Chemical and Process Engineering

Chemical and Process Engineering

Edited by **Serena Gibson**

Willford Press
New York

Published by Willford Press,
118-35 Queens Blvd., Suite 400,
Forest Hills, NY 11375, USA
www.willfordpress.com

Chemical and Process Engineering
Edited by Serena Gibson

International Standard Book Number: 978-1-68285-174-6 (Hardback)

Printed in China.

Contents

Preface

This book aims to highlight the current researches and provides a platform to further the scope of innovations in this area. This book is a product of the combined efforts of many researchers and scientists from different parts of the world. The objective of this book is to provide the readers with the latest information in the field.

Chemical and process engineering deals with designing of processes that aim to transform chemicals and other substances into useful products. This innovative and comprehensive book integrates the well-developed theory and practical applications of this field through topics such as polymer science, thermodynamics, rheology, process modeling, nanofabrication, catalysis, industrial chemistry, etc. This book aims to equip students and experts with the advanced topics and upcoming concepts in this area.

I would like to express my sincere thanks to the authors for their dedicated efforts in the completion of this book. I acknowledge the efforts of the publisher for providing constant support. Lastly, I would like to thank my family for their support in all academic endeavors.

Editor

Absorption of Phenol on Nitrogen-Enriched Activated Carbon from Wood Fiberboard Waste with Chemical Activation by Potassium Carbonate

Xiao-Juan Jin* and Yue Mei Zhu

MOE Key Laboratory of Wooden Material Science and Application, Beijing Key Laboratory of Lignocellulosic Chemistry, MOE Engineering Research Center of Forestry Biomass Materials and Bioener, Beijing Forestry University, Beijing, China

Abstract

Preparation of nitrogen-enriched activated carbons from wood fiberboard waste was studied. Activated carbons were obtained with an impregnation ratio (gram chemical agent/gram wood fiberboard waste) of 3 with 50 % potassium carbonate solution in 800°C activation temperature carbonized for 1h. Optimum adsorption conditions were determined as a function of effects of contact time, pH, and dosage of activated carbon and temperature of solution for phenol adsorption. Adsorption equilibrium was achieved with in 120 min at the given phenol concentration of 250 mg/L. When 0.3 g of the carbon absorbent and 100 mL of phenol solution at 250 mg/L were used, maximum adsorption capacity of phenol on activated carbon can reach 208 mg/g. In the adsorption isotherm .the Langmuir model fit better than the Freundlich model in phenol adsorption. The kinetics of phenol adsorption followed nicely the pseudo-second-order rate expression.

Keywords: Phenol adsorption; Activated carbon; Wood fiberboard waste

Introduction

Phenols are pollutants of high priority concerns because of their toxicity and possible accumulation in the environment. Phenols are introduced into surface water from industrial effluents such as those from the coal tar, gasoline, plastic, rubber proofing, disinfectant, pharmaceutical and steel industries and domestic wastewaters, agricultural run-off and chemical spills. Various methods have been proposed for the treatment of wastewaters containing organic and inorganic pollutants. These processes are based on the principles of precipitation and coagulation, chemical oxidation, sedimentation, filtration, adsorption, osmosis, ion exchange, etc. [1,2].

Adsorption technology has been widely studied for removal of organic and inorganic micro pollutants from aqueous solutions. There are many absorbents in use. Activated carbon is the most widely used for the removal of a variety of organics from waters, but the disadvantage associated with it is the high regeneration cost, due to the use of non-renewable and relatively expensive starting material such as coal, a major economic consideration. This has prompted a growing research interest in the production of activated carbons from locally available materials especially for application concerning waste wood [3]. Several wood processing residues and agricultural residues have been found to be suitable precursors for production of activated carbon. Because the biomass obtained from these materials are cheaper, renewable and abundantly available and have high carbon and low ash contents [4].

In the past few years, enhanced adsorption capacity toward phenol on activated carbons has currently receiving great attention [5,6]. Because introducing nitrogen-containing surface group makes activated carbon more alkaline, increased adsorption of acidic agents is expected. Therefore, nitrogen-enriched activated carbon could increase ability to adsorb phenolic compounds [7]. However, nitrogen content in commercially used activated carbon is very low. There are several methods for preparing activated carbon with enhanced levels of nitrogen. The methods used for the synthesis of nitrogen-containing porous carbons can be divided into two groups. First, it is the treatment of conventional activated carbon at an elevated temperature with ammonia, ammonia-air or ammonia-steam mixtures [8-10]. The second approach consists in the condensed phase pyrolysis of nitrogen-containing precursors [11-13] or co-pyrolysis of suitable nitrogen carrier with a common carbon precursor, followed by the activation of the resultant chars [14-17]. Among above-mentioned methods for making nitrogen-containing activated carbon, it is quite necessary to find more cost-effective ways of production. There is great potential for developing activated carbon products from wood waste. In waste wood fiberboard, nitrogen is introduced to the wood matrix from the use of nitrogen containing amino plastic adhesives such as urea-formaldehyde and melamine-formaldehyde resins. Nitrogen-containing adhesives are dispersed through wood fiberboard. Accordingly, preparation of nitrogen-enriched activated carbon from waste wood and its adsorption behavior on phenol would be an interesting research subject.

The aim of this work is to illustrate adsorption of phenol on nitrogen-enriched activated carbon prepared from wood fiberboard waste. A series of important factors such as contact time, adsorbent dose, pH, and temperature were investigated. Also, adsorption isotherms and adsorption kinetic studies were reviewed.

Materials and Methods

Materials

Waste fiberboard (WFB) was provided by Beijing Forestry University wood factory. WFB consisted of poplar wood and urea–formaldehyde resin (10 %).Phenol, Potassium carbonate (K_2CO_3) and hydrochloric acid (HCI) were analytical grade and were purchased from Beijing

***Corresponding author:** Xiao-Juan Jin, Associate Professor, MOE Key Laboratory of Wooden Material Science and Application, Beijing Key Laboratory of Lignocellulosic Chemistry, MOE Engineering Research Center of Forestry Biomass Materials and Bioener, Beijing Forestry University, Beijing, China

Lanyi chemical (Beijing, China). Double distilled water was used for preparation of all required solutions.

Preparation of the activated carbons

The preparation of activated carbon is divided into two steps. The first step is carbonation, WFB was cut into 250×20×5 mm pieces, There pieces were heated to a carbonization temperature of 500°C at a heating rate of 100°C/h and were maintained at this temperature for 1 h. The samples were then grounded and screened with sieves. The fraction in the particle diameter ranged from 40-60 mesh. Particle samples were dried in a105°C oven for 4 h.

In the activation step, 3 g of oven-dried samples was soaked in 50 % K_2CO_3 solution for 16 h at the investigated impregnation (gram chemical agent/gram WFB waste) of 3. The soaked samples were then put in an electric furnace and heated up to about 800°C at a heating rate of 10°C /min^{-1}.Samples were held at this temperature for 50 min before cooling down and followed by washing sequentially with a 0.5 N HCI solution. Activated carbons were repeatedly washed with hot distilled water until pH of the solution reached about 6-7. Finally, these activated carbons were dried in the 105°C oven for 4 h.

Phenol adsorption process

Adsorption kinetics and equilibrium studies were conducted at ambient temperature. A known amount of the adsorbent was put in flasks containing100 mL of aqueous phenol solution at predetermined concentration [18]. The flasks were hermetically closed, not only for avoiding the loss of phenol by volatilization, but also for preventing the entry of oxygen, which is known to lead to irreversible phenol adsorption through oxidative coupling [19]. Then the flasks were shaken for 100 min. At the end of the adsorption process, adsorbent particles were filtered out using 0.45 mm membranes. The residual phenol concentration was measured by UV absorption in a UV-2102c spectrophotometer (Unico, USA) with a wavelength of λ max=270 nm [20].

The amount of adsorption at equilibrium, q_e (mg/g), was calculated by:

Fiberboard	Properties	Activated carbon from WFB
1	C content (wt %)	90.46
2	H content (wt %)	0.59
3	N content (wt %)	1.36
4	Iodine number (mg/g)	1370
5	BET surface area (m²/g)	1576
6	Total pore volume (cm³/g)	0.953

WFB, waste fiber board; BET, Brunauer-Emmett-Telle

Table 1. Characteristics of activated carbon from fiberboard.

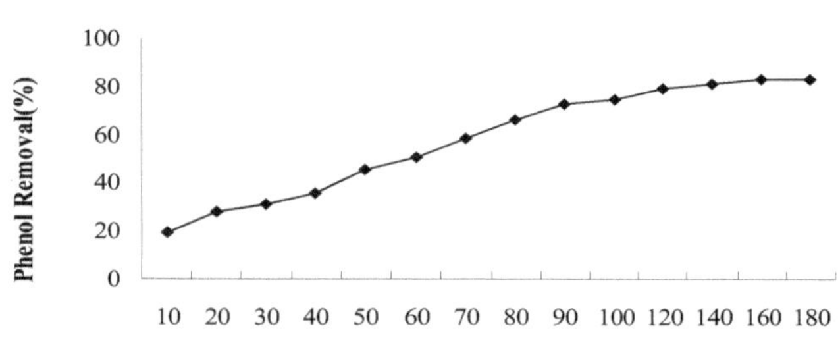

Figure 1: Effect of contact time on adsorption of phenol (carbon dosage = 0.1 g; phenol concentration = 250 mg/L).

$$q_e=(c_0 - c_e)\times v/w \qquad (1)$$

Where c_0 and c_e (mg/L) are the concentrations of phenol at initial stage and equilibrium stage, respectively. *V* is the volume of the solution (L) and *w* is the mass of adsorbent used (g).

Results and Discussion

Characteristics of the as-prepared activated carbons

The properties of the prepared activated carbon were determined by standard methods as summarized in Table 1. Elemental analysis (contents in carbon, hydrogen and nitrogen) of the prepared activated carbon w as made in a CHNS analyzer (Flash, EA.1112 series, Thermo Finnigan San Jose, CA). Iodine number was determined at 303 K based on ASTM D4607-94 (ASTM 2011). Pore structure characteristics of activated carbon were measured by N_2 adsorption isotherm at -196°C using an ASAP 2010 analyzer (Micro meritic Norcross, GA). The Brunauer-Emmett-Teller (BET) surface areas were calculated from N_2 adsorption isotherms by using the BET equation.

Effect of contact time

The 0.1 g activated carbon sample was put into 250 mL flasks containing 100 mL of buffered aqueous phenol solution. The initial concentration of phenol was 250 mg/L. The result is shown in Figure 1.

Adsorbent were filled and the adsorption of phenol became slower gradually and reached a plateau. The phenol removals versus time curves were single, smooth and continuous leading to saturation, suggesting the possible monolayer coverage of phenol on the surface of the adsorbents [21].

In this experiment, the equilibrium time was 120 min, which the maximum surface areas were obtained with the activation of 50 % K_2CO_3 impregnated samples as 1576 m²/g. meantime, the capacity of the activated carbon from WFB was 208 mg/g and the phenol removal could get 83 %. Tancredi et al. [22] reported that phenol adsorption capacity of activated carbon from Eucalyptus wood was 64 mg/g in 750 min and phenol removal could reach 32%. In a study reported by Mohanty et al. [23], capacity of activated carbon from Sterculia alata nutshell could reach 140 mg/g and phenol removal achieved 80% in 300 min. Compared with results reported in these two studies, phenol removal was higher and equilibrium time shorter in this study. Because introduction of nitrogen-containing surface groups made activated carbon more alkaline, increased adsorption of phenol was expected.

Since the adsorption rate became constant at about 120 minutes, there was no significant growth in the adsorption when the contact time increased from 120 minutes to 180 minutes. Therefore, the optimum contact time for all the further experiments was set as 120 min.

Effect of dosage of activated carbon

For investigating of effect of adsorbent dosage on adsorption, the experiments were conducted with the initial concentration of phenol was maintained at 250 mg/L and samples having different adsorbent dosage ranging from 0.01 to 0.30 g (0.01 g, 0.02 g, 0.05 g, 0.15 g 0.20 g 0.25 g 0.30 g) in 100 mL unbuffered aqueous phenol solution .The result was illustrated in Figure 2.

It could be observed from the Figure 2. At the dosage of activated carbon level is 0.25 g, the phenol adsorption dose is at its maximum, and then as dosage of active carbon increased, the unit

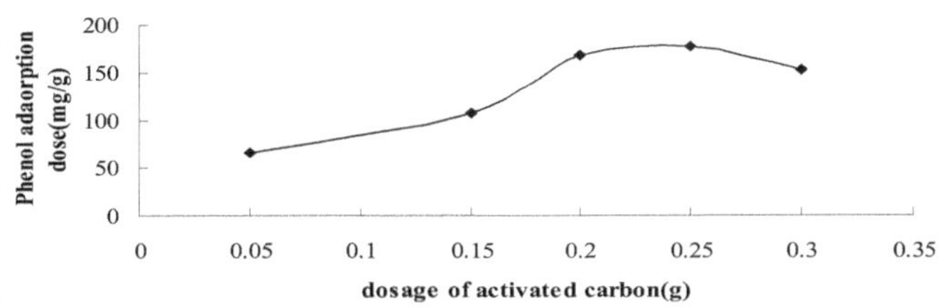

Figure 2: Effect of dosage of activated carbon on adsorption of phenol (phenol concentration = 250 mg/L, equilibration = 120 min).

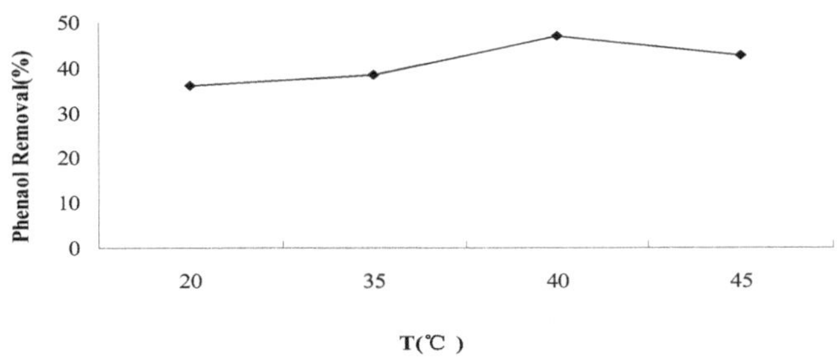

Figure 3: Effect of temperature on adsorption of phenol (carbon dosage = 0.1 g; phenol concentration = 250 mg/L; equilibration time = 120 min).

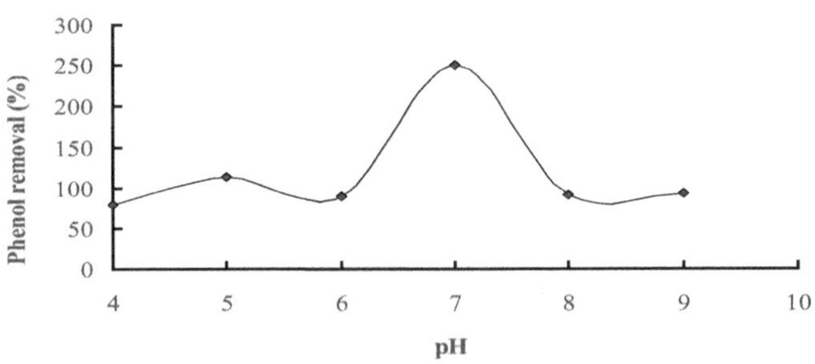

Figure 4: Effect of pH on adsorption of phenol (carbon dosage = 0.1 g; phenol concentration = 250 mg/L; equilibration time = 120 min).

adsorption volume has dropped significantly. The phenol removal a sharp drop as the adsorbents dose increased. When the adsorbent dose further increased from 0.1 g to 0.3 g, a significant increase in phenol adsorption was observed. In any case, because introducing nitrogen-containing surface groups makes activated carbon more alkaline. The increase in the percent adsorption of phenol was due to the availability of larger surface area with more active functional groups [24]. As a result, the optimum dosage of activated carbon was 0.25g.

Similar phenomenon was also observed by Murat Kilic et al. [25], reported that that the adsorption behavior of phenol onto two carbons are generally similar. The adsorption capacity increased with increasing adsorbent amount and then became constant indicating that 0.2 and 0.4 g/L were sufficient for the optimum removal of phenol from aqueous solution, respectively. Therefore, the dosages of 0.2 and 0.4 g/L were used for the following phenol adsorption experiments.

Effect of temperature

The adsorptions of phenol on activated carbon removal as a function of temperature value were verified in the range of 20-45°C. The initial concentration of phenol was 250 mg/L and the contact time for adsorption was kept at 120 min to ensure the complete equilibrium. The result was seen in the Figure 3.

With the temperature varied from 20°C to 35°C, the phenol removal slowly increased, and as the temperature further increased from 35°C to 40°C, the phenol removal rate increased from 38.47 % to 47.05 %. In the paper, the adsorption capacities of 90.00, 96.18, 117.64 and 106.73 mg/g were reported at 20, 35, 40, 45 °C respectively. Similar phenomenon was also shown by Karabacakoglu et al. [26], Hazelnut Bagasse carbon activated with zinc chloride was examined for the removal of phenol from water. The results showed that the adsorption capacities of 97.36, 91.32 and 99.27 mg/g were found at 25, 35 and 45°C respectively. Compared with results reported by Karabacakoglu et al. [26] and in this paper, phenol removal was higher at the same temperature. The reason was that introduction of nitrogen-containing surface groups made activated carbon more alkaline, increased adsorption of phenol was expected.

As the temperature further increased beyond 40°C, the phenol removal rate decreased sharply. The result indicated that the maximum adsorption for phenol was obtained at 40°C for the activated carbon prepared from WFB.

Effect of pH

The effect of pH on the equilibrium adsorption of phenol on prepared activated carbon was examined over a pH range of 4-9. The initial concentration of phenol was 250 mg/L and the adsorbent dose was 0.1 g. The pH of unbuffered aqueous phenol solution was 7.0, and then pH values were adjusted with 0.1 M HCI and 0.1 M K_2CO_3

Figure 4 is show that the maximum removal of phenol at the given absorbent dosage and given phenol concentration is observed at pH of 7.0 and adsorption capacity was 208mg/g. The potential of nutshells of Sterculia alata, a forest waste to prepare activated carbons by zinc chloride activation to remove phenol from aqueous solution was evaluated Mohanty et al. [23] Results were shown that the maximum uptake of phenol was found at pH 3.5 and adsorption capacity was 2.8 mg/g.

This trend could be explained considering the nature of the adsorbent at different pH in phenol adsorption. Due to the phenol itself is soluble in acidic water solution (pKa=9189), and phenol in acidic or neutral aqueous solution exists mainly in molecular state, with affinity is bigger, the surface of activated carbon to adsorption; While in alkaline aqueous solution exists mainly in ion state, large and is not conducive to the affinity adsorption with water. When the pH>7, dissociation forms of phenol are more likely to be soluble in solution, and only binding force strong chemical bonds between adsorption and water is broken, only possible adsorption, dissociation and activated carbon adsorption in the form of the increase of repulsion between the surface also increases the adsorption force of smaller. In the higher pH range phenol forms salts which readily ionize leaving negative charge on the phonetic group. At the same time the presence of OH- ions on the adsorbent prevents the uptake of phenol [27,28]. Therefore, the optimum pH was 7.0.

Adsorption isotherm

The Langmuir and Freundlich models were the most common models to describe the adsorption isotherms.

Langmuir adsorption isotherm was given by the following equation [28].

$$c_e/q_e = 1/(q_m \times b) + (1/q_m) \times c_e \qquad (2)$$

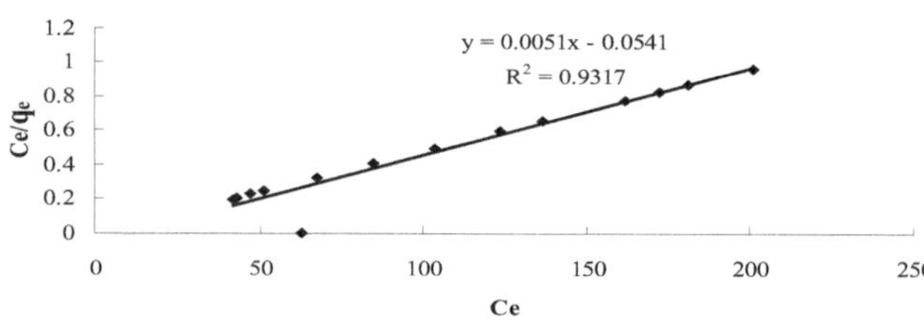

Figure 5: Linear plot from Langmuir equation of phenol adsorption on activated carbon.

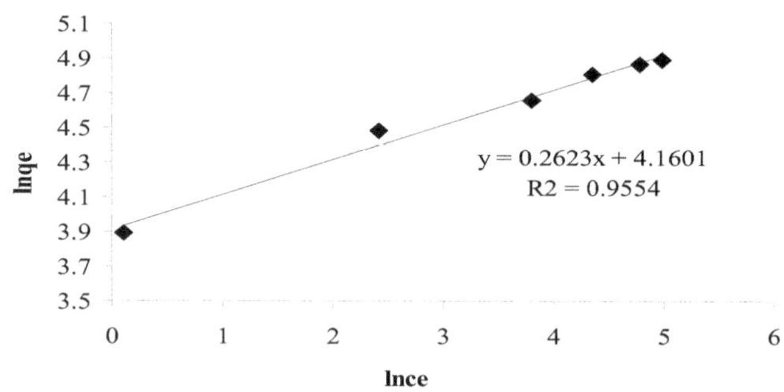

Figure 6: The linear from Freundlich equation of adsorption of phenol on the AC.

Langmuir			Freundlich		
qm (mg/g)	B	R^2	n	K_f	R^2
213,46	0.23	0.93	3. 51	64.05	0.96

Table 2: Parameters of different isotherms of adsorption of phenol on activated carbon.

Where q_m (mg/g) is the amount of adsorption corresponding to complete monolayer coverage and b (1/mg) is the Langmuir constant related to the energy or net enthalpy of adsorption. The linear plots of c_e/q_e versus c_e were shown in Figure 5.

Freundlich adsorption isotherm was given by the following equation [29].

$$q_e = K_f C_e^{1/n} \qquad (3)$$

Where K_f and n are Freundlich constants related to adsorption capacity and adsorption intensity. Usually, for a good adsorbent 1<n<10. The phenol adsorption isotherm followed the linearized Freundlich model as shown in Figure 6.

The values of Langmuir and Freundlich constant obtained were given in Table 2. Table 2 Parameters of different isotherms of adsorption of phenol on activated carbon. It could see that the Langmuir model fitted the results better then the Freundlich model. Since the value of *1/n* was less than 1, it indicated a favorable adsorption. Similar trends were found in the article of Fierro et al. [5], which the Langmuir fitted better in the adsorption of phenol.

Adsorption kinetic studies

The rate constant of adsorption was studied under the pseudo-first-order equation and the pseudo-second-order equation.

The pseudo-first-order equation was represented by the following equation.

$$\ln(q_e - q) = \ln q_e - k_1 \times t \qquad (4)$$

Where q (mg/g) is the amount of phenol absorbed at time t (min); q_e (mg/g) is the amount of phenol absorbed at equilibrium and k_1 (1/min) is the first-order adsorption rate constant. The linear plot of $\ln(q_e\text{-}q) \vee t$ was shown in Figure 7.

The pseudo-second-order equation [30] based on equilibrium adsorption was expressed as:

$$t/q = 1/(k_2 \times q_e^2) + t/q_e \qquad (5)$$

Where k_2(g.mg.min^{-1}) is the rate constant of second-order adsorption. Also, the linear plot was shown in Figure 8.

The constants and R^2 values obtained from the two linear plots were summarized in Table 3.

The correlation coefficients R^2, showed that the pseudo-second-order equation was more suitable for the experimental data. The data shows that good agreement between the experimental and the calculated q_e values was observed by using the pseudo-second-order equation. As a result, it is concluded that the adsorption of phenol on the activated carbon prepared from WFB follows the pseudo-second-order model. This may be expected that the overall rate of the adsorption process was controlled by chemisorptions [31].

Conclusion

The study results showed that the nitrogen-enriched activated carbon made from chemical activation of WFB with K_2CO_3, which had a significantly higher phenol adsorption capacity contrasted with other non-nitrogen containing activated carbon. Compared to a series effects of contact time; pH, dosage of activated carbon, and temperature on

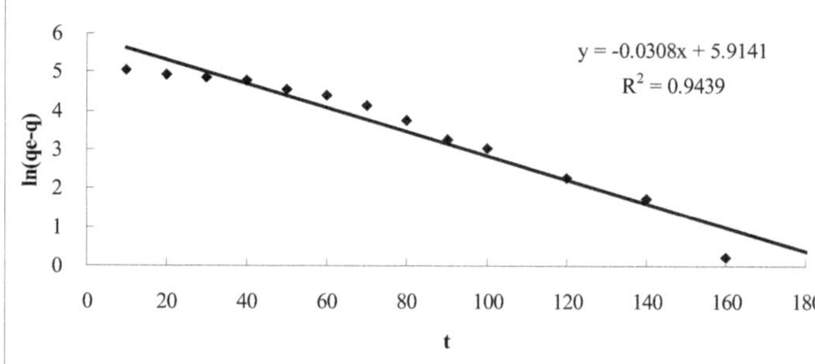

Figure 7: First-order adsorption rate of phenol on activated carbon.

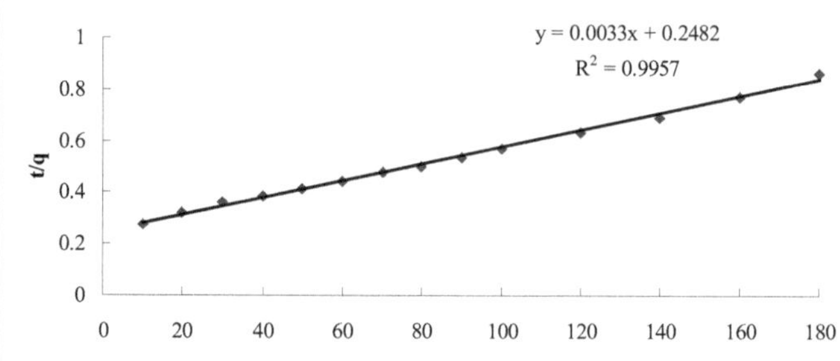

Figure 8: The second-order adsorption rate of phenol on activated carbon.

Pseudo-first-order equation			Pseudo-second-order equation		
q^e(mg/g)	k_1(min^{-1})	R^2	q^e(mg/g)	k_2(g. mg .min^{-1})	R^2
208.36	0.03	0.94	208.37	0.0070	0.99

Table 3: Parameters of different adsorption rate of phenol on activated carbon.

phenol adsorption capacity of activated carbons; we concluded that the phenol removal became higher and the equilibrium time was shorter. Because introducing of N-containing surface groups made activated carbon more alkaline, increased adsorption of phenol was expected. The kinetics of phenol adsorption rate followed the pseudo-second-order, and the Langmuir model fitted better than Freundlich model in phenol adsorption in the study of adsorption isotherm.

Acknowledgment

This study was funded by State Forestry Administration under Project 201204807: the study on the technology and mechanism of the activated carbon electrode preparation from waste hard board.

References

1. Cermola F, DellaGreca M, Iesce MR, Montella S, Pollio A, et al. (2004) A mild photochemical approach to the degradation of phenols from olive oil mill wastewater. Chemosphere 55: 1035-1041.

2. Rengaraj S, Moon SH, Sivabalan R, Arabindoo B, Murugesan V (2002) Agricultural solid waste for the removal of organics: adsorption of phenol from water and wastewater by palm seed coat activated carbon. Waste Manag 22: 543-548.

3. Hameed BH, Tan IAW, Ahmad AL (2008) Adsorption isotherm, kinetic modeling and Mechanism of 2, 4, 6-trichlorophenol on coconut husk-based activated carbon. Chem Eng J 144: 235-244.

4. Mohd Din AT, Hameed BH, Ahmad AL (2009) Batch adsorption of phenol onto physiochemical-activated coconut shell. J Hazard Mater 161: 1522-1529.

5. Fierro V, Torné-Fernández V, Montané D, Celzard A (2008) Adsorption of phenol onto activated carbons having different textural and surface properties. Micro porous and Mesoporous Materials 111: 276-284.

6. Ozkaya B (2006) Adsorption and desorption of phenol on activated carbon and a comparison of isotherm models. J Hazard Mater 129: 158-163.

7. Boudou JP, Chehimi M, Broniek E, Siemieniewska T, J Bimer (2003) Adsorption of H2S or SO2 on an activated carbon cloth modified by ammonia treatment. Carbon 41:1999-2007.

8. Mangun CL, Benak KR, Economy J, Foster KL (2001) Surface chemistry, pore sizes and adsorption properties of activated carbon fibers and precursors treated with ammonia. Carbon 39:1809-1820.

9. Jansen RJJ, van Bekkum H (1994) Amination and ammoxidation of activated carbons. Carbon 32: 1507-1516.

10. Boudou JP, Paredes JI, Cuesta A, Martinez-Alonso A, Tascon JMD (2003) Oxygen plasma modification of pitch-based isotropic carbon fibres. Carbon 41: 41-56.

11. Laszlo K, Tombacz E, Josepovits K (2001) Effect of activation on the surface chemistry of carbons from polymer precursors. Carbon 39: 1217-1228.

12. Stan´czyk K, Dziembaj R, Piwowarska Z, Witkowski S (1995) Transformation of nitrogen structures in carbonization of model compounds determined by XPS. Carbon 33: 1383-1392.

13. Lahaye J, Nanse G, Bagreev A, Strelko V (1999) Porous structure and surface chemistry of nitrogen containing carbons from polymers. Carbon 37: 585-590.

14. Huang MC, Teng H (2003) Nitrogen-containing carbons from phenol–formaldehyde resins and their catalytic activity in NO reduction with NH3. Carbon 41: 951-957.

15. Kapteijn F, Moulijn JA, Matzner S, Boehm HP (1999) The development of nitrogen functionality in model chars during gasification in CO2 and O2. Carbon 37: 1143-1150.

16. Raymundo-Pinero E, Cazorla-Amoros D, Linares-Solano A, Find J, Wild U, et al. (2002) Structural characterization of N-containing activated carbon fibers prepared from a low softening point petroleum pitch and a melamine resin. Carbon 40: 597-608.

17. Machnikowski J, Grzyb B, Machnikowska H, Weber JV (2005) Surface chemistry of porous carbons from N-polymers and their blends with pitch. Micropor Mesopor Mater 82: 113-120.

18. Bagheri N, Abedi J (2009) Preparation of high surface area activated carbon from corn by chemical activation using potassium hydroxide. Chem Eng Res Des 87: 1059-1064.

19. Mohanty K, Jha M, Meikap BC, Biswas MN (2005) Removal of chromium (VI) from dilute aqueous solutions by activated carbon developed from terminalia arjuna nuts activated with zinc chloride. Chem Eng Sci 60: 3049-3059.

20. Terzyk A (2007) The impact of carbon surface chemical composition on the adsorption of phenol determined at the real oxic and anoxic conditions. Appl Surf Sci 253: 5752-5755.

21. Altenor S, Carene B, Emmanuel E, Lambert J, Ehrhardt JJ, et al. (2009) Adsorption studies of methylene blue and phenol onto vetiver roots activated carbon prepared by chemical activation. J Hazard Mater 165: 1029-1039.

22. Tancredi N, Medero N, Möller F, Píriz J, Plada C, et al. (2004) Phenol adsorption onto powdered and granular activated carbon, prepared from Eucalyptus wood. J Colloid Interface Sci 279: 357-363.

23. Mohanty K, Das D, Biswas MN (2006) Preparation and characterization of activated carbons from Sterculia alata nutshell by chemical activation with zinc chloride to remove phenol from waste water. Adsorption 12: 119-132.

24. Girods P, Dufour A, Fierro V, Rogaume Y, Rogaume C, et al. (2009) Activated carbons prepared from wood particleboard wastes: characterisation and phenol adsorption capacities. J Hazard Mater 166: 491-501.

25. Kilic M, Apaydin-Varol E, Pütün AE (2011) Adsorptive removal of phenol from aqueous solutions on activated carbon prepared from tobacco residues: Equilibrium, kinetics and thermodynamics. Journal of Hazardous Materials 189: 397-403.

26. Karabacakoglu B, Tumsek F, Demiral H, Demiral I (2008) Liquid Phase Adsorption of Phenol by Activated Carbon Derived From Hazelnut Bagasse. J Int Environ Applic Sci 3: 373-380.

27. Ania CO, Parra JB, Pis JJ (2002) Effect of texture and surface chemistry on adsorptive capacities of activated carbons for phenolic compounds removal. Fuel Processing Technology 77-78: 337-343.

28. Namasivayam C, Ranganathan K (1995) Removal of Cd(II) from wastewater by adsorption on waste Fe(III)/Cr(III) hydroxide. Water Res 29: 1737-1744.

29. Bimer J, Salbut PD, Berlozecki S, Boudou JP, Broniek E, et al. (1998) Modified active carbons from precursors enriched with nitrogen functions: Sulphur removal capabilities. Fuel 77: 519-525.

30. Ho YS, McKay G (1999) Pseudo-second order model for sorption processes. Process Biochem 34: 451-465.

31. Singh BK, Mishra NM, Rawat NS (1994) Sorption characteristics of phenols on fly ash and impregnated fly ash. Indian J Environ Hlth 36: 1-7.

2

Change Detection in a Distillation Column Based on the Generalized Likelihood Ratio Approach

Yahya Chetouani*

Département Génie Chimique, Rue Lavoisier, 76821, Mont Saint Aignan Cedex, France

Abstract

With increasing demands for efficiency, product quality, reliability and process safety, the field of fault detection (FD) plays an important role in chemical industries. This paper deals with a FD method based on the combination of Generalized Likelihood Ration Test (GLRT) and Artificial Neural Networks (ANNs). A reliable neural model in normal conditions, under all regimes (i.e. steady-state and dynamic conditions), is found by means of a NARX (Nonlinear Auto-Regressive with eXogenous input) model and by an experimental design. The efficiency of the combination of these two approaches used for detecting faults has been tested under real anomalous conditions on a real plant as a distillation column. From the experimental results, it is observed that the proposed FD is able to detect the process status effectively.

Keywords: Fault detection; Reliability; Safety; GLRT, ANNs; Distillation column

Introduction

Early detection of occurrence of an abnormal event, fault or failure in an operating plant is vital for ensuring plant safety and maintaining product quality in chemical industries. The necessity to improve FD methods is further underlined by the finding that about 70% of industrial accidents are caused by human errors [1]. Traditionally, the most usually implemented FD methods have been based on model-based approaches. However, in modern process industry, there is a demand for data-based methods because of the complexity and the limited availability of the nonlinear models in chemical units.

Artificial Neural Networks (ANNs) have an adaptive behavior i.e. they are able to adjust and to modify their behavior according to nonlinear dynamics of processes. ANNs can be trained to learn new associations, complex modelling, functional dependencies and new patterns [2]. Owing to their inherent nature to model and learn complexities, ANNs have been successfully applied in medicine and biomedical studies [3]. Also they have found wide applications in various areas of chemical engineering [4,5].

This paper proposes a fault detection approach, which combines a statistical test as GLRT with ANNs to quickly detect faults in a distillation column. Firstly, this study consists to obtain a reduced and reliable model of this chemical process in steady-state and dynamic conditions. The chosen model is NARX model for forecasting the process dynamics. The performance of this neural model was then evaluated using the performance criteria. Then, the abnormal behaviour of the process is inspected when it is submitted to faults. Fault detection results show that the statistical test based on the GLRT is a powerful tool to detect changes in the behaviour of the distillation column.

This paper is organized as follows: Section 2 describes the neural training and selection of neural structure, Section 3 describes the fault detection strategy, in Section 4 the experimental set-up is introduced and presents the experimental results with the proposed FD method, and Section 5 ends the paper.

Network training modelling

In many fields, there are several models representing systems that are dominated by nonlinear characteristics. Many physical processes which have nonlinear behavior can be well represented by a polynomial representation, Volterra or Wiener series, or other nonlinear techniques. It is showed that the Autoregressive with eXternal input (ARX) model can be easily identified and outperform linear models [6]. Also, Nonlinear AutoRegressive Moving Average model with exogenous inputs (NARMAX) model [7] can provide a unified and simple representation for a wide class of discrete-time nonlinear stochastic systems. The NARMAX model is an input-output recursive model where the current output depends on lagged inputs, outputs and noise terms through a suitable nonlinear function. The simpler Nonlinear AutoRegressive with eXternal input (NARX) model is often preferred, although the absence of a disturbance model may result in bias problems i.e. a NARMAX model with the noise terms excluded. Structure determination of the NARX model is to choose sufficient and necessary terms for capturing the system dynamics. This model could represent a nonlinear model with a smaller number of parameters [4,8,9]. In this study, the effectiveness of this technique for black-box nonlinear modelling is described.

The ANN propagates the error from the output layer (y_i) to the hidden layer to update the weight matrix (w_i). Each node produces an output signal, which is a function of the sum of its inputs where $y(k)$ is the Auto-Regressive (AR) variable or system output; $u(k)$ is the eXogenous (X) variable or system inputs:

$$y_i = \phi_{ANN}(\sum x_i w_i) \quad (1)$$

Where $\varphi_{ANN}(\cdot)$ is the activation function. In this study, the neural model is defined as follows

***Corresponding author:** Université de Rouen, Département Génie Chimique, Rue Lavoisier, 76821, Mont Saint Aignan Cedex, France

(10,11):

$$y_i = \phi_{ANN}(y(t-1),...,y(t-n_y), u_i(t-n_k),...,u_i(t-n_k-n_u)) \quad 1 \le i \le m \tag{2}$$

where y(k) is the Auto-Regressive (AR) variable or system output; u(k) is the eXogenous (X) variable or system input. ny and nu are the AR and X orders, respectively. m is the number of the used inputs. nk is the time delay between u and y.

For smaller number of hidden layers and nodes, the performance of the ANN may not be satisfactory, while with too many hidden nodes, there is the risk of over-fitting the training data. In this case, the ANN generalization on the new data is poor. Different methods, both heuristic and systematic, can be chosen in order to select the number of hidden layers and the nodes. There are a large number of selection methods of the ANN. Two most popular ways of significant structures are in general used. We start with a small number of nodes in the hidden layer and add new ones when a certain criterion is met (growing algorithms). On the other hand, we can start with a large number of nodes and delete some of them under certain conditions (pruning algorithms).

In this study, the hidden layer nodes have log-sigmoid transfer hyperbolic function as the activation function and the output have linear activation function. The ANN multi-layer feed forward network is trained to capture the underlying relationship between the input *u(k)* and output *y(k)* using the training data. The input data are presented to the network via the input layer. These data are propagated through the network to the output layer to obtain the network output. The network error (generally the mean square error function) is then determined by comparing the network output with the actual output. If the error is not smaller than a desired performance, the weights are adjusted and the training data are presented to the network again to determine a new network error. The aim of this step is to find the appropriate weights which minimize the cost function. This is usually done using an iterative procedure. One of the best known learning mechanisms for neural networks is the training algorithm of Levenberg-Marquardt [12], which is used along with back-propagation (BP). In this case the ANN is trained iteratively using the training dataset to minimize the performance function of mean square error (MSE) between the network outputs and the corresponding target values. The Levenberg-Marquardt Algorithm (LMA) shows the fastest convergence during the training process based on gradient descent methods because it performs as a compromise between the stability of the first-order optimization methods (steepest-descent method) and the fast convergence properties of the second-order optimization methods (Gauss-Newton method). In many cases, LMA is able to obtain lower mean square errors than any of the other algorithms [12]. For weights initialization, the Nguyen- Widrow initialization method [13] is best suited for the use with the sigmoid/linear network which is often used for function approximation. After training, the networks thus developed are tested with the test data set to assess the generalization capability of each developed network. The best ANN model developed is trained off-line and then used on-line for detecting faults in the separation unit. All computations have been made on Matlab® 7.0.4.

Fault detection method: GLRT

GLR test is applied in order to establish an adaptive system, which achieves three important operations; fault detection, estimation and magnitude compensation of jumps generated by instantaneous changes of the process. GLRT is used for fault detection in signals and dynamic systems [14], geophysical signal segmentation [13], signals, incident fault detection on freeways [16], missiles trajectory [17], changes detection issue in a stochastic system [18]. Nikiforov [19] used GLRT when the model parameters after the change are unknown. He introduced an alternative approach to reduce the computational burden of the GR scheme. The idea of this solution is to decompose a given parameter space into several sub-sets so chosen that in each subset the fault detection problem can be solved with loss of a small part of optimalit y by a recursive change detection algorithm. This test is also proposed for estimating the time instant of muscle contraction onset via electromyographic (EMG) signal processing [20]. The proposed algorithm proves to be reasonably accurate even for low levels of EMG activity; the improved behavior of the GLRT comes with just a modest increase in the computational complexity. Keller et al. [21] developed a new GLRT based method for detecting, identifying, and estimating gross errors in steady state processes. This method improves GLRT and can be applied to all types of gross errors including leaks and biases. Boukai [22] used the GLRT to detect a local shift of parameter of a distribution function occurring at unknown point of time between consecutive independent observations. Three parametric models of distributions are considered; the one parameter exponential family, the location parameter family and the scale parameter family. The class of considered tests is based on GLRTs, appropriately adapted for such a change-point problem. Asymptotic techniques are used to obtain the limiting distribution of the test statistics, under both the null hypothesis of no change and the change-point alternative. Micera et al. [23] utilized a hybrid approach to EMG pattern analysis for classification of movements. The hybrid model included GLRT and the principal component analysis. They show that the classification method classify correctly all patterns related to the selected movements. Kay et al. [24] derived a GLRT detector for the detection of a sinusoid in complex environment where the noise variance and the signal amplitude, phase and frequency are all unknown. They derived a detector that gives an upper performance bound for the GLRT applied to this problem.

In this study, the recursive algorithm of the GLRT is established according to normal conditions of the operating plant (null hypothesis) vs. an alternate hypothesis. This letter is normally distributed (where variance (σ^2) is unknown). The problem is described in terms of a hypothesis test H_0, with the null hypothesis denoted by with plant behaviour belonging to the normal region Ω_o, and the alternate hypothesis H_1 denoted by with this behaviour belonging to Ω_1. One way to distinguish between these two hypotheses is to build the likelihood ratio test where P(l≤λ/H_0 is true)=α. The main aim is to detect a change in the behaviour of a conditional probability distribution (P_0), from the known value of the parameter (θ_0) to the unknown value (θ_1) which occurs at an unknown time of fault (t_f). The detection rule of GLRT is defined as following:

$$lk - \max_{1 \le j \le k} \sup_{\theta_1} \sum_{i=j}^{k} \log\left(\frac{P_{\theta_1}(\gamma_i)}{P_{\theta_0}(\gamma_i)}\right) \tag{3}$$

With t_f = min (k: $k \ge \alpha$)

In the case of the Equ 2, the GLRT test rule is given as follows:

$$lk - \max_{1 \le j \le k} \sup_{\theta_1} \left[\frac{1}{2}\log(\sigma_0^2 / \sigma_1^2) + \frac{(\gamma_i^o)^2}{2\sigma_0^2} - \frac{(\gamma_i^1)^2}{2\sigma_1^2}\right] \tag{4}$$

where γ (k) = $y(k) - y^{\wedge}(k)$ is the sequence of the residuals between the actual and predicted output of the system. During a normal operation, innovations fluctuation γ (k) = $y(k) - y^{\wedge}(k)$ is low and corresponding to random variations in measurements. Moreover, when a fault occurs, the innovation evolves differently according to the type of fault and its influence on the general state of the plant. In order to limit the computing time, it is decided to detect changes in a window of fixed size (*M*). The window size is chosen according to the process dynamics,GLRT sensitivity, and to the delay in detection.

Experimental Results

Experimental set-up

The proposed FD scheme is applied to a distillation unit. The feed tank contains a mixture to be separated (toluene-methylcyclohexane) with amass composition at 23% in methylcyclohexane. The operation in continuous mode involves charging the still with the mixture to be separated, bringing the column to equilibrium under total reflux. The product is introduced through the optimal feed tray so that the light components are volatilized, while the heavy part goes down again with the reflux in the column reboiler. The quality of the collected top product of the column depends on the reflux flow rate. The reflux ratio is varied through the magnetic valve by changing the relative quantities of material returning to the column and flowing to product storage. Feed preheating system is constituted by three elements of 250 W each one. In addition it has a low liquid level switch in order to avoid the running if the level is excessively low. The reciprocating feed pump is constituted by a membrane allowing firstly the suction of the mixture and the discharge towards the tank with a flow capacity F = 4.32 L.h-1. The column has also a reboiler of 2 liters hold-up capacity, an immersion heater of a power Qb = 3.3 kW and of a level liquid switch sensor which allows the automatic stop of heating if the level is insufficient. The column can be used in atmospheric (Patm) or in vacuum conditions (Pr). The stirring of the mixture in the reboiler is ensured by the boiling mixture. The internal packing is made of Multiknit stainless 316L which enhances the mass transfer between the vapor and liquid phases. In order to approach the adiabatic conditions, a heat-insulating made of glass wool is laid around the column. A condenser is placed at the column overhead in order to condense the entire vapor coming out from the column. The cooling (Q_c) used in exchangers is water. The heat transfer area of the total overhead condenser is 0.08 m^2. Moreover the reflux timer (R_t) allows to control the reflux ratio (R_r). It is monitored by the overhead product temperature (T_d). When the required distillate temperature (T_d) is attained, the reflux timer opens. In the opposite case, it remains closed. Distillation supervision control system allows to modify the parameters and to follow their evolution such as the pressure drop (ΔP), the flow or the temperatures at different points of the distillation column. This control system, therefore, must hold product compositions as near the set points as possible. The thermocouples are coupled to a calibrated amplification circuit (4-20 mA, 0-150°C) whose signals are inputted to the computer on-line, which permits the bottom and top temperatures to be obtained. The unit has twelve sensors which measure continuously the temperature throughout the column.

Selection of the reliable and reduced structure of the ANN model

For ANNs learning, it is very important to appropriately choose the input and output variables, determination of the relative importance of inputs and time delay [25]. For the experimental application as this used distillation column, it is often faced with the challenge of pre-processing a huge set of possible inputs. However, many variables possess redundant information and not all variables are relevant or even necessary. The strategy adopted here is the most relevant input variables are those preferentially define the most influential rules appearing in the model. For this reason, each input variable is modified and the variation relevance on the output variables is examined. It has to be bear in mind that this analysis of the relative importance is based on an experimental work but it takes the real physical nature into account as well. This analysis is based mainly on the physical knowledge of the industrial processes. The fact of changing an input and only one allows to visually observing the delay between this input and the outputs of the process. If the delay exists, therefore there is dependence between this input and the observed output. Otherwise, there is no correlation between these quantities. For this particular reason, the distillation unit operates at T_d = 95°C from the isobar diagram of the toluene-methylcyclohexane by using the Wilson thermo-dynamic model for vapor-liquid equilibrium (VLE). In order to define this nominal mode, all the regulation systems of the column are put in a closed-loop configuration. When the steady-state regime is achieved, the column was operated for approximately four hours in steady-state mode (T_d = 95°C) in order to collect a sufficient number of data. The data are taken every 11 s. The set of the most important measurable variables of the process are (R_t, Q_b, P_r, ΔP, Q_f, T_f, F, Q_c). Amongst the remaining variables, an experimental analysis of the process is carried out in order to observe the influence of each input variable on the output variables. This analysis aims at reducing more the system. Table 1 gives the values of the measurable variables calculated on average from the nominal steady-state regime.

Influence of all input variables

This part summarizes the experimental variation of all inputs as seen previously in 4.2. Table 2 collects the mean time delay measured between each input variable and the output variable [25]. The lower time delay is, the higher the relative importance is, because if a variation of the input variable occurs the output variable would be disturbed afterwards a time delay. On another side, it is important to highlight that the time delay between the reflux timer and the distillate temperature is 44 s. The feed flow rate has an influence on (T_d); however the response of this temperature is especially low to a feed variation (210.6 s). Therefore, the feed flow rate has less influence on (T_d) than the other input variables. When a time delay is called infinite (represented by "∞"), it means that there is no influence of the input variable upon the output variables.

Pattern generation

This study aims at modelling the overhead product temperature (T_d) according to the input variables of the process such as ΔP, R_t, Q_b, Q_f, Q_c, T_f, P_r and F. To have this database which is rich in amplitudes and in frequencies, the column behaviour is modified in both in a steady- state and dynamics conditions in the chosen temperature range ($T_d = 93.4 \rightarrow T_d = 95.9$°C).

The test data are a set of independent data used to verify the consistency of the efficiency of the model. For more legibility of figures, the parameters of the distillation column are represented in Figures 1 and 2. The operation duration of the distillation column is 12 hours continuously.

Data normalization

To maintain the influence of smaller data values in comparison to higher input values, the generated experimental data are no-directly introduced in the network as training patterns. For this reason, the

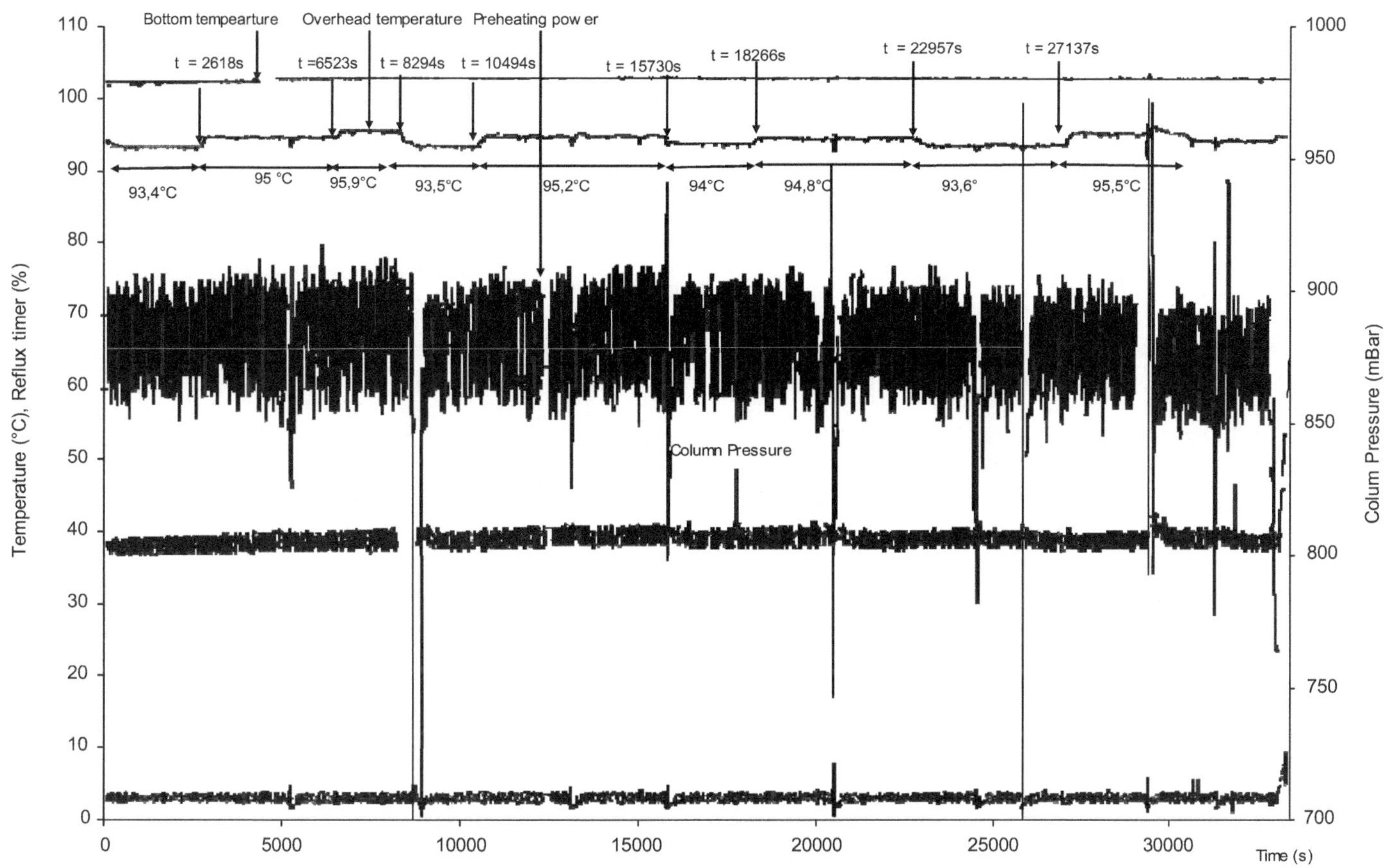

Figure 1: The operating conditions of the nominal steady-state regime.

experimental data are normalized before being presented to the ANN which gives equal priority to all the inputs variables. In our study, data normalization compresses the range of training data between 0.1 and 0.9.

Reliability of modelling of the overhead product temperature (T_d)

The number of nodes in the input and the output layers depend on the number of input and output variables, respectively. In this study, the adopted strategy is chosen as follows; the initial model has a low number of parameters and hidden nodes were gradually added during learning until the optimal result is achieved in the test subset. In the present work, the number of hidden nodes was modified from 1 to 12. Also, one hidden layer was used. The complete training process of networks took approximately 80,000 epochs using the Levenberg-Marquardt algorithm. In this case, the model composed by the set of inputs (ΔP, R_t, Q_b, Q_f, Q_f, T_f, P_r and F) and the output (T_d) is reduced according the equation 2 and tables 1 and 2.

$$T_d(t) = \varphi(T_d(t-1), T_f(t-3), R_t(t-4), \Delta P(t-8), Q_t(t-8) \quad (13)$$

The difficult trade-off between model accuracy and complexity can be clarified be clarified by using model parsimony indices from linear estimation theory (Ljung, 1999) such as Akaike's Information Criterion (AIC), Final Prediction Error (FPE) and Bayesian Information Criterion(BIC). A strict application of the indices would select a number Nh=5 because it exhibits the lowest of three indices for all the model structures compared. In conclusion, the developed network architecture used consisted of 5 nodes in the input layer and 5 nodes for the hidden layer. This reduced neural model is considered as a reliable one for describing the dynamic behaviour of the studied distillation column. In conclusion, the identified model is reduced from a (9-9-1) neural structure to (5-5-1)

Application of the developed fault detection

Once the identified ANN is trained and tested, it is ready for detecting faults. This neural model developed in part 4.5. is used both for the prediction and the FD procedure using the analysis of residuals. The proposed method is based on the GLRT as described in Section 3. This test should be processed to detect any real fault condition, rejecting any false alarms caused by noise. In this section, the proposed FD is applied to an experimental fault to verify its workability and effectiveness under real measurement conditions.

Statistics concerning the incidents which occurred in the distillation column showed that the most frequent faults are due to an inadequate heating power, feed pump, feed preheating, reflux or pressure used in the column. Consequently, the choice was carried out on this type of faults. To illustrate the adopted approach for the fault detection, it was decided to attempt among these faults to detect a sudden closing of the reflux timer (R_t=0%). This fault is frequent in the distillation operation and introduces a large deviation in comparison with the normal behaviour. It is important to notice that this fault occurs at 8173 s causes a large decrease of the overheat temperature (T_d). This evolution is showed in Figure 4.

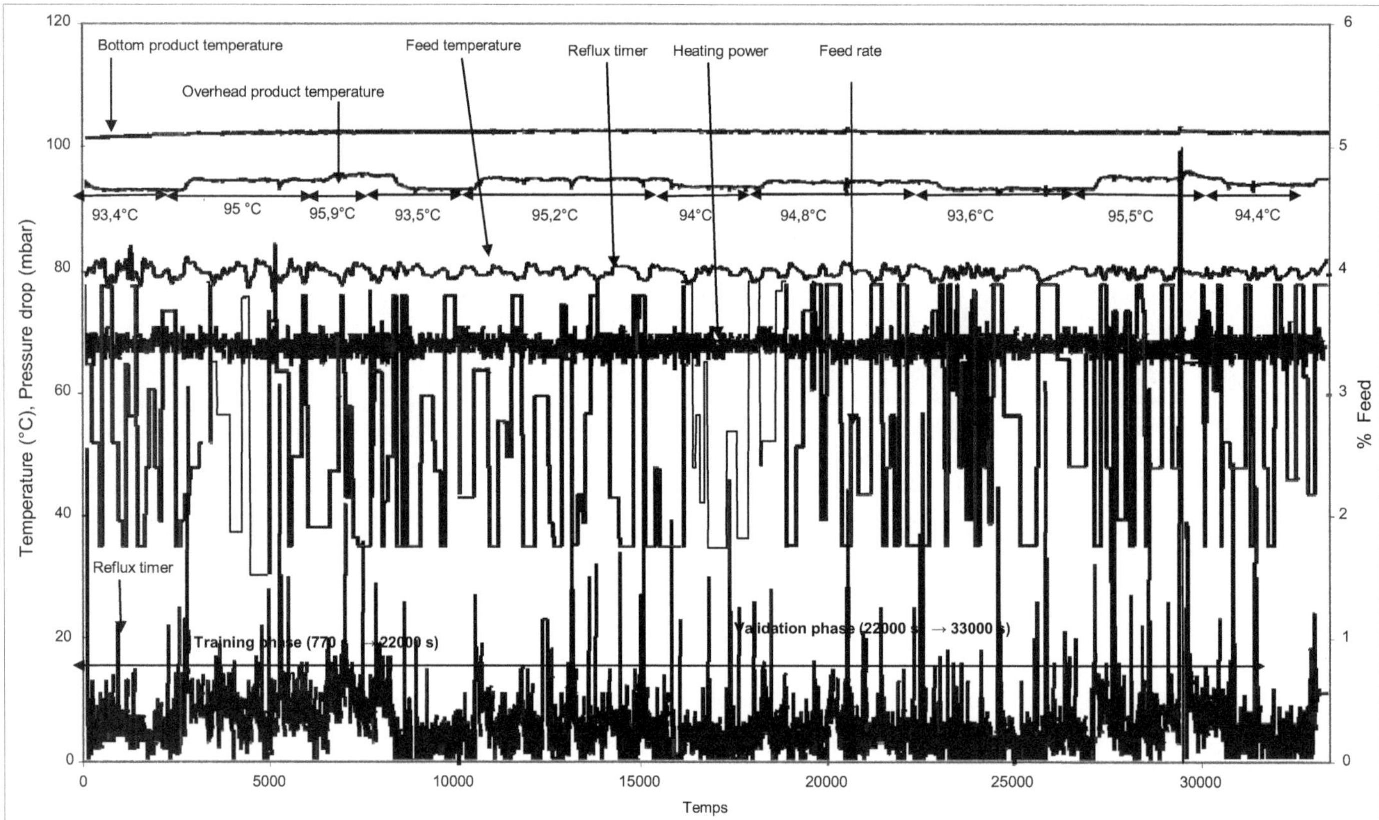

Figure 2: Mean time delay (seconds) between input and output variables.

Rt	*Qb*	*Tf*	*Td*	*F*	*Qc*	*ΔP*	*Pr*	*xb*	*xF*	*Qf*
7 %	68 %	80.1 °C	95.15 °C	80 %	250 $L.h^{-1}$	3.02 mbar	800 mbar	0.25 [-]	0.25 [-]	16 %

Table 1: The operating conditions of the nominal steady-state regime.

Input / Output	*Rt*	*Qb*	*ΔP*	*Qf*	*Tf*	*F*	*Qc*
Td	44	104.5	93.5	95.3	29.3	201.6	∞

Table 2: Mean time delay (seconds) between input and output variables.

This decrease should be detected by the GLRT because it exceeds a statistical threshold (3.841). The statistical threshold of decision is chosen according to the physical description of the process. It considers the different noises resulting from the process and the imperfections. Of its modeling. Indeed, it must be chosen too high in order to avoid generating false alarms due to the variations of the measurement equipment. By analyzing the various rates of false alarms probability, a rate of 5% represents a good threshold of decision. This statistical acceptance threshold is shown clearly in Figure 5 which represents the evolution of the GLRT. The size of the window is chosen according to the system dynamics. The analysis of the GLRT algorithm shows that the size value M =10 is a judicious compromise between the process dynamics, the delay time in detection and the GLRT sensitivity.

Figure 5 shows two operating zones: a fault region and a confidence one. It is important to notice that the fault which occurs at 8173 s, is detected at 8195 s i.e. with a delay of 33 s which corresponds to a difference ($\Delta T \approx 0.5$°C) between the temperature of the desired overhead temperature (T_d =95°C) and the fault one (T_d =94.5°C). This delay in detection defined as the difference between the time of the fault occurrence and the time of its detection, depends essentially on the evolution of the process dynamics.

Conclusion

Faults that change the system dynamics by causing surges of drifts of components, abnormal measurements, sudden shifts in the measurement channel, and other difficulties such as the decrease of instrument accuracy, an increase of background noise, reduction in actuator effectiveness etc., effect the characteristics of the likelihood ratio. This study shows that the combination of the ANN and the GLRT solves efficiently the FD problem. The experimental results show that the one-layer perceptron network provides promising assignments

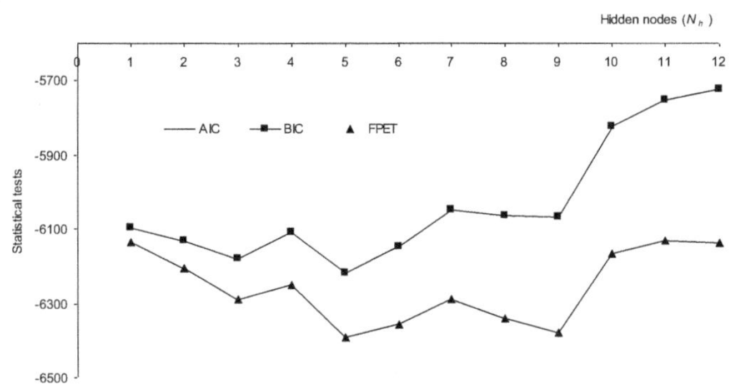

Figure 3: Evolution of the statistical criteria.

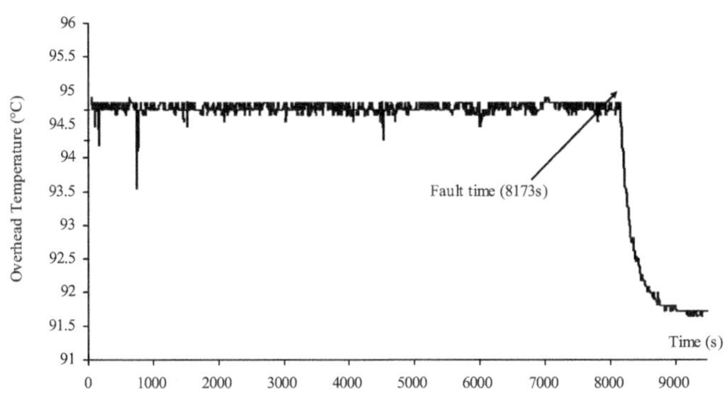

Figure 4: Evolution of the overheat temperature (*Td*) caused by the chosen fault.

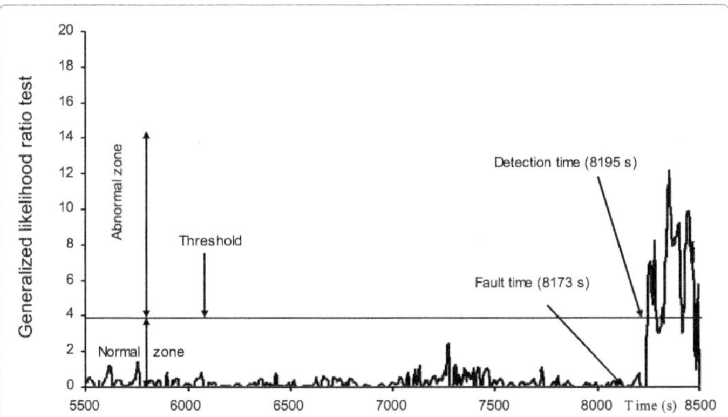

Figure 5: Fault detection test based on GLRT.

to normal and faulty states of the investigated reference process. The dedicated example indicates the strength of the proposed approach for reliability of models, prediction of the future data and fault detection. It makes this combination very attractive for solving the modelling issues and fault detection in real time operation of complex plants as a distillation column. GLRT is used to advise the operator of an abnormal behavior of the process by setting-on the suitable alarm. Such alarm could be used in order to initiate the back-up procedure, to stop the operating plant or even to begin the diagnosis algorithm of the fault physical origin.

References

1. Venkatasubramanian V, Rengaswamy R, Yin K, Kavuri S (2003) A review of process fault detection and diagnosis Part I: Quantitative model-based methods. Computers and Chemical Engineering 27: 293-311.
2. Fuller R (2000) Introduction to Neuro-Fuzzy Systems. Springer, New York 2: 289.
3. Zurada J, Karwowski W, Marras WS (1997) A neural network-based system for classification of industrial jobs with respect to the risk of low back disorders due to workplace design. Applied Ergonomics 28: 49-58.
4. Sharma R, Singh K, Singhal D, Ghosh R (2004) Neural network applications for detecting process faults in packed towers. Chemical Engineering and Processing 43: 841-847.
5. Himmelblau DM (2000) Applications of artificial neural networks in chemical engineering. Korean J Chem Eng 17: 373-392.
6. Söderström T, Stoica P (1989) System Identification. Englewood Cliffs NJ: Prentice-Hall.
7. Leontaritis IJ, Billings SA (1985) Input–output parametric models for nonlinear systems part I: deterministic nonlinear systems. Int J Control 41: 303-328.
8. Chetouani Y (2007) Use of Cumulative Sum (CUSUM) test for detecting abrupt changes in the process dynamics. International Journal of Reliability, Quality and Safety Engineering14: 65-80.
9. Previdi F (2002) Identification of black-box nonlinear models for lower limb movement control using functional electrical stimulation. Control Engineering Practice 10: 91-99.
10. Ljung L (1999) System Identification, Theory for the User, Prentice Hall, Englewood Cliffs, NJ.
11. Zaknich A (2003) Neural networks for intelligent signal processing. Singapore.
12. Chen TC, Han DJ, Au FTK, Tham LG (2003) Acceleration of Levenberg-Marquardt training of neural networks with variable decay rate. In: Neural networks. Proceedings of the international joint conference 3:1873-1878.
13. Nguyen D, Widrow B (1990) Improving the learning speed of 2-layer neural networks by choosing initial values of the adaptive weights. International Joint Conference of neural networks 3: 21-26.
14. Gustafsson F (1996) The marginalized likelihood ratio test for detecting abrupt changes. IEEE Trans. on Automatic Control 41: 66-78.
15. Basseville M, Beneveniste A (1987) Detection and diagnosis of changes in the eigenstructure of nonstationnary multivariable systems. Automatica 23: 479-489.
16. Willsky AS, Chow EY, Gershwin SB, Greene CS, Houpt PK, et al. (1980) Dynamic model-based techniques for the detection of incidents on freeways. IEEE Tran Aut Control 25: 347-360.
17. Dowdle JR (1982) Nonlinear generalized likelihood ratio algorithms for manoeuver detection and estimation. American Control Conf 985-987.
18. Sanderson AC, Segen T (1980) Hierarchical modelling of EEG signals. IEEE Trans Pattern Analysis and Machine Intelligence 2: 405-414.
19. Nikiforov IV (2001) A simple change detection scheme. Signal Processing 81: 149-172.
20. Micera S, Sabatini AM, Dario P (1998) An algorithm for detecting the onset of muscle contraction by EMG signal processing. Medical Engineering & Physics 20: 211-215.
21. Keller JY, Darouach M, Krzakala G (1994) Fault detection of multiple biases or process leaks in linear steady state systems. Computers & Chemical Engineering 18: 1001-1004.
22. Boukai B (1990) The asymptotic power function of GLR tests for local change of parameter. Journal of Statistical Planning and Inference 26: 291-303.
23. Micera S, Sabatini AM, Dario P (2000) On automatic identification of upper-limb movements using small-sized training sets of EMG signals. Medical Engineering and Physics 22: 527-533.
24. Kay SM, Gabriel JR (2002) Optimal invariant detection of a sinusoid with unknown parameters. IEEE Trans. Signal Process 50: 27-40.
25. Chetouani Y (2008) Using neural networks and statistical tests for detecting changes in the process dynamics. International Journal of Modelling Identification and Control 3: 113-123.

Determination by Modeling of the Interactions during the Adsorption Equilibrium of N-Propanol and Ortho Xylene on Activated Carbon Surface

Abdouelilah Hachimi*

Research Institute of Catalysis and Environment of Lyon (IRCELYON), UMR 5256 CNRS, Claude University, Bernard Lyon I, Villeurbanne, France

Abstract

Temperature Programmed Adsorption Equilibrium (TPAE) procedure was used to determine the heats of adsorption of ortho xylene and 1-propanol on activated carbon at 300 K. The evolutions of the surface coverage with the adsorption temperature (θ_e=f(T_a)) were compared to the Temkin model in order to determine the heats of adsorption and reveal the affinity to the surface of activated carbon regarding the polarity of each adsorbate. It has been shown that the heats of adsorption of ortho xylene were $E_{\theta 0}$=100 kJ/mol and $E_{\theta 1}$=62 kJ/mol at 185 Pa, respectively, at low and high surface coverage. Whereas the heats of adsorption of 1-propanol at the adsorption equilibrium was found to be dependent to the surface coverage with $E_{\theta 0}$=67 kJ/mol and $E_{\theta 1}$=58 kJ/mol at 475 Pa. It was shown that ortho-xylene molecules have more affinity to the activated carbon surface than 1-propanol regarding the difference in the heats of adsorption.

Keywords: Adsorption equilibrium; Heat of adsorption; Temkin model; Langmuir Model; VOCs removal; Adsorption modeling

Introduction

Alcohol and benzene derived compounds among others volatile organic compounds (VOCs) are known to participate to the air quality alteration. These compounds are generally emitted from petrochemical process industries, vehicles and building/construction materials. Their abatement constitutes a priority to anticipate their effects on human health [1]. The adsorption-desorption/destruction constitutes the most processes currently used to prevent the VOCs emission in the atmosphere [2-5]. Their severe reduction to lowest level needs sorbents with large surface area and pore volume. Zeolites [2], metal-organic framework (MOF) [4] and activated carbons [3,6-8], are the most frequently utilized adsorbents, having great adsorption capacity for VOCs.

Activated carbon has been the subject of several studies for the removal of VOCs from air [9] and water [10]. The gas adsorption on activated carbon is dominated by the forces of dispersion and repulsion of Van Der Waals [11]. The electrical charges on the activated carbon are low, which leads to a low potential gradient on surface [12]. The existence of small pore sizes plays an important role in the adsorption capacity of gases on activated carbons [13]. Furthermore, the surface groups have a significant effect in the adsorption of gas polar. Thus, the adsorbed amount increases with the existence of oxidic sites on surface such as carboxyl, lactone, carbonyl, quinone, phenol, ether, carboxylic anhydride, lactol…etc [11-13].

Activated carbon has hydrophilic and hydrophobic adsorption sites that promote adsorption of non-polar and polar molecules in the micropores [14]. However, condensation can occur in the micropores and the mesopores in the adsorption pressure once approaches the critical pressure [15]. Dobruskin [16] modeled the adsorption of benzene on a microporous activated carbon (BET=805 m^2/g, V_{micr}=0.42 cm^3/g), he concluded that the inflection point after the saturation of the monolayer is obtained at a relative pressure of 0.17. For alcohols, Salame and Bandosz [17] studied the adsorption of methanol on several microporous activated carbons; they noticed that the methanol begins to condense at the relative pressure of 0.3. Pre et al. [11] conducted a quantitative study to predict the heat of adsorption and activation energy of desorption of VOCs on activated carbon using micro-calorimetry technique. They studied several parameters having a relationship with charcoal-VOCs interactions, including dipole moment, ionization potential, the dielectric constant, polarizability and surface tension. They showed linear regression between these variables and that the adsorption/desorption varies inversely with the potential of ionization and linearly with the polarizability. The heats of adsorption found in the range 40-80 kJ/mol.

The goal of the present investigation was to study the kinetic parameters (heats of adsorption and surface coverage) involved in the adsorption equilibrium of 1-propanol and ortho-xylene on activated carbon using temperature programmed adsorption equilibrium method (TPAE). Therefore, it is of great scientific and practical interest to study the adsorption properties of a commercial activated carbon with respect to different adsorbates (polar and/or non-polar). Such investigations could provide evidence for the importance of some adsorbent parameters including the surface heterogeneity and polarity of adsorbate and leading to measure by modeling the interactions in the adsorbate/adsorbent system. In addition, discussions about the determination the heats of adsorption of alcohol and benzene-derived compounds on activated carbon are limited in the open literature.

Experimental

Materials

Activated carbon (denoted AC) from Prolabo has been selected for two main interests: (a) it is a micro-porous solid and (b) it has non-reactivity with OX or PrOH excepted the adsorption. The AC samples were pretreated before each adsorption in order to remove eventual trace of adsorbates, as follow:

***Corresponding author:** Abdouelilah Hachimi, Research Institute of Catalysis and Environment of Lyon (IRCELYON), UMR 5256 CNRS, Claude University, Bernard Lyon I, Villeurbanne, France

He, 300 K → He, 723 K (2 h) → He, 300 K.

For the adsorption experiments, ortho di-methyl benzene (Alfa Aesar, 99%) and 1-Propanol (VWR BDH Prolabo, 99%) were used as adsorbates. Dry helium was used as carrier gas with purity greater than 99.999%.

Sorbent characterization

The textural properties (specific surface area and pores volume) of AC were determined by exploiting the isotherm data of the nitrogen physisorption at 77 K. The adsorption of nitrogen was performed on an automated Micrometrics ASAP 2010 V5.02H analyzer after out degassing the samples at 873 K in a vacuum of a relative pressure $<10^{-7}$ to ensure a dry clean surface and free sites of adspecies.

Figure 1 shows the nitrogen adsorption and desorption isotherms of the AC sample. It is obvious that the adsorption isotherm of N_2 on the AC is of typical type-I profile with secondary uptakes, which implies the presence of micro and mesopores. The BET and Langmuir surface areas and pores volume are given in the table 1.

Analytical procedures

The temperature programmed adsorption equilibrium (TPAE), previously developed for hydrogen chemisorption on Pt/Al_2O_3 catalysts [18], was used in the present work in order to determine the kinetic parameters (heats of adsorption and surface coverage) involved in the adsorption equilibrium of heavy molecules such as PrOH and OX on AC, and to evaluate its efficiency of VOCs removing from bulk stream. TPAE experiments were performed in dynamic conditions using a purpose-built device described above. In brief, it consisted of a gas chromatography (GC) with flame ionization detector (FID) for transient experiments. The assembly permitted (i) to prepare gas mixture in the reactor inlet and to control feeding flow rate, (ii) to perform in-situ pre-treatment, adsorption and desorption steps in quartz micro reactor (1 cm^3) and (iii) to monitor OX and PrOH molar fractions in the reactor outlet by mean of GC-FID system.

TPAE experiments

The principle of the TPAE experiments was reported elsewhere [5,19]. Briefly, around 0.100 g of AC was treated as described above and cooled down under helium to 300 K in order to remove moisture and adspecies from the surface. Afterward, under atmospheric pressure, a switch He x% PrOH/He or y% OX/He (x and y fixed by saturator/condenser system using the well-known coefficient table of Antoine Equation) was performed to start the adsorption step and the mixture in the reactor outlet was analyzed by GC-FID at the frequency of 1/180 Hz. Once molar fractions at the reactor inlet and outlet are equal, the adsorption equilibrium was considered established. The adsorption temperature (T_a) was slightly increased (0.5-1 K/min) under quasi-constant partial pressure of the adsorbate (PrOH or OX) leading to desorption of the adspecies associated to positive peaks. An isothermal desorption is carried out under helium to verify the total desorption of adspecies once the molar fractions are again equal at high temperature. Finally, the AC was cooled down to the room temperature. The experimental data was presented as breakthrough curves showing the evolution of molar fractions with time.

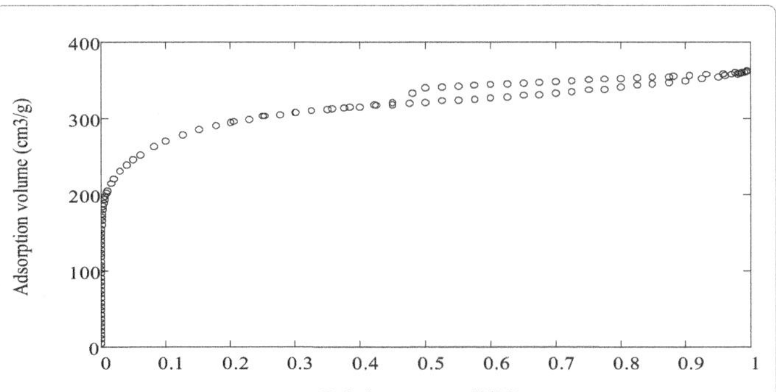

Figure 1: N_2 sorption isotherms at 77 K of the AC. P/Po is the ratio of gas pressure (P) to saturation pressure (Po=101325 Pa).

Results and Discussion

Mathematique models

Surface coverage (θ) variation with temperature (T_a) and pressure (P_a) of adsorption allows measuring the interaction strengths implied in the adsorbate/adsorbent systems. The heats of adsorption (denoted by E_θ) of the adspecies and their evolutions with the surface coverage θ (with $0 \le \theta \le 1$) constitute a fundamental characterization of the adsorbate/adsorbent systems. Temkin model [20] among numerous mathematical models describes the curve of θ = f (Ta) using statistical thermodynamics. This model assumes that (i) the adspecies are localized, (ii) the heat of adsorption linearly decreases with increase in surface coverage, (iii) heterogeneous surface sites and (iv) and indirect interactions between adspecies. The Temkin model is given by the following equation:

$$\theta_e = \frac{RT_a}{\Delta E} \ln\left(\frac{1+K_{\theta 0}P_a}{1+K_{\theta 1}P_a}\right) \quad (1)$$

Where R is the gas constant, $K_{\theta 0}$ and $K_{\theta 1}$ are the coefficient of adsorption respectively at low (θ =0) and high coverage (θ =1) with:

$$K_{\theta 0} = \frac{h^3}{(2\pi M)^{\frac{3}{2}}.(kT_a)^{\frac{5}{2}}} \exp\left(\frac{E_{\theta 0}}{RT_a}\right), \quad K_{\theta 1} = \frac{h^3}{(2\pi M)^{\frac{3}{2}}.(kT_a)^{\frac{5}{2}}} \exp\left(\frac{E_{\theta 1}}{RT_a}\right)$$

and $\Delta E = E_{\theta 0} - E_{\theta 1}$ Where $E_{\theta 0}$ and $E_{\theta 1}$ are the heats of adsorption respectively at low (θ = 0) and high (θ = 1) coverage. The heats of adsorption ($E_\theta = E_a - E_d$) are the difference between the activation energy of adsorption (E_a) and desorption (E_d) at the adsorption equilibrium.

The theoretical curve obtained from Eq. 1 provides the evolution of the surface coverage with the adsorption temperature (T_a) by supposing $E_{\theta 0}$ and $E_{\theta 1}$, and its comparison with the experimental surface coverage variation permits the determination of the adsorption heats ($E_{\theta 0}$ and $E_{\theta 1}$). It could be noted that during TPAE step, the slight perturbation of the equilibrium by increasing the temperature of adsorption leads to a slight variation of adsorption pressure of about 20% that conducts to the variation of heat of adsorption of around 3 kJ/mol. This value was considered as the measurement error of the heat of adsorption determined by TPAE.

Reference of AC	Particles Size (mm)	S_{BET} (m^2/g)	S_{ext} (m^2/g)	S_{meso} (m^2/g)	S_{micr} (m^2/g)	V_{micr} (cm^3/g)	V_{meso} (cm^3/g)	V_p (cm^3/g)	D_p (A°)
N°89067	1 to 5	1055	125	250	680	0.36	0.13	0.56	10.6

Table 1: Measured Textural parameters of activated carbon with particle diameter (Dp), volumes: total pore (V_p), mesopores (V_{meso}), micropores (V_{micr}), surfaces: BET (S_{BET}), external (S_{ext}), microporous (S_{micr}) and mesoporous (S_{meso}).

At the equilibrium, the surface coverage (θ_e (T_a, P_a)) can be determined by the ratio of the adsorbed amount (Q_a) at T_a and P_a and the amount forming the monolayer (Q_{ML}):

$$\theta_e(T_a) = \frac{Q_a}{Q_{ML}} \quad (2)$$

The Q_{ML} can be calculated considering surface area of the solid (S = 1055 m^2/g) and the cross-section area σ (m^2/molecule) of an adsorbed molecule (Eq. 3):

$$Q_{ML} = \frac{S.10^{20}.10^{6}}{\sigma.N_A} \text{ (µmol/g)} \quad (3)$$

Where N_A is the Avogadro number (N_A= 6.02210[23] molecule. mol^{-1}), S is the surface area (m^2/g) and σ(m^2) is the cross-section between a molecule and an adsorbent. This later has been determined experimentally by McCLellan and Harnsberger [21]. They found that the cross-section of PrOH and OX is around 31 and 48 ($Å^2$/molecule), respectively. Considering Eq. 3 the Q_{ML} was determined to around 5650 and 3650 µmol/g, respectively, for PrOH and OX. The adsorption pressures were chosen to be significantly less than those obtained by the vapor at the saturation at the temperature of adsorption (T_a=300 K), 0.75% and 2.3%, respectively, for OX and PrOH. Those values were calculated considering the molar fractions at the peaks of TPAE, which are linked to the adsorbate partial pressure that increases with 20% during the TPAE step. At the equilibrium, the surface coverage evolution, during TPAE step, can be determined using the following equations:

$$\theta(T_a) = \frac{Q_a - \frac{F}{W}\int_{t_{eq}}^{t}\left(C_t - C_{t_{eq}}\right)dt}{Q_{ML}} \quad (4)$$

$$\text{With } Q_a(T_a, P_a) = \frac{F}{W}\int_{t_0}^{t_{eq}}\left(C_{t_{eq}} - C_{t_0}\right)dt \quad (5)$$

Where: Q_a is the amount (mol/g) of an adsorbate at T_a and P_a. F is the flow rate (cm^3/min) of adsorbate/He. W is the weight (g) of the adsorbent. C_{t0}, C_{teq} and C_t are the adsorbate's molar fraction at the reactor outlet, respectively, at the start time of the adsorption (t_0), at the time at which equilibrium is established (t_{eq}) and at any time during TPAE experiment (t > t_{eq}). Q_{ML} is the amount (mol/g) of an adsorbate forming monolayer.

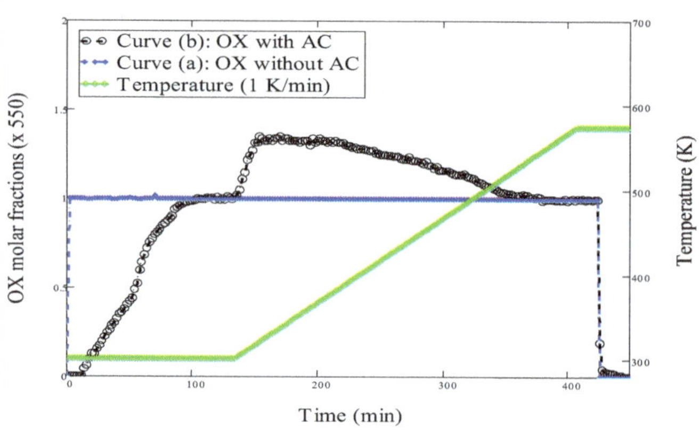

Figure 2: Molar fractions evolutions of OX (185 Pa) adsorption at 300 K and 10^5 Pa, TPAE and isothermal desorption processes in the reactor outlet with and without activated carbon inside.

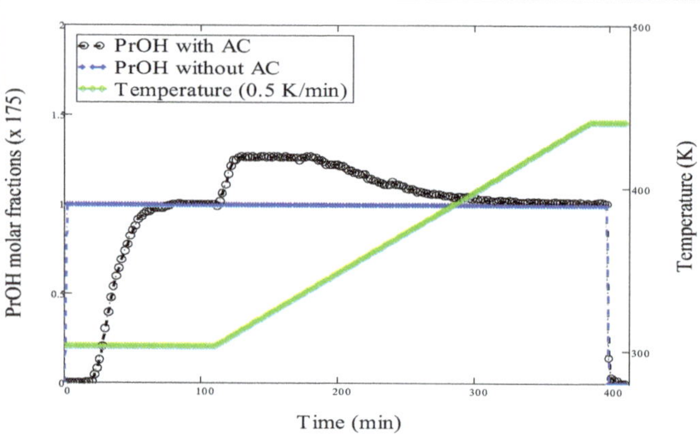

Figure 3: Molar fractions evolutions of PrOH (475 Pa) during the adsorption at 300 K and 10^5 Pa, TPAE and isothermal desorption processes in the reactor outlet with and without activated carbon inside.

Characterization of the OX adsorption equilibrium on AC

The breakthrough curve permitted the monitoring of adsorbate loading on adsorbent as function of time. Figure 2 shows the evolution of molar fractions of the 0.183%OX/He mixture at the reactor outlet in presence (curve (b)) and absence of 0.100 g of AC (curve (a)) during the TPAE experiment steps. The adsorption step of OX on AC was carried out using an isobaric (P_a=185 Pa) of OX at T_a=300 K. The equilibrium was established after 100 minutes with AC in the reactor whereas the breakthrough time for the empty reactor is less than minute. The amount of adsorbed OX was determined using Eq. 5. The subtraction of the area of (curve (b)) from the area of the empty reactor (curve (a)) permitted to assess the adsorbed amount of OX. This later was found to be 4160 µmol/g corresponding to θ_e=114% of the AC surface coverage. This result can be explained by the fact that, at 185 Pa and 300 K, the pores were partially filled with OX molecules because of the formation of multilayers on the surface of AC [22]. This assumption could be verified with comparing the adsorbed volume with the total pores volume. Indeed, such result is consistent since the total pores volume of the AC is 0.56 cm^3/g which is close to the corresponding adsorbed volume of OX (0.51 cm^3/g) even though the adsorption pressure (185 Pa) is lower than the OX saturation pressure at 300 K (970 Pa) with a relative adsorption pressure of 0.19. This result is consistent with the work of Dobruskin [16]. The author [16] has modeled the adsorption of benzene on a microporous activated carbon (BET = 805 m^2/g, V_{micr}=0.42 cm^3/g), he concluded that the inflection point after the saturation of the monolayer is obtained at a relative pressure of 0.17.

Once the equilibrium established (OX molar fractions are equal at reactor inlet and outlet), the temperature programmed (1 K/min) was performed under 0.183%OX/He mixture resulting a positive and asymmetric peak of desorbed OX adspecies. The amount of the OX adspecies desorbed during TPAE was determined using the Eq. 5 by subtracting areas of the curve (b) from the curve (a) in the temperature range 300-550 K and found to be 4130 µmol/g. While the amount of OX desorbed during the isotherm at high temperature, was 30 µmol/g.

Similar results were obtained by applying high adsorption pressure (680 Pa) until equilibrium using 0.100 g of AC and under 0.68% OX/He mixture at 300 K. The amount of OX adsorbed was around 4655 µmol/g which corresponds to the adsorbed OX volume of 0.56 cm^3/g that matches with the total pores volume of AC and then the saturation of AC. However, the adsorption potential depends not only on the characteristics of the adsorbent but also on those of the adsorbate. Materials having narrow micropores adsorb more than those with wide or non-porous. The adsorption capacity of AC for OX could be compared with those given for the adsorption of xylene on activated carbon AC40 (1300 m^2/g) [23] and much higher than those obtained for non-porous silica (220 m^2/g) [5].

Characterization of the PrOH adsorption equilibrium on AC

Similar experiments to the adsorption of OX were carried out on the adsorption of PrOH on AC (Figure 3). The adsorption equilibrium was obtained during 70 min using 0.147 g of AC and 0.46% PrOH/He at 300 K. Same methods were used for determination of the PrOH adsorbed amounts as for OX on AC. The amounts of the PrOH uptake during the adsorption, TPAE and desorption under helium steps were, 6050, 5950 and 100 μmol/g, respectively. The adsorbed amount corresponds to 0.45 cm^3/g which represents 80 vol. % of pores filling and θ_e=112 % of the surface coverage at the relative pressure of 0.16 at 300 K. The effect of the adsorption pressure was studied in order to reveal the saturation limit of the AC. The experiment revealed that the amount of PrOH adsorbed on AC increases with the adsorption pressure. In fact, at the saturation pressure and T_a=300 K, the uptake amount was found to be 7550 μmol/g which corresponds to the volume of 0.56 cm^3/g of PrOH adsorbed. These results can be explained by formation of multilayers on the surface and condensation of PrOH molecules in the pores of AC. However. These results are unlike to those found by Salame and Bandosz [17]. They noticed in similar studies that the adsorption of methanol on several microporous activated carbons leads to the interparticulate and the pores capillary condensation that starts with applying relative pressure from 0.3. The difference, in the condensation at relative pressure, could be linked to the chemical properties of the adsorbents used in both studies. On the other hand, the AC was completely regenerated at 420 K and 550 K for PrOH and OX, respectively; unlikely to that found in a previous work performing similar studies on non-porous silica sorbent [5]. This later was found readily regenerated at around 400 K for both adsorbates.

Heats of adsorption of OX and PrOH on AC under adsorption equilibrium

The experimental curves in Figure 4, show the plot of θ_e=f (T_a) obtained from the data in Figure 2 by using (Eq. 4) and (Eq. 5) and provides the evolution of OX surface coverage on AC for T_a > 300 K and P_a=185 Pa (OX) during TPAE experiment. The equilibrium surface

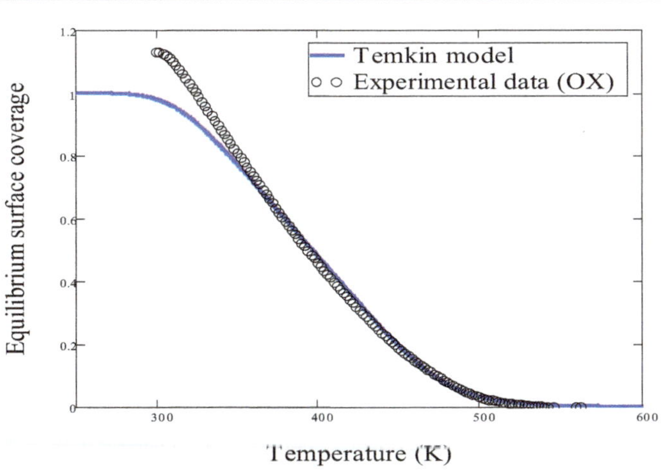

Figure 4: Modeling of the experimental surface coverage of OX on AC under quasi-equilibrium with Temkin model using the heats of adsorption ($E_{\theta 0}$= 100 kJ/mol and $E_{\theta 1}$=62 kJ/mol) at 185 Pa.

coverage was obtained using the ratio of the amount of desorbed molecules at a given T_a and P_a during TPAE step to the amount of OX necessary for the formation of monolayer on the AC surface (Eq. 4). At 300 K and 185 Pa, the equilibrium surface coverage was θ_e=1.14 and its variation θ_e=f (T_a) (experimental curve, Figure 4) was modeled using Temkin model (Eq. 1) taking into account that the heats of adsorption depend to the surface coverage with $E_{\theta 0}$ = 100 kJ/mol, $E_{\theta 1}$ = 62 kJ/mol and P_a=185 Pa.

It could be noticed that the experimental data deviates of the theoretical model in the temperature range 300-360 °C. In fact, at this temperature range the surface coverage drops from 1.14 to 0.73 because of desorption of the OX adspecies forming multilayers and/or probably from the evaporation of condensed OX molecules into AC pores [24]. However, at T_a>360 °C, the model Temkin fitted perfectly the experimental data curve. This result can be explained by desorption of the OX adspecies forming the monolayer of AC surface; in agreement with Hill [24] work in which the author reported that the multilayer formation occurred at surface coverage θ>0.70. On the other hand, the heats of adsorption of OX on AC determined using Temkin model parameters at high and low surface coverage under adsorption equilibrium found to be close to those found by Pre et al. [11] for the adsorption of toluene on activated carbon (63 to 71 kJ/mol), and to those found by Dollimore et al. [25] using micro-calorimetric technique and isosteric heats of adsorption for meta-xylene on non-porous graphite (60 kJ/mol). The authors [25] showed that at low surface coverage the heats of adsorption were much higher than at high surface coverage; they explained their results by the fact that the graphite surface was heterogeneous and the high values of the heats of adsorption at low surface coverage were associated to the oxygen chemisorbed on the surface. Hockey and Pethica [26] studied the adsorption of benzene on silica in various forms, and they have shown that micro porosity and lateral interactions increases the heats of adsorption at low surface coverage of silica. If their theory is correct, the high values of the heats of adsorption found at low surface coverage ($E_{\theta 0}$=100 kJ/mol) are due to the presence of micropores in AC, and at high surface coverage ($E_{\theta 1}$=62 kJ/mol) are due to the hydrogen physisorption on sites with low energy. The very first molecules adsorbed will be onto highly energetic sites and the molecules will remain fixed. Molecules arriving at high surface coverage will release less energy upon adsorption and may form a dilute mobile film due to the condensation.

On the other hand, similar results were obtained during the adsorption of PrOH on AC. Figure 5 depicts the variation of the surface coverage of PrOH on AC at 475 Pa with the adsorption temperature (T_a>300 K). The fitting of the Temkin model on the experimental curve permitted determination of the heats of adsorption. These later were found to be $E_{\theta 0}$=67 kJ/mol, $E_{\theta 1}$=58 kJ/mol at low and high surface coverage, respectively. Same remark could be noted on deviation of the experimental curve at high coverage. This could be related to the PrOH desorption that adsorbed physically on sites with low energy. However, it could be noticed that the heats of adsorption for polar molecules (PrOH) were found to be lower than those for non-polar molecules (OX), and the large difference in the heats of adsorption between low and high surface coverage for OX (100 kJ/mol) against that for PrOH (67 kJ/mol) let's assume that the heats of adsorption of OX could not be associated in major part to the microporosity of AC but to the chemisorption phenomena on specific sites in contrast to the PrOH adsorbate. The difference in the interactions between both adsorbates and AC can be associated to the difference in their dipolar moments. Indeed, the OX and PrOH molecules are characterized by different dipole moment 0.64 D and 1.68 D, respectively [27,28]. This

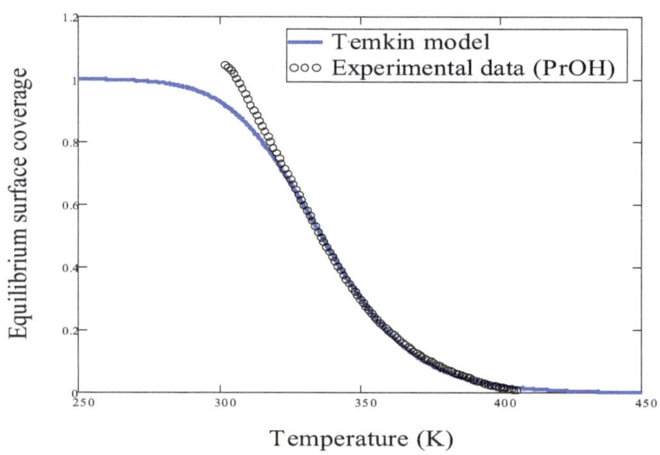

Figure 5: Modeling of the experimental surface coverage of PrOH on AC under quasi-equilibrium with Temkin model using the heats of adsorption ($E_{\theta 0}$=67 kJ/mol and $E_{\theta 1}$=58 kJ/mol) at 475 Pa.

characteristic may lead to different affinities to the AC surface sites and subsequently promotes their adsorption.

The heats of adsorption of PrOH on activated carbon obtained by using the Temkin model and adsorption equilibrium that were found to be consistent with the isosteric heats obtained for the adsorption of alcohol compounds on AC. Herry et al. [29] have found similar heats of adsorption for ethanol (54.78 kJ/mol), 1-propanol (59.47 kJ/mol) and butanol (76.87 kJ/mol), while Nastaj et al. [30] and Salame Bandosz [17] have found, 60.08 kJ/mol for the adsorption of 2-propanol and 55 kJ/mol for ethanol on AC. However, the heats of adsorption determined by microcalorimetric analysis which is an accurate tool were found to be close to those fond by TPAE method. Pre et al. [11] have found, in similar investigation on AC, that of the adsorption of ethanol (54.7 kJ / mol) and isopropanol (55 kJ / mol) whereas Popescu et al. [31] found similar results for butanol (63 kJ / mol).

Note that heats of adsorption given in literature seem to be an average values and did not take into account the surface coverage factor; in contrast, TPAE method gives variation of the heats of adsorption of PrOH with the surface coverage of AC.

In the present investigation, the heat of adsorption obtained indicates that weak interactions between PrOH and AC surface that may be due to the nonspecific interactions (Van Der Waals forces, London forces) and present good agreement with the fact that the PrOH adsorption is reversible.

Conclusions

The main objective of the present work was to demonstrate the interest of an easy and efficient method to determine to heats of adsorption. Polar (Propanol) and non- polar (ortho xylene) molecules were selected for the adsorption on a microporous adsorbent (activated carbon). The adsorption capacities for OX and PrOH showed that both adsorbates were condensed into the micropores of AC at 300 K at relative adsorption pressures of 0.16 and 0.19, respectively, of PrOH and OX. The evolutions of the surface coverage (θ_e) at the equilibrium of adspecies with the temperature of adsorption (Ta) have been determined by using temperature programmed adsorption equilibrium (TPAE) method. The heats of adsorption have been determined by modeling the experimental curves (θ_e=f(Ta)) with the theoretical model of Temkin considering that PrOH and OX adspecies were localized and the heats of adsorption found to depend to the surface coverage. The heats of adsorption of OX on AC found to be $E_{\theta 0}$=100 kJ/mol at low coverage and $E_{\theta 1}$=62 kJ/mol at high coverage. Whereas, the heats of adsorption of PrOH on AC were found to be lower than for OX with $E_{\theta 0}$=67 kJ/mol, $E_{\theta 1}$=58 kJ/mol at low and high surface coverage, respectively. It was concluded that OX molecules have more affinity to the AC surface than PrOH regarding the difference in the heats of adsorption. It could be concluded that TPAE experiment could be used into the characterization of the interactions involved in the adsorption of polar or non-polar molecules on microporous adsorbents.

Acknowledgment

I gratefully acknowledge the funding of the Institute of Research on Catalysis and Environment of Lyon (IRCELyon), France, for this study.

References

1. Larson RA, Weber EJ (1994) Reaction mechanisms in environmental organic chemistry. Boca Raton ed., Lewis Publishers.
2. Takeuchi Y, Iwamoto H, Miyata N, Asano S, Harada M (1995) Adsorption of 1-butanol and p-xylene vapor and their mixtures with high silica zeolites. Separations Technology 5: 23-34.
3. Lillo-Ŕdenas MA, Fletcher AJ, Thomas KM, Cazorla-Amor ́s D, Linares-Solano A (2006) Competitive adsorption of a benzene-toluene mixture on activated carbons at low concentration. Carbon 44: 1455-1463.
4. Yang K, Xue F, Sun Q, Yue R, Lin D (2013) Adsorption of volatile organic compounds by metal-organic frameworks MOF-177. Journal of Environmental Chemical Engineering 1: 713-718.
5. Hachimi A (2014) Adsorption models and heat of adsorption of adsorbed ortho di-methyl benzene and 1-propanol species in competitive adsorption on silica by using temperature programmed adsorption equilibrium methods. Journal of Environmental Chemical Engineering, 2 377-385.
6. Fuertes AB, Marbán G, Nevskaia DM (2003) Adsorption of volatile organic compounds by means of activated carbon fibre-based monoliths. Carbon 41: 87-96.
7. Mirzaei A, Ebadi A, Khajavi P (2013) Kinetic and equilibrium modeling of single and binary adsorption of methyl tert-butyl ether (MTBE) and tert-butyl alcohol (TBA) onto nano-perfluorooctyl alumina. Chemical Engineering Journal 231: 550-560.
8. Sing KSW (2014) 10-Adsorption by Active Carbons. In: F. Rouquerol, J. Rouquerol, K.S.W. Sing, P. Llewellyn, G. Maurin (Eds.) Adsorption by Powders and Porous Solids (Second Edition). Academic Press, Oxford 321-391.
9. Gallego E, Roca FJ, Perales JF, Guardino X (2013) Experimental evaluation of VOC removal efficiency of a coconut shell activated carbon filter for indoor air quality enhancement. Building and Environment 67: 14-25.
10. Marbán G, Fuertes AB (2004) Co-adsorption of n-butane/water vapour mixtures on activated carbon fibre-based monoliths. Carbon 42: 71-81.
11. Pré P, Delage F, Faur-Brasquet C, Le Cloirec P (2002) Quantitative structure–activity relationships for the prediction of VOCs adsorption and desorption energies onto activated carbon. Fuel Processing Technology 77-78: 345-351.
12. Tanada S, Nakamura T, Kawasaki N, Torii Y, Kitayama S (1999) Removal of Aromatic Hydrocarbon Compounds by Hydroxypropyl-cyclodextrin. See comment in PubMed Commons below J Colloid Interface Sci 217: 417-419.
13. Daifullah AAM, Girgis BS (2003) Impact of surface characteristics of activated carbon on adsorption of BTEX. Colloids and Surfaces A: Physicochemical and Engineering Aspects 214: 181-193.
14. Bae JS, Do DD (2006) On the equilibrium and dynamic behavior of alcohol vapors in activated carbon. Chemical Engineering Science 61: 6468-6477.
15. Dobruskin VK (2001) Correlation between two- and three-dimensional condensations in cylindrical mesopores of active carbons. Carbon 39: 583-591.
16. Dobruskin VKh (2002) Micropore structure of activated carbons and predictions of adsorption equilibrium. Carbon 40: 1003-1010.
17. Salame II, Bandosz TJ (2000) Adsorption of Water and Methanol on Micro- and Mesoporous Wood-Based Activated Carbons. Langmuir 16: 5435-5440.

18. Derrouiche S, Bianchi D (2004) Heats of adsorption using temperature programmed adsorption equilibrium: application to the adsorption of CO on Cu/A12O3 and H2 on Pt/Al2O3. See comment in PubMed Commons below Langmuir 20: 4489-4497.

19. Hachimi A, Chafik T, Bianchi D (2008) Adsorption models and heat of adsorption of adsorbed ortho di-methyl benzene species on silica by using Temperature Programmed Adsorption Equilibrium methods. Applied Catalysis A: General 335: 220-229.

20. Tompkins FC, Gundry PM (1960) Chemisorption of Gases on Metals. Harcourt Brace Jovanovich.

21. McCLellan AL, Harnsberger HF (1967) Cross-sectional Areas of Molecules Adsorbed on Solid Surfaces. Journal of Colloid and Interface Science 23: 577-599.

22. Rouquerol J, Llewellyn P, Rouquerol F (2007) Studies in Surface Science and Catalysis. Elsevier 49-56.

23. Benkhedda J, Jaubert JN, Barth D, Perrin L, Bailly M (2000) Adsorption isotherms of m-xylene on activated carbon: measurements and correlation with different models. The Journal of Chemical Thermodynamics 32: 401-411.

24. Hill TL (1946) Theory of Multimolecular Adsorption from a Mixture of Gases. Journal of Chemical Physics 14: 441.

25. Dollimore D, Heal GR, Martin DR (1972) Thermodynamics of adsorption of a series of related organic molecules on graphite and a carbon black. Part 1.-Heat of adsorption. J Chem Soc Faraday Trans 68: 832-838.

26. Hockey JA, Pethica BA (1962) Thermodynamic studies of adsorption on characterized silicas. Part 1.-Benzene. Transactions of the Faraday Society 58: 2017-2026.

27. Wilson RD, Barton DG, Baertsch CD, Iglesia E (2000) Reaction and Deactivation Pathways in Xylene Isomerization on Zirconia Modified by Tungsten Oxide. Journal of Catalysis 194: 175-187.

28. Zheng S, Jentys A, Lercher JA (2006) Xylene isomerization with surface-modified HZSM-5 zeolite catalysts: An in situ IR study. Journal of Catalysis 241: 304-311.

29. Herry C, Baudu M, Raveau D (2001) Estimation of the influence of structural elements of activated carbons on the energetic components of adsorption. Carbon 39: 1879-1889.

30. Nastaj JF, Ambrożek B, Rudnicka J (2006) Simulation studies of a vacuum and temperature swing adsorption process for the removal of VOC from waste air streams. International Communications in Heat and Mass Transfer 33: 80-86.

31. Popescu M, Joly JP, Carré J, Danatoiu C (2003) Dynamical adsorption and temperature-programmed desorption of VOCs (toluene, butyl acetate and butanol) on activated carbons. Carbon 41: 739-748.

4

Photodegradation study of Congo Red in Aqueous Solution using ZnO/UV-A: Effect of pH And Band Gap of other Semiconductor Groups

Laouedj Nadjia[1], Elaziouti Abdelkader[2]* and Bekka Ahmed[2]

[1]DR. Moulay Tahar University, Saida, Algeria
[2]LCPCE Laboratory, Faculty of sciences, Department of industrial Chemistry, University of the Science and the technology of Oran (USTO M.B). BP 1505 El M'naouar 31000 Oran, Algeria

Abstract

The photocatalytic degradation of Congo red in aqueous solution was investigated under UVA light at different operating conditions, including irradiation time, pH solution and band gap of MO_2 ($M=Ti^{+4}$ and Ce^{+4}) and M'_2O_3 ($M'=Al^{+3}$ and Fe^{+3}) semiconductor groups by UV-spectrophotometric monitoring. The results showed slight stability of 597nm-band of CR over the pH range of 6-7, while, it almost completely disappeared at pH higher than 7. Maximum photo degradation was obtained at pH 8 as a result of 95.02% degradation efficiency of CR for 60 min of irradiation time. The photo decomposition efficiency of MO_2 ($M=Ti^{+4}$ and Ce^{+4}) semiconductor group was higher, compared to that of M'_2O_3 ($M'=Al^{+3}$ and Fe^{+3}). Photodecomposition reactions correlated with pseudo-first-order kinetic model. These findings can support the design of remediation processes and also assist in predict their fate in the environment.

Keywords: Kinetics; Photocatalysis; Congo red; Zinc oxide

Introduction

Heterogeneous photocatalysis oxidations performed with light irradiated semiconductors dispersions has been extensively investigated owing to their highly efficiency to completely mineralize the harmful organic and inorganic ions species to CO_2 and water [1]. Most researches has been focalized on the heterogenic systems based on high dispersion TiO_2 with a crystalline modification of anatase (Degussa P25, Hombriat UV-100, Aldrich, etc.) as a result of their high photocatalytic activity and widespread uses for large-scale water treatment. However, the relatively elevated intrinsic band gap of anatase TiO_2 (3.2 eV), limited their efficiencies under solar light, so that the effective utilization of solar energy is limited to about 4% of total solar spectrum. In order to meet the requirement of future environment and energy technologies, it is necessary to develop highly efficient, non toxic and chemically stable photocatalyst. Various semiconductor catalysts such as MO_2 ($M= CeO_2$, ZrO_2, SnO_2) and M'_2O_3 ($M'=\alpha\text{-}Fe_2O_3$, Bi_2O_3, Al_2O_3, Sb_2O_3 etc...) metal oxides and DS (D=Zn, Cd, Pb ect...) metal chalcogenide groups were investigated, but their practical uses have been constrained by their low photocatalytic activity under solar light, short-term stability against photo- and chemical corrosion as well as potential toxicity [2]. Many attempts have been made to study ZnO-mediated photocatalytic degradation of organic compounds [3-6]. Semiconductor, on irradiation with photon of sufficient energy, greater than or equal to the band gap energy of the semiconductor ($h\nu \geq Eg$), a free electron (e^-) and electronic vacancy-a hole (h^+) are generated and recombine or migrate in the semiconductor surface being partially localized on structural defective centers of its crystalline lattice Equation (1). The photogenerated electrons take part in the reduction reaction with dissolved oxygen, producing superoxide anion ($O^{\cdot-}_{2ads}$), hydroperoxide (HO_{2ads}) radicals and hydrogen peroxide (H_2O_{2ads}) Equation (2-4), while the photo generated holes can oxidize either the organic compound directly Equation (5) or both hydroxylic ions and water molecules adsorbed on the photo catalyst surface Equation (6-7) forming the organic cation-radicals ($R^{\cdot+}_{ads}$) and hydroxylic radicals ($HO^{\cdot}_{ads}$). The stepwise photo catalytic mechanism is illustrated below:

$$ZnO + h\nu \rightarrow ZnO\,(e^-_{(CB)} + h^+_{(VB)}) \quad (1)$$

$$O_{2ads} + e^- \rightarrow O^{\cdot-}_{2ads} \quad (2)$$

$$O^{\cdot-}_{2ads} + H^+ \rightarrow HO_{2\,ads} \quad (3)$$

$$O^{\cdot-}_{2ads} + 2H^+ + e^- \rightarrow H_2O_{2ads} \quad (4)$$

$$R_{ads} + h^+ \rightarrow R^{\cdot}{}^{+}_{ads} \quad (5)$$

$$HO^-_{ads} + h^+ \rightarrow {}^{\cdot}OH^{+}_{ads} \quad (6)$$

$$H_2O_{ads} + h^+ \rightarrow {}^{\cdot}OH_{ads} + H^+ \quad (7)$$

The hydroxylic, peroxide and hydroperoxide radicals are the main oxidizing agents in the heterogeneous photocatalytic systems used in the water treatment technologies. The heterogeneous photocatalytic processes substantially depend on a variety of environment conditions such as surface charge and electronic structure of catalyst, the nature of surface-active center, the localization degree of photogenerated charges, the amphoteric properties of photocatalyst, pH solution, temperature of system, nature of pollutant, photocongeners, crystalline structure, synthesis method and photoreactor dimension [7]. Aggregation is one of the features of dyes in solution. Based on literature, Congo red tends to aggregate in aqueous and organic solutions leading to dimer formation and sometimes even higher order aggregates due to hydrophobic interactions between aromatic rings of dye molecules. This aggregation phenomenon is more noticed for high Congo red concentrations, at high salinity and/or low pH. The formed aggregates (micelles) separate and precipitate onto solid surfaces.

***Corresponding author:** Elaziouti Abdelkader, LCPCE Laboratory, Faculty of sciences, Department of industrial Chemistry, University of the Science and the technology of Oran (USTO M.B). BP 1505 El M'naouar 31000 Oran, Algeria

In the present study, the stability of optical properties of the Congo red (CR) in aqueous solutions and upon photodegradation process was investigated at various pH values using UV spectroscopic monitoring. The potential ability of ZnO-assisted photocatalytic degradation of CR, was assessed in terms of evolution of the adsorption and photodecomposition efficiency at different operating parameters such as, pH solution and band gap of MO_2 (M=Ti^{+4} and Ce^{+4}) and M'_2O_3 (M'=Al^{+3} and Fe^{+3}) semiconductor groups as a function of irradiation time. The experimental data were quantified by applying the pseudo-first order kinetic.

Materials and Methods

Nanoparticle semiconductors ZnO (BET surface area, S= 10 m^2/g and particle size D=60 nm, 99.99%), TiO_2 (anatase 99.99%), Al_2O_3 (99.99%), CeO_2 (99.99%), and Fe_2O_3 (99.99%), were obtained from Merck, and were used without further purification. Congo red (C.I. 22020, MW = 696.67 g mol^{-1}, $C_{32}H_{24}N_6O_6S_2$.2Na, pKa=4 and λ_{max}=497 nm). Distilled water was used for preparation of various solutions. The molecular structure of the dye is illustrated in Figure 1.

Photocatalytic reactions were carried out inside a (BLX-E365) photoreactor equipped with 6UV-A lamps with an emission maximum at λ of 365 nm. The suspension was irradiated perpendicularly to the surface of solution, and the distance between the UV source and vessel containing reaction mixture was fixed at 15 cm. The experiments were performed at 298K. In all photocatalytic degradation experiments, 300mL CR solution of appropriate concentration was taken in photocatalytic reactor vessel of 600 ml capacity. A known quantity of semiconductor was added and mixture was stirred to obtain uniform suspension. The suspension pH values were previous adjusted using NaOH/ H_2SO_4 solutions *via* pH meter. Before irradiation, photocatalyst/ substrate suspension was stirred in the dark for 30 minutes at 298K to ensure the adsorption equilibrium was established. Next, the lamp was switched on to initiate the photocatalytic degradation reaction. During irradiation, agitation was maintained by a magnetic stirrer to keep the suspension homogeneous. The suspension was sampled at regular intervals of time and immediately centrifuged using (EBA-Hetlich) at 3500 rpm for 15 min to completely remove photocatalyst particles. The residual concentration of the solution samples was monitored using UV-Vis Spectrophotometer (Shimadzu UV mini-1240) at λ = 497 nm as a function of irradiation time.

The effect of initial pH on the photocatalytic degradation of Congo red was researched over a range of pH values from 6 to 10 for avoiding dye aggregation. The experiments were also performed by replacing ZnO catalyst by MO_2 (M=Ti^{+4} and Ce^{+4}) and M'_2O_3 (M'=Al^{+3} and Fe^{+3})

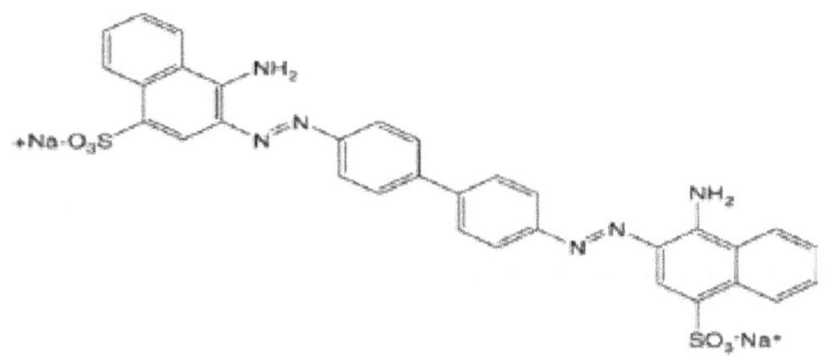

Figure 1: Molecular structure of Congo red (CR).

semiconductor groups. The data obtained from the photocatalytic degradation experiments were then used to calculate the degradation efficiency η (%) of the substrate Equation (8):

$$\eta(\%) = \left[\frac{(c_i - c_f)}{c_i}\right]100 \quad (8)$$

where C_i: dye initial concentration (mg·L^{-1}) and C_f : dye residual concentration after certain intervals (mg·L^{-1}). To calculate the corresponding energy at UV-A wavelength. The energy of an electro-volt, *E* (eV), at a given wavelength, λ (nm), is given by Equation (9):

$$E(eV) = \frac{hc}{\lambda j} \quad (9)$$

where *h* is Planck's constant (6.626 10^{-34} J s); *c* is the speed of light (3 10^8 m/s); and *j* is the number of electro-volt per joule (joule=1.6 10^{-19} electro-volt). The corresponding light energy at UV-A wavelength (λ_{UV-A}=365 nm) was estimated to $E_{\lambda\,UV-A}$ = 3.4 eV.

The photocatalytic degradation efficiency of ZnO catalyst for the degradation CR was quantified by measurement of dye apparent first order rate constants under operating parameters.

Surface catalyzed reactions can often be adequately described by a monomolecular Langmuir–Hinshelwood mechanism, in which an adsorbed substrate with fractional surface coverage θ is consumed at an initial rate given as follow Equation (10) [8]:

$$-\left[\frac{dC}{dt}\right] = r_0 = K_{app}\theta = \frac{K_1 K_2 C_0}{1 + K_1 C_0} \quad (10)$$

where K_1 is a specific rate constant that changes with photocatalytic activity, K_2 the adsorption equilibrium constant, and C_o is the initial concentration of the substrate (Benzopurpurine 4B in our case). Inversion of the above rate equation is given by Equation (11):

$$\frac{1}{K_{app}C_0} = \frac{1}{K_1 K_2} + \frac{C_0}{K_1} \quad (11)$$

Thus, a plot of reciprocal of the apparent first order rate constant 1/ K_{app} against initial concentration of the dye C_o should be a straight line with a slope of 1/K_1 and an intercept of 1/K_1K_2. Such analysis allows one to quantify the photocatalytic activity of ZnO catalyst through the specific rate constant K_1 (with larger K_1 values corresponding to higher photocatalytic activity) and adsorption equilibrium constant K_2 (K_2 expresses the equilibrium constant for fast adsorption-desorption processes between surface of catalyst and substrates). The integrated form of the above equation (Equation 10) yields to the following Equation (12):

$$t = \frac{1}{K_1 K_2}\ln\frac{C_0}{K_2} + \frac{1}{K_2}(C_0 - C) \quad (12)$$

where *t* is the time in minutes required for the initial concentration of the dye C_o to decrease to *C*. Since the dye concentration is very low, the second term of the expression becomes small when compared with the first one and under these conditions the above equation reduces to Equation (13).

$$\ln\frac{C_0}{C} \approx K_1 K_2 t = K_{app}t \quad (13)$$

where k_{app} is the apparent pseudo-first order rate constant, C and C_o are the concentration at time 't' and 't=0', respectively. The plot of ln C_o/C against irradiation time t should give straight lines, whose slope is equal to K_{app}.

The half-life of dye degradation at various process parameters was raised from Equation (14).

$$t_{1/2} = \frac{0.5C_0}{K_2} + \frac{0.693}{K_1K_2} \quad (14)$$

where half-life time, $t_{1/2}$, is defined as the amount of time required for the photocatalytic degradation of 50% of CR dye in a aqueous solution by ZnO catalyst.

Results

Effect of UV irradiation and ZnO nanoparticle

Figure 2 illustrated the photocatalytic degradation kinetics of 20 mg/L of CR in aqueous solution under different experimental conditions under UV-A alone, dark/ ZnO and UV-A/ZnO. The degradation rate was found to increase with increase in irradiation time and 95.02% of degradation were achieved within 60 min at pH of 8 (curve CR/ZnO/UV-A). When 20 mg/L of both dye along with ZnO were magnetically stirred for the same optimum irradiation times in the absence of light, lower (20.78 %) degradation was observed (curve CR/ZnO), whereas, disappearance of dyes was negligible (0.49%) in the direct photolysis (curve CR/UV-A) indicating that the observed high decomposition of the dye in the UV/ZnO process is exclusively attributed to the photocatalytic reaction of the semiconductor particles. Similar results have been reported for ZnO-assisted photocatalytic degradation of azo dyes such as Congo red [9] and Reactive Black 5 [10].

Effect of pH solution

Qualitative study: The evolution of maxima absorption bands of CR in aqueous solution and upon ZnO-mediated photocatalytic degradation under UV-A light at various pH: Figure 3 illustrated UV visible spectra of CR in aqueous solution at various pH solutions. The position and the evolution of the maxima absorption bands of CR in aqueous solution and upon ZnO-mediated photocatalytic degradation of substrate CR in aqueous solution were displayed in Table 1. Congo red in aqueous solution exhibited a main band at 497nm (C_0=20 mg/L and natural pH=8), assigned to the absorption band of anionic monomer, and associated with two absorption bands in the ultraviolet region appeared at 235 and 347 nm, attributed to the benzoic and naphthalene rings, respectively. As seen in Table 1, the CR monomers have nearly constant absorption bands and absorbance intensities over a range of pH values of 6 to10, while they are most sensitive to the pH solution under acidic medium (pH<6). The main band of CR monomer at 497nm (in aqueous solution) gradually shifted to the longer wavelengths, almost reaching 595.5 and 537.5 nm at pH 2 and 4 respectively. The intensity decrease and red shift of CR monomer bands are attributed to the partially self-association of CR monomers as anionic dimers in face-to-face arrangement to minimize their hydrophobic interaction with water. For that reason, the adsorption and photocatalytic degradation conditions were conducted only within neutral and alkali pH range.

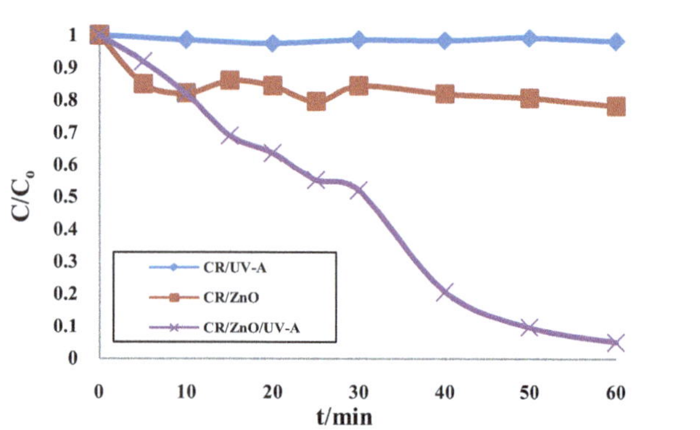

Figure 2: Photocatalytic degradation kinetics of CR under different experimental conditions ([ZnO]= 0.5g/L, [CR]=20mg/L, pH=8, T=298K, λ_{max}=365 nm and I=90 J /cm²).

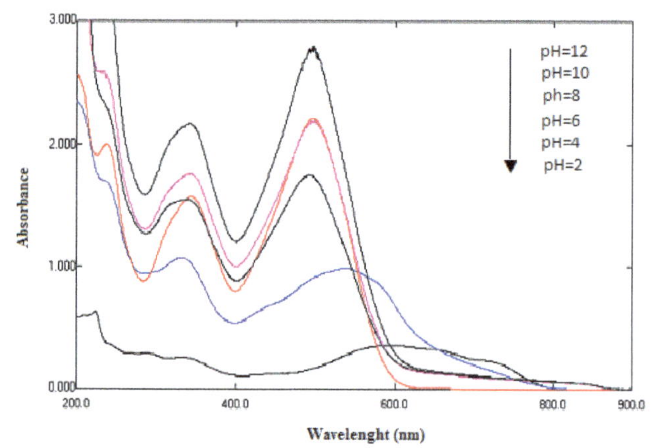

Figure 3: UV visible spectra of CR in aqueous solution as a function of pH ([CR]=20mg/l and T=298K).

The adsorption equilibrium of CR on ZnO catalyst in dark for 30 min at different pH values (Table 1) showed that CR monomers have nearly constant absorption bands, while a slight decrease in the absorbance intensities over the pH values in the range of 6 and 10 was observed.

On the other hand, the behavior of the maxima absorption bands of CR during the ZnO-mediated photocatalytic degradation as a function of pH solution differ in certain aspects from those of CR in aqueous solution and adsorbed on ZnO catalyst surface. As elucidated Table 1, the photodecomposition profile of CR was accompanied by a blue shift of the main bands of CR monomers, compared to those in aqueous solutions.

The maxima bands became visible at 497, 481, 492, 481 and 488 nm at pH 6, 7, 8, 9 and 10 respectively as the light illumination time was increased up to 40 minutes. However, they remained nearly constant at pH 6 and 7 and completely disappeared within 60 minutes of irradiation time at pH beyond of 7. The shift does not exceed 10nm over the pH in the range of 6 and 10. These results are consistent with the destruction of CR chromophoric structure in the vicinity of the azo-linkage. Additionally, this shift was accompanied by a parallel decrease in the intensities of bands in the ultraviolet region located at 235 and 347 nm, ascribed to the benzoic and naphthalene rings, respectively. No new absorption bands appeared in either the visible or the ultraviolet spectra regions.

Initial pH	Photolyse (CR /Aqueous solution) t=0 min A	Species λ (nm) $(CR^-)_2$	Species λ (nm) (CR^-)	Adsorption (CR/ZnO) t=30 min A	Species λ (nm) (CR^-)	AE (%)	Photocatalytic degradation (CR/ZnO/UV-A) t=20 min A	Species λ (nm) (CR^-)	t=40 min ABS	Species λ (nm) (CR^-)	t=60 min A	Species λ (nm) (CR^-)	PDE (%)
2	0.359	595.5	-	-	-	-	-	-	-	-	-	-	-
4	0.986	537.5	-	-	-	-	-	-	-	-	-	-	-
6	0.83	-	497	0.679	497	18.19	0.605	497	0.415	497	0.202	492	70.25
8	0.628	-	497	0.562	498	10.51	0.357	496	0.116	492	0.028	D	95.02
10	0.83	-	497	0.706	499	14.94	0.222	481	0.146	481	0.075	D	89.38
12	0.83	-	498	0.732	497	11.81	0.281	488	0.184	488	0.1	D	86.34

CR^- and $(CR^-)_2$: Monomer and dimer anion forms of Congo red respectively; A : Absorbance; AE : Adsorption efficiency; PDE: Photocatalytic degradation efficiency; D: Disappearance

Table1: The position of the absorption bands of CR in aqueous solution and during the ZnO-mediated photocatalytic degradation of dye as a function of pH ([ZnO] =0.5g/L, [CR]=20mg/l, T=298K, λ_{max}=365 nm, I=90 J /cm^2 and irradiation time =60min).

Quantitative study: The evolution of photocatalytic degradation rate of CR in aqueous solution by ZnO under UV-A irradiation at various pH: It is commonly accepted that in photocatalyst/aqueous systems, the potential of the surface charge is determined by the activity of ions (e.g. H^+ or pH). A convenient index of the tendency of a surface to become either positively or negatively charged as a function of pH is the value of the pH required to give zero net charge (pH_{ZPC}) [11-12]. pH_{ZPC} is a critical value for determining the sign and magnitude of the net charge carried on the photocatalyst surface during adsorption and photocatalytic degradation process. Most of the semiconductor oxides are amphoteric in nature, can associate (Equation 15) or dissociate (Equation 16) proton. To explain the relationship between the layer charge density and the adsorption, so-called models of surface complexation (SCM) was developed [13], which consequently affects the sorption–desorption processes and the separation and transfer of the photogenerated electron–hole pairs at the surface of the semiconductor particles. In the 2-pK approach we assume two reactions for surface protonation. The zero point charge pH_{ZPC} for ZnO is 9.0. For pH values lower than the pH_{ZPC} of ZnO, the surface becomes positively charged, according to the following reaction Equation (15):

$$Zn\text{-}OH + H^+ \rightarrow ZnOH_2^+ \quad pH < pH_{ZPC} \qquad (15)$$

ZnO surface becomes negatively charged for pH values higher than pH_{pzc}, given by the following reaction Equation (16):

$$Zn\text{-}OH + OH^- \rightarrow ZnO^- + H_2O \quad pH > pH_{ZPC} \qquad (16)$$

The experimental data revealed that higher degradation rate of CR was observed in acidic medium (Figure 4 and Table 1). Since CR is an anionic dye, its adsorption mainly performed via an electrostatic interactions between the positive ZnO surface and CR anions, leading to a maximum extent at pH 6. Thus, a high adsorption (18.19%) can lead to a decrease in the active centers on the catalyst surface, which results in decrease in the absorption of the light quanta by the catalyst and consequently to a reducing of the kinetic reaction. As a result, the high degradation efficiency cannot be ascribed to the photocatalytic oxidation of the CR dye, but to catalyst behavior under strong acid pH. ZnO nanoparticles can undergo photo-corrosion through self-oxidation at pH lower than 4 Equation (17). In particular, ZnO can be photo-oxidized with decreasing the pH Equation (18).

$$ZnO + 2h^+_{(VB)} \rightarrow Zn^{2+} + 1/2O_2 \qquad (17)$$

$$ZnO + 2H^+ \rightarrow Zn^{2+} + H_2O \text{ (acidic dissolution)} \qquad (18)$$

Photocatalytic activity of anionic dyes (mainly sulfonated groups) such as CR reaches a maximum value (95.02%) in lower pH_{ZPC} (i.e. pH=8). At alkaline mediums, excess of hydroxyl anions facilitate photogeneration of •OH radicals which is accepted as primary oxidizing species responsible for photocatalytic degradation, resulting in enhancement of the efficiency of the process. Furthermore we found that, where the adsorption of dye was weak, degradation scarcely occurred. The adsorption affects strongly the accessibility of the surface reducing species to the CR reduction kinetics. However, adsorption is not the only factor that controls the photocatalytic degradation of dye. Although the adsorption extents of the dye were lower, the degradation rates were in the reverse order (Figure 4 and Table 2).

At pH higher than pH_{ZPC} value (i.e. pH=9-10), a dramatically decrease in the degradation efficiency could be explained on the basis of amphoteric behaviors of ZnO catalyst. The negatively surface of ZnO catalyst (highly concentration of hydroxide ions) and the great negatively charged CR dye anions results in electrostatic repulsion electrostatic.

Moreover, the stability of ZnO may not be guaranteed at this high pH due to possibility of alkaline dissolution of ZnO Equation (19):

$$ZnO + H_2O + 2OH^- \rightarrow Zn(OH)_4^+ \text{ alkaline dissolution} \qquad (19)$$

The linearity of the plots of $\ln C/C_0$ against irradiation time *t* for the photocatalytic degradation process of CR by ZnO catalyst at low dye

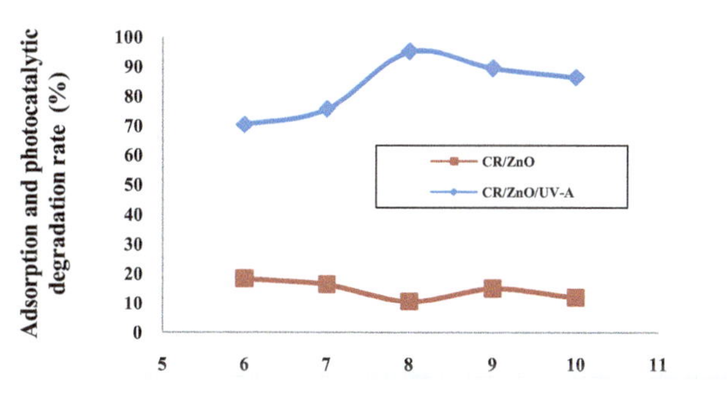

Figure 4: Adsorption and photocatalytic degradation of CR as a function of initial pH ([ZnO]= 0.5g/L, [CR]= 20mg/L, T=298K, λ_{max}=365 nm and I=90 J / cm^2).

concentrations and under various pH solutions adsorption (Figure 5, Table 2) suggest that the adsorption data were satisfactorily described by pseudo-second order model (R^2=0.747-0.968).

Effect of band gap of semiconductor

The photocatalytic degradation reactions were further performed in two semiconductor groups MO_2 (M=Ti^{+4} and Ce^{+4}) and M'_2O_3 (M'=Al^{+3} and Fe^{+3}) having different band gap values. Figure 5 depict the plot of C/C_0 versus irradiation time CR on ZnO catalys. It is evident that the photocatalytic degradation of dye greatly depends on the electronic structure and properties of semiconductor surface/solvent. The results exposed in Figure 6 shows that the highest degradation efficiency of 95.02% was obtained for ZnO catalyst. The photo decomposition process profiles under the same optimized conditions (pH=5, C=20 mg/L CR, m/v=0.5g/L ZnO and irradiation time of 60min) were different for both MO_2 and M'_2O_3 semiconductor groups in comparison with that of ZnO catalyst. As elucidated Figure 5, the best result of photocatalytic degradation process was observed for MO_2 group. TiO_2 catalyst exhibited high degradation efficiency of 93.34 %, whereas, only 14.95% of CR was degraded for CeO_2 semiconductor. On the other hand, the M'_2O_3 (M'=Al^{+3} and Fe^{+3}) semiconductors group showed lower (4.20 and 5.15 % for Al_2O_3 and Fe_2O_3 respectively) degradation of CR.

Experimental parameters		Experimental results: Adsorption efficiency (%)	Experimental results: Photocatalytic degradation efficiency (%)	Pseudo-first order model: K_{app} (min^{-1})	$t_{1/2}$ (min)	R^2 (%)
Initial pH	6	18.19	70.25	0.014	49.511	79.8
	7	16.26	75.68	0.025	27.726	96.8
	8	10.51	95.02	0.044	15.753	94.4
	9	14.94	89.38	0.038	18.241	79.0
	10	11.81	86.34	0.035	19.804	74.7

Table2: Kinetic parameters of the photocatalytic degradation of CR as a function of initial pH ([ZnO]= 0.5g/L, [CR]= 20mg/l, T=298K, λ_{max}=365 nm and I=90 J /cm²).

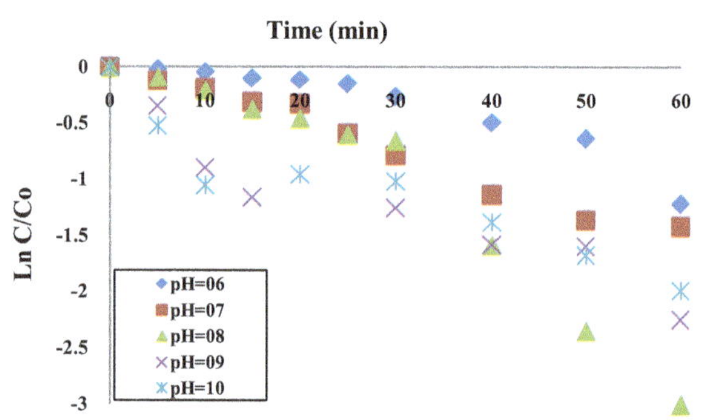

Figure 5: Pseudo first order kinetics of photocatalytic degradation of CR as a function of initial pH ([ZnO]= 0.5g/L, [CR]= 20mg/L, T=298K, λ_{max}=365 nm and I=90 J /cm²).

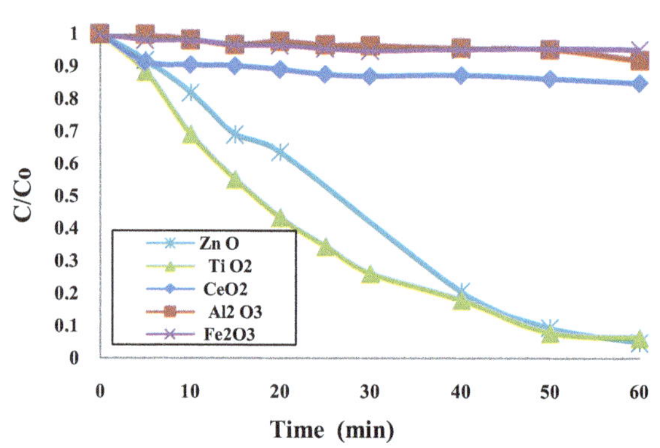

Figure 6: Photocatalytic degradation kinetics of CR as a function of band gap of semiconductors ([CR]= 20mg/L, [Catalyst]= 0.5g/L, pH =8, T=298K, λ_{max}=365 nm and I=90 J /cm²).

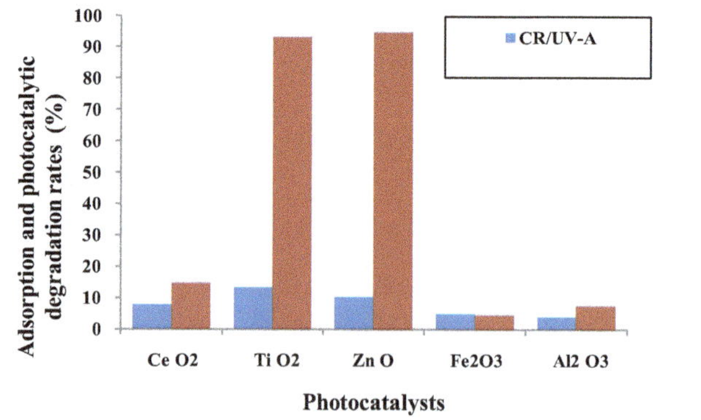

Figure 7: Adsorption and photocatalytic degradation rates of CR a function of band gap of semiconductors ([CR]= 20mg/L, [Catalyst]= 0.5g/L, pH=8, T=298K, λ_{max}=365 nm and I=90J/cm²).

Experimental parameters		Experimental results: Adsorption efficiency (%)	Experimental results: Photocatalytic degradation efficiency (%)	Pseudo-first order model: K_{app} (min^{-1})	$t_{1/2}$ (min)	R^2 (%)
Band gap of semiconductors (eV)	Zn O (3.2eV)	10.510	95.018	0.044	15.753	94.4
	Ti O_2 (3.3eV)	13.479	93.342	0.045	15.403	98.5
	Ce O_2(2.6eV)	8.000	14.928	0.003	231.049	49.0
	Al_2O_3 (5.6eV)	4.207	7.905	0.001	693.147	88.7
	Fe_2O_3(3.7eV)	5.151	4.723	0.001	693.147	34.0

Table3: Kinetic parameters of photocatalytic degradation of CR as a function of band gap of semiconductors ([CR]= 20mg/L, [Catalyst]= 0.5g/L, pH=8, T=298K, λ_{max}=365 nm and I=90 J /cm²).

Generally, semiconductor on irradiation with light energy greater or equal than band gap energy of the semiconductor (hν≥Eg), a free electron (e^-) and electronic vacancy-a hole (h^+) are formed and recombine or migrate in the semiconductor surface. Probability of electron transfer in the semiconductor/substrrate system is determined by a relative position of the valence band, the photocatalyst conductance band and the value of the oxidation-reduction potential (ORP) of the oxidant and the substrate. The photogeneration of electrical charge is in dynamic equilibrium with their recombination substantially reducing the quantum yield of the photocatalytic process. The ORP of water oxidation, hydroxylic ions, and most of organic compounds below of reducing photogenerated holes within a wide interval of the pH due to which the formation of hydroxylic radicals and organic cation-radicals of photocatalyst surface are thermodynamically possible processes. It has already reported that semiconductors such as ZnO and TiO_2 having band gaps larger than 3 eV are excellent photocatalysts. Obviously, MO_2 (M=Ti^{+4}) semiconductor exhibits a higher degradation activity

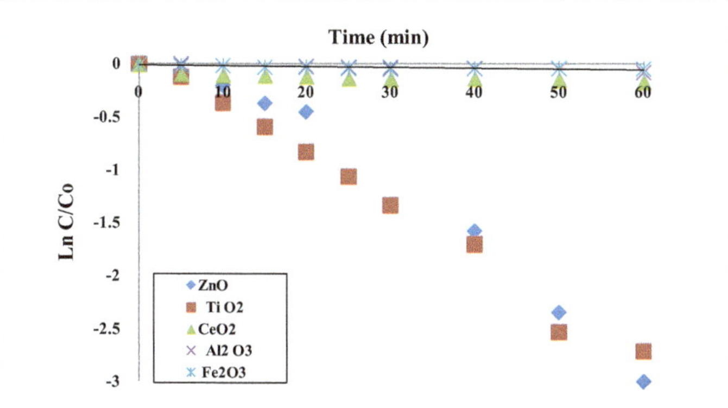

Figure 8: Pseudo first order kinetics of photocatalytic degradation of CR as a function of band gap of semiconductors ([CR]= 20mg/L, [Catalyst]= 0.5g/L, pH=8, T=298K, λ_{max}=365 nm and I=90 J /cm^2).

Catalyst composition	Preparation method	Catalyst loading	Pollutant concentration	pH	UV source (nm)	Processing time	Photo degradation efficiency (%)	Ref.
TiO_2		0.5 g/L	20 mg/L	8	365	298k for 60 min	93.34	This work
TiO_2(P25)		1 g/L	55 mg/L	-	254 (O_2)	480min	>99	[16]
TiO_2	Hydrothermal	1 g/L	10 mg/L	-	350			[17]
Mesoporous TiO2		0.5 g/L	5ppm	-	254	90min	>76	[18]
ZnO		0.5 g/L	20 mg/L	8	365	298K for 60 min	95.02	This work
ZnO		0.5 g/L	5ppm	8	254 (O_2)	11min	> 99	[8]

Table 4: Comparative chart of photocatalytic degradation of Congo red on various catalysts.

than that of the other systems because its band gap (Eg = 3.3 eV) is slightly equal to that of UV-A irradiation source (E_{UV-A} =3.4 eV). The photogenerated electron (e^-) hole (h^+) pairs can be easily separated and transferred to the semiconductor/adsorbate interface efficiently, consequently improving the photocatalytic activity. The minimum energy required for excitation of an electron from the valence band to conductance band for the semiconductor such as MO_2 ($M=Ce^{+4}$) is 2.6 eV. The photodecomposition process of CR by CeO_2 system was much lower than that of the other catalysts, although it band gap being smaller than of E_{UV-A} light energy. The CeO_2 system might reduce the life of electron-hole pairs, and enhance the opportunities of their recombination [14]. On the other hand, M'_2O_3 ($M'=Al^{+3}$ and Fe^{+3}) catalysts showed lower activity for the photocatalytic of CR than ZnO since their conductance bands of 5.6 and 3.7 eV respectively, are much higher than of the E_{UV-A} light irradiation as a result of low light energy conversion efficiency, so the photogenerated electrical charge in semiconductor cannot efficiently transfer in the surface and are lost due to recombination.

Table 3 records the kinetic parameters of photodecomposition of CR by ZnO in aqueous solution at various band gap of semiconductors. The results showed that the decolorization process for ZnO as well as TiO_2 catalysts could be described by the pseudo-first-order kinetic model. The plots of the experimental data ($\ln C/C_o$ against t, Figure 8) yielding to a straight line with relatively high regression coefficients R^2=0.944 and 0.985 for ZnO and TiO_2 respectively. However, the regression coefficients, R^2, (given in Table 3) for the pseudo-first-order model for semiconductors such as CeO_2, Al_2O_3 and Fe_2O_3 did not exceed 0.49, 0.887 and 0.34 respectively. These results suggested that the photodegradation process of CR by CeO_2, Al_2O_3 and Fe_2O_3 semiconductors did not follow the pseudo-first-order.

Comparative study

Table 4 gives clear assessment chart of photocatalytic degradation of Congo red on ZnO and TiO_2 catalysts compared to those of prior work of ZnO and TiO_2 photocatalytic systems for dye degradation applications.

Conclusion

The photocatalytic degradation of Congo red in aqueous solution was investigated under UVA light at different operating conditions, including irradiation time, pH solution and band gap of other semiconductor groups by UV-spectrophotometric monitoring. The results showed slight stability of 597nm-band of CR over the pH range of 6-7, while, it almost completely disappeared beyond pH 7. The highest photodecomposition was obtained at pH 8 as a result of 95.02% degradation efficiency of CR for 60 min exposure of irradiation time. The photodegradation rate of MO_2 ($M=Ti^{+4}$ and Ce^{+4}) semiconductor group was higher, compared to that of M'_2O_3 ($M'=Al^{+3}$ and Fe^{+3}). Photodecomposition process profiles were satisfactory correlated with pseudo-first-order kinetic model. These findings suggest that the degradation of Congo red mainly occurred by the attack of OH radical [15].

Acknowledgments

We greatly acknowledge the material support obtained from DR T.M. University

References

1. Eslami A, Nasseri S, Yadollahi B, Mesdaghinia A, Vaezi F, et al. (2008) Photocatalytic degradation of methyl tert-butyl ether (MTBE) in contaminated water by ZnO nanoparticles. Journal Chemical and Technology Biotechnology 83: 1447-1453.

2. Lingzhi L, Bing Y (2009) CeO_2-Bi_2O_3 nanocomposite: Two step synthesis, microstructure and photocatalytic activity. Journal of Non-Crystalline Solids 355: 776-779.

3. Djurišić AB, Chan Y, Li EH (2002) Progress in the room temperature optical functions of semiconductors. Material Science Engineering R 38: 237-293.

4. Sushil K, Kansal NK, Sukhmehar S (2009) Photocatalytic Degradation of Two Commercial Reactive Dyes in Aqueous Phase Using Nanophotocatalysts. Nanoscale Research Letter 4:709-716.

5. Kansal SK, Singh M, Sud D (2007) Studies on photodegradation of two commercial dyes in aqueous phase using different photocatalysts. Journal of Hazardous Materials 141:581-590.

6. Poulios I, Kositzi M, Kouras A (1998) Photocatalytic decomposition of triclopyr over aqueous Semiconductor suspensions. Journal of Photochemestry and Photobiology A: Chem 115: 175-183.

7. Soboleva NM, Nosovich AA, Goncharuk VV (2007) The heterogenic photocatalysis in water treatment processes Journal of Water Chemistry and Technology 29:72-89.

8. Movahedi M, Mahjoub AR, Janitabar-Darzi S (2009) Photodegradation of Congo Red in Aqueous Solution on ZnO as an Alternative Catalyst to TiO_2. Journal Iranian of Chemical Society 6: 570-577.

9. Habibi MH, Hassanzadeh A, Zeini-Isfahani A (2006) Spectroscopic studies of Solophenyl red 3BL polyazo dye tautomerism in different solvents using UV–visible, ^{1}H NMR and steady-state fluorescence techniques. Dyes and Pigments 69: 93-101.

10. Akyol A, Yatmaz HC, Bayramoglu M (2004) Photocatalytic decolorization of Remazol Red RR in aqueous ZnO suspensions. Applied Catalysis B: Environment 54: 19-24.

11. Zhang F, Zhao J, Shan T, Hidaka H, Pelizzetti E, et al. (1998) TiO_2-assisted photodegradation of dye pollutants : II. Adsorption and degradation kinetics of eosin in TiO_2 dispersions under visible light irradiation Applied Catalysis B: Environment 15: 147-150.

12. Yates DE, Levine S, Healy TW (1974) Site-binding model of the electrical double layer at the oxide/water interface. Journal of Chemical Society Faraday Trans 70: 1807-1818.

13. Fernandez J, Kiwi J, Lizama C, Freer J, Baeza J, et al. (2002) Factorial experimental design of Orange II photocatalytic decolouration. Journal of Photochemistry and Photobiology A: Chem 151: 213-219.

14. Pradhan GK, Parida KM (2010) Fabrication of iron-cerium mixed oxide: an efficient photocatalyst for dye degradation International. Journal of Engineering Science and Technology. 2: 53-65.

15. Zhang LS, Wong KH, Zhang DQ, Hu C, Yu JC, et al. (2009) Zn: In(OH)ySz solid solution nanoplates: synthesis, characterization, and photocatalytic mechanism. Environ Sci Technol 43: 7883-7888.

16. Curkovic L, Ljubas D, Juretic H (2010) Photocatalytic decolorization kinetics of diazo dye Congo Red aqueous solution by UV/TiO_2 nanoparticles. Reaction Kinetic and Mechanism Catalysis 99: 201–208.

17. Rajeev KW, William WY, Yunping L, Michelle LM, Joshua CF, et al. (2005) Photodegradation of Congo Red catalyzed by nanosized TiO_2. Journal of Molecular Catalysis A: Chemical 242: 48–56

18. Darzi JS, Mahjoub AR (2010) Synthesis of Spongelike Mesoporous Anatase and its Photocatalytic Properties Iran. Journal of Chemistry and Chemical Engineering 29: 2.

Immobilization of Molybdenum in Ordinary Portland Cement

Minocha AK and Manish Kumar Goyal*

Environmental Science and Technology Division, Central Building Research Institute, Roorkee- 247667(UA) & NIPER Mohali, Punjab-160062, India

Abstract

In this paper we report the results of the studies concerning the addition of varying concentrations of molybdenum on physical, chemical and engineering properties of 43 grade ordinary Portland cement. The studies regarding initial and final setting time of cement, bulk density, compressive strength, microscopic and spectroscopic properties have been carried out and the result of these studies are presented and discussed. Different building products containing varying concentrations of added molybdenum were prepared and subjected to hydrologic environment to investigate the leaching behavior of molybdate ions. The efforts have also been made to establish a quantitative co-relation between the concentration of added molybdenum and the intensity of any effect in above properties of the cement. Also the possible fixation mechanism of molybdenum in cement is suggested. The X-ray diffraction analysis of molybdenum added cement sample was carried out and it was observed that the insoluble molybdenum is distributed in Portland cement in the form of powellite ($CaMoO_4$).

Keywords: Cement; Heavy metals; Molybdenum; Setting time; Waste management

Introduction

The protection of public health and the environment from the hazardous pollutants is a major concern of industrialized nations [1-3]. One of the most toxic forms of inorganic industrial waste is that contains toxic heavy metals. This type of waste is generated from metallurgical, mining, chemical, leather industry, distillery, sugar, battery, electroplating and pigment industry. Generally the metal concentration in the waste is too low for economical recovery but high enough to represent toxicity hazards. Therefore, these industries discharge their waste without treatment or with improper treatment in the landfills. Its disposal in hydrologic environment can cause environmental risk due to the mobility of metal ions [4]. These metal ions get mixed in surface water and ground water system, which may be detrimental to human being as well to the environment. Molybdenum is frequently used in alloy industry and in the manufacturing of electronic devices owing to its high melting point, high strength at elevated temperature, high thermal conductivity and good resistant to corrosion. It is also used as a catalyst in many petrochemical and inorganic reactions in solutions. Molybdenum generally occurs in form of molybdate ions ($MoO^4{}_2$-) at about pH 5 and above. Various technologies have been developed which purport to render a waste or to reduce the potential for the release of toxic species into the environment. Molybdenum containing waste can be immobilized in cement based binder system [5-7]. This technique is used to transform potentially hazardous liquid or solid waste into less hazardous and or non-hazardous solid before disposal in a landfill, thus preventing the waste from contaminating the environment [7-13]. The United State Environmental Protection Agency also recognizes cementitious Solidification/Stabilization as "the Best Demonstrated Available Technology for land disposal of most toxic elements [14]. Many formulations have been developed for the S/S process according to the kinds of wastes. Portland cement can be modified for suitable S/S process using fly ash, lime slag soluble silicates etc. [15-23]. One of the most difficult problems in this process is that the hydration of cementitious materials is too retarded to set and harden enough due to the inhibition of hydration reaction of heavy metals in a landfill area. It seems that Solidification/Stabilization (S/S) process would be the best practical technology to treat the molybdenum containing waste. There is very less data reported in the literature about the effects of addition of molybdenum on various properties of cement. Therefore, efforts have been made to fill void in this data. The results of these studies on the effects of addition of molybdenum on various properties of 43 grade ordinary Portland cement are presented and discussed in this paper. A quantitative co relation between the concentration of molybdenum added and the intensity of any effect has been established.

Materials, Apparatus and Methods

Materials

Sodium molybdate (GR) from Loba, India was used for this study. Commercial ordinary Portland cement 43 grade was used. Molybdenum has been added to the ordinary Portland cement in the different concentrations of molybdate ions. All solutions were prepared in double distilled water. Fly ash is also used as a filler material with the ordinary Portland cement to make mortar samples as fly ash is treated as industrial waste which does not have any cementing property itself but in the presence of lime and water it behaves like cement.

Apparatus

Atomic Absorption Spectrophotometer (AAS) from Hitachi (Z-7000) was used to determine the metal ion concentrations. Hazardous waste filtration system from Millipore (YT-30142 HW) was used to carry out Toxicity Characteristic Leaching Procedure as recommended by United State Environment Protection Agency (USEPA); UV-Visible spectrophotometer (Aquamate) from Thermo Corporation was used for spectrophotometric studies. Vicat apparatus was used to determine initial and final setting time of control cement as well as metal ion doped cement samples. Compressive strength testing machine from Central Scientific Instruments Company was used to determine the compressive

***Corresponding author:** Manish Kumar Goyal, Environmental Science and Technology Division, Central Building Research Institute, Roorkee-247667 (UA), NIPER Mohali, Punjab-160062, India

strength of mortar samples. Scanning Electron Microscope (SEM) from LEO 438VP, UK was used for the microstructure visualization of fractured cement surfaces. X-Ray powder diffraction analysis was carried out by using the instrument from Rikagu, UK to identify the crystalline phases present in the control as well as in metal containing cement samples.

Methods

The chemical analysis of the cement was carried out according to Indian standard specifications IS: 4032:1985 guidelines. The physical properties were tested according to the Indian standard specification IS: 4031:1996 guidelines. The chemical composition and physical properties of Ordinary Portland Cement used are summarized in Tables 1 and 2 respectively.

Preparation of cement pastes and mortars: The Initial and final setting times of control as well as the cement samples, containing varying concentrations of molybdenum, were determined by following the methods as described in IS: 4031:1996. The results obtained for controls were compared with those obtained for molybdenum containing samples to know the effect of addition of molybdenum on the setting time of cement paste. To investigate the effect of addition of molybdenum on engineering properties like compressive strength, bulk density of mortar samples containing cement and fly ash, were cast in 2.78" cubic iron molds. The samples were demolded after 24 hours and were dipped into water for curing the compressive strength

Constituent	Weight percentage
Silica	20.80
Aluminum oxide	4.40
Iron oxide	3.79
Calcium oxide	66.10
Magnesium oxide	3.30
Anhyd. Sulfuric acid	3.00
Sodium oxide	0.20
Potassium oxide	0.70

Table 1: Chemical composition of cement.

Physical parameter	Results
Loss on ignition	0.7 %
Consistency	31.58 %
Soundness	1.0 mm
Bulk density	1.421g/cm3
Initial setting time	175 min.
Final setting time	300 min

Table 2: Physical properties of cement.

S. No.	Concentration of metal ion (ppm)	IST	FST
1.	--	175	300
2.	1	180	300
3.	5	180	300
4.	10	170	300
5.	100	170	310
6.	1000	170	300
7.	1500	155	275
8.	2000	135	265
9.	3000	135	265
10.	5000	135	265

IST- Initial setting time of cement in min.
FST- Final setting time of cement in min.

Table 3: Effect of fixation of molybdenum on initial and final setting time of cement.

of the cubes were determined on 3, 7, 28, 60, 90, 180 and 360 day of curing. Six samples were taken each time and the average value of these results was reported.

X-ray diffraction analysis: It is possible to find and quantify crystalline components which have over 1% abundance in a sample If the correct experimental technique are used. XRD can give a semi quantitative or quantitative analysis of the components of the crystalline fraction. Thus XRD gives most information from the crystalline components of the material under investigation. X-ray diffraction analysis was performed with a Rigaku, Japan (Dmax 2200 VK/PC) automated X-ray diffractometer. The samples were crushed to a fine powder with porcelain and sieved through a 45 mm sieve. The XRD scan was made with copper ka radiation from 3°-70° 2θ with 0.02° step width and 1 to 3 s counting time.

Scanning electron microscopy: Scanning Electron Microscopy (SEM) is a technique used to know the fractured surface of cementitious material and of three dimensional particle level morphology. It is

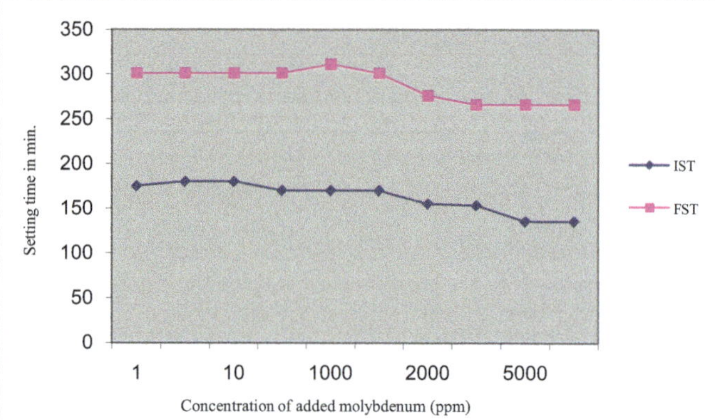

Figure 1: Effect of fixation of molybdenum on the setting time of cement.

Binder system	No. of days	Average bulk density (+ 0.001 g/cm³) Molybdenum			
		Control	1000 ppm	1500 ppm	2000 ppm
Cement + fly ash	1	2.303	2.385	2.393	2.398
	28	2.381	2.385	2.393	2.399

Table 4: Average bulk density values of solidified products on 1 and 28 days of curing.

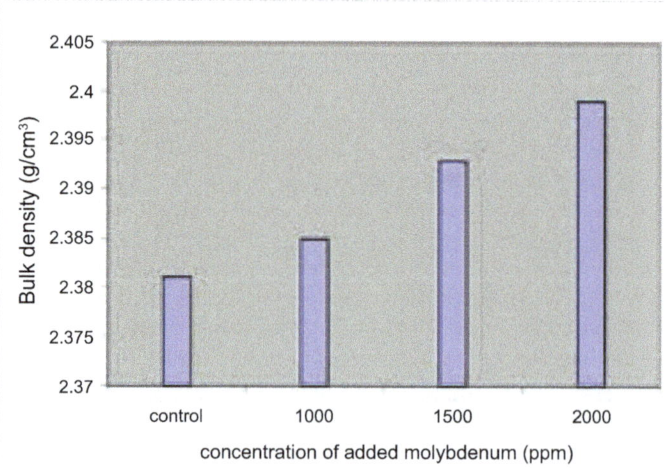

Figure 2: Bulk density of control and molybdenum on 28 day of curing.

Added molybdenum (ppm)	No. of days						% change in compressive strength in comparison of control					
	3	7	28	60	90	180	3	7	28	60	90	180
Control	323.4	436.0	486.0	570.0	570.0	620.0						
1000	345.0	410.0	550.0	555.0	555.0	606.6	106.6	93.9	113.0	97.3	97.3	97.8
1500	354.0	423.4	580.0	592.0	640.0	660.0	96.9	119.1	103.8	112.2	106.4	109.4
2000	354.0	426.6	590.0	564.0	640.0	700.0	109.4	97.7	121.2	98.9	112.2	112.9

Table 5: Compressive strength (kg/cm²) of mortar samples.

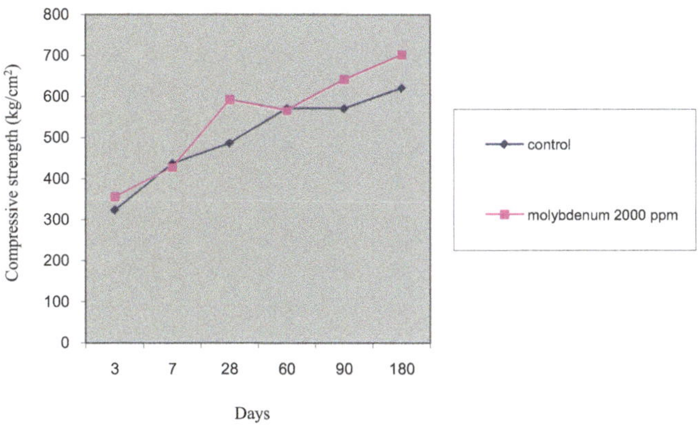

Figure 3: Effect of addition of 2000 ppm molybdenum on the compressive strength of cement.

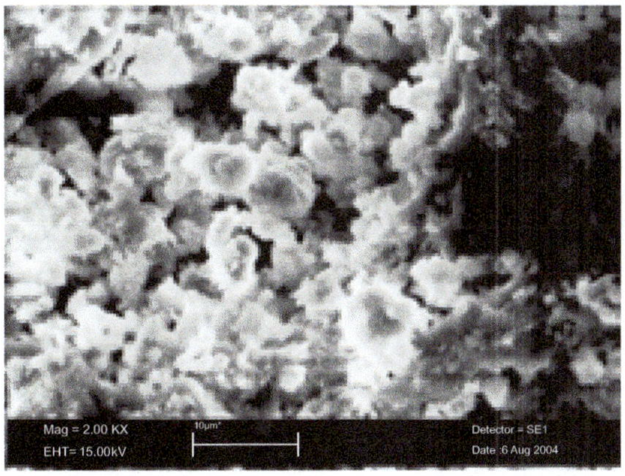

Figure 6: SEM image of molybdenum 2000 ppm containing cement sample on 28 day of curing.

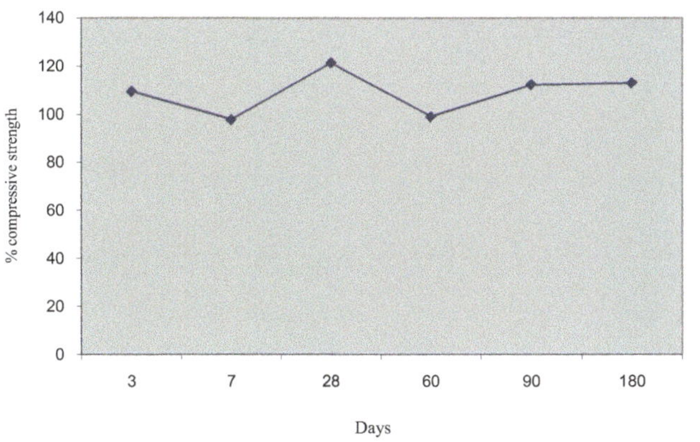

Figure 4: Percentage change in compressive strength of molybdenum 2000 ppm containing samples in comparison with control.

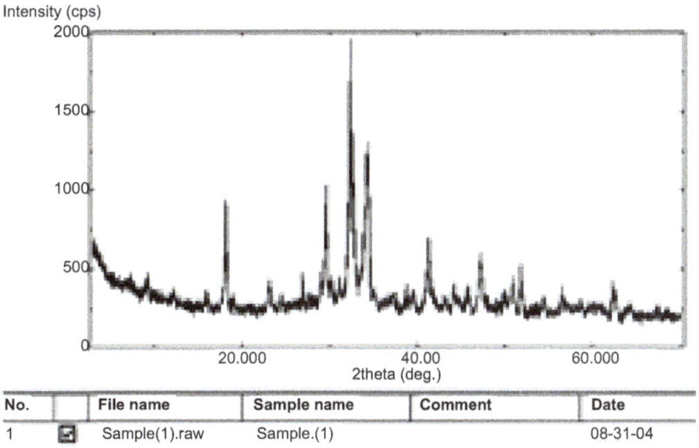

Figure 7: X-Ray diffractogram of molybdenum 2000 ppm containing cement sample.

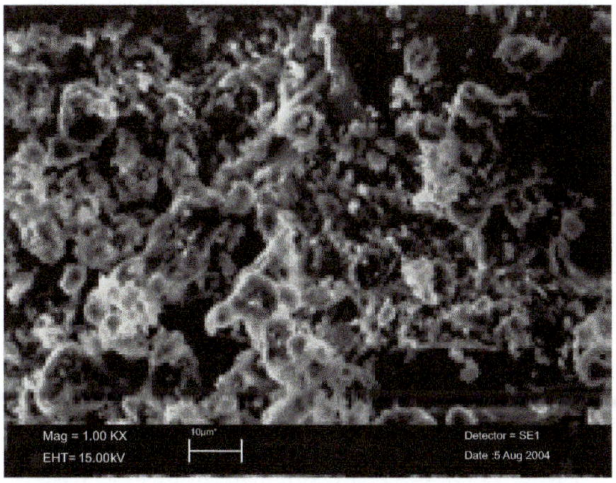

Figure 5: SEM image of pure cement sample on 28 day of curing.

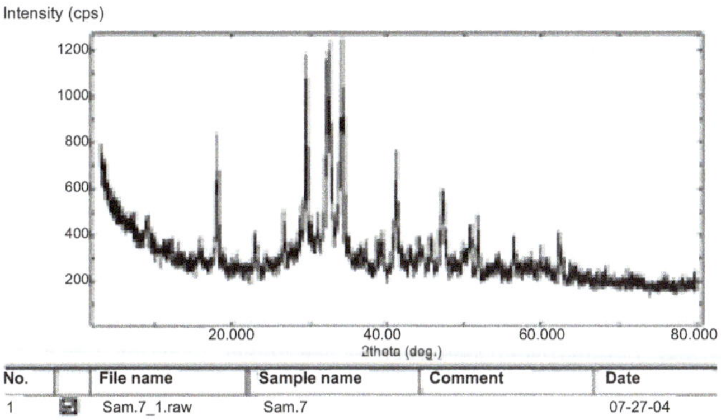

Figure 8: X-Ray diffractogram of Blank cement sample.

molybdenum-2000 ppm

PEAK: 21-pts/Parabolic Filter, Threshold=3.0, Cutoff=0.1%, BG=3/1.0, Peak-Top=Summit

2-Theta d(A)	BG	Height	I%	Area	I%	FWHM	
7.131	12.3865	223	52	5.4	1137	7.4	0.372
7.339	12.0347	228	55	5.7	449	2.9	0.139
9.223	9.5807	194	95	9.9	1507	9.8	0.270
12.244	7.2229	169	63	6.6	958	6.2	0.259
15.939	5.5556	145	67	7.0	1404	9.1	0.356
18.142	4.8858	151	415	43.4	5679	36.9	0.233
19.019	4.6624	153	58	6.1	464	3.0	0.136
23.081	3.8502	144	113	11.8	1727	11.2	0.260
24.516	3.6280	153	46	4.8	542	3.5	0.200
26.822	3.3212	160	126	13.2	820	5.3	0.111
27.591	3.2303	160	55	5.7	823	5.3	0.254
28.839	**3.0932**	**203**	**87**	**9.1**	**685**	**4.4**	**0.134**
29.539	3.0215	183	433	45.2	7693	50.0	0.302
30.200	2.9569	205	73	7.6	494	3.2	0.115
31.182	2.8660	203	69	7.2	347	2.3	0.085
32.318	2.7678	220	957	100.0	15399	100.0	0.274
32.660	2.7395	225	586	61.2	11339	73.6	0.329
33.680	2.6589	222	85	8.9	3446	22.4	0.689
34.419	**2.6035**	**213**	**568**	**59.4**	**13572**	**88.1**	**0.406**
36.818	2.4392	155	52	5.4	1825	11.9	0.597
37.460	2.3988	157	56	5.9	938	6.1	0.285
38.836	2.3169	148	93	9.7	1145	7.4	0.209
39.503	2.2793	152	69	7.2	1037	6.7	0.255
41.258	2.1863	157	261	27.3	6028	39.1	0.393
43.344	2.0858	152	44	4.6	453	2.9	0.175
44.303	2.0429	149	87	9.1	2099	13.6	0.410
45.846	1.9776	155	83	8.7	1044	6.8	0.214
47.281	**1.9209**	**149**	**211**	**22.0**	**5459**	**35.5**	**0.440**
50.020	1.8220	161	60	6.3	591	3.8	0.167
50.919	1.7919	151	107	11.2	1923	12.5	0.306
51.821	1.7628	136	179	18.7	2104	13.7	0.200
54.456	**1.6836**	**140**	**47**	**4.9**	**476**	**3.1**	**0.172**
56.503	1.6273	140	96	10.0	1545	10.0	0.274
58.802	1.5690	128	57	6.0	1223	7.9	0.365
59.467	1.5531	131	46	4.8	798	5.2	0.295
62.302	1.4891	118	134	14.0	2962	19.2	0.376
62.518	1.4844	117	103	10.8	2949	19.2	0.487
64.278	1.4480	119	54	5.6	1291	8.4	0.406
67.453	1.3873	120	46	4.8	558	3.6	0.206

Table 6: Peak Search Report (39 Peaks, Max P/N = 13.9)

generally used to magnify an image. It can magnify image up to three lack times. There are Three types of rays emitted from the instruments that is back scattered electron rays, secondary rays. Both are used for imaging. And X-rays are used for elemental analysis of the sample. It is used for conducting material for non-conducting and biological material like cement coat a conducting material on it. It may be carbon or gold.

Leaching studies: A suitable way to know the effectiveness of the immobilization of contaminants after solidification process is to perform the extraction test. Usually this test is performed under the chosen condition of leaching medium. Standard method No.1311 recommended by United State Environment Protection Agency is followed. Crushed solid material has been taken in hazardous waste filtration system with zero head space extractor. A measured volume of water was added and this assembly has been shaken for 18 hours continuously in agitation assembly. Filtered extract was collected in a closed vessel and analyzed for metal concentration by using Atomic Absorption Spectrophotometer. The results were compared with that of blank samples to know the actual enhancement of the particular metal in the leachate. It was observed that the leachate of the samples doped with 1000, 1500 and 2000 ppm molybdenum contain 2.5, 3.0 and 11.0 ppm molybdenum respectively. It is clear that the presence of all molybdenum in the leachate is due to the molybdenum containing samples because molybdenum concentration level in leachate of control cement samples is below the detection limit.

Results and Discussion

Although hydration of cement begins immediately as water is added into it but there is a period of time-several hours in which cement remains in fluid like state before setting as a rigid, load bearing material. Setting refers the change of state from fluid paste to rigid form. According to IS: 8112:1996. The initial setting time of a 43 grade ordinary Portland cement should be at least 30 min. and final setting time should be within 600 min. The initial and final setting time of control as well as molybdenum containing samples were determined and presented in Table 3. Molybdenum was added in the concentrations 1, 5, 10, 100, 1000, 1500, 2000, 3000 and 5000 ppm respectively. A perusal of data shows that there is no considerable effect of molybdenum addition on initial as well as final setting time up to the concentration 1000 ppm. The maximum change in this parameter was observed within the concentration range 1000-2000 ppm molybdenum (Figure 1). The setting time became stable beyond this concentration limit. The effect of molybdenum addition on setting time of cement was recorded directly proportional to the concentration of molybdenum added. All the experiments were carried out in triplicate to assure accuracy and reproducibility. All the samples were observed for any visual change before the compressive strength testing each time. There was no significant change found in colour and shape except the surface of molybdenum containing samples was smoother than that of control samples.

Bulk density of control samples as well as molybdenum containing samples was determined on 28 day of curing. Six samples were taken each time and the average value of these observations was reported. There was no significant change in this parameter for molybdenum containing sample in comparison with control (Table 4). Bulk density of control samples was 2.381 ± 0.001 g/cm^3, while it changed to 2.385, 2.393 and 2.399 ± 0.001 g/cm^3 for 1000, 1500 and 2000 ppm molybdenum containing samples respectively (Figure 2). The figure shows that bulk density values of molybdenum containing samples increased with the increase in concentration of molybdenum although the change is in significant. Compressive strength of control mortar samples as well as molybdenum containing samples was determined on 3, 7, 28, 60, 90, 180 and 360 days of curing. All the values have been taken in triplicate and the average values have been reported. The

[Sam.5_1.raw] Sam. Blank

PEAK: 23-pts/Parabolic Filter, Threshold=3.0, Cutoff=0.1%, BG=3/1.0, Peak-Top=Summit

2-Theta d(A)	BG	Height	I%	Area	I%	FWHM	
9.081	9.7305	215	77	14.6	1728	13.5	0.382
12.272	7.2065	186	47	8.9	337	2.6	0.122
13.090	6.7578	168	56	10.6	459	3.6	0.139
14.835	5.9667	153	47	8.9	278	2.2	0.101
15.819	5.5976	153	81	15.4	1576	12.4	0.331
18.058	4.9083	168	335	63.7	4605	36.1	0.234
18.961	4.6766	158	49	9.3	560	4.4	0.194
23.001	3.8634	142	97	18.4	2045	16.0	0.358
26.658	3.3411	172	129	24.5	1773	13.9	0.234
27.514	3.2391	177	47	8.9	465	3.6	0.168
29.420	3.0335	208	500	95.1	7966	62.4	0.271
31.043	2.8785	212	69	13.1	745	5.8	0.184
32.160	2.7810	224	491	93.3	12756	100.0	0.442
32.579	2.7462	243	500	95.1	10704	83.9	0.364
34.121	2.6255	222	526	100.0	11840	92.8	0.383
36.636	2.4508	161	59	11.2	1696	13.3	0.489
37.242	2.4123	158	48	9.1	592	4.6	0.210
38.741	2.3224	158	80	15.2	630	4.9	0.134
39.421	2.2839	144	105	20.0	2290	18.0	0.371
41.242	2.1871	154	303	57.6	6325	49.6	0.355
41.645	2.1669	164	69	13.1	1658	13.0	0.408
44.081	2.0526	152	83	15.8	2253	17.7	0.461
44.679	2.0266	156	58	11.0	1749	13.7	0.513
45.798	1.9796	150	92	17.5	1599	12.5	0.295
47.140	1.9263	148	208	39.5	6343	49.7	0.518
49.865	1.8273	169	55	10.5	458	3.6	0.142
50.803	1.7957	164	103	19.6	1485	11.6	0.245
51.724	1.7659	137	156	29.7	1981	15.5	0.216
54.516	1.6819	139	57	10.8	751	5.9	0.224
56.453	1.6287	149	90	17.1	1114	8.7	0.210
58.593	1.5741	147	53	10.1	447	3.5	0.143
59.912	1.5426	146	69	13.1	450	3.5	0.111
62.380	1.4874	123	120	22.8	2553	20.0	0.362
63.710	1.4595	122	51	9.7	434	3.4	0.145
64.265	1.4482	122	40	7.6	786	6.2	0.334
71.050	1.3257	115	40	7.6	271	2.1	0.115
62.518	1.4844	117	103	10.8	2949	19.2	0.487
64.278	1.4480	119	54	5.6	1291	8.4	0.406
67.453	1.3873	120	46	4.8	558	3.6	0.206

Table 7: Peak Search Report (36 Peaks, Max P/N = 9.6).

results of this study are presented in Table 5. It is important to mention that according to IS:1882:1989 the compressive strength of 43 grade ordinary Portland cement should be at least 230 kg/cm^2, 330 kg/cm^2 and 430 kg/cm^2 for 3, 7 and 28 days of curing. Compressive Strength is the most common performance measures by compression testing machine. The compressive strength is calculated from the failure load divided by the cross section area resisting the load. It is reported in units of kg/cm^2.

The addition of molybdenum caused considerable change on the rate of strength attainment as well as on the compressive strength of the binder system. Utmost care was taken to reduce variability associated with batch preparation step and reagent addition to avoid any substantial variability within a batch. A perusal of data presented in Table 5, exhibits that compressive strength of molybdenum containing samples increases significantly in comparison with control sample. The effect intensifies as the concentration of added molybdenum is increased (Figures 3 and 4).

Microscopic and spectroscopic analysis of the cement paste

The microstructures of hardened control cement paste as well as molybdate containing cement paste were investigated by Scanning Electron Microscopy (Figures 5 and 6). The microstructure of all these samples was very compact with gel like C-S-H phase as a main component. It is clear from scanning electron microscopy images that in molybdenum dopped samples the C-S-H phase is more densified than that of control samples. The XRD diffractogram of sample containing molybdenum and blank sample are presented in Figures 7 and 8. It has been found that the intensity of powellite mineral is clearly indicated in the diffractogram. The presence of molybdenum reduced the leaching of Ca through the solidification process. Molybdenum is mostly dissolved in the interstitial phases while very small amount is found in the silicate phase The X-Ray diffraction study shows that the insoluble molybdenum found in the cement as powellite form $CaMoO_4$. The enclosed peak search reports shows that peaks of Powellite mineral ($CaMoO_4$) are present in the X-ray pattern characteristics (Tables 6 and 7).

Conclusions

The effect of fixation of molybdenum on physical, chemical and engineering properties of 43-grade Ordinary Portland Cement was studied. A quantitative correlation between the concentrations of molybdenum used and the intensity of any change in the properties of 43 grade Ordinary Portland Cement like initial and final setting time, compressive strength, bulk density, leaching studies, microscopic and spectroscopic studies etc. has been established. Molybdenum was used in the concentration range 1 ppm-5000 ppm. It is evident from the results obtained that molybdenum addition improves setting and hardening characteristics of the cement. The compressive strength of the cement improves to 110.5% on 28 day of curing for 2000 ppm of molybdenum addition. SEM studies show that the density of C-S-H gel is more in molybdenum doped cement samples in comparison with control. The presence of peaks of powellite ($CaMoO_4$) in X-ray diffractogram of molybdenum doped cement samples suggested that molybdate ions combine with calcium ions to give $CaMoO_4$ and get fixed chemically.

Acknowledgement

The authors are thankful to the Director, Central Building Research Institute, Roorkee. The help of Dr. L.P. Singh regarding SEM and XRD analysis is gratefully acknowledged.

References

1. Park CK (2000) Hydration and solidification of Hazardous waste containing heavy metals using modified cementitious materials. Cem Conc Res 30: 429-435.
2. Cote' PO, Bridle TR, Hamilton DP (1984) Evaluation of Pollutant Release from Solidified Aqueous Waste Using a Dynamic Leaching Test. Hazardous Waste and Environmental Emergencies, Houston TX.
3. Glasser FP (1993) Chemistry of cement solidified waste forms chemistry and microstructure of solid waste forms, Lewis publishers Chelsea, MI-P.I.
4. Nocun W, Goteborg JM (1997) Studies on Immobilization of heavy metals in Cement Paste-C-S-H leaching behavior. Proceedings of the 10th international Congress on the chemistry of cement, Sweden.
5. Kindness A, Lachowski EE, Minocha AK, Glasser FP (1994) Immobilization and Fixation of Molybdenum (VI) by Portland cement. Waste Management 14: 97-102.
6. Kasselouri V, Ftikos Ch (1997) The effect of MoO_3 on the C3S and C3A formation. Cem Conc Res 27: 917-923.
7. Duchesne J, Laforest G (2004) Evaluation of the degree of Cr ions immobilization by different binders. Cem Conc Res 34: 1173-1177.
8. Mayers TE, Eappi ME (1992) Laboratory evaluation of Stabilization/ Solidification technology for reducing the mobility of heavy metals in New Bedford harbor superfund site sediment stabilization of hazardous radioactive and mixed wastes, 2nd Edn. ASTM publication Philadelphia PA: 304.
9. Adaska WA, Tresouthick SW, West PB (1991) Solidification and Stabilization of wastes using Portland cement. Portland cement Association Skokie IL.
10. Deja J (2002) Immobilization of Cr6+, Cd2+, Zn2+ and Pb2+ in alkali activated slag binders. Cem Concr Res 32: 1971-1979.
11. Hills CD, Pollard SJT (1997) The influence of interference effects on the mechanical microstructural and fixation Characteristics of cement-solidified hazardous waste forms. J Hazard Mater 52: 171-191.
12. Wang S, Vipulanandan C (2000) Solidification/Stabilization of Cr (VI) with cement Leachability and XRD analysis. Cem Concr Res 30: 385-389.
13. Macias A, Kindness A, Glasser FP (1997) Impact of carbon dioxide on the immobilization potential of cemented wastes: Chromium. Cem Concr Res 27: 215-225.
14. U.S. Environmental Protection Agency (1986) Test Method for Evaluating Solid Wastes, SW-846, 3rd ed., Office of Solid Waste and Emergency Response, Washington DC.
15. Alunno Rosseti V, Medici F (1995) Inertization of toxic metals in cement matrices: effects on hydration and hardening. Cem Concr Res 25: 1147-1152.
16. Corner JR (1990) Chemical Fixation and Solidification of Hazardous wastes. Van Nostrand- Reinhold, New York.
17. Taylor HFW (1964) The Chemistry of Cement, Vol. (II) Academic press, New York.
18. Trezza MA, Ferraiuelo MF (2003) Hydration study of limestone blended cement in the presence of hazardous wastes containing Cr(VI). Cem Concr Res 33: 1039-1045.
19. Minocha AK, Jain N, Verma CL (2003) Effect of Inorganic Materials on the Solidification of Heavy metal Sludge. Cem Concr Res 33: 1695-1701.
20. Halim CE, Amal R, Beydaun D, Scott JA, Low G (2004) Implication of the structure of cementitious wastes containing Pb(II), Cd(II), As(V) and Cr(VI) on the leaching of metals. Cem Concr Res 34: 1093-1102.
21. Kolovas K, Tsivilis S, Kakali G (2002) The effect of foreign ions on the reactivity of the CaO-SiO2-Al2O3–Fe2O3 system Part (II): Cations. Cem Concr Res 32: 463-469.
22. Olmo IF, Chacon E, Irabien A (2001) Influence of Lead, Zinc, Fe(III) and Cr(III) oxide on the setting time and strength development of Portland cement. Cem Concr Res 31: 1213-1219.
23. Roy A, Eaton HC, Cartledge FK, Tittlebaum ME (1991) Solidification/ Stabilization of a Heavy metal sludge by a Portland cement/fly ash binding mixture. Hazardous waste and Hazardous materials 8: 33-41.

6

Application of Multi-Objective Optimization by S and R* Optimal Combination Criteria to Determine the Freeze Drying Mode of *Penaeus monodon*

Nguyen Tan Dzung[1]*, Nguyen Quoc Dzung[1], Trinh Van Dzung[2] and Le Xuan Hai[2]

[1]HCMC University of Technical Education, 01-Vo Van Ngan Street, Thu Duc District, Viet Nam
[2]HoChiMinh City University of Technology, Viet Nam National University, HoChiMinh City

Abstract

The establishment of the freeze drying mode of *Penaeus monodon* was based on the solution to multi-objective optimization problem. Experiments was carried out to set up the objective functions describing the influence of technological factors (temperature and pressure of freeze drying chamber, and time of freeze drying) to the freeze drying process. The restricted area method with R*(Z) optimal combination criterion was applied to solve the multi-objective optimization problem, determining the optimal technological mode of freeze drying process (correspondingly 33.96°C, 0.008 mmHg and 13.21h) in order that the objective functions reached the minimum value in terms of the finished product, including the energy consumption of 68.77 kWh/kg, the residual water content of 4.76%, the anti-rehydration capacity of 7.82%, the volume contraction of 8.82% and the loss of vitamin C of 1.91%.

Keywords: Multi-objective optimization; Freeze drying; *Penaeus monodon*

Introduction

Freeze drying is a complicated technique including three consecutive main stages which are represented in the process diagram of Figure 17. The first stage is freezing the material; the second and the third stage are respectively the sublimation drying and vacuum drying. Both these final stages determine the quality of the product [6,7].

According to Figure 1, the determination of the freeze drying mode required the outputs to reach the minimal level, including the energy consumption per weight (y_1,kWh/kg), the residual water content (y_2,%), the anti-rehydration capacity (y_3,%), the volume contraction (y_4,%) and the loss of vitamin C (y_5,%) of the freeze-dried product (finished product). It should be emphasized that these 5 outputs were affected by the 3 technological factors: temperature of freeze drying chamber (Z_1°C), pressure of freeze drying chamber (Z_2,mmHg) and time of freeze drying (Z_3,h).

However, the simultaneous consideration of all these outputs above to reach the minimal level resulted in the standard solution to the multi-objective optimization problem [4,11]. This problem regularly appears in reality and in different fields. The answer to the multi-objective optimization problem was found in the case of the application of the R*(Z) optimal combination criterion (also known as the restricted area method) for the freeze drying process of *Penaeus monodon*. By solving the heat and mass transfer model of the freeze drying dehydration [2,3], the multi-objective optimization results were used to establish the freeze drying mode of *Penaeus monodon* which was the closest to the utopian point but the furthest from the restricted area C, [4,11,12].

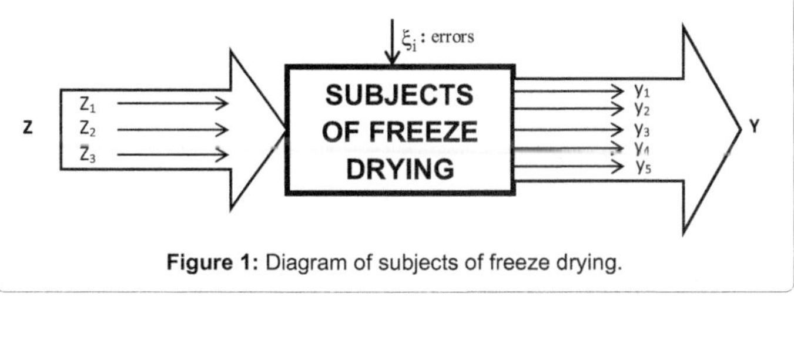

Figure 1: Diagram of subjects of freeze drying.

The fundamental of multi-objective optimization by S And R* optimal combination criteria

Basic concepts: The technological subjects including m objective functions $f_1(Z), f_2(Z), ..., f_m(Z)$ form the vector of these functions $f(Z) = \{f_j(Z)\} = \{f_1(Z), f_2(Z), ..., f_m(Z)\}$, where $j = 1 \div m$. Every objective function $f_j(Z)$ will be affected by n variables $Z_1, Z_2, ..., Z_n$ which form the Z variable vector $Z = \{Z_i\} = (Z_1, Z_2, ..., Z_n)$, where $i = 1 \div n$. These variables vary in the identified domain Ω_Z and the function values form the domain of the objective function Ω_f (in the two-objective optimization problem, the domain can be performed geometrically in the closed curve **A – f(ZS) – f(ZR) – B – N – M**, Figure 2), [4,11,12].

Every objective function $f_j(Z)$ with Z variable vector $Z = \{Z_i\} = (Z_1, Z_2, ..., Z_n)$, where $i = 1 \div n$, is considered as the one-objective optimization problem. Hence, the m-objective optimization problem can be simply transformed into the problem to find the minimum value for the set of m one-objective optimization problems, [4,11,12]:

$$f_{jmin} = f_j(Z_1^{j\,opt}, Z_2^{j\,opt}, ..., Z_n^{j\,opt}) = \text{Min}\, f_j(Z_1, Z_2, ..., Z_n) \quad (1)$$

$$Z = \{Z_i\} = (Z_1, Z_2, ..., Z_n) \in \Omega_Z \quad (2)$$

$$j = 1 \div m;\ i = 1 \div n \quad (3)$$

The utopian plan and the utopian effect: If the variable vector $Z^{UT} = \{Z_i^{UT}\} = (Z_1^{UT}, Z_2^{UT}, ..., Z_n^{UT}) \in \Omega_Z$ is the test for all one-objective

***Corresponding author:** Nguyen Tan Dzung, HCMC University of Technical Education, 01-Vo Van NganStreet, Thu Duc District, Viet Nam

optimization problems (1) + (2) + (3), it means that $Z_i^{UT} = Z_i^{jopt}$ with i = 1 ÷ n. Thus, Z_i^{UT} is called the utopian plan or the utopian test of the m-objective optimization problem, [4,11,12].

In reality, Z_i^{UT} does not exist. However, every one-objective optimization problem (1) + (2) + (3) has its own f_{jmin} (with j = 1 ÷ m) respectively, so $f^{UT} = (f_{1min}, f_{2min}, ..., f_{mmin})$ does exist. Then, $f^{UT} = (f_{1min}, f_{2min}, ..., f_{mmin})$ is called the utopian effect or the utopian point. According to Figure 2, the utopian point f^{UT} of the two-objetive optimization problem exists but lies outside the identified domain Ω_f, i.e. the utopian test does not exist.

The dominant plan and the dominated plan: It is assumed that there are two variable vectors ZQ = {ZQ_i} and ZV = {ZV_i} with i = 1 ÷ n. Then, there exist respectivly two function vectors f(ZQ) = {$f_j(ZQ)$} and f(ZV) = {$f_j(ZV)$} with j = 1 ÷ m.

If with all j: $f_j(ZQ) < f_j(ZV)$, ZQ is called the dominant plan (or the dominant test) over ZV, symbolizing: ZQ '>' ZV; and ZV is called the dominated plan (or the dominated test), symbolizing: ZV '<' ZQ, [4,11,12].

The optimal paréto plan

The ZP plan is called the optimal Paréto plan in condition that ZP cannot be dominated by any other plans dependable on the identified domain Ω_Z. Then, f(ZP) would be called an optimal Paréto effect in the set of the optimal Paréto effects ΩfP. Figure 2 performs the set of the optimal Paréto effects ΩfP as the curve **A – f(ZS) – f(ZR) – B,** [4,11,12].

Theorem 1: (Theorem Paréto): If the multi-objective optimization problem has the test which is the so-called optimal one according to some definition, this test received has to be the optimal Paréto plan without the dependence on the chosen definition, [4,11,12].

Proof: If optimal Z test of the multi-objective optimization problem is not the optimal Paréto plan, it is certainly able to find at least one plan dominating Z. This proves that Z can not be recognized as the optimal test and leads to the conflict with the assumption that Z is the optimal test. Thus, Z must be a plan over which cannot be dominant, i.e. the Paréto plan is optimal.

Therefore, one test of the multi-objective optimization problem (1) + (2) + (3) found by any method, to be recognized as the optimal by the method chosen, must in advance be certified as the optimal Paréto plan.

Multi-objective optimization by the utopian point method with S(Z) optimal combination criterion

Considering the m-objective optimization problem (1) + (2) + (3): The optimal values $f_{1min}, f_{2min}, ..., f_{mmin}$ can be determined after solving each problems, and the fact that the utopian test (the test for the whole system) does not exist still identifies the utopian point $f^{UT} = (f_{1min}, f_{2min}, ..., f_{mmin})$. A S(Z) optimal combination criterion is defined by the following expression [4,12]:

$$S(Z) = \sqrt{\left[\sum_{j=1}^{m} s_j^2(Z)\right]} = \sqrt{\left[\sum_{j=1}^{m}\left(f_j(Z) - f_{jmin}\right)^2\right]} \qquad (4)$$

It is obvious that S(Z) is the distance from f(Z) to f^{UT}. Choosing S(Z) optimal combination criterion as an objective function, the m-objective optimization problem are restated as:

Find ZS = ($Z_1S, Z_2S, ..., Z_nS$) ∈ Ω_Z in order that the objective function S(Z) reaches the minimum value:

$$S_{min} = S(ZS) = \min\{S(Z)\} = \min\left\{\sqrt{\left[\sum_{j=1}^{m}\left(f_j(Z) - f_{jmin}\right)^2\right]}\right\} \qquad (5)$$

$\forall Z = (Z_1, ..., Z_n)\ \Omega_Z$

Theorem 2: (Theorem Paréto): If ZS of the optimization problem (5) does exist, ZS is the optimal Paréto test of the m-objective optimization problem (1) + (2) + (3), [4,11].

Proof: It is assumed that ZS is not optimal Paréto test. Then, it will be found that ZS^* is dominant over ZS. By definition, ZS^* must have at least an effect $f_k(ZS^*)$, where $1 \le k \le m$ in order that $f_k(ZS^*) < f_k(ZS)$. As a result, $S(ZS^*) < S(ZS)$. This contradicts the assumption that ZS is the optimal test (5). Hence, there does not exist any other dominant tests over ZS. Therefore, ZS must be an optimal Paréto test [4,11].

Symbol: f(ZS) = fPS = ($f_1PS, f_2PS, ..., f_mPS$). With the utopian point method (i.e. the m-objective optimization problem convert into the S optimal combination criterion), the optimal Paréto test ZS will be found to have the optimal Paréto effect f(ZS) = fPS closest to the utopian point $f^{UT} = (f_{1min}, f_{2min}, ..., f_{mmin})$. The case m = 2 (two objectives) is illustrated in Figure 2.

Optimizing the multi-objective functions by the restricted area method with R*(Z) optimal combination criterion

In fact, every objective function $f_j(Z)$ is restricted by the conditions set up by technology. Such as:

a) Case 1: The obligatory conditions $f_j(Z) < C_j$, $\forall j = 1 \div m$, $\forall Z \in \Omega_Z$ (6)

From (6), the restricted area would be made: C = {$f_j(Z) \ge C_j$}, with $f_j(Z)$ (7)

The restricted area method suggests the solution to the m-objective optimization problem (1) + (2) + (3) by R*(Z) optimal combination criterion, defined as, [4,12]:

$$R^*(Z) = \sqrt[m]{r_1(Z).\ r_2(Z)...\ r_m(Z)} = \sqrt[m]{\prod_{j=1}^{m} r_j(Z)} \qquad (8)$$

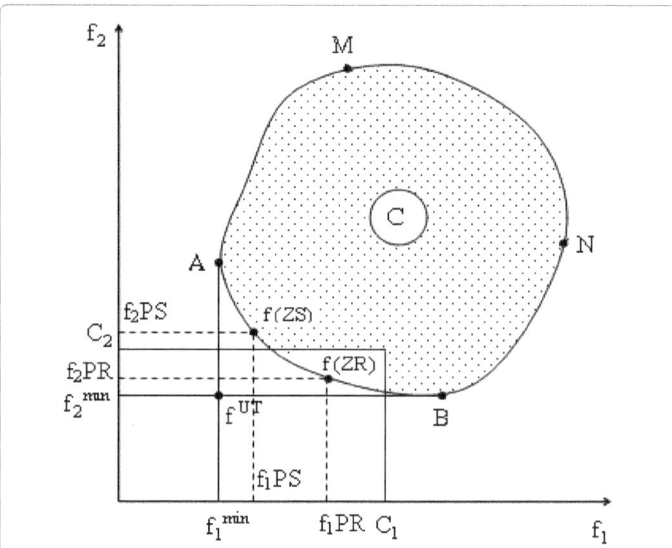

Figure 2: Dimension of objective functions of the two-objective optimization problem.

With $r_j(Z)=\left(\frac{C_j-f_j(Z)}{C_j-f_{jmin}}\right)$ when $f_j(Z)<C_j$ (9)

$r_j(Z)=0$ when $f_j(Z)\geq C_j$ (10)

According to (9), if $f_j(Z) \to f_{j\,min}$ and $\forall f_j(Z) < C_j$, $r_j(Z) \to r_{jmax} = 1$.

By choosing R*(Z) as the objective function, the m-objective optimization problem is restated as:

Find $ZR = (Z_1R, Z_2R... Z_nR) \in \Omega_Z$ in order that R*(Z) reaches the maximum value.

$$R^*_{max}=R^*(ZR)=\max\left\{R^*(Z)\right\}=\max\left\{\sqrt[m]{\left[\prod_{j=1}^{m} r_j(Z)\right]}\right\} \quad (11)$$

$\forall Z = (Z_1... Z_n) \in \Omega_Z$

From (9), it can be seen: $0 \leq R^*(ZR) \leq 1$. If $R^*(ZR) = 1$, $ZR = Z^{UT}$ – the utopian test. If $R^*(ZR) = 0$, one of the values of $f_j(Z)$ violates (6), which means that $f_j(Z)$ belongs to the restricted area C (7).

Theorem 3: If the multi-objective optimization problem (11) has it own test ZR, this test ZR is also the optimal Paréto test of the m-objective optimization problem (1) + (2) + (3), [4,11].

Symbol: $f(ZR) = fPR = (f_1PR, f_2PR, ..., f_mPR)$. With the optimal ZR, the optimal Paréto effect $fPR = (f_1PR, f_2PR... f_mPR)$ would be the closest to the utopian point and the furthest from the restricted area C.

b) Case 2: The obligatory conditions $a_j < f_j(Z) < b_j$, $\forall j = 1 \div m$, $\forall Z \in \Omega_Z$ (12)

Then, setting a new objective function:

$$I_j(Z)=\left[f_j(Z)-\frac{a_j+b_j}{2}\right]^2 \quad (13)$$

From (13), it can be seen: if $I_j(Z) \to I_{j\,min} = 0$, $f_j(Z) \to f_{jmin} = (a_j + b_j)/2$, $\forall Z\ \Omega_Z$.

When $\frac{a_j+b_j}{2} < f_j(Z) < b_j$, then

$$0 < \left[f_j(Z)-\frac{a_j+b_j}{2}\right] < \frac{b_j-a_j}{2} \quad (14)$$

When $a_j < f_j(Z) \leq \frac{a_j+b_j}{2}$, then

$$-\frac{b_j-a_j}{2} < \left[f_j(Z)-\frac{a_j+b_j}{2}\right] \leq 0 \quad (15)$$

From (14) and (15), we have: $0 \leq \left|f_j(Z)-\frac{a_j+b_j}{2}\right| < \frac{b_j-a_j}{2}$ (16)

Combining (13) and (16), we have:

$$I_j(Z)=\left[f_j(Z)-\frac{a_j+b_j}{2}\right]^2 < \left(\frac{b_j-a_j}{2}\right)^2 = C_j \quad (17)$$

From (17), the restricted area of the new objective function is defined as:

$$C = \{I_j(Z) \geq C_j\}, \text{ with } C_j=\left(\frac{b_j-a_j}{2}\right)^2 \quad (18)$$

As a result of the fact that the restricted area (18) is the same as (6), the multi-objective optimization problem is solved similarly as case 1.

Materials and Methods

Materials

The material used for the freeze drying experiment was *Penaeus monodon* which had the approximate weight of (41÷50) prawns/pound, with the size coefficient K = 11, [1]. Blanched at 70°C during (15 ÷ 30) seconds, the material was processed minimally by peeling off the shells and cutting off the heads.

Apparatus

- The freeze drying machine DS-3 was controlled automatically by computer (Figure 3)

- The tools determining these factors such as energy consumption, residual water content, rehydration capacity, volume contraction, and loss of vitamin C could be referred to by [1,5].

Methods

Determining the energy consumption (**y_1, kWh/kg** product) for 1 kg finished product by Watt meter, [1,5].

$$y_1=\frac{U.I.\tau.\cos\varphi}{G} \quad (19)$$

Where: G [kg] – mass finished product;

U [V] – number of Voltmeter;

I [A] – munber of Amperemeter;

τ [s] – times

Determining the residual water content of the finished product (**y_2**, %) by mass sensor through computer, [1,5].

$$y_2=100-\frac{G_i}{G_e}(100-W_i) \quad (20)$$

Determining the anti-rehydration capacity of the finished product

Figure 3: The freeze drying system DS-3 with the auto-freezing (-50 ÷ - 45)°C.

(y_3, %) indirectly by IR [%], which is the rehydration capacity of the finished product: $y_3 = 100 - IR$, [1,5].

$$IR = \frac{G_1 - G_e}{G_i - G_e}.100\% \quad (21)$$

$$y_3 = 100 - IR = \frac{G_i - G_1}{G_i - G_e}100\% \quad (22)$$

where: G_i [kg] – weight of the initial material used for freeze drying; G_e [kg] - weight of the finished product, G_1 [kg] – weight of the finished product which was soaked into the water at 25°C until the constant mass (the saturation of the water content), W_i [%] – initial water content of the material.

The ideal rehyration capacity of the product means that the in-water content is equal to the out-water content of the product, i.e. $G_1 = G_i$ and $IR_{max} = 1 = 100\%$, $y_{3min} = 0$. In fact, $y_3 > 0$.

Determining the volume contraction (**y_4**, %) by the volume of the initial material (V_1) and of the finished product after freeze drying (V_2), [1,5]:

$$y_4 = \frac{V_1 - V_2}{V_1}100\% = \frac{\Delta V}{V_1}100\% \quad (23)$$

The fact that the surface of the product is not rough and not contracted means $y_{4min} = 0$. In fact, $y_4 > 0$.

Determining the loss of vitamin C of the finished product (**y_5**,%) according to the method TCVN 4715 – 89, [1,5].

$$y_5 = \frac{m_1 - m_2}{m_1}100\% = \frac{\Delta m}{m_1}100\% \quad (24)$$

where: m_1 and m_2 [mg%] – the vitamin C content of the material before and after freeze drying respectively. The fact that the product achieves the best quality means $y_{5min} = 0$. In fact, $y_5>0$.

1. Determining the temperature and pressure by sensors in the machine.
2. Quadratic orthogonal experimental planning method, [4,10].
3. Establishing and solving 5-objective optimization problem by the restricted area method.

Results and Discussion

Establishing the constituent objective functions of the multi-objective problem

The constituent objective functions of the optimal freeze drying (y_1, y_2, y_3, y_4, y_5) depended on the parameters, including: temperature of freeze drying chamber (**Z_1, °C**), pressure of freeze drying chamber (**Z_2, mmHg**), time of freeze drying (**Z_3, h**), and were determined by the experimental planning method with the quadratic orthogonal experimental matrix ($k = 3$, $n_0 = 4$, carrying out 18 experiments). These variables x_1, x_2, x_3 were coded variables of Z_1, Z_2, Z_3. The value of the star point $\alpha = 1.414$.

The experimental parameters investigated by establishing and solving the heat and mass transfer model in the freeze drying [2,3] to reach the suitable residual water content of the finished product were summarized in Table 1.

The values of the objective functions y_1, y_2, y_3, y_4, y_5 were shown in Table 2 by carrying out the experiments with these parameters in Table 1.

The regression equations below were obtained after processing the experimental data, calculating the coefficients, testing the significance of the coefficients by the Student test, and testing the regression equations for the fitness of the experimental results by Fisher test.

$$y_1 = f_1(x_1, x_2, x_3) = 75.758 + 1.883x_1 + 10.167x_3 - 2.491x_2^2 + 2.487x_3^2 \quad (25)$$

$$y_2 = f_2(x_1, x_2, x_3) = 4.17 - 0.39x_1 - 0.614x_{39} - 0.226x_1x_3 + 0.27.x_1^2 + 0.245x_2^2 + 0.142x_3^2 \quad (26)$$

$$y_3 = f_3(x_1, x_2, x_3) = 7.578 + 2.069x_1 + 0.575x_2 + 1.187x_3 + 0.851x_1^2 + 1.205x_3^2 \quad (27)$$

$$y_4 = f_4(x_1, x_2, x_3) = 8.307 + 1.45x_1 + 0.789x_2 + 0.76x_3 - 0.484x_2x_3 + 0.429x_1^2 + 0.607x_2^2 \quad (28)$$

$$y_5 = f_5(x_1, x_2, x_3) = 2.363 + 0.393x_1 + 0.375x_2 + 0.205x_3 - 0.178x_3^2 \quad (29)$$

Solving the one-objective optimization problems

These one-objective optimization problems were found to achieve: $y_{1min} = \min f_1(x_1, x_2, x_3)$; $y_{2min} = \min f_2(x_1, x_2, x_3)$; $y_{3min} = \min f_3(x_1, x_2, x_3)$; $y_{4min} = \min f_4(x_1, x_2, x_3)$; $y_{5min} = \min f_5(x_1, x_2, x_3)$, with the identified domain $\Omega_x = \{-1.414 \le x_1, x_2, x_3 \le 1.414\}$. By using the Excel – Solver software, the results of the optimal parameters of every objective function (25), (26), (27), (28) and (29) limited in the experimental domain were summarized in Table 3, [4,11,12]:

According to the Table 3, the utopian points were indentified: $f^{UT} = (f_{1min}, f_{2min}, f_{3min}, f_{4min}, f_{5min}) = (58.71, 3.119, 5.215, 5.163, 0.7497)$. However, the utopian plan did not exist, because of $x^{jopt} = (x_1^{jopt}, x_2^{jopt}, x_3^{jopt}) \neq x^{kopt} = (x_1^{kopt}, x_2^{kopt}, x_3^{kopt})$ with $j, k = 1 \div 5$, $j \neq k$.

Solving the multi-objective optimization problem by the restricted area method

The purpose of the experiment was to reach the targets of the freeze drying process which were expressed by 5 regression equations (25), (26), (27), (28) and (29), but the tests satisfying all function values (y_{1min}, y_{2min}, y_{3min}, y_{4min}, y_{5min}) could not be found. Hence, the idea of the multi-objective optimization problem was to find the optimal Paréto test for $yPR = (y_1PS, y_2PR, y_3PR, y_4PR, y_5PR)$ closest to the utopian point and the furthest from the restricted area, but $y_j = y_j(x) = f_j(x_1, x_2, x_3)$ must satisfy technological conditions with the initial requirements, [4,11,12]:

$$y_1 < C_1 = 86.21;\ 2 = a < y_2 < b = 6;\ y_3 < C_3 = 11.56;$$
$$y_4 < C_4 = 10.91;\ y_5 < C_5 = 2.98 \quad (30)$$

Setting the new objective functions as the followings:

$I_1(x) = y_1(x)$ with $I_1(x) < C_1 = 86.21$ (31a)

$I_2(x) = [y_2(x) - (a + b)/2]^2 = [y_2(x) - 4]^2$ with $I_2(x) < C_2 = [(b - a)/2]^2 = 4$ (31b)

$I_3(x) = y_3(x)$ with $I_3(x) < C_3 = 11.56$ (31c)

$I_4(x) = y_4(x)$ with $I_4(x) < C_4 = 10.91$ (31d)

$I_5(x) = y_5(x)$ with $I_5(x) < C_5 = 2.98$ (31e)

Parameters	Levels					Deviation ΔZ_i
	$-\alpha$ (-1.414)	Low (-1)	Central (0)	High (+1)	$+\alpha$ (1.414)	
Z_1 [^{0}C]	20.102	23	30	37	39.898	7
Z_2 [mmHg]	0.008	0.094	0.3	0.507	0.592	0.2065
Z_3 [h]	11.172	12	14	16	16.828	2

Table 1: Parameter level design.

	N	X_0	X_1	X_2	X_3	X_1X_2	X_1X_3	X_2X_3	X_1^2-2/3	X_2^2-2/3	X_3^2-2/3	y_1	Y_2	Y_3	Y_4	Y_5
2^k	1	1	1	1	1	1	1	1	0.333	0.333	0.333	80.42	3.72	12.45	10.91	3.253
	2	1	-1	1	1	-1	-1	1	0.333	0.333	0.333	83.51	4.81	10.01	8.978	2.554
	3	1	1	-1	1	-1	1	-1	0.333	0.333	0.333	89.71	3.51	11.71	10.92	2.161
	4	1	-1	-1	1	1	-1	-1	0.333	0.333	0.333	84.34	5.04	7.73	8.525	1.192
	5	1	1	1	-1	1	-1	-1	0.333	0.333	0.333	66.52	5.36	10.94	10.31	2.581
	6	1	-1	1	-1	-1	1	-1	0.333	0.333	0.333	65.78	5.69	7.78	8.671	2.247
	7	1	1	-1	-1	-1	-1	1	0.333	0.333	0.333	65.89	5.39	10.09	8.478	1.628
	8	1	-1	-1	-1	1	1	1	0.333	0.333	0.333	64.82	5.87	5.97	6.19	1.287
2k	9	1	1.414	0	0	0	0	0	1.333	-0.667	-0.667	81.24	4.08	13.32	12.85	3.388
	10	1	-1.41	0	0	0	0	0	1.333	-0.667	-0.667	68.15	4.96	5.45	6.379	1.707
	11	1	0	1.414	0	0	0	0	-0.667	1.333	-0.667	70.62	4.93	8.61	11.63	2.748
	12	1	0	-1.41	0	0	0	0	-0.667	1.333	-0.667	73.73	4.01	7.75	8.307	2.653
	13	1	0	0	1.41	0	0	0	-0.667	-0.667	1.333	98.71	3.51	12.61	10.13	2.521
	14	1	0	0	-1.41	0	0	0	-0.667	-0.667	1.333	65.45	5.02	7.57	7.697	1.787
n_0	15	1	0	0	0	0	0	0	-0.667	-0.667	-0.667	77.31	4.41	7.08	7.858	2.281
	16	1	0	0	0	0	0	0	-0.667	-0.667	-0.667	75.52	4.21	7.12	8.018	2.263
	17	1	0	0	0	0	0	0	-0.667	-0.667	-0.667	78.24	4.29	7.16	8.449	2.127
	18	1	0	0	0	0	0	0	-0.667	-0.667	-0.667	73.63	4.13	7.74	7.659	2.011

Table 2: The orthogonal experimental matrix level 2, k = 3, n_0 = 4.

Drying material	j	y_{jmin}	$x_1^{j\,opt}$	$x_2^{j\,opt}$	$x_3^{j\,opt}$
Penaeus monodon	1	58.71	-1.414	1.414	-1.414
	2	3.119	1.314	-6.11E-08	1.414
	3	5.215	-1.216	-1.414	-0.4925
	4	5.163	-1.394	-1.2137	-1.405
	5	0.7497	-1.241	-1.414	-1.341

Table 3: Minimum tests of each one-objective optimization problem.

$$\forall x = (x_1, x_2, x_3) \in \Omega_x \quad (31f)$$

From the system of equations (31a), (31b), (31c), (31d), (31e), (31f) and table 3, it can be easily found:

$I_{1min} = y_{1min} = 58.71;$

$I_{2min} = 0.00;$

$I_{3min} = y_{3min} = 5.215;$

$I_{4min} = y_{4min} = 5.163;$

$I_{5min} = y_{5min} = 0.7497;$

Establishing the R*-objective combination function $R^*(I_1, I_2, I_3, I_4, I_5) = R^*(y_1, y_2, y_3, y_4, y_5) = R^*(x_1, x_2, x_3) = R^*(x)$ as the followings [4,11, 12]:

$$\begin{cases} R^*(x) = R^*(x_1, x_2, x_3) = \sqrt[5]{\prod_{j=1}^{5} r_j(x_1, x_2, x_3)} = \sqrt[5]{\prod_{j=1}^{5} r_j(x)} \\ \Omega_x = \{-1,414 \le x_1,\ x_2,\ x_3 \le 1,414\};\ x = (x_1, x_2, x_3) \end{cases} \quad (32)$$

With: $$r_j(x) = \left(\frac{C_j - I_j(x)}{C_j - I_{jmin}}\right) \text{ when } I_j(x) < C_j \quad (33)$$

$$r_j(x) = 0 \text{ when } I_j(x) \ge C_j \quad (34)$$

The five-objective optimization problem needed to indentify xR = $(x_1R, x_2R, x_3R) \in \Omega_x$ in order that $R^*(x_1R, x_2R, x_3R) = Max\{R^*(x_1, x_2, x_3)\}$. The maximum value of (32) was determined by using the Excel – Solver sofware:

$R^*(x)_{max} = Max\{R^*(x_1, x_2, x_3)\} = R^*(x_1R, x_2R, x_3R) = 0.876$

With: $x_1R = 0.5659$; $x_2R = -1.414$; $x_3R = -0.3936$; Then, transforming into real variables:

$Z_1^{opt} = 33.96^{\circ}C$; $Z_2^{opt} = 0.008mmHg$; $Z_3^{opt} = 13.21h$

Substituting x_1R, x_2R, x_3R into these equations (31a), (31b), (31c), (31d), (31e) and (31f), the results were obtained as:

$I_1PR = 68.226;$

$I_2PR = 0, 7056;$

$I_3PR = 7.928;$

$I_4PR = 8.795$;

$I_5PR = 1.9$

Substituting x_1R, x_2R, x_3R into these equations (25), (26), (27), (28) and (29), the results were obtained as:

$y_1PR = 68.226$;

$y_2PR = 4.84$;

$y_3PR = 7.928$;

$y_4PR = 8.795$;

$y_5PR = 1.947$

The rehydration capacity of the product was determined as:

$IR = 100 - y_3PR = 92.072$

where: $xR = (x_1R, x_2R, x_3R)$ called optimal Paréto test; $yPR = (y_1PR, y_2PR, y_3PR, y_4PR, y_5PR)$ called the optimal Paréto effect.

As a result, through the calculation from the experimental models (25), (26), (27), (28) and (29), the parameters of the freeze drying process which satisfied the maximum R*-Optimal combination criterion were determined as: temperature of freeze drying chamber was $Z_1^{opt} = 33.96^0C$, pressure of freeze drying chamber was $Z_2^{opt} = 0.008mmHg$, time of freeze drying was $Z_3^{opt} = 13.21h$. The total energy consumption per weight of the product was $y_1PR = 68.22$ kWh/kg; the residual water content of the product was $y_2PR = 4.84\%$ (acceptable with the initial requirements of $2 \div 6$ %); the rehydration capacity of the product was $IR = 100 - y_3PR = 92.07$ %; the volume contraction of the product was $y_4PR = 8.79$ % and the loss of vitamin C of the product was $y_5PR = 1.94\%$. Compared with the experimental results from the table 2, these results above were suitable and satisfying with the objectives of the problem.

Experiment to test the results of multi-objective optimization problem

Carrying out the freeze drying process of *Penaeus monodon* at the optimal Paréto test: temperature of freeze drying chamber of $Z_1^{opt} = 33.96^0C$, pressure of freeze drying chamber of $Z_2^{opt} = 0.008mmHg$, and time of freeze drying $Z_3^{opt} = 13.21$ hours, the experimental results were determined as: the energy consumption per product weight of $y_1 = 68.77$ kWh/kg, the residual water content of $y_2 = 4.76\%$, the rehydration capacity of $IR = 100 - y_3 = 92.17\%$ (the anti-rehydration capacity of $y_3 = 7.82\%$), the volume contraction of $y_4 = 8.82\%$ and the loss of vitamin C of $y_5 = 1.91\%$.

Consequently, it was very noticeable that the results from the optimization problems of the freeze drying process had the fitness for the experimental results.

When the pressure of freeze drying chamber was fixed: $x_2 = -1.414$, respectively $Z_2 = 0.008$ mmHg, the relationship between y_1, y_2, y_3, y_4, y_5 and S combination function with 2 variables x_1, x_3 was performed geometrically in 3D (Figures 4, 5, 6, 7, 8, 9). When x_3 was fixed with constant values, the variation of x_1 was shown in Figures 10, 11, 12, 13, 14, 15.

Establishing the freeze drying mode

By the optimal Paréto test: the temperature, pressure of freeze drying chamber and the total drying time of 2 stages (2 and 3) respectively $Z_1^{opt} = T_f = 33.96^0C$, $Z_2^{opt} = P_m = 0.008mmHg$ and $Z_3^{opt} = 13.21h$, and by the temperature of the freezing chamber $T_e = -45^0C$,

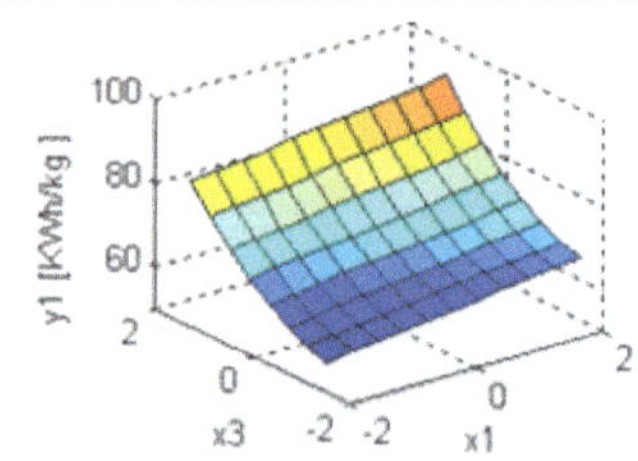

Figure 4: Energy consumption, x2 = -1.414.

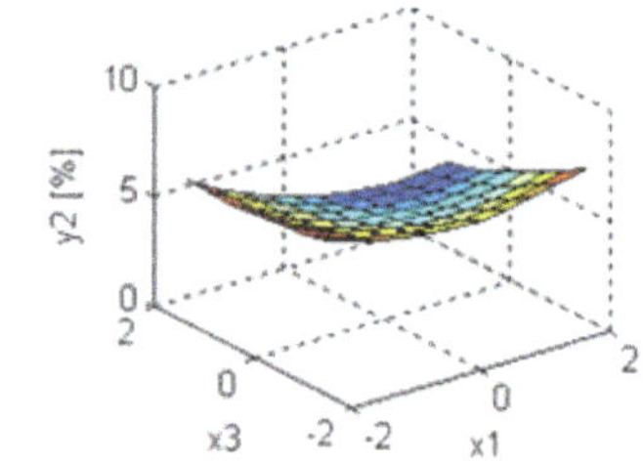

Figure 5: Water content of product, x2 = -1.414.

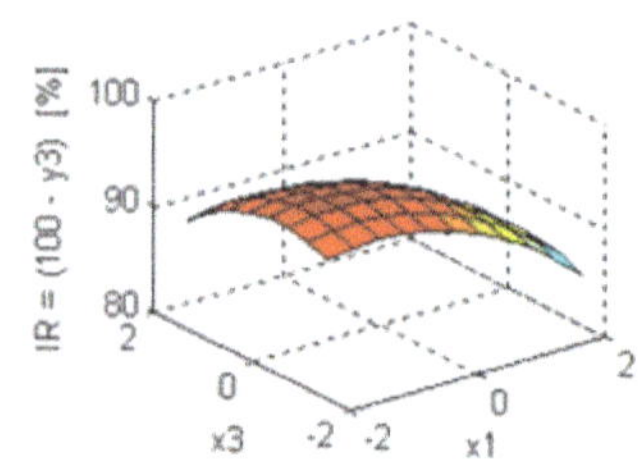

Figure 6: Rehudration capacity, x2 = -1.414.

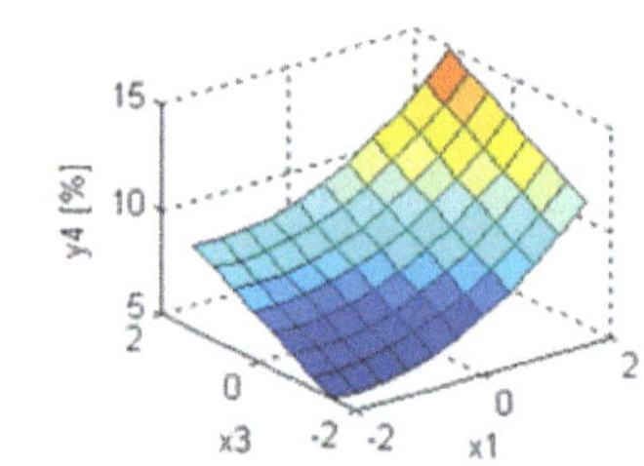

Figure 7: Volume contraction, x2 = -1.414.

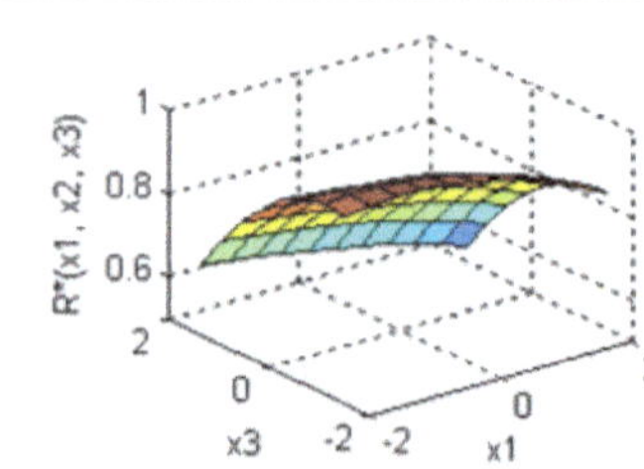

Figure 8: Loss of vitamine C, x2 = -1.414.

the crystallization temperature of the water in *Penaeus monodon* $T_{kt} = -1.21^{\circ}C$ [1], and the thermo-physical parameters of *Penaeus monodon* [1,13,14], the heat transfer model of the freezing process [2] and the heat and mass transfer model of the freeze drying process [3,13,14] were solved to determine the optimal freezing temperature, the time of stage 1 (τ_1, **h**), the time of sublimation drying at stage 2 (τ_2, **h**), and the time of vacuum drying at stage 3 (τ_3, **h**). The results were shown in Figure 17.

Stage 1: The frozen material was carried out at the optimal technological mode [2,3,13] (water completely frozen in the material) with the temperature of the freezing chamber of T_e = -45°C, the surface material temperature of T_s = -34°C, the central material temerature of T_c = -11.78°C and the average material temperature of T_m = -25.11°C, the frozen time of τ_1 = 2.5h.

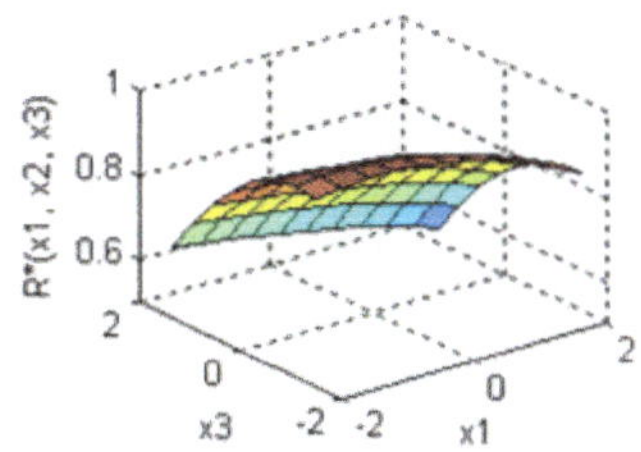

Figure 9: Conbination function R¹, x2 = -1.414.

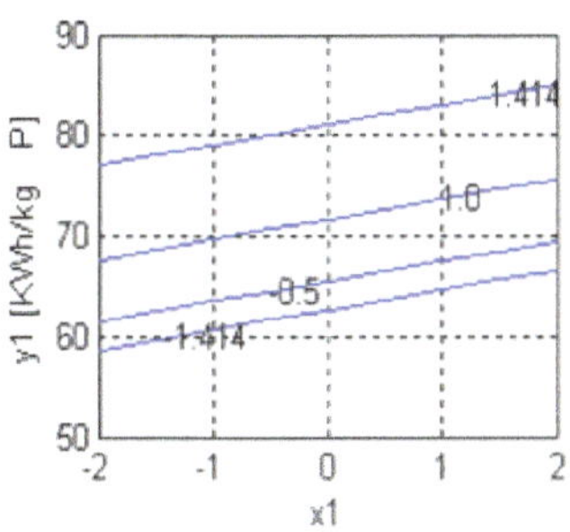

Figure 10: Energy consumption, x2 = -1.414.

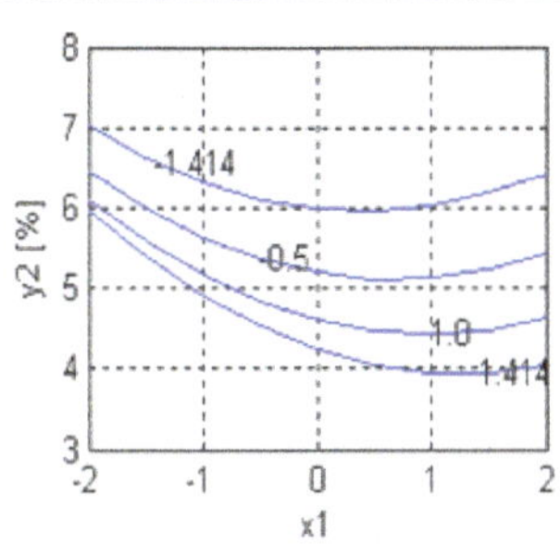

Figure 11: Water content of product, x2 = -1.414.

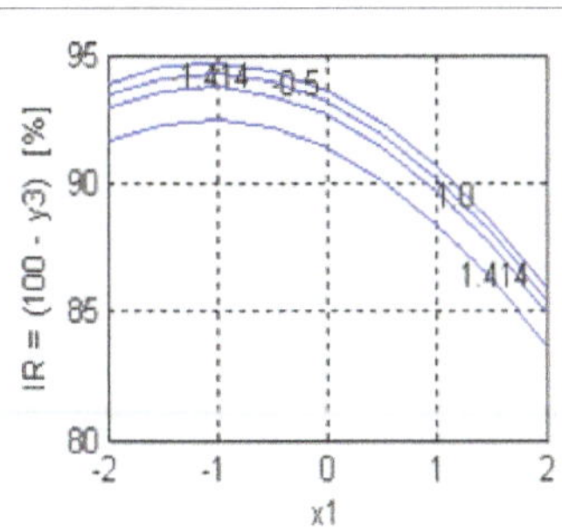

Figure 12: Rehudration capacity, x2 = -1.414.

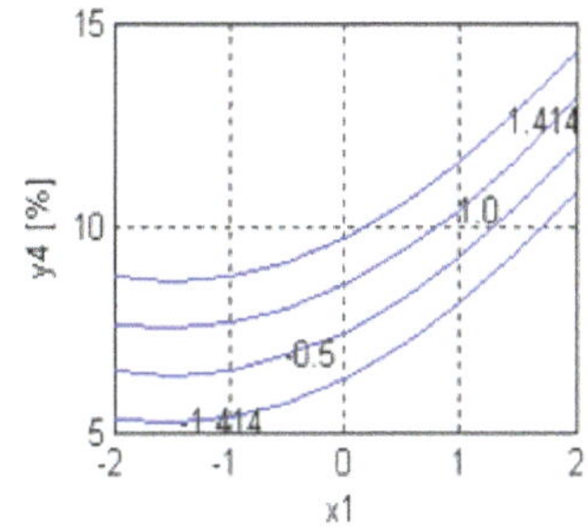

Figure 13: Volume contraction, x2 = -1.414.

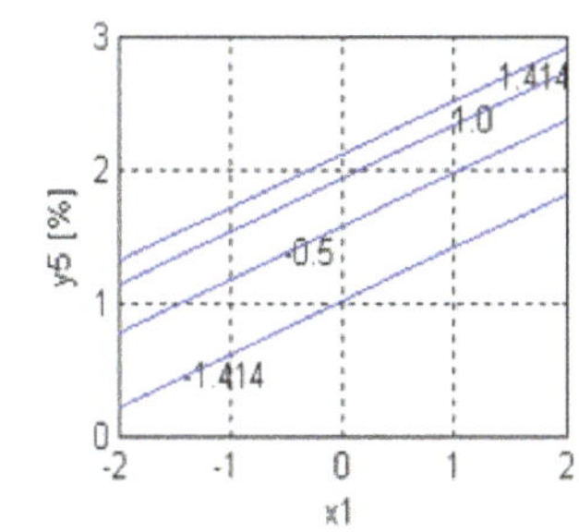

Figure 14: Loss of vitamine C, x2 = -1.414.

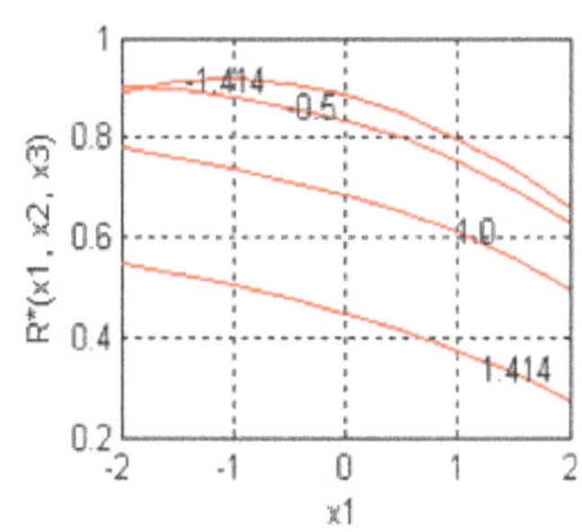

Figure 15: Conbination function R¹, x2 = -1.414.

Figure 16: The freeze-dried *Penaeus monodon*.

Stage 2: The sublimation drying was carried out with the temperature of freeze drying chamber of $Z_1^{opt} = T_f$ = 33.96°C, the pressure of freeze drying chamber of $Z_2^{opt} = P_m$ = 0.008mmHg, the sublimation temperature of the water frozen in the material of T_m = -25.11°C, the central material temperature of $T_c \leq$ -1.21°C, and the drying time of τ_2 =11.46h, [2,3,13].

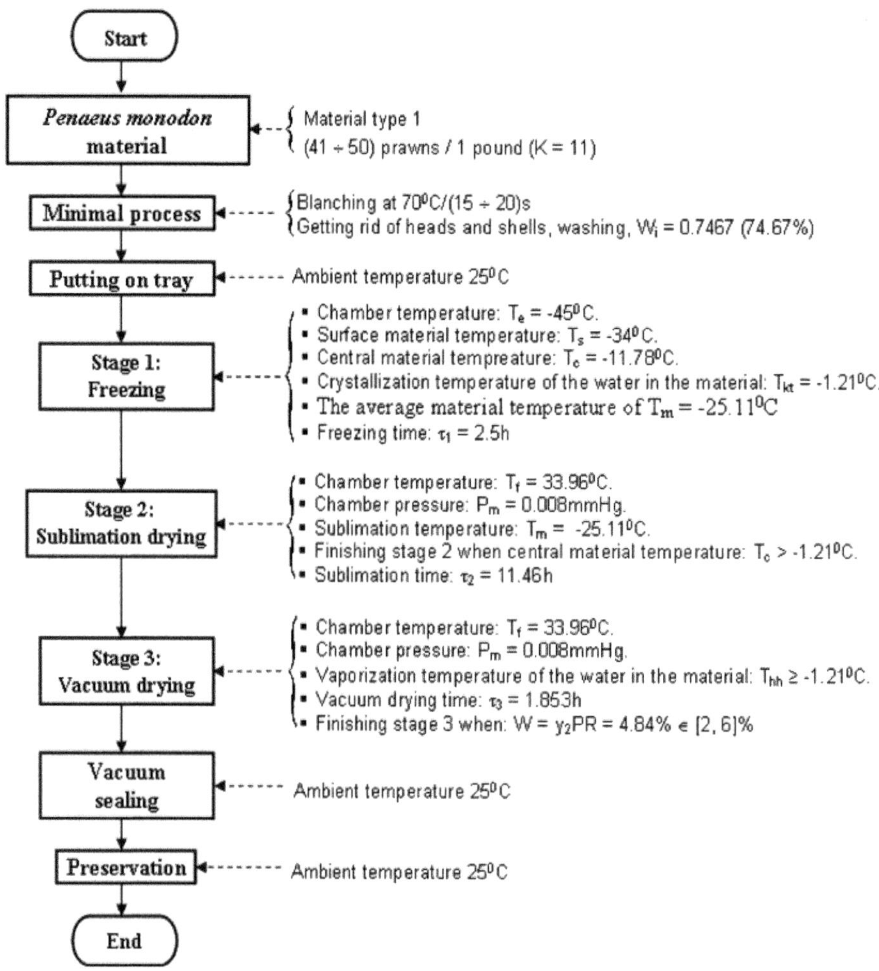

Figure 17: The freeze drying process diagram of *Penaeus monodon*.

Stage 3: The vacuum drying was carried out with the temperature of vacuum drying chamber of $Z_1^{opt} = T_f = 33.96^{\circ}C$, the pressure of vacuum drying chamber of $Z_2^{opt} = P_m = 0.008mmHg$, the central material temperature of $T_c > -1.21^{\circ}C$, and the drying time of $\tau_3 = 1.75h$. The residual water content of the finished product was $W = y_2PR = 4.84\%$, [2,3,13].

Thus, the total drying time of 2 stage was $Z_3^{opt} = \tau_2 + \tau_3 = 13.21h$, the total time of 3 stage was $\tau = \tau_1 + \tau_2 + \tau_3 = 2.5 + 11.46 + 1.75 = 15.71h$

The quality of the finished product was displayed in Figure 16 when all the parameters of the freeze drying process were applied as the above. The energy consumption per weight was the same as the determined in priority.

Conclusion

The results showed that the restricted area method with R*(Z) optimal combination criterion solving the five-objective optimization problem (y_1, y_2, y_3, y_4, y_5) determined three optimal technological parameters: $\mathbf{Z_1^{opt}}$ - temperature of freeze drying chamber, $\mathbf{Z_2^{opt}}$ - pressure of freeze drying chamber, $\mathbf{Z_3^{opt}}$ - time of freeze drying, which were entirely consistent with the experiments, [4,12].

The results also demonstrated that optimal Paréto test ZR = (Z_1^{opt} = 33.96°C; Z_2^{opt} = 0.008mmHg; Z_3^{opt} = 13.21h) for optimal Paréto effect yPR = (y_1PR = 68.226; y_2PR = 4.84; y_3PR = 7.928; y_4PR = 8.795; y_5PR = 1.947) was the closest to the utopian point but the furthest from the restricted area.

The freeze drying process was researched systematically, combining mathematical method and experimental design.

The regression equations (25), (26), (27), (28) and (29) obtained from the experiments were the experimental statistical models which could well describe the impact of the freeze drying chamber temperature, the freeze drying chamber pressure and the freeze drying time on the energy consumption, the residual water content, the rehydration capacity, the volume contraction and the loss of vitamin C of the finished product.

The freeze drying mode was determined by the restricted area method in order to minimize the energy consumption per product weight and to maximize the quality of the finished product.

References

1. Dzung NT, Dzung TV, Ba TD (2008) Survery thermophysical properties of crustacean group (*Penaeus monodon, Penaeus Merguiensis, Penaeus Vannamei*) affected heat and mass transfer in processing freeze drying. Journal of Fisheries Science and Technology 3: 46-54.

2. Dzung NT, Dzung TV, Ba TD (2008) Researched the method to determine ice ratio in materials freeze and optimal temperature of freeze in stage 1 of processing freeze – drying. Journal of Science and Technology National University 11: 62-73.

3. Dzung NT, Dzung TV, Ba TD (2009) Research for the establishment and solution mathematical model of heat and mass transfer at the same time in processing freeze drying. Journal of Science and Technology National University11: 82-94.

4. Hai LX, Lan NT (2008) Multi-objective optimization with combination criteria applied to extraction of the high colour degree anthocyanin. Science and Technology development 11: 69-76.

5. Figura LO, Teixeira AA (2007) Food Physics: Physical properties Measurement and Application. Germany.

6. Pikal MJ, Shah ML (1984) Mass and Heat transfer in vial freeze drying of pharmaceuticals: role of the vial. J Pharm Sci 73: 1224-1237.

7. Liapis I, Bruttini R, Pikal MJ (1996) Research and development needs and opportunities in freeze drying. Journal of Food Engineering 14: 1265-1300.

8. Pikal MJ, Shah S, Roy ML, Putman R (1989) The secondary drying stage of Freeze drying: drying kinetics as a function of temperature and chamber pressure. Journal of Food Engineering 60: 203-207.

9. Khalloufi S, Robert JL, Ratti (2005) Solid foods Freeze drying simulation and experimental data. Journal of Food Engineering 28: 107-132.

10. Hai LX, Nguyen NV (2008) Multi-objective optimization for grouper (e.coioides) compounded feed formulation. Journal of Science and Technology 48: 123-136.

11. Dzung NT, Hai LX, Dzung TV, Ba TD (2010) Multi-objective optimization applied to determine regime technological freeze drying of penaeus vannamei. Journal of Fisheries Science and Technology 1: 82-93.

12. Dzung NT, Hai LX, Dzung TV (2010) Multi-objective optimization with optimal standard combination of R applied to determine regime technological freeze drying of *penaeus merguiensis*. Journal the Development of Science and Technology 13: 56-68.

13. Dzung NT, Dzung TV, Hai LX (2010) Building and solving a mathematical model of heat transfer and humidity seperation in conditions freeze drying by the finite element method. Journal of Science and Technology 48: 23-35.

14. Dzung NT, Dzung TV (2009) Building a mathematical model to determine the relationship between heat emission coefficient and pressure of the freeze drying environment of solid materials. Proceedings of the 11th Conference on Science and Technology, Viet Nam.

Synthesis of Palladium-Platinum Bimetallic Nanoparticles and their Catalytic Activity towards the Hydrogenation Reaction of Palm Olein

Salimah Shafii*, Wu Lihua, M. Ridzuan Nordin and Liew Kong Yong

Faculty of Industrial Science and Technology, University Malaysia Pahang, 26300, Kuantan, Pahang, Malaysia

Abstract

Bimetallic Pd-Pt catalyst was successfully prepared via conventional heating (CH), microwave (MW), and ultrasonic irradiation (US) methods. Bimetallic Pd-Pt nanoparticles stabilized with polyvinylpyrrolidone (PVP) were prepared with molar ratio of PVP to metal, 40:1. Smaller particles sizes with narrow distribution obtained after 30min and 10min with US and MW methods respectively. All the particles are spherical or near spherical in shape.

The average particles size and size distribution for Pd-Pt (US) nanoparticles was 1.28 ± 0.29 nm, while for Pd-Pt (MW) nanoparticles was 1.05 ± 0.21 nm. The average particles sizes of these bimetallic nanoparticles are comparable with the classical CH method, which was 1.23 ± 0.25 nm. Based from the XPS analysis, surface compositions of Pd and Pt show that PdPt(CH) and PdPt(MW) catalyst were enriched with Pd, meanwhile PdPt (US) composition was comparable and it might be in alloy structure formation. The activity of these three catalysts for the hydrogenation reaction of palm olein was studied. The reaction conducted under ambient temperature and atmospheric pressure. A molar ratio of palm olein to catalyst of 25000:1 has been used in the hydrogenation reaction to determine the conversion of linoleate, selectivity of trans-isomer and iodine value (IV) in the hydrogenated palm

olein. After 180 min of reaction, full conversion achieved with Pd-Pt (MW) catalyst, while only 91.4% and 80.6% conversion achieved with Pd-Pt (US) and Pd-Pt (CH) respectively. However, the Pd-Pt (MW) catalyst shows the highest elaidate selectivity after 180 min, which was up to 13.1%, meanwhile the elaidate selectivity of 11.82% and

8.1% for both Pd-Pt (US) and Pd-Pt (CH) respectively. Even though higher conversion of linoleate achieved in short time and lower trans selectivity produced with Pd-Pt bimetallic catalyst. The IV calculated shows that Pd-Pt (MW) catalyst has the lowest IV then followed by Pd-Pt (US) and Pd-Pt (CH) catalysts.

Keywords: Polyvinylpyrrolidone; Platinum; Palladium; Hydrogenation; Palm olein

Introduction

The partial hydrogenation of vegetables is an industrial process that use to obtain a more stable product (which no oxidation on storage), together with a suitable texture and melting-temperature range at human mouth conditions for use as margarine, edible shortenings and baking applications [1,2]. During the hydrogenation process, the carbon double bonds are partially or fully saturated. Unfortunately, the catalytic isomerization of naturally occurring cis isomer to trans isomer fatty acids takes place [3]. In recent years, however it has been published that the intake of trans fatty acids (TFA) adversely affects blood lipid levels. Metabolic and epidemiologic studies confirm the potential role of TFA in increasing the risk of coronary heart disease. Furthermore, on each gram basis the negative health impact of TFA appears to be stronger than that of saturated fatty acids [4,5]. A numbers of alternative have been studied to figure out the ways for substantial reduction of inevitable TFA content in the hydrogenated edible oils such as the electro-catalytic hydrogenation, supercritical fluid hydrogenation, membrane reactor technology, catalytic transfer hydrogenation, and modification including supported precious metals [3].The investigated catalysts offer various advantages but also have considerable shortcomings.

In this part of work, such modification of precious metals included the use of microwave and ultrasonic irradiation selected and the effects of the catalyst on hydrogenation of palm olein were studied. Obviously, the use of microwave energy to heat chemical reactions has attracted a considerable amount of attention, due to its numerous successful applications in organic synthesis, polymer chemistry, material sciences, nanotechnology, and biochemical processes [6-10]. A huge number of paper published, had discovered that by using microwave technology can synthesis nanoparticles successfully. The motivation for the use of microwave energy has mainly been to design faster, cleaner and economically more viable methods for synthesis. The super rapid heating and sometimes extreme temperatures observable in microwave chemistry generally lead to faster processes. The transformations that normally required several hours when performed in a solvent at reflux temperature in an oil bath however may reach completion in a few minutes or even seconds using superheated solvent in a seal vessel or microwave reactor [11]. These unique features explained that this heating method has gained its popularity in many different field of chemistry. Mispa and co workers in their previous research in 2010 determined that the electric field applies a force on charged particles, resulted the charged particles started to migrate or rotate. Due to the movement of charged particles, further polarization of polar particles takes place. The concerted force applied by the electric and magnetic

***Corresponding author:** Salimah Shafii, Faculty of Industrial Science and Technology, University Malaysia Pahang, 26300, Kuantan, Pahang, Malaysia

components of microwaves are rapidly changing in direction, which creates friction and collisions of the molecules, claimed effects of the microwave irradiation include thermal and non-thermal effects [12]. Many papers published about microwave compared with conventional method, shows that microwave synthesis has the advantages of short reaction time, small particles size and high purity [8,9,11,13,14].

On the other hand, the ultrasonic irradiation in particular the acoustic techniques is one of the most useful approaches for acceleration of chemical processes. Acoustic waves with frequencies of more than 20 kHz, can cause structural changes and accelerate chemical reactions [15]. The initiation of most of the sonochemical reactions in aqueous solution applying acoustic vibration was caused by cavitation. Acoustic cavitation is a disturbance of the continuity of a liquid, which connected to the creation, growth, oscillation, and collapse of steam-to-gas bubbles inside the liquid [16,17]. The development of cavitation bubbles follows the sound field in a liquid during the cycles of compression and expansion and stimulated by time-varying pressure. Bubbles either oscillate around their equilibrium position over several expansion or compression cycles or can grow over one or more acoustic cycles to double their initial size and then finally collapse violently [16-18]. Radziuk D et al. [18] report that the tremendous heating or cooling rates of bubbles and transient of high temperature inside hotspot in heterogeneous medium which contained metal nanoparticles, caused it not only to accelerate catalytic activity of metal nanoparticles but also produced smaller size particles that are highly dispersed. Another comprehensive study was on the sonochemical synthesis of colloidal bimetallic Pd-Sn nanoparticles and conducted by Kim et al. [19]. Metals salt of Pd $(NH_4)_2Cl_4$ and $SnCl_2$ dissolved in an aqueous ethanol solution containing citric acid, which acts as the stabilizing agent for the bimetallic nanoparticles. Ultrasonic irradiation performed with collimated 20 kHz bean from a ceramic transducer with a titanium amplifying horn directly immersed in the solution and operated with an input power of 42 Wcm-2 for 2 hours. The bimetallic nanoparticles shape and size were characterized with TEM. Highly dispersed Pd-Sn nanoparticles with sizes of 3-5 nm in average were obtained. Thus, implies that the ultrasonic irradiation method is an effective to form nanoparticles.

The aim of the present work is to synthesized bimetallic nanoparticles of Pd-Pt under ultrasonic irradiation and microwave heating and compared with the conventional heating method. As a lot papers published about the interesting characteristic of bimetallic catalyst compared to monometallic catalyst, it is not only helps to enhance catalytic activity but also increasing the yield of desired product [20,21]. Thus, Pd-Pt catalysts were used in hydrogenation reaction of palm olein to determine the effect of Pd-Pt catalyst especially on catalytic activity and the formation of elaidate (trans fatty acid).

Materials and Methods

Chemicals

Analytical grade or equivalent chemicals: Polyvinylpyrrolidone (PVP), (Mw 40000); methyl oleate (>99%) and methyl elaidate (>99%) (Sigma Aldrich); chloroplatinic acid hexahydrate (Merck, Germany); methanol and hexane AR grade (Merck, Germany); Butan-1-ol (Merck, Germany); anhydrous sodium sulfide and sodium hydroxide (BDH, UK) were used without purification. Palm olein (100% refined palm oil) was obtained from a local supermarket. The fatty acid compositions of the olein were determined to be: 36.11% palmitic, 4.04% stearic, 45.21% oleic, 11.85% linoleic and 0.39% linolenic acid.

Preparation of PVP-stabilized bimetallic Pd-Pt colloidal nanoparticles

PVP-stabilized Pd-Pt bimetal colloids was prepared where 0.0222g $PdCl_2$ (1.25×10^{-4} mol) were converted to $H_2PdCl_4{\cdot}nH_2O$ by adding 100μl of concentrated HCl, then 0.065g $H_2PtCl_6{\cdot}6H_2O$ (1.25×10^{-4} mol) and 1.11g PVP (1.00×10^{-2} mol) was added to the mixture of methanol (130 ml) — distilled water (150 ml) which act as solvent. The solution was mixed in the 500 mL 3-neck round bottom flask. Then, 20 ml of 0.1 M NaOH in methanolic was added drop wise under vigorous stirring. Then the solution was divided into three portions of 100 ml. Each portion was synthesized under different method to produced colloidal metal nanoparticles. Brief descriptions of the different method condition were shown in the Table 1 below. Finally, the final colloidal solution resulted in a dark brown color will be stored in a dark bottle at 4°C [18].

Hydrogenation of palm oil

A constant gas flow rate of 36-37 ml/min of hydrogen gas was used for the whole experiments. The gas flow was first stabilized for 30 min before calibrated thrice. The hydrogenation reaction was performed at ambient temperature and atmospheric pressure. Colloidal catalyst of 0.5 ml (4.17×10^{-7} mol) and 48 ml of butan-1-ol was fed into the reactor. Next, hydrogen gas flown inside the reaction vessel continuously and the catalyst was activated for 60 min. Finally, palm olein (0.01mol) was injected into the reaction vessel and the reaction started immediately for 180 min. The hydrogenation reactions were done under conventional stirring at the speed of 200 rpm. Partially hydrogenated palm olein were sampled at selected interval; 10 min, 20 min, 30 min, 40 min, 50 min, 60 min, 90 min, 120 min, 150 min, and 180 min.

Analysis method

UV-Visible spectrophotometer (UV-Vis): UV-Vis analysis was using Hitachi U-1800 UV-Visible spectrophotometer. All spectra were taken in the range of 200-700 nm with scan speed of 400 nm/min.

Transmission electron microscopy (TEM): Characterization of the particles size and dispersion were done by using the TEM with LaB_6 source Tecnai G^2 20 S-TWIN. Sample for TEM were prepared by placing a drop of a colloidal catalyst onto a perforated carbon copper grid, then the solvent was evaporated for at least 3 min. The average particle size and standard deviation was determined based on that of 250-300 particles, from enlarged photographs via Gatan micrograph software Choo et al. [22].

X-ray diffractometer (XRD): X-Ray diffraction is a strong method to investigate the solid structure ofmetal nanoparticles. The X-ray diffraction (XRD) patterns of bimetallic Pd-Pt nanoparticles were

Type of method	Synthesized process condition
Conventional heating, Pd-Pt(CH)	This method was done by using oil bath to maintain the refluxing process condition was heated at reflux temperature for 3hr [22,23].
Microwave irradiation, Pd-Pt(MW)	Power of 100 watts was used. Time for the heating was dependent to the color of colloidal solution until it is started to change to dark brown.
Ultrasonic irradiation, Pd-Pt(US)	Ultrasonic bath (ELMA D-78224, type S60H) with frequency of 37kHz was used for this experiment. Temperature of 80°C as it was the maximum temperature for this equipment. Time for the heating was same with the microwave method.

Table 1: Different method and their synthesized process condition.

determined by a Bruker-D8 diffractormeter using monochromatized Cu-Kα radiation (λ = 1.54056 Å) operated at 40 kV and 40 Ma and collected by a Vantec-1 detector.

X-Ray photoelectron spectroscopy (XPS): The XPS measurements of the ITO (indium-tin-oxide) substrates were carried out in a VG ESCALAB MK II spectrometer, using a monochromatic Al Kα 1486.60 eV as X-ray source. The vacuum in the analysis chamber was maintained at approximately 10^{-8} Pa or lower. All binding energies were referenced to the binding energy of the carbon C 1s peak at 284.0 eV.

Gas Chromatography with FID detector (GC-FID): The hydrogenated palm olein was methylated before analyzed with Agilent-6890 gas chromatograph (GC) equipped with FID detector and a HP-88 capillary column (*L*, 100m; i.d., 0.25 mm; thickness of film, 0.25 μm). The column temperature was programmed from 185 to 210°C at

2°C.min^{-1} and held at 210°C for 2 min and then raised to 230°C at 3°C.min^{-1} and held for 3 min with helium as carrier gas. Both the injector and detector temperatures were set at 250°C. Methyl oleate and methyl elaidate served as standard for determination of cis- and trans- isomers. Below are the formulas used for calculating linoleate conversion, trans selectivity and iodine value (IV):

Calculations for C18:2 conversions (Brito, 2007)

The C18:2 conversions (linoleate) were determined directly from the GC fatty acid composition using the following formula (Brito, 2007):

$$Conversion\,C_{18:2} = \frac{C18:2_i - C18:2_0}{C18:2_0} \times 100\%$$

(C18:2_0 = initial value; C18:2_i current value)

Calculations for trans selectivity

The trans (elaidate) selectivity was determined directly from the GC fatty acid composition using the following formula (D. Pham Minh, 2009):

$$Trans\,selectivity = \frac{C18:1_{Trans}}{C18:1_{Trans} + C18:1_{Cis}} \times 100\%$$

Calculations for Iodine Values

The iodine value for oil samples was determined directly from the GC fatty acid composition using the following formula (Xiao, 2007):

Iodine value = [(% C18:1 + % trans C18:1) x 0.86] + [(%C18:2 + % trans C18:2) x 1.732] + [(% C18:3 + % trans C18:3) x 2.616]

Results and Discussion

In this part, Pd-Pt bimetallic nanoparticles were synthesized using chemical reduction method as mentioned in the literature. Then, all the metal nanoparticles were reduced by conventional heating (CH), microwave (MW) and ultrasonic irradiation (US). Refluxing-reduction method was done by repeating previous work that has been done before by Yu and Liu [23]. The same method uses to reproduce the previous work and compared with the new preparation method of microwave and ultrasonic irradiation. Comparison was not only based on times required to reduce metal nanoparticles but also the effect on the size and particles distribution, but also its activity towards hydrogenation reaction of palm olein.

Table 2 below shows the time required to produce colloidal nanoparticles for all different methods. For the CH method, it took about 3 hr to produce a homogeneous dark brown colloidal solution of metal nanoparticles [22,23]. On the other hand, ultrasonic and microwave method does not required long time to produce a homogeneous colloidal nanoparticles solution. The result in the Table 2 below shows that bimetallic nanoparticles Pd-Pt took only 10 min and 30 min for microwave and ultrasonic irradiation method respectively. It show that the microwave resulted faster reduction to produced colloidal bimetallic Pd-Pt nanoparticles, followed by ultrasonic irradiation and then the classical heating method.

Method	Time
Refluxing*	3 hr
Ultrasonic	30 min
Microwave	10 min

*Following methods from literature [22].

Table 2: Time required for the formation of bimetallic Pd-Pt nanoparticles

UV-Vis analysis

Even though a large number of researches have been reported on the synthesizing bimetallic nanoparticles [11,19,24-26], but the preparation of Pd-Pt stabilized by PVP via microwave and ultrasonic irradiation method has never been reported yet. The UV-Vis spectrums of Pd-Pt (CH), Pd-Pt (MW) and Pd-Pt (US) under different reduction process method shows in Figure 1. All samples including before and after reduction were taken for analysis to determine the changes of reduction processes for both Pd and Pt ions in the solution. A very broad peak was observed before the reduction occurred which indicated that an abundance of metal ions contained in the solution [27]. After the reduction process was completed where it can be determined via visual color changed, the broad peak totally disappeared. It is because of the synthesis of this bimetallic are using the co-reduction method. Where both metal salt of $PdCl_2$ and $H_2PtCl_6{\cdot}6H_2O$, stabilizer and reducing agent are added at the same time and started the reduction process. Besides that, Pd-Pt bimetallic nanoparticles dispersion shows that at shorter wavelengths, the absorbance are the highest, which was described that the formation of smaller size and dispersed bimetallic colloidal nanoparticles [28,29].

TEM analysis

Among the techniques commonly used, transmission electron microscopy (TEM) is indispensable for metal nanoparticles study. Toshima and Yonezawa [30] stated that in their previous work, where noble metal nanoparticles such as Pd and Pt, give high contrast when particles are dispersed on the carbon supported copper grids, thus, the TEM image are much easier to analyze and clearly seen.

Typical micrographs and distribution histograms were shown in Figure 2A-C. The particles are of nanometer sizes with narrow distribution. Microwave-reduction method shows that it has resulted in smallest average particles size than the other methods, followed by ultrasonic irradiation and refluxing method. The average particles size are 1.05 ± 0.21 nm, 1.28 ± 0.29 and 1.23 ± 0.25 nm for MW, US, and CH method respectively. All the TEM images show spherical and nearly spherical in shape and are highly dispersed. Both microwave and ultrasonic irradiation have shown well-improved method in synthesizing metal nanoparticles especially in chemical reduction technique that has been used in this research.

As mentioned previously, that MW method shows smallest

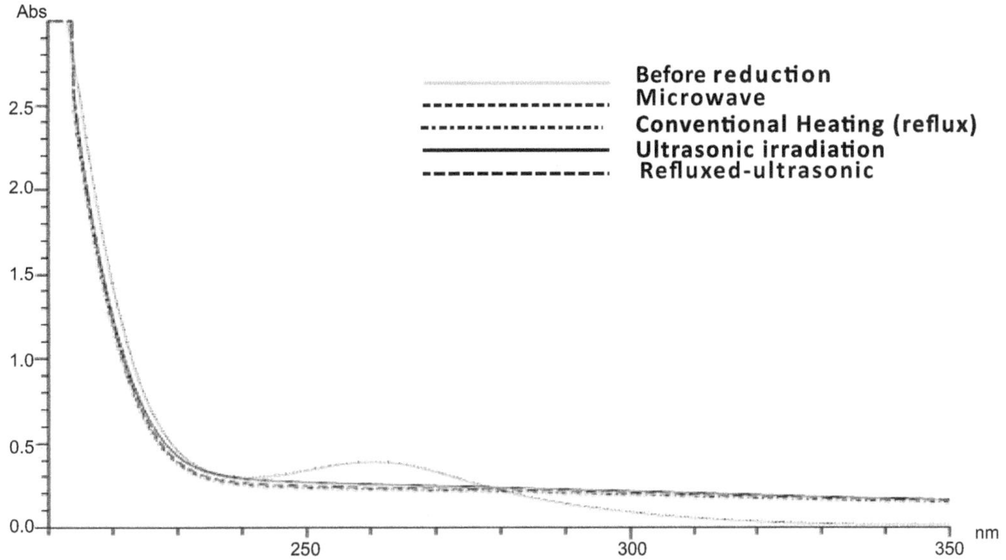

Figure 1: UV-Vis spectrum for Pt-Pd bimetallic nanoparticles.

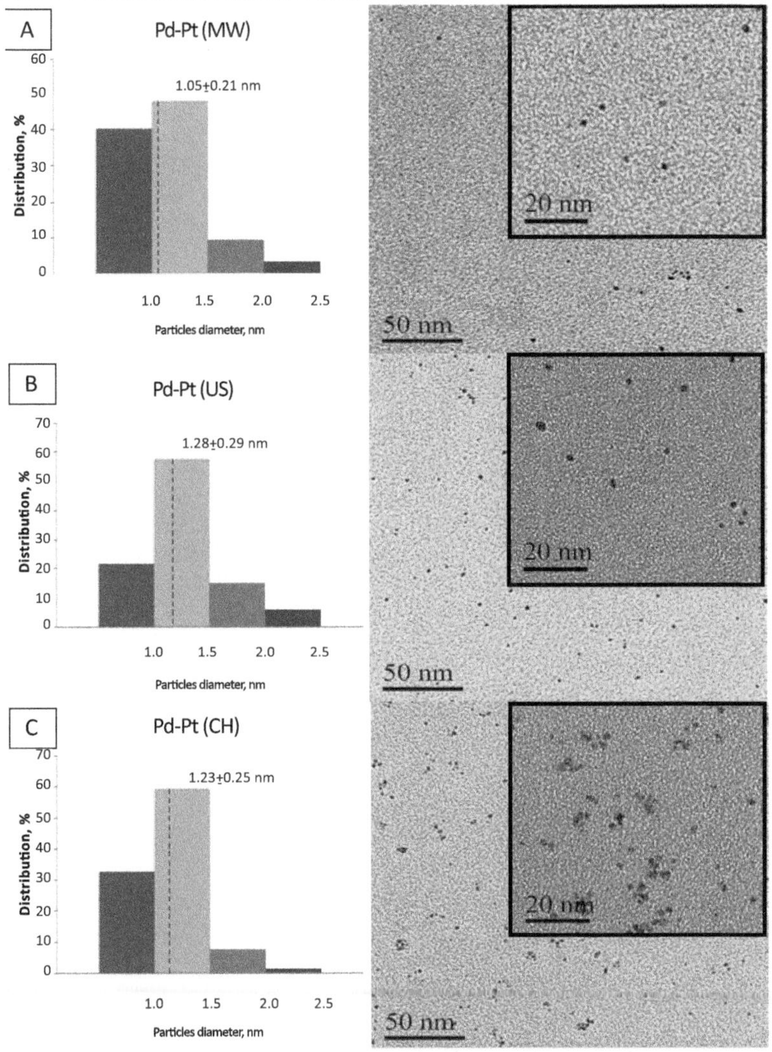

Figure 2: TEM image of Pd-Pt bimetallic nanoparticles with the histogram chart of particles size distribution that has been prepared via; (A) microwave (MW) ;(B) ultrasonic irradiation (US) and, (C) conventional heating (CH)..

particles size compared to the other method. It can be explained by the microwave heating which allows an instantaneous volumetric and therefore a more rapid heating in comparison with the conventional heating process [11]. The MW technique has demonstrated as a better alternative in synthesizing bimetallic nanoparticles.

On the other hand, it was determined that particles are spherical or near spherical in shape and highly dispersed were also produced by ultrasonic irradiation method. Cavitation bubbles produced the shock wave increased the momentum of metal nanoparticles in the solution and caused them to collide with great force [31,32]. The particles are fracture upon collusions, leading to an overall decrease in the average particles sizes [33].

XPS analysis study

All the Pd-Pt (CH), Pd-Pt (MW), and Pd-Pt (US) samples were analyzed by XPS in order to identify the effect of all different method prepared on the surface composition of both Pd and Pt elements in bimetallic nanoparticles. Figure 3 (A-C) indicates the XPS spectrum for both Pd and Pt elements for each method of catalysts preparation. Both Pd and Pt element spectrums have been deconvoluted. Binding energy (BE) for each element and the Pd-Pt surface composition ratio of Pd to Pt was presented in the Table 3 below.

Pd-Pt (CH) sample shows that Pd spectrums have two peaks in Figure 3 (A1-A2). The binding energy for both peaks are, 335.1 and 341.4 eV assigned to electron form $Pd3d_{5/2}$ and $Pd3d_{3/2}$ orbital respectively. On the other hand,this spectrum can be curve-fitted with a spin orbit-split doublet by having BE at about 342.9 eV ($3d_{3/2}$) and 337.9 eV ($3d_{5/2}$) for Pd^{2+} species [34]. At the same table, the BEs for Pt element was also presented. The binding energy signals obtained are 70.9 and 74.2 eV. The Pt0 species was found at binding energy signal of 70.9 and 74.2 eV for the $4f_{7/2}$ and $4f_{5/2}$ components respectively [35]. The same XPS spectrum also indicated the existence of Pt^{2+} species at BEs of 73.1 ($4f_{7/2}$) and 76.4eV ($4f_{5/2}$). Next, the ratio of surface composition of Pd to Pt calculated was 2.7. As the composition of Pd was almost three times higher than Pt, it shows that the Pd enrichment might be to the polymer in anchoring Pt nanoparticles and thus allowing Pd to coalesce onto the Pt nanoparticles. The result shows that this bimetallic catalyst formation described that Pt metal atom had been covered by the Pd metal atom.

The Pd-Pt (MW) sample was presented in single element spectrum of Pd and Pt in Figure 3 (B1-B2). The binding energy of Pd3d5/2 and Pd3d3/2 electrons amounts are 334 and 340.3 eV respectively, that virtually coincides with the data for palladium in zero-valent metal state [36,37]. However Pd^{2+} species was also observed at BE of 339.3 eV (3d3/2) [38]. This same XPS spectrum was also indicated the presence of oxidized Pt^{2+} species at about Bes of 73.4 and 76.5eV for the ($4f_{7/2}$) and ($4f_{5/2}$) component respectively [34,35]. Then the ratio of Pd to Pt surface composition for this sample was 2.5. Like Pd-Pt (CH) samples, it can shows that the Pd has enriched onto the Pt nanoparticles, and resulted higher composition of Pd on the surface of the bimetallic catalyst [30].

The last sample of bimetallic Pd-Pt (US) that is prepared under ultrasonic irradiation was also analyzed with XPS 3(C1-C2). For the Pd metal state element, deconvoluted XPS peaks of are recorded at BEs of 335.1 eV ($Pd3d_{5/2}$) and 340.5 eV ($Pd3d_{3/2}$), while Pd^{2+} species was spotted at BEs about 336.2 eV for Pd3d5/2 and 342.9 eV for Pd3d3/2 components [37]. The zero-valent state of Pt metal species was shown at binding energy signal of 70.7 eV and 74.1 eV for the $4f_{7/2}$ and $4f_{5/2}$

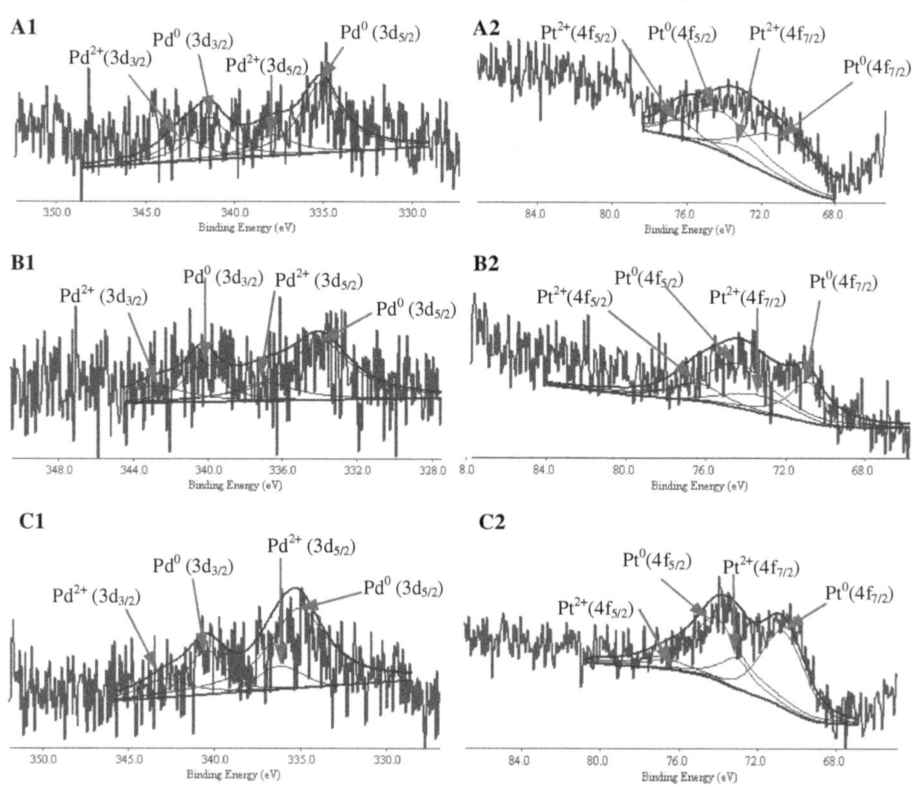

Figure 3: XPS spectrum of Pd and Pt element for on the surface (A1-A2) PdPt-CH; (B1-B2) PdPt-MW; (C1-C2) PdPt-US catalysts.

component respectively. Whereas unreduced Pt2+ ions was seen at 73.1 eV for Pt4f7/2 and 76.4 eV for $Pt4f_{5/2}$ components [35]. Unlike Pd-Pt (CH) and Pd-Pt (MW) results, the ratio of Pd to Pt surface composition in this sample was 1.2. The surface composition ratio is almost identical; it suggested the alloy formation of bimetallic Pd-Pt (US). The alloy structure formation might be because due to the high intensity of the ultrasonic irradiation that caused the destruction of the bimetallic particles and thus rearrangement of both Pd and Pt atom to form an alloy structure [18,38].

Hydrogenation reaction of palm olein

It is well established that the double bonds in vegetables oils are naturally in cis-conformation. Upon hydrogenation reaction the trans isomer that is thermodynamically more stable are produced. The trans isomer of the fatty acid is unhealthy for consumption and hence their formation in edible fats during hydrogenation must be suppressed as far as possible [2]. Thus, the hydrogenation reaction has been carried out at ambient condition using a direct flow of hydrogen gas (99.5% purity) at a constant flow rate of 36 ml/min through the reaction vessel. This reaction has been done to demonstrate the effect of the Pd-Pt bimetallic catalyst on the composition of partially hydrogenated palm olein.

Figure 4A shows the conversion of C18:2 and selectivity of elaidate (trans-oleic) versus time of reaction for all catalysts of Pd-Pt (CH), Pd-Pt (MW) and Pd-Pt (US) shows the conversion of linoleate and selectivity of elaidate after hydrogenation reaction was conducted with bimetallic PdPt catalyst under conventional stirring method. After reaction stopped at 180 min, full conversion was achieved with PdPt (MW) catalyst, meanwhile only 91.4 % and 80.6 % conversion were achieved with PdPt (US) and PdPt (CH) respectively. All catalyst shows similar trend, where the conversion of linoleate increased gradually.

On the other hand, it can be seen from the TEM image that although both PdPt (CH) and PdPt (US) catalysts have comparable particles size but PdPt (US) had shows lower linoleate composition compared to PdPt (CH). However, the conversions of linoleate between PdPt (CH) and PdPt (US) from the chart are significantly different. This result occurred maybe due to the well-dispersed of PdPt (US) nanoparticles which in contrary with

PdPt (CH) nanoparticles formation (refer to Figure 4C for TEM study). This aggregated PdPt (CH) nanoparticles caused it to have lower in activity and thus less amount of linoleate hydrogenated [23].

The PdPt(MW) catalyst shows the highest elaidate selectivity after 180 min which up to 13.1%, meanwhile the elaidate selectivity of 11.82% and 8.1% for both PdPt (US) and PdPt (CH) respectively. It can be seen a slightly difference of the elaidate selectivity between catalysts, where the highest selectivity of elaidate obtained with PdPt (MW), followed by PdPt (US) and PdPt (CH) catalyst. Bimetallic PdPt catalyst shows to be less favorable to isomerizes from cis (oleate) to trans (elaidate) conformation thus lower elaidate selectivity obtained for all catalyst.

This may be due to the combination of Pd and Pt characteristics that produced such catalytic activity result. From the previous research,

Sample	Pd3d5/2	Pd3d3/2	Pt4f7/2	Pt4f5/2	Surface ratio (Pd/Pt)
(CH)	335.1(Pd^0) 337.9(Pd^{2+})	341.4(Pd^0) 342.9(Pd^{2+})	70.9(Pt^0) 73.1(Pt^{2+})	70.9(Pt^0) 73.1(Pt^{2+})	2.7
(MW)	334.0(Pd^0) 336.9(Pd^{2+})	340.3(Pd^0) 342.8(Pd^{2+})	70.9(Pt^0) 73.4(Pt^{2+})	74.4(Pt^0) 76.5(Pt^{2+})	2.5
(US)	335.1(Pd^0) 336.2(Pd^{2+})	340.5(Pd^0) 342.9(Pd^{2+})	70.7(Pt^0) 73.1(Pt^{2+})	74.1(Pt^0) 76.4(Pt^{2+})	1.2

Table 3: Binding energy and raw area for each material detected with XPS.

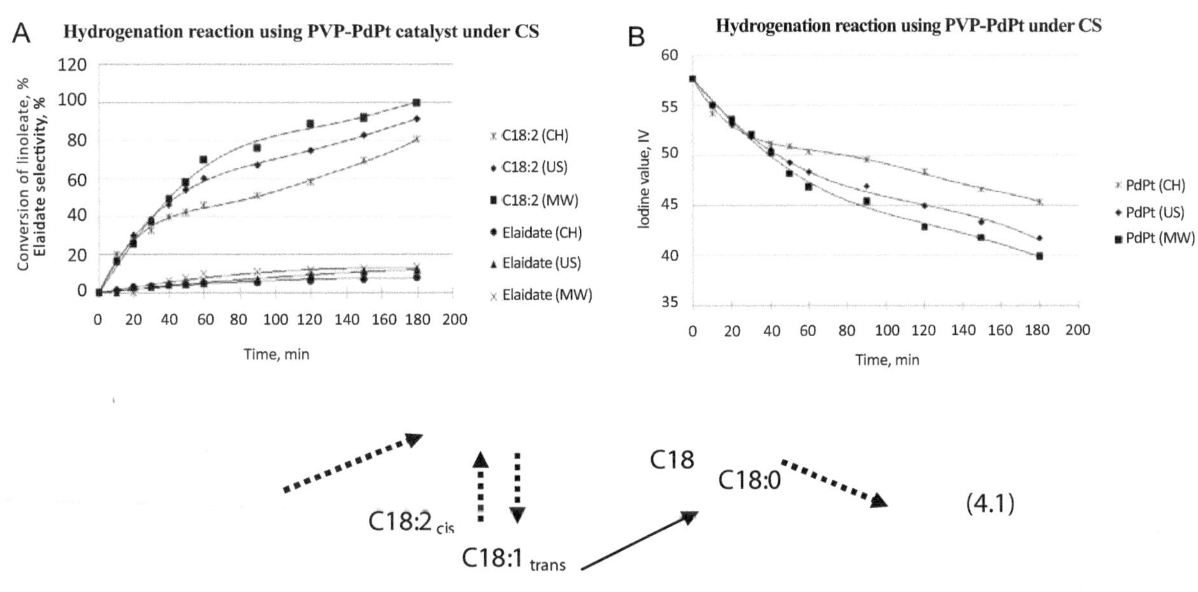

Figure 4: (A) Graph of the C18:2 conversion and selectivity of elaidate versus time of reactions; (B) Graph of IV value versus time of reaction.

Pd shows to be highly active and selective towards monoenes formation rather than Pt catalyst. However, Pd also exhibits higher selectivity towards TFA production than Pt [39]. Thus, in the context of no-TFA production, Pt was viewed as much better for hydrogenation of vegetable oils, as a consequences Pt are favor to produce non-desirable of saturated fatty acid. As a result, the combination of Pd and Pt bimetallic catalyst had shows an interesting result in this part of work. As we know that previously, Pd are highly active on hydrogenation reaction of linoleate, but it is also produced higher trans isomer [23]. However, Pt has the opposite characteristic with Pd catalyst. Thus, the combination of these two noble metals resulted higher activity but lower selectivity of trans isomer.

Figure 4B shows the IV versus time of reaction for all catalyst, Pd-Pt (CH), Pd-Pt (MW) and Pd-Pt (US). The IV for palm olein before hydrogenation reaction was 57. When the reactions proceed, the IV started to decrease slowly as the increasing numbers of double bond were hydrogenated. Based on the Figure 4B, all the catalysts show a similar trend of IV reduction. However, Pd-Pt (CH) and Pd-Pt (US) show a comparable result between each other and both of it have a slightly lower IV than Pd-Pt (MW). After the reaction stopped at 180 min, the IV for Pd- Pt (CH), Pd-Pt (US), and Pd-Pt (MW) are 45, 42, and 40 respectively.

A summary of the reaction pathway with the bimetallic PdPt is described in scheme 4.1 below. The PdPt catalyst favors the hydrogenation for both dienes and monoenes, which was describes by the dash line arrow in the scheme 4.1. Besides that, hydrogenation and isomerization of C18:1 is slightly favorable.

Conclusion

PVP stabilized Pd-Pt bimetallic nanoparticles with average particle sizes range at 1-2 nm were successfully prepared. Synthesis of bimetallic Pd-Pt with microwave and ultrasonic irradiation shows better alternative as a lot of time can be save and small and highly dispersed particles obtained. Interestingly, almost similar nanoparticles characterization achieved compared to classical heating method. The study of catalytic reaction of bimetallic Pd-Pt shows an impressive result. The combination of two noble metals, improved the compositions of partially hydrogenated palm olein, where higher conversion of C18:2 achieved in short time and in addition the elaidate selectivity are also very low. The IV is highly decreased as the reaction proceeds. Further work will be continued to determine the effect of ultrasonic irradiation on the hydrogenation reaction to compare with the conventional stirring.

References

1. Santana A, Larrayoz M, Ramirez E, Nistal J, Recasens F (2007) Sunflower oil hydrogenation on Pd in supercritical solvents: Kinetics and selectivities. J Supercrit Fluids 41: 391-403.
2. Singh D, Rezac ME, Pfromm PH (2009) Partial Hydrogenation of Soybean Oil with Minimal Trans Fat Production Using a Pt-Decorated Polymeric Membrane Reactor. J Am Oil Chem Soc 86: 93-101.
3. Stanković M, Gabrovska M, Krstić J, Tzvetkov P, Shopska M, et al. (2009) Effect of silver modification on structure and catalytic performance of Ni-Mg/ diatomite catalysts for edible oil hydrogenation. Journal of Molecular Catalysis 297: 54-62.
4. Tonetto G, Sanchezm J, Ferreira M, Damiani D (2009) Partial hydrogenation of sunflower oil: Use of edible modifiers of the cis/trans-selectivity. Journal of Molecular Catalysis 299: 88-92.
5. Xiao H (2007) Low trans fatty acid containing hydrogenated edible oils. Miami University: 69.
6. Wu ZL, Ondruschka B, Cravotto G (2008) Degradation of Phenol under Combined Irradiation of Microwaves and Ultrasound. Environ Sci Technol 42: 8083-8087.
7. Daugaard AE, Hansen TS, Larsen NB, Hvilsted S (2011) Microwave assisted click chemistry on a conductive polymer film. Synth Met 161: 812-816.
8. Ayala A, Fetter G, Palomares E, Bosch P (2011) CuNi/Al hydrotalcites synthesized in presence of microwave irradiation. Mater Lett 65: 1663-1665.
9. MacKenzie K, Dunens O, Harris AT (2009) A review of carbon nanotube purification by microwave assisted acid digestion. Sep Purif Technol 66: 209-222.
10. Pasquet V, Chérouvrier JR, Farhat F, Thiéry V, Piot JM, et al. (2011) Study on the microalgal pigments extraction process: Performance of microwave assisted extraction. Process Biochem 46: 59-67.
11. Pein A, Baghbanzadeh M, Rath T, Haas W, Maier E, et al. (2011) Investigation of the Formation of CuInS2Nanoparticles by the Oleylamine Route: Comparison of Microwave-Assisted and Conventional Syntheses. Inorg Chem 50: 193-200.
12. Mispa MKJ, Subramaniam DP, Murugesan DR (2010) Microwave- assisted route for synthesis of nanosized metal sulphides. Chalcogenide Lett 7: 335-340.
13. Xu W, Liew KY, Liu H, Huang T, Sun C, et al. (2008) Microwave-assisted synthesis of nickel nanoparticles. Mater Lett 62: 2571-2573.
14. Cano M, Benito A, Maser WK, Urriolabeitia EP (2011) One-step microwave synthesis of palladium–carbon nanotube hybrids with improved catalytic performance. Carbon 49: 652-658.
15. Mason TJ (1986) Use of ultrasound in chemical synthesis. Ultrasonics 24: 245-253.
16. Ruan Q, Zhu Y, Zeng Y, Qian H, Xiao J, et al. (2009) Ultrasonic-Irradiation-Assisted Oriented Assembly of Ordered Monetite Nanosheets Stacking. J Phys Chem B 113: 1100-1106.
17. Joseph CG, Li Puma G, Bono A, Krishnaiah D (2009) Sonophotocatalysis in advanced oxidation process: A short review. Ultrason Sonochem 16: 583-589.
18. Radziuk D, Mohwald H, Shchukin D (2008) Ultrasonic Activation of Platinum Catalysts. J Phys Chem C 112: 19257-19262.
19. Kim J, Momma T, Osaka T (2009) Synthesis of carbon-supported Pd-Sn catalyst by ultrasonic irradiation for oxygen reduction reaction. J Power Sources 189: 909-915.
20. Li X, Li B, Cheng M, Du Y, Wang X, et al. (2008) Catalytic hydrogenation of phenyl aldehydes using bimetallic Pt/Pd and Pt/Au nanoparticles stabilized by cubic silsesquioxanes. J Mol Catal A Chem 284: 1-7.
21. Li T, Zhang W, Lee R, Zhong Q (2009) Nickel–boron alloy catalysts reduce the formation of Trans fatty acids in hydrogenated soybean oil. Food Chem 114: 447-452.
22. Choo HP, Liew KY, Liu H, Seng CE, Mahmood WAK, et al. (2003) Activity and selectivity of noble metal colloids for the hydrogenation of polyunsaturated soybean oil. J Mol Catal A Chem 191: 113-121.
23. Yu WW, Liu H (2006) Singular modification effects of metal cations and metal complex ions on the catalytic properties of metal colloidal nanocatalysts. J Mol Catal A Chem 243: 120-141.
24. Long NV, Asaka T, Matsubara T, Nogami M (2011) Shape-controlled synthesis of Pt–Pd core–shell nanoparticles exhibiting polyhedral morphologies by modified polyol method. Acta Mater.
25. Kumar PSS, Manivel A, Anandan S, Zhou M, Grieser F, et al. (2010) Sonochemical synthesis and characterization of gold–ruthenium bimetallic nanoparticles. Colloid Surface A 356: 140-144.
26. Patel K, Kapoor S, Dave DP, Mukherjee T (2005) Synthesis of Pt, Pd, Pt/Ag and Pd/Ag nanoparticles by microwave-polyol method. J Chem Sci 117: 311-316.

27. Taguchi N, Iwase A, Maeda N, Kojima,T, Taniguchi R, et al. (2009) Synthesis of Au–Pd bimetallic nanoparticles under energetic irradiation fields. Radiation Physics and Chemistry 78: 1049-1053.

28. Zhang H, Okuni J, Toshima N (2011) One-pot synthesis of Ag-Au bimetallic nanoparticles with Au shell and their high catalytic activity for aerobic glucose oxidation. J Colloid Interface Sci 354: 131-138.

29. He W, Liu J, Qiao Y, Zou Z, Zhang X, et al. (2010) Simple preparation of Pd–Pt nanoalloy catalysts for methanoltolerant oxygen reduction. J Power Sources 195: 1046-1050.

30. Toshima N, Yonezawa T (1998) Bimetallic nanoparticles novel materials for chemical and physical applications. New J Chem 22: 1179-1201.

31. Toukoniitty B, Kuusisto J, Mikkola J-P, Salmi, T, Murzin D Y (2005) Effect of Ultrasound on Catalytic Hydrogenation of D-Fructose to D-Mannitol. Ind Eng Chem Res 44: 9370-9375.

32. Radziuk D, Grigoriev D, Zhang W, Su D, Mo¨hwald H, et al. (2010) Ultrasound-Assisted Fusion of Preformed Gold Nanoparticles. J Phys Chem C 114: 1835-1843.

33. Gogate PR, Kabadi AM (2009) A review of applications of cavitation in biochemical engineering/biotechnology. Biochem Eng J 44: 60-72.

34. Wang W, Neoh K, Kang E, Lim S, Yuan D (2004) Metal ion reduction and resultant deposition on viologenfunctionalized LDPE films and viologen-containing microporous membranes. J Colloid Interface Sci 279: 391-398.

35. Zhou J, Zhou X, Sun X, Li R, Murphy M, et al. (2007) Interaction between Pt nanoparticles and carbon nanotubes – An X-ray absorption near edge structures (XANES) study. Chem Phys Lett 437: 229-232.

36. Coronado E, Ribera A, García-Martínez J, Linares N, Liz-Marzán L M (2008) Synthesis, characterization and magnetism of monodispersed water soluble palladium nanoparticles. J Mater Chem 18: 5682-5688.

37. Pisarevskaya EY, Zolotarevskiy VI, Kazanskiy LP, Ovsyannikova EV and Alpatova NM (2009) Double-stage polyo- phenylenediamine modification with palladium nanoparticles. Synth Met 159: 304-310.

38. Nemamcha A, Rehspringer JL (2008) Morphology of dispersed and aggregated pvp-pd nanoparticles prepared by ultrasonic irradiation of Pd $(No_3)_2$ solutions in ethylene glycol. Rev Adv Mater Sci 18: 685-688.

39. Belkacemi K, Boulmerka A, Arul J, Hamoudi S (2006) Hydrogenation of Vegetable Oils with Minimum trans and Saturated Fatty Acid Formation Over a New Generation of Pd-catalyst. Topics in Catalysis 37: 113-120.

40. Sohn Y, Pradhan D, Leung K (2010) Electrochemical Pd Nanodeposits on a Au Nanoisland Template Supported on Si (100): Formation of Pd-Au Alloy and Interfacial Electronic Structures. ACS Nano 4: 5111-5120.

41. Shafii S, Lihua W, Nordin MR, Liew KY (2010) Effect of Ultrasonic Irradiation on the Catalytic Activity of Pt and Pd Nanoparticles for hydrogenation of Palm Olein. National Science Officer Conference.

8

Fixed bed Column Adsorption of Cu (II) onto Maize Tassel-PVA Beads

Mahlatse M. Sekhula[1], Jonathan O. Okonkwo[1*], Caliphs M. Zvinowanda[1], Nana N. Agyei[2] and Abdul J.Chaudhary[3]

[1]*Department of Environmental, Water & Earth Sciences, Faculty of Science, Tshwane University of Technology, 175 Nelson Mandela Drive, Arcadia Campus, Pretoria, South Africa*

[2]*Department of Chemistry, University of Limpopo, MEDUNSA Campus, Pretoria, South Africa*

[3]*Institute for the Environment, Brunel University, Kingston Lane, Uxbridge, Middlesex, UB8 3PH, UK*

Abstract

The intention of this study was to explore the efficacy and feasibility for Cu (II) adsorption onto fixed bed column of maize tassel-PVA beads. The effects of flow rate and bed height were explored. The Thomas, Adams and Bohart and Yoon-Nelson models were analysed to evaluate the column adsorption performance. The adsorption rate constant and correlation coefficient associated to each model for column adsorption was calculated. Thomas model indicated that increase in column height and flow rate increased the values of k_{Th} *and* q_o. Thereafter, the value of q_o decreased with decreasing bed height. With Adam-Bohart model, the values of k_{AB} increased as the flow rate and the bed height increased and continued to increase with decrease in bed height. However, the value of N_o first increased with increase in flow rate and bed height, but later decreased with decrease in bed height. This suggests that the overall system kinetics may have been influenced by external mass transfer, particularly in the initial part of adsorption in the column. In the case of Yoon-Nelson model, the rate constant k_{YN} and τ increased with increase in bed height and flow rate. However, decrease in bed height at a flow rate of 2.33×10^{-3} L min^{-1} resulted in an increase in τ while the values of k_{YN} decreased. The three models gave high values of R^2 (0.9694-0.98920), although the Adam-Bohart model values were, on average, the least. Therefore, it can be said that both the Thomas and Yoon-Nelson models describe the behavior of the adsorption of Cu (II) in a fixed-bed column better than Adam-Bohart.

Keywords: Maize tassel; Polyvinyl alcohol; Beads; Cu(Ii) removal; Column studies

Introduction

Effective removal of trace metal ions from aqueous solution is important in order to protect the environment and to improve the quality of public health. Trace metals such as mercury, lead and cadmium are not biodegradable and are known to bio-accumulate in the food chain. Therefore, the removal of trace metals from the environment, particularly from water, is very important since water is one of the most valuable natural resources. Copper is an essential element in mammalian nutrition as a component of metallo enzymes in which it acts as an electron donor or acceptor. However, exposure to high levels of copper can result in a number of adverse health effects [1]. Hence, the study of removal of copper ions in aqueous system is essential.

Physical and chemical methods have been employed in the removal of trace metals from contaminated water through processes such as chemical precipitation, membrane filtration, ion exchange, and adsorption. One of the new developments in recent years to remove toxic trace metals from aqueous solutions is the use of adsorbents of biological origin, including alginate, dead and living biomass, chitosan, lignin, and others [2]. Agricultural waste products such as maize tassel, rice husks, corn cobs, olive mill products and polymerized orange skin have been reported for their removal of toxic metals from aqueous solutions [3-9].

Maize tassel, a waste biomaterial, is the 'male' flower of the maize plant that forms at the top of the stem. The tassel grows at the apex of the maize stalk and end up as agricultural waste product after being involved in fertilisation. It is discarded by farmers in large quantities with the rest of the plant once the cobs have been harvested. In our earlier studies, we reported the potential of maize tassel to remove trace metals from aqueous solution, physicochemical characterisation of the tassel for texture, microstructure, elemental composition and thermal stability [8]. However, earlier attempt to use ground maize tassel in column experiments showed that the material was soft and, therefore, clogged columns thus preventing free flow of water. In order to apply this waste material effectively in column experiments, it became necessary to immobilize onto an inert substance which can provide support without the maize tassel losing its capacity to remove metals from aqueous solution. Immobilisation of such material will create material with mechanical strength and rigidity. For this purpose, materials such as Polyvinyl Alcohol (PVA), has been used [10,11]. To date, the immobilisation of maize tassel-PVA for the removal of heavy metals from aqueous solution has not been reported.

Batch experiments are usually conducted to evaluate the ability of a material to adsorb as well as the adsorption capacity of the material. The data obtained from batch experiments are, in most cases, limited to laboratory scale and thus do not provide data which can be accurately applied in industrial systems. Column experiments, on the other hand, are necessary to provide data which can be applied for industrial purposes [12].

In this study, the efficacy and feasibility for Cu (II) adsorption onto fixed bed column of maize tassel immobilised in polyvinyl alcohol matrix to form beads was explored. The effects of flow rate and bed

***Corresponding author:** Jonathan O. Okonkwo, Department of Environmental, Water & Earth Sciences, Faculty of Science, Tshwane University of Technology, 175 Nelson Mandela Drive, Arcadia Campus, Pretoria, South Africa

height were investigated. The Thomas, Adam-Bohart and Yoon-Nelson models were employed to evaluate the column adsorption performance. Also scanning electron microscopy was carried out on maize tassel-PVA beads for morphology identification.

Methods

Materials and equipment

All chemicals used were of analytical grade and were used without further purification. Polyvinyl alcohol (PVA, 99% hydrolyzed, average molecular weight 89,000-98,000) was obtained from Aldrich Chemical. For column experiments, aqueous samples were pumped through the column with a Watson Marlow 101U/R peristaltic pump (Watson-Marlow limited, England). Microstructure was determined using a JSM-5800LV (JEOL, Japan) scanning electron microscope and a JSM-7500F (JEOL, Japan). Vernier calliper was used to measure the diameter of the maize tassel-PVA beads.

Preparation of maize tassel powder

Maize tassel collected from the Tshwane University of Technology research farm in Pretoria, South Africa, was first air dried for two days in a fume cupboard and then thoroughly rinsed with copious amounts of deionised water. The material was then placed in an air powered drying hood to remove water. Finally, the tassel was dried in an oven at 105°C for 24 h to expel any residual moisture. The dried material was then milled by a hammer mill model Laboratory Mill 3 100 (Stockholm, Sweden). Thereafter, the milled tassel was sieved to a particle size of 150-500 μm.

Preparation of maize tassel-PVA beads

The method of preparation was as described by Jin and Bai [10]. Briefly, about 8.50 g of PVA powder was weighed into a 0.25 L beaker containing 0.10 L deionised water. The mixture was stirred at a speed of 800 rpm at 80°C for 5 h and 10 g of sieved tassel powder added to the mixture with stirring. The mixture was stirred for another 1 h until the tassel powder blended well with the PVA and the temperature of the mixture reduced to 50°C. The gel formed was, thereafter, cooled and extruded into small beads. The beads were placed into a beaker containing 99% methanol for 1 h, and later rinsed with deionised water, air dried in a fume cupboard and stored in airtight container before use. The diameters of the beads (5.50 ± 0.48 mm) were measured using a vernier calliper. The PVA beads were prepared in the same but without the addition of maize tassel powder.

Microstructure analysis

About 1.0 g of maize tassel-PVA beads and 1.0 g of tassel powder were separately mounted onto double sided tape and placed on a special SEM stage. The samples were sputter coated with carbon to avoid sputtering. The surface morphology of the samples was obtained with SEM mode and later changed to EDX mode to obtain the elemental composition of the samples.

Batch experiment

Batch experiment was first conducted with polyvinyl alcohol beads (PVA), maize tassel powder (MT) and maize tassel-polyvinyl beads (MT PVA) in order to determine the potential of PVA to adsorb Cu (II) from aqueous solution. About 5.0 g of the aforementioned adsorbents were weighed and added into separate 250 mL solutions containing 20 mg L^{-1} of Cu (II) and the pH of the mixture adjusted to 4 with 0.1 M HCl. The mixtures were shaken in an orbital shaker at a speed of 200 rpm under room temperature for 1-2 h. Samples were taken at predetermined time intervals for the analysis of Cu (II) ion concentrations in the solution until adsorption equilibrium was reached. Cu (II) concentrations were analyzed by using a Varian 220 FS Flame Atomic Absorption Spectrometer (FAAS) coupled to an SPS5 (sample preparation system) and a Sample Introduction Pump System (SIPS). The adsorbed amount of Cu (II) per unit weight of beads and powder were calculated from equation 1.

$$Q_e = \frac{(Ci - Ce)V}{m} \tag{1}$$

where Q_e is the metal adsorbed in mg g^{-1}; Ci and Ce are the initial and equilibrium concentrations of metal (mg L^{-1}); V is the volume of the solution (L), and m is the mass of the biosorbent (g).

Colum experiments

A glass column of internal diameters (I.D) of 25 mm and 20 cm in length, 20 mg L^{-1} of Cu (II) solutions at flow rates of 0.80 mL min^{-1} and 2.33 mL min^{-1} and column bed heights of 5 cm and 10 cm were used for the experiment. The columns were packed by placing weighed layers of glass wool before and after packing the maize tassel-PVA beads. All experiments were conducted at room temperature and the direction of flow was from bottom to top. The treated metal solution was collected at the outlet at different time intervals and analysed. The column was run at a predetermined pH of 4.0 as established in our previous study [8]. The experimental conditions for the column runs are shown in Table 1.

Kinetic models

In this study, three models namely Thomas [13], Adam-Bohart [14] and Yoon-Nelson [15] models were used for kinetic studies. The Thomas solution is one of the most widely used models in assessing column performance. The expression by Thomas for an adsorption column is given as follows:

$$\ln\left(\frac{C_o}{C_t} - 1\right) = \frac{k_{TH} q_o X}{Q} - \frac{k_{TH} C_o}{Q} V_{eff} \tag{2}$$

where C_t is the effluent concentration at time t (mg L^{-1}), C_0 the influent concentration (mg L^{-1}), k_{Th} is the Thomas rate constant (L min^{-1} mg^{-1}), q_0 the equilibrium uptake per g of the adsorbent (mg g^{-1}), x the amount of adsorbent in the column (g), V eff the effluent volume (L) and v is the flowrate (L min^{-1}). The kinetic coefficient k_{Th} and the adsorption capacity of the column q_0 can be determined from a plot of C_t/C_0 against t at a given flow rate using linear regression.

The Adam-Bohart model is used for the description of the initial part of the breakthrough curve. The data obtained in column in continuous mode studies was used to calculate maximum solid phase concentration of Cu (II) on maize tassel-PVA beads. The Adam-Bohart expression is given as:

Column No	Height (cm)	Column ID (mm)	pH	Flow rate (x 10^{-3} L min^{-1})	Feed Concentration (mg L^{-1})
1	5.0	25	4	0.8	20.0
2	0.0	25	4	2.33	20.0
3	5.0	25	4	2.33	20.0

Table 1: Experimental conditions for column runs.

$$\ln\left(\frac{C_o}{C_t}-1\right)=k_{AB}N_o\frac{Z}{F}-k_{AB}C_o t \tag{3}$$

where C_t and C_0 are the effluent and influent concentrations (mg L^{-1}) at time t and zero, k_{AB} is the kinetic constant (L mg^{-1} min^{-1}), t is time (min), N_0 is the saturation concentration (mg L^{-1}) and Z is the bed depth of column (cm), F is the linear flow rate (L min^{-1}). Parameters describing the characteristic operations of the column (k_{AB} and N_0) were determined from a linear plot of ln (C_t/C_0) against time (t), as the intercept and slope respectively.

Yoon and Nelson model is based on the assumption that the rate of decrease in the probability of adsorption of adsorbate molecule is proportional to the probability of the adsorbate adsorption and the adsorbate breakthrough on the adsorbent. The Yoon-Nelson model for a single component system is expressed as:

$$\ln\left(\frac{C_o}{C_t}-1\right)=\tau k_{YN}-k_{YN}t \tag{4}$$

where C_t and C_o are the effluent and influent concentrations (mg L^{-1}) at time t and zero, k_{YN} (L min^{-1}) is the Yoon-Nelson constant, τ (min) is the time required for 50% adsorbate breakthrough. From a linear plot of $ln[Ct/(C_o-C_t)]$ against sampling time (t), values of k_{YN} and τ were determined from the intercept and slope of the plot.

Results and Discussions

Maize tassel-PVA beads

The beads formed from the immobilization of powdered maize tassel in polyvinyl alcohol matrix were reasonably spherical in shape, and their diameters were in the range of 5.50 ± 0.48 mm.

Microstructure analysis

Figure 1 shows scanning electron micrographs of powdered tassel (a) and maize tassel-PVA beads (b). From Figure 2, maize tassel powder appeared as elongated fibrous particles which are bound tightly to PVA, although the elongated shape was still maintained. The SEM images verified that the addition of PVA made the maize tassel become much denser, as shown in Figure 1.

Batch experiment

The results of the batch experiments with polyvinyl alcohol beads, maize tassel powder and maize tassel-polyvinyl alcohol beads are shown in Figure 2. From Figure 2, the amount of Cu (II) ions adsorbed by PVA beads was very insignificant compared to the adsorption exhibited by maize tassel powder and maize tassel-PVA beads. The amounts of Cu (II) adsorbed by maize tassel powder and maize tassel-PVA beads were 0.50 mg g^{-1} and 0.52 mg g^{-1} respectively. Further experiments were conducted using only maize tassel-PVA beads.

Effect of influent flow rate on Cu (II) adsorption by maize tassel-PVA beads

The column adsorption study was conducted at different influent flow rates of 8.0×10^{-4} and 2.33×10^{-3} L min^{-1} using initial Cu (II) concentration of 20 mg L^{-1}, bed height of 5 cm, pH 4 and room temperature of 25°C. Figure 3 shows the resultant breakthrough curve in which the breakthrough occurred faster at higher flow rate of 2.33×10^{-3} L min^{-1} than at lower flow rate of 8.0×10^{-4} L min^{-1}. The faster breakthrough curve exhibited by flow rate 2.33×10^{-3} L min^{-1}, was attributed to faster movement of the adsorption zone along the bed, thus reducing the contact time between the influent, Cu (II) and the adsorbent, maize tassel-PVA beads. On the other hand, the gradual breakthrough curve for flow rate of 8.0×10^{-4} L min^{-1} suggested longer residence time of the influent in the column. Similar trends have been reported by researchers using activated rice husk to remove Cu (II) and dye from aqueous solution [16, 17]. It can also be observed from Figure 3, that the maximum C/C_0 value for flow rate of 2.33×10^{-3} L min^{-1} is lower than that of 8.0×10^{-4} L min^{-1}. This can be attributed to the reduction of adsorption efficiency since the contact time of the influent in the column was low.

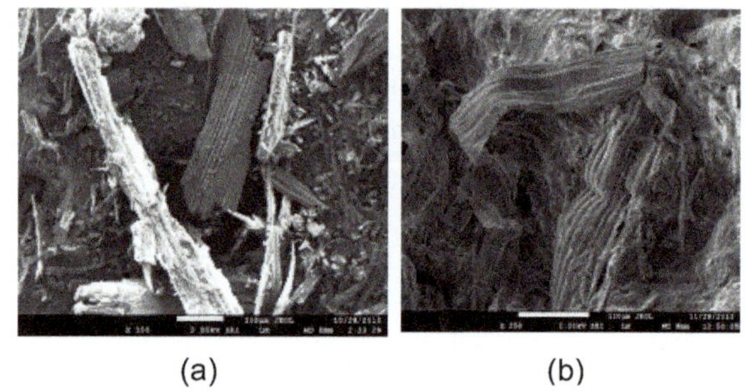

Figure 1: (a) SEM micrographs showing the surface morphologies of maize tassel powder and (b) Maize tassel PVA beads.

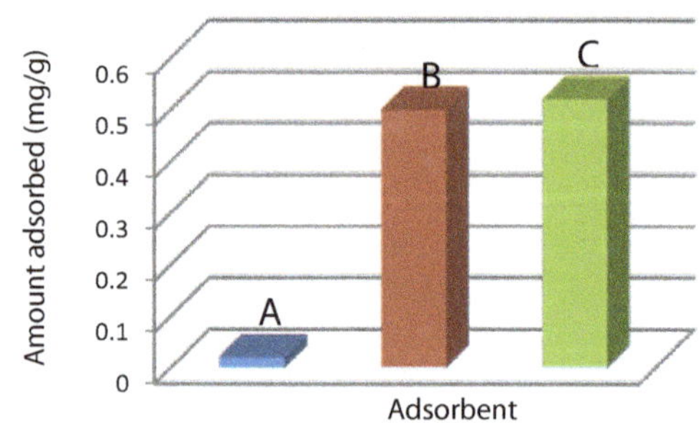

Figure 2: (A) Adsorption of Cu(II) by polyvinyl alcohol only, (B) Maize tassel powder only and (C) Maize tassel-PVA beads.

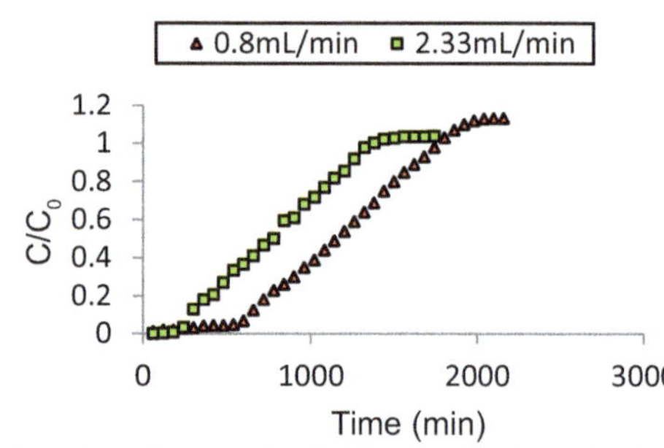

Figure 3: Breakthrough curves for Cu(II) adsorption by maize tassel-PVA beads at different flow rates (Cu(II) concentration 20 mg L^{-1}, column height 5 cm, pH 4 and 25°C).

Effect of bed height on Cu (II) adsorption by maize tassel-PVA beads

The effects of bed heights of 5.0 cm and 10.0 cm were studied at influent concentration of 20 mg L^{-1}, 2.33×10^{-3} L min^{-1} flow rate and pH 4. From Figure 4, the breakthrough time increased with bed height.

This was attributed to the fact that as the bed height increased from 5 to 10 cm, Cu (II) had more contact time with the maize tassel beads which resulted in higher removal of Cu (II) ions from the column. Furthermore, the slope of the breakthrough curves decreased with increase in bed height as a result of broadened influent movement zone. Higher Cu (II) removal was observed for bed height of 10.0 cm because of the increase in the bed surface which provided more binding sites for the adsorption of Cu (II) onto maize tassel-PVA beads. It should be noted that the breakthrough curves for bed height 5 cm in Figure 4 and the 2.33×10^{-3} L min^{-1} in Figure 3 were performed under identical conditions. However, the dissimilarities observed on the curves may be due to variability of flow rates during experimentation as well as the non-reproducibility in the production of bead used in this study.

Thomas model

The behaviour of the column was modelled using the Thomas model. The influences of flow rate and bed height on the adsorption of Cu (II) with maize tassel-PVA were studied. The results are shown in Table 2. The Thomas model is suitable for adsorption processes where the external and internal diffusions will not be the limiting step [17]. The column data were fitted to the Thomas model to determine the Thomas rate constant (k_{Th}) and maximum solid-phase concentration (*qo*). The determined coefficients and relative constants were obtained using linear regression analysis according to equation (2) and the results are shown in Table 2. As can be seen in Table 2, as the column height and flow rate increased the values of k_{Th} and q_o increased. Thereafter, the value of q_o decreased with decreasing bed height. The values of k_{Th} obtained in the present study are within the range reported in a fixed bed study for the removal of Cu (II) from aqueous solution using rice

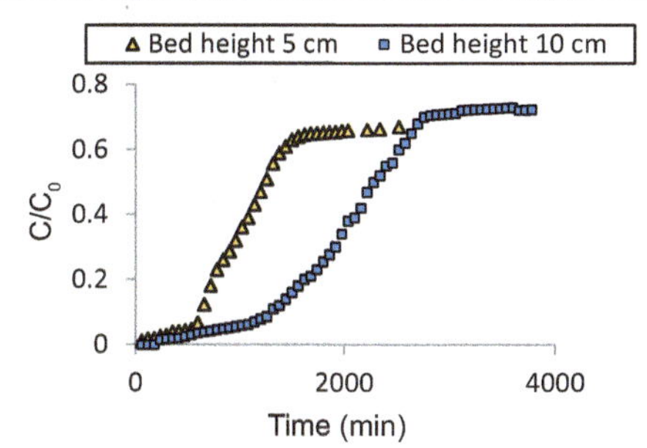

Figure 4: Breakthrough curves for Cu(II) adsorption by maize tassel-PVA beads at different bed heights (Cu(II) concentration 20 mg L^{-1}, flow rate 2.33×10^{-3} L min^{-1}, pH 4 and 25°C).

Inlet Concentration (mg g^{-1})	pH	Bed height (cm)	Flow rate (x10^{-3} L min^{-1})	k_{Th} (x10^{-3} L min^{-1}mg^{-1})	Q_0 (mg g^{-1})	R^2
20	4.0	5.0	0.80	0.20304	3.98	0.9806
20	4.0	10.0	2.33	0.20468	7.766	0.9858
20	4.0	5.0	2.33	1.0995	3.102	0.9745

Table 2: Thomas model parameters at different conditions using linear regression analysis.

Inlet Concentration (mg g^{-1})	pH	Bed height (cm)	Flow rate (x10^{-3} L min^{-1})	k_{AB} (x 10^{-4} L mg^{-1}min)	N_0 (mg L^{-1})	R^2
20	4.0	5.0	0.80	1.86	2670	0.9827
20	4.0	10.0	2.33	2.35	9570	0.9826
20	4.0	5.0	2.33	9.25	116	0.9694

Table 3: Adam-Bohart model parameters at different conditions using linear regression analysis.

Inlet Concentration (mg g^{-1})	pH	Bed height (cm)	Flow rate (x10^{-3} L min^{-1})	k_{YN} (L min^{-1})	τ (mg g^{-1})	R^2
20	4.0	5.0	0.8	0.004	1176	0.9893
20	4.0	10.0	2.33	0.048	1513	0.9806
20	4.0	5.0	2.33	0.0194	3601	0.9727

Table 4: Yoon-Nelson model parameters at different conditions using linear regression analysis.

husk based activated carbon [16], ammonium ion removal by zeolite 13X [18] and adsorption of Acid Yellow 17 Dye onto Tamarind seed powder [19], but significantly lower than the values reported for the removal of methylene blue from aqueous solution using rice husk [20]. However, the q_o values obtained in the present study are significantly lower that the values reported [17,18,19]. The R^2 values ranged from 0.9745-0.9858, indicating good linearity. It can, therefore, be said that the high R^2 value indicates that the Thomas model equations of linear regression analysis describes the breakthrough data under the studied conditions.

Adams-Bohart model

This model assumes that the adsorption process is continuous and that equilibrium is not attained instantaneously. The Adam-Bohart adsorption model was applied to experimental data for the breakthrough curve. After applying equation (3) to the experimental data, a linear relationship was found for the time for breakthrough. For all breakthrough curves, respective values of N_o and k_{AB} were calculated and presented in Table 3. From Table 3, it can be seen that the values of k_{AB} increased as the flow rate and the bed height increased and continued to increase with decrease in bed height. However, the value of N_o first increased with increase in flow rate and bed height, but later decreased with decrease in bed height. This suggests that the overall system kinetics may have been influence by external mass transfer, particularly in the initial part of adsorption in the column [20]. The values of saturation concentration, $N_{o,}$ obtained in the present study are significantly higher than the values reported [16]. The correlation coefficient, R^2 values were between 0.9694-0.9827.

Yoon-Nelson model

The values of k_{YN} and τ are listed in Table 4. From Table 4, the rate constant k_{YN} and τ increased with increase in bed height and flow rate. However, decrease in bed height at a flow rate of 2.33 x 10^{-3} L min^{-1} resulted in an increase in τ while the values of k_{YN} decreased. These observations are in agreement with a fixed bed column adsorption of Acid Yellow 17 Dye onto Tamarind seed powder [16].

Conclusion

The present study demonstrated the ability of maize tassel-PVA to remove Cu (II) from aqueous solution in a column model. The change in flow rate and column height influenced the amount of Cu (II) adsorbed. The results obtained with maize tassel-PVA were in agreement with studies reported in the literature. The adsorption models applied to evaluate the performance of the column describe the fixed bed column well. However, the value of R^2 for Adam-Bohart model was generally lower than Thomas and Yoon-Nelson models under the same experimental conditions. Therefore, it can be concluded that both the Thomas and Yoon-Nelson models describe the behaviour of the adsorption of Cu (II) in a fixed-bed column better than Adam-Bohart.

Acknowledgements

This Research was supported by the International Cooperation Unit (facilitated by Dr B. Kgarebe) of the Organisation for the Prohibition of Chemical Weapon (OPCW), Tshwane University of Technology and National Research Foundation of South Africa.

References

1. Stern BR, Solioz M, Krewski D, Aggett P, Aw TC, et al. (2007) Copper and human health: biochemistry, genetics, and strategies for modeling dose-response relationships. J Toxicol Environ Health B Crit Rev 10: 157-222.

2. Bailey SE, Olin TJ, Bricka RM, Adrian DD (1999) A review of potentially low-cost sorbents for heavy metals. Water Research 33: 2469-2479.

3. Lokatos J, Brown SD, Snape CE (2002) Coals as sorbents for the removal and reduction of hexavalent chromium from aqueous waste streams. Fuel 81: 691-698.

4. Doyurum S, Celik A (2006) Pb (II) and Cd (II) removal from aqueous solutions by olive cake. J Hazard Mater 138: 22-28.

5. Ahluwalia SS, Goyal D (2007) Microbial and plant derived biomass for removal of heavy metals from wastewater. Bioresour Technol 98: 2243-2257.

6. Mohan D, Pittman CU Jr (2006) Activated carbons and low cost adsorbents for remediation of tri- and hexavalent chromium from water. J Hazard Mater 137: 762-811.

7. Šćiban M, Klašnja M, Škrbić B (2008) Adsorption of copper ions from water by modified agficultural by-products. Desalination 229: 170-180.

8. Zvinowanda CM, Okonkwo OJ, Agyei NM, Shabalala PN (2009) Physicochemical characterization of maize tassel as an adsorbent. I. Surface texture, microstructure, and thermal stability. J Appl Polymer Sci 111: 1923-1930.

9. Zvinowanda CM, Okonkwo JO, Sekhula MM, Agyei NM, Sadiku R (2009) Application of maize tassel for the removal of Pb, Se, Sr, U and V from borehole water contaminated with mine wastewater in the presence of alkaline metals. J Hazard Mater 164: 884-891.

10. Jin L, Bai R (2002) Mechanisms of lead adsorption on chitosan /PVA hydrogel beads. Langmuir 18: 9765- 9770.

11. Khoo K, Ting Y (2001) Biosorption of Gold by immobilized fungal biomass. Biochem Eng J 8: 51-59.

12. Wang S, Li H, Xie S, Liu S, Xu L (2006) Physical and chemical regeneration of zeolitic adsorbents for dye removal in wastewater treatment. Chemosphere 65: 82-87.

13. Thomas HC (1944) Heterogeneous ion exchange in a flowing system. J Am Chem Soc 66: 1664-1666.

14. Bohart G, Adams EQ (1920) Some aspects of the behaviour of charcoal with respect to chlorine. J Am Chem Soc 42: 523-544.

15. Yoon YH, Nelson JH (1984) Application of gas adsorption kinetics--II. A theoretical model for respirator cartridge service life and its practical applications. Am Ind Hyg Assoc J 45: 517-524.

16. Patel H, Vashi RT (2012) Fixed bed column adsorption of ACID Yellow 17 dye onto Tamarind Seed Powder. Can J Chem Eng 90: 180-185.

17. Yahaya NKEM, Abustana I, Latiffa MFPM, Bello OS, Ahmad MA (2011) Fixed-bed column study for Cu (II) removal from aqueous solutions using rice husk based activated carbon. International Journal of Engineering & Technology 11: 248-252.

18. Han R, Wang Y, Yu W, Zou W, Shi J, et al. (2007) Biosorption of methylene blue from aqueous solution by rice husk in a fixed-bed column. J Hazard Mater 141: 713-718.

19. Zheng H, Han L, Ma H, Zheng Y, Zhang H, et al. (2008) Adsorption characteristics of ammonium ion by zeolite 13X. J Hazard Mater 158: 577-584.

20. Guibal E, Lorenzelli R, Vincent T, Cloirec PL (1995) Application of silica gel to metal ion sorption: static and dynamic removal of uranyl ions. Environmental Technology 16: 101-114.

Effect of Particle Size on Detergent Powders Flowability and Tabletability

Abarasi Hart*

School of Chemical Engineering, University of Birmingham, Edgbaston, Birmingham, B15 2TT, UK

Abstract

Detergent powders are sometimes processed into compact tablets for cleaning purposes based on the advantage of dosage, minimised dusting and packaging its offers over their powder counterpart. The flow-ability greatly influences handling. The particle size within the powder plays a key role on handling, flow-ability and the strength of the produced compact detergent tablet. The Ariel powder used in this study was separated into different size ranges via sieving; thereafter flowability, compression and compaction were carried out. The compressibility of the powder under 2 KN compaction force was analysed using Heckel equation. It was found that at the same compaction pressure, smaller size particle powders formed compact of greater tensile strength than larger particle. However, the powder with size range of 125-180 μm showed higher flowability compared to other size range.

Keywords: Detergent powder; Particle size; Flowability; Tabletability

Introduction

It is well know that particle size and size distribution is one of the many factors that greatly impacts on the flowability of powder during handling, processing and tableting. Flowability is a vital parameter for efficient and effective transport, storage and handling of detergent powder in the detergent industries. It is the ability of the powder to flow in desired manner in a given processing or handling piece of equipment [1]. Flowability remains a crucial factor that affects the design and processing of detergent powder in handling equipment such as hoppers, silos, filling and packaging operations, conveying, etc. [2]. There are many techniques to assess the flowability of powders which includes; measuring the time required to discharge a given amount of powder from a flowmeter, carr's compressibility index, Hausner ratio, angle of repose and flow function [3,4]. However, the Jenike flow function is mostly used to classify the flowability of powders. The higher the flow function of the powder, the better the flowability. As poor powder flow can lead to sticking or caking during storage, prone to cohesion, rat-holing, arching, poor content uniformity and poor solubility [4].

On the other hand, detergent powders are currently been processed into compact powder and detergent tablets for cleaning purposes based on the advantages it present such as easy dosage, minimised dusting, easy packaging and safer for consumers [2,5]. A tablet is a defined assembly of discrete particles with each particles bonded to its closest neighbours. The strength of the tablet formed depends on particle size and particle rearrangement as well as densification and interaction between the particles by bonding during compaction [6]. Therefore, among the mechanical properties of interest in detergent tablet, the tensile strength is of importance to the detergent industries as it determines the tablet potential to sustain loading encountered during handing, filling and processing. The tabletability of the detergent tablet is described in *XY*-plot of tensile strength of the tablet versus compaction pressure. The tensile strength of the compacted tablets is determined by diametric compression. In diametrical compression test, the tablets are generally placed vertically between two platens and compressed until the tablet breaks; the value of the maximum force is used to calculate the tensile strength [7].

To form compact tablet however uniaxial compaction is used to consolidate the powder bed by reducing the voids and creating particle-to-particle bonding. During the tabletting, the applied stress causes the particles to rearrange themselves leading to increased bulk density, fragmentation, plastic and elastic deformations. The strength of the resulting tablet depends on the particle-to-particle bond formed during the compaction. The strength of the particle-to-particle bond in the tablet depends on the solid pressure (i.e., particle-to-particle pressure), which is the difference between the total applied pressure and interstitial air pressure in between the punches in the die. The tensile strength of the formed tablet depends on the strength of the particle-to-particle bonding, the porosity, and the resultant pressure. However, powder flow properties can be utilised to predict uniaxial compaction behaviour [1]. The coordination number within the tablet depends on the particle size and size distribution and on the solid fraction (or porosity) of the tablet [4]. Therefore, particle size and solid fraction (or porosity) greatly impact on the tensile strength of the tablet compared to the compaction pressure used in producing the compact. In addition, the solid fraction (or porosity) of the tablet is the direct result of the applied compression pressure [8].

Besides particle size distribution (*PSD*), shape, morphology and surface chemistry can affect the flow properties of powder and tabletability significantly. The flowability of powder is a multifunctional parameter which depends on particle size and size distribution, shape, particle interactions and moisture content [9,10]. Mullarney and leyva [11] investigated the possibility of predicting powder flowability based on the particle size distribution data. In this study, however, the effect of powder particle size on flowability and tensile strength of formed detergent powder tablet was investigated. The compressibility of the powder was evaluated using Heckel equation. The effect of particle sizes on deformation and rearrangement under compaction pressure was explored. However, the influence of particle size on the detergent powder activity and dissolution of compacted tablet are outside the scope of this study.

***Corresponding author:** Abarasi Hart, School of Chemical Engineering, University of Birmingham, Edgbaston, Birmingham, B15 2TT, UK

Material and Methods

The detergent powder used in this study is a Professional Ariel® Regular Powder supplied by Procter & Gamble, United Kingdom. The composition of the detergent power are thus oxygen-based bleaching agent, anionic surfactants, non-ionic surfactants, phosphate, polycarboxylate, zeolites, enzymes, perfumes, optical brighteners, butylphenyl methlpropional, hexyl cinnamal and linalool. The properties of the detergent powder are presented in Table 1. The true density of the detergent powder was measured with a helium gas pycnometer (AccuPyc™ II 1340, Micrometritics®, Germany).

The particle size distribution (PSD) of the as-received detergent powder was obtained using sieve analysis with a stack of sieves arranged in decreasing mesh size (2000, 1400, 1000, 710, 500, 355, 250, 180 and 125 μm, respectively) and shaken for 15 min at an amplitude of 1 mm on the sieve tower (Retsch AS200, Germany) to separate the detergent powder into various size range and to obtain a distribution by mass.

An important processability parameter is the flowability of the powder. The flow properties of the detergent powder and separated size ranges were measured using the Schulze ring shear cell tester RST-4 (RST-XS, Dietmar Schulze, Wolfenbuttel, Germany). The powder was loaded into the annular space between the concentric rings of the shear cell tester. The shear stress is applied through the pins at the corners of the shear lid. The powder was first pre-sheared under a consolidation stress of 10 kPa until a steady consolidated state is reached. To construct a yield-locus, the samples were pre-sheared with four other normal stresses 8.0, 6.0, 4.0, and 2.0 kPa. The consolidation stress (σ_1) of the detergent powder and the corresponding unconfined yield strength (σ_c) were determined by constructing Mohr's circles [12], from which the effective angle of internal friction (φ_e) and powder flow function (ff_c) (i.e., the ratio of σ_1/σ_c) was determined.

A confined uniaxial compression test (Figure 1a) was performed with compression forces of 2 kN using Zwick® universal material testing machine (Zwick-Roell Z030, Germany). The powder was compressed with flat-faced punches 13 mm in diameter with an approximate sample mass of 1.15 g at a constant compaction rate of 60 mm/min to form compact tablets. The compression data were registered during the experiments with a Zwick-testXpert II software. The compaction pressure was calculated from the applied force and the cross-sectional area of the punch.

The thickness of the compacted detergent tablets were measured with a digital calliper, before subjected to diametrical compression test (Figure 1b) at a compression speed of 0.5mm/min. The maximum force required to break the tablet under diametrical compression was recorded and converted into a tensile strength value, σ_t, using equation (1). Each test was carried in triplicate.

$$\sigma_t = \frac{2F}{\pi D t} \quad (1)$$

where F is the force required to break the detergent tablet, D is the diameter of the detergent tablet and t is the tablet thickness.

Parameter	Value
True density (g.cm^{-3})	2.0147 ± 0.01
PSD d_{50} (μm)	469.5 ± 19.5
Moisture content (%)	7.37 ± 1.35
Angle of repose (°)	32 ± 0.24
Adhesive strength (Pa)	238 ± 18.83

Table 1: Properties of the detergent powder.

Results and Discussion

Flowability

Detergent powders have wide particle size distribution ranges giving to problems in the course of their handling. Operations such as fillings, blending, transportation and processing need reliable flowability to avoid stoppage, downtime and wastage. To evaluate the influence of particle size on flowability, the flow function, ff_c, of each particle size ranges of the powder were measured using Schulze ring shear cell tester. The plot of the flow function as a function particle size range of the detergent powder is shown in Figure 2. It is clear that the particle size distribution have a significant effect on the flowability of the powder. The detergent powder with particle size ranges 125-180 μm and 180-250 μm show superior bulk flow properties compared to larger particles because their sphericity make them roll or slide over one another when shear stress is applied. However, the particles (<125μm) have high surface area to volume ratio, causing particle-particle interaction forces to inhibit flow when stress is applied [11].

The flowability of the powder with size ranges 125-355 μm and 710-1000 μm are easy flowing, while the other size ranges showed poor flowability. It has been reported that rough surfaces and irregular shape causes particles to interlock and resist bulk flow when shear stress is applied [11]. However, the poor flow behaviour of the smaller size range particles <125 μm can be attributed to the cohesive nature of the particles as they tend to agglomerate upon shear application. Smallness of the particle size of the powder could tend to reduce powder flow due to the increase surface area per unit mass has provided a greater surface area for surface cohesive forces to interact resulting in more cohesive flow. Powder of size range 1400 to greater than 2000 μm was not analyzed because the separated amounts were less than necessary for the test. The processability and handling of the detergent powder therefore will be improved if the particle size range can be narrowed to the range of 125 to 250 μm compared to a wider size range distribution.

Compressibility

Compressibility is the densification and reduction in volume of a powder bed in response to applied pressure. Heckel [13] proposed that the reduction of the powder bed porosity follows a first-order response with applied pressure. The force-displacement data obtained from the uniaxial compaction experiments was mathematically transformed according to equation (2) to describe the compressibility of the detergent powder.

$$\ln\left(\frac{1}{1-D_{rel}}\right) = kP + A \quad (2)$$

$$D_{rel} = \frac{\rho_b}{\rho_t} \quad (3)$$

$$\rho_b = \frac{m}{\frac{\pi D^2}{4}(h-d_i)} \quad (4)$$

$$\varepsilon = 1 - D_{rel} \quad (5)$$

$$k = \frac{1}{3}Y \quad (6)$$

where; D_{rel} is the relative density (i.e. solid fraction) of the powder bed at applied compaction pressure, ρ_b and ρ_t are bulk and true density of the powder, m is the mass of the powder, h is the initial height of the powder bed, d_i displacement, D is the diameter of punch, Y is the

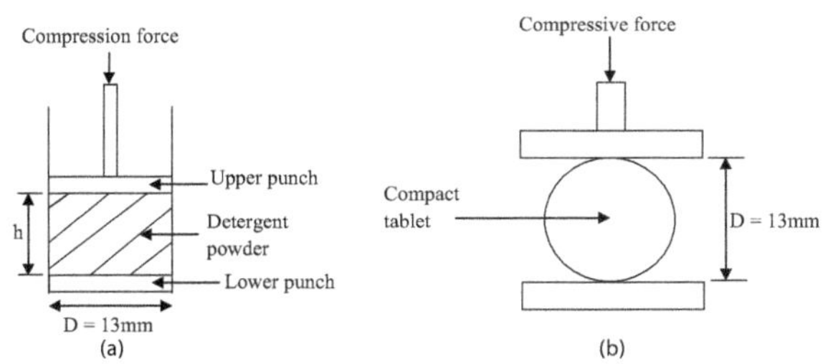

Figure 1: Schematic of (a) Confined uniaxial and (b) diametrical compression tests.

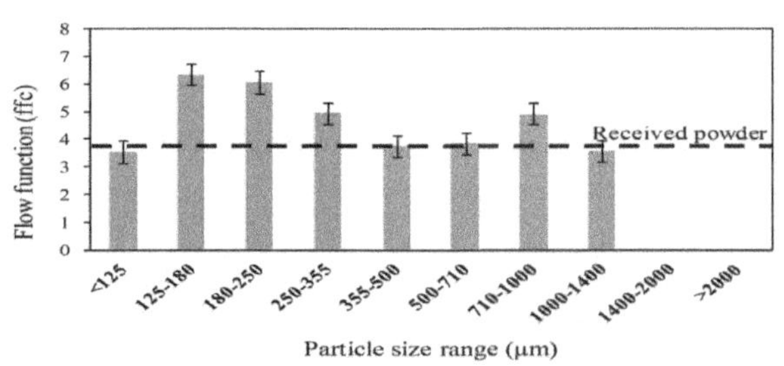

Figure 2: Effect detergent powder particle size on flowability.

yield strength and ε is the porosity. The constant *k* is a measure of the plasticity of the material and the constant *A* is related to the die filling and particle rearrangement before deformation and bonding of the discrete particles.

In Figure 3, the Heckel plots at the compression force of 2 kN for the received detergent powder and the separated powder of different particle size ranges using sieve are presented. The initial curvature describes particle fragmentation and re-arrangement without inter-particle bonding. The linear section of the plots represents plastic deformation.

Notably, the powder deforms more for the as-received detergent powder than the separated powder with size ranges 125-180, 355-500, and 500-710 μm, respectively. However, when compression pressure is greater than 15 MPa the densification is almost 100% (Figure 3) as the compacts are close to a solid continuum. This is as a result of the changes in material properties due to broken capsules containing perfumes and surfactant within particles as pressure increases during strain hardening. This results in negative porosities and the undefined parts of the Heckel plot beyond 15 MPa [14].

The linear section of the Heckel plots obtained for the different particle size range powders are linearly fitted to determine the constants *k* and *A*. It has been observed that the particle size is related to the deformation and fragmentation behaviour of powder under compression during compaction to form tablet [15]. The Heckel constants k and A describe the structure of the powder bed changes and consolidation mainly due to particle rearrangement, plastic deformation, and fragmentation. The effect of particle size on *k* and *A* are presented in Figure 4a and 4b.

The *k* values increases up to particle size 250-355μm and decreases sharply with increasing particle size. This implies that the plastic deformation for the material decreases with increasing particle size. However, the Heckel constant *A* increases with increasing particle size, expect for the powder with particle size range of 710-100 μm. This shows that for larger particle size powder there is greater particle rearrangement and fragmentation before bonding. In this regard, a greater compression pressure is required to create effective particle bonding. Hence, particle size of the powder will influence the final compact porosity, tensile strength and subsequently the compact dissolution.

Compatibility

The variation of tensile strength of the detergent with particle size range at 2kN compression force is shown in Figure 5. It is clear that the tensile strength of the tablet decreases with increasing particle size. Notably, at the same applied compaction pressure, smaller particles formed tablets of higher tensile strength than larger size counterparts. Additionally, the density of the compressed power increases with decreasing particle size, thereby reducing the porosity of the formed table which correlates with the increasing tensile strength with particle size decrease. This is as a result of particle size increment and the smoothness of the larger particles which reduces particles interlocking and bonding.

The tensile strengths of the size ranges 125-180, 180-250, and 250-355 μm are approximately the same, before significant decrease in tensile strength was noticed as the particle size increase from 355 to 1400 μm. Therefore, smaller particles provide a larger surface area for interparticle bonding than larger particles, hence the resulting tablets after compaction is stronger. However, Abdel-Hamid et al. [6] has reported the size of particles plays a vital role in the interaction as well as available surface area and bond propensity during compaction of powder material. A similar observation in the increase of tensile strength of tablet with decreasing particle size for L-lysine monohydrochloride dehydrate powder was reported by Sun and Grant [16]. The porosity of the compacted tablet decreases with decreasing powder particle size. Additionally, reducing powder particle size increase its surface area per unit mass and the feed powder quantities and this result correspond to increase tensile strength [17]. Smaller particle size of powders provided a greater surface area for surface cohesive forces to interact thereby making the tablet to increase in tensile strength

Statistical tools such as correlation and linear regression were used to test the relation between flowability and tabletability for the separated detergent power particle size ranges. The correlation coefficient (R^2) is a vital estimate of the strength of the linear regression analysis. The R^2 value of 0.273 was obtained and the association is significant at the P value of 0.554. The P value is therefore not significant at $P < 0.05$. Hence, the tensile strength of the detergent tablet and the flowability of the separated powder size ranges are poorly correlated. Flowability and tabletability are functionally dependent on the interparticle interaction. The interparticle interactions are strongly affected by surface properties such as texture, surface chemistry, and contact area. In compaction, the dominating interparticle interactions are identified as solid bridges, intermolecular forces, and mechanical interlocking. However, flowability is influenced by particle size, size distribution, shape, surface texture, surface energy, chemical composition, moisture content, particle geometry, and other factors [18]. Therefore, the flowability and tabletability are multifunctional parameters strongly influenced by geometrical, material, physical, chemical and mechanical properties of the particle.

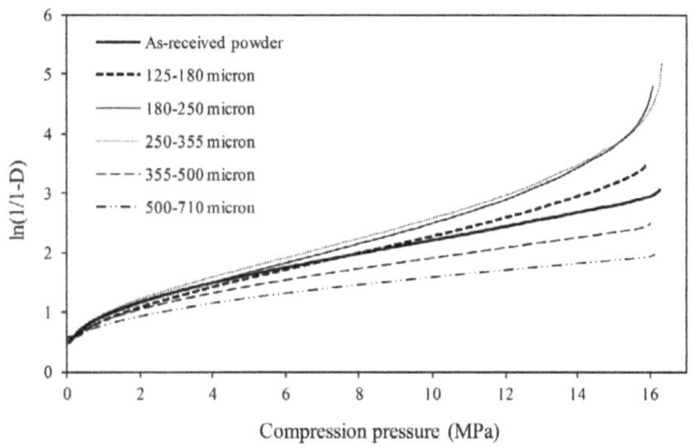

Figure 3: Heckel plots for the received detergent powder and separated particle size ranges (2 kN compression force).

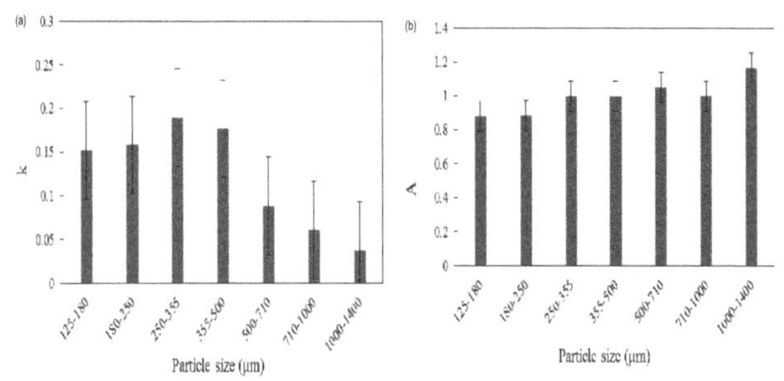

Figure 4: Effect of particle size on Heckel contacts (a) k and (b) A at 2 kN compression force.

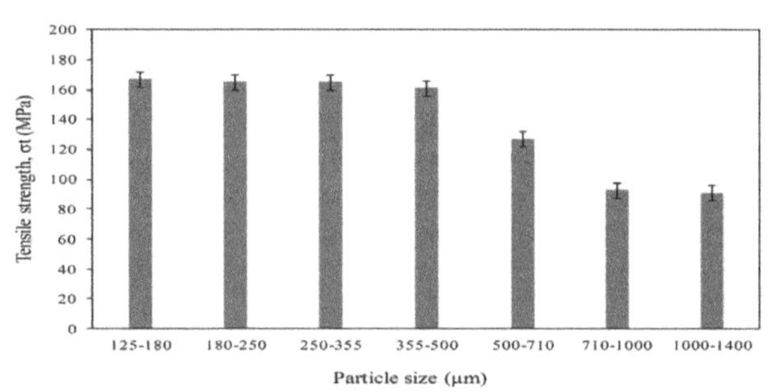

Figure 5: Variation of tensile strength with particles size for detergent powder (2 kN compression force).

Conclusions

The effect of particle size on detergent powder flowability and tensile strength of compacted tablet was investigated using Ariel detergent powder. Also investigated was the compressibility and compactibility of the as-received powder and the different separated particle size range powder using Zwick® universal material testing machine. It was found that the strength of the produced tablet was dependent on the particle size of the powder. Powder of smaller sizes showed higher tensile strength than larger particles because of their large surface area available for bonding. However, powder with size range 125-180 µm showed the highest flowability compared to other size ranges after sieving. This study is relevant with regard to the compressibility and compactibility useful to improve the production of cleaning and laundry detergent tablet with optimum strength to resist fracture on handling and transportation with the level of compaction that will enhance dissolution. However, the influence of particle size on activity of the detergent powder and the dissolution of the compacted detergent tablets require further investigation.

References

1. Prescott KJ, Barnum AR (2000) On powder flowability. Pharmaceutical Technology 60-84.
2. Hart A, Wu CY (2011) The Impact of Dry Granulation on Detergent Powder Properties. In: Wu CY, Ge W (eds.) Particulate Materials: Synthesis, Characterisation, Processing and Modelling. RSC Publishing 233-239.
3. Schwedes J (2003) Review on testers for measuring flow properties of bulk solids. Granular Matters 5: 1-43.
4. Sun CC (2010) Setting the bar for powder flow properties in successful high speed tableting. Powder Technology. 201: 106-108.
5. Chateau ME, Laurence G, Yannick S, Jacques F (2005) Processing a detergent powder formulation: Direct compression, and high shear wet granulation followed by compression. Powder Technology 157: 191-198.
6. Abdel-Hamid S, Alshihabi F, Betz G (2011) Investigating the effect of particle size and shape on high speed tableting through radial die-wall pressure monitoring. Int J Pharm 413: 29-35.
7. Fell JT, Newton JM (1970) Determination of tablet strength by the diametral-compression test. J Pharm Sci 59: 688-691.
8. Tye CK, Sun CC, Amidon EG (2005) Evaluation of the effects of tableting speed on the relationships between compaction pressure, tablet tensile strength, and tablet solid fraction. Journal of Pharmaceutical Sciences 94: 465-472.
9. Ganesan V, Rosentrater KA, Muthukumarappan K (2008) Flowability and handling characteristics of bulk solids and powders - a review with implication for DDGS. Biosystems Engineering 101: 425-435.
10. Jiang Y, Matsusaka S, Masuda H, Qian Y (2009) Development of measurement system for powder flowability based on vibrating capillary method. Powder Technology 188: 242-247.
11. Mullarney PM, Leyva N (2009) Modeling Pharmaceutical powder-flow performance using particle-size distribution. Pharmaceutical Technology 33: 126-134.
12. Carson WJ, Wilms H (2006) Development of international standard for shear testing. Powder Technology 167: 1-9.
13. Heckel RW (1961) Density-pressure relationships in powder compaction. Trans Metall Soc AIME 221: 671-675.
14. Krumme M, Schwabe L, Frömming K (2000) Development of computerised procedures for the characterisation of the tableting properties with eccentric machines: extended Heckel analysis. Eur J Pharm Biopharm 49: 275-286.
15. Sebhatu T, Alderborn G (1999) Relationships between the effective interparticulate contact area and the tensile strength of tablets of amorphous and crystalline lactose of varying particle size. Eur J Pharm Sci 8: 235-242.
16. Sun C, Grant DJ (2001) Effects of initial particle size on the tableting properties of L-lysine monohydrochloride dihydrate powder. Int J Pharm 215: 221-228.
17. Yusof YA, Smith CA, Briscoe BJ (2010) Uniaxial Die Compaction of food powders. Journal of the Institution of Engineers Malaysia 70: 41-48.
18. Li Q, Rudolph V, Weigl B, Earl A (2004) Interparticle van der Waals force in powder flowability and compactibility. Int J Pharm 280: 77-93.

The Degradation Mechanisms Study of the Organoclay by Using the Nitrogen Absorption/Desorption and X-ray Diffraction

Yuan-Cheng Cao*, Wenjun Wei and Jiyan Liu

Key Laboratory of Optoelectronic Chemical Materials and Devices of Ministry of Education, Jianghan University, Wuhan 430056, China

Abstract

Surfactants modified organoclays are known to undergo degradation during the melt process. In this work nitrogen absorption/desorption and X-ray diffraction were applied to study the mechanic of the degradation by employing the thermal treated commercial organoclay 93A. The 93A powder samples were heated in the Muffle furnace for 2 minutes at temperature ranged from 200 Celsius from 600 Celsius (200°C, 250°C, 300°C, 350°C and 600°C). The N2 absorption/desorption results showed the BET surface area is 10.402 m2/g, 9.455 m2/g, 14.169 m2/g, 8.860 m2/g and 9.198 m2/g respectively. The XRD results showed the d-space [001] 2θ=3.8 for all the samples (200°C -350°C) which did not change compared with the original 93A while the 600°C this unique reflection disappeared. The results indicate that the silicate clay skeletons thermal resistant ability can protect the surfactant molecules from degradation in the test temperature range low 350°C and the surfactant (which intercalated in the galleries) at the edge part of the galleries is first to degrade. High heating temperature can accelerates the vapour of the surfactants in the degradation process.

Keywords: Organoclay; Degradation; Surface area; Clay porosity; X-Ray diffraction

Introduction

The exploration of layered materials such as clay related polymer nanocomposites are one of the most attractive research fields for the novel materials designers which merged with the development of nanoscience and nanotechnology during the recent decades [1-5]. And the organoclay based polymer fillers, which composed of sodium montmorillonite (MMT) cationically exchanged with alkyl ammonium surfactants which expand the interlayer distance of the clay, have attracted considerable interest in many properties, such as mechanic properties, gas barrier properties, fire retardancy, conductivity, thermal stability, and degradability [1-14]. And with the incorporation of nanofillers/nano-reinforcements into polymer matrices, the clay based nanofillers are considered to be the most promising area which could lead to a significant enhancement compared to conventional composites [1]. These improvements can be realized without significantly increasing the density of the polymer or changing its optical properties at very low loading level. All these advantages make this kind of materials loads of interesting prospects for a wide range of applications in microbiology, nanotechnology, electronics, food packaging and so on [2-4].

Generally, the clay surface is hydrophilic and the surfactants such as tertiary or quaternary alkyl ammonium salts are bipolar molecules with hydrophobic tails [1,3,7]. For the clay nanocomposites, the significant property enhancements are rely on the affinity which to some extent is determined by the polarity of the polymer matrix and the type of organic modifier. And the affinity of the polymer with the surface of the clay and organoclay is essential for the favorable interactions within these materials. And the intercalated surfactant in the organoclay galleries is the bridge for the affinity of the clay layers and the polymer matrix and in some cases it also provides the resultant materials additional property such as antibacterial ability [7]. And also research reports that the compatibility of the polymer matrix and the organoclays can be greatly improved by the modifying the surfactant alky tails (branched or chain) which can provide more or less coverage during the exfoliation [2-4,12]. Thus, to obtain high levels of exfoliation is the key approach to achieve the high aspect ratio and surface area which is essential for the affinity. Therefore, the dispersion, or exfoliation of the individual organoclay layers within the polymer matrix is very important in the significant property enhancements.

Even the raw clay such as montmorillonite is not thermally stable at high temperatures, the organoclays filled with the surfactants may undergo degradation reactions at temperatures below the typical melt processing temperatures for the matrix polymers [3-6]. The instability of the surfactants in the clay galleries might be the first to degradation during the melt exfoliation process which will be resulted in the failure of the effective affinity function for each individual layer in the polymer matrix [11-13]. And the lack of thermally stable organoclays is one of the difficulties in the clay/polymer nanocomposites development. Thus, organoclay degradation becomes an issue that must be considered when forming polymer nanocomposites by melt processing. Although there are loads of the reports about the organoclay degradations during the melt process [8-21], the details about the mechanic of degradation are still unclear which is of prime importance to the understanding of organoclay degradation.

Thus, this study is focused on the thermal process effects on the organoclays during the melt process by monitoring the external surface areas and the interlayer spaces. These results provide the insights into the understanding of the organoclay degradation and the direction for stable organoclay designing and also provide the clue to improve the quality of organoclay based polymer nanocomposites in the thermal process.

***Corresponding author:** Yuan-Cheng Cao, Key Laboratory of Optoelectronic Chemical Materials and Devices of Ministry of Education, Jianghan University, Wuhan 430056, China

Experimental

Materials

The 93A organoclay (Organic Modifier: methyl dihydrogenated tallow ammonium; X-Ray Diffraction d-Spacing (001)=2.36 nm) was purchased form Southern Clay Products (USA). All other chemicals used in this experiment were used as received and the water used in this work was distilled water.

Samples thermal degradation

The organoclay degradation process was carried out by heating the 93A in the Muffle furnace for 2 minutes at different temperature ranged from 200 Celsius from 600 Celsius (200°C, 250°C, 300°C, 350°C and 600°C). And during the experiment the sample weight was recorded before and after the heating at each temperature.

Absorption/Desorption

The surface area, isotherm, and pore size distribution were measured by nitrogen adsorption and desorption at liquid nitrogen temperature (77 K) on a Quantachrome Autoabsorb. The tubes used in this experiment were 9 mm and 12 mm in diameter which were completely cleaned before each measurement using acetone and water in turn and dried in the oven at 70°C over-night. Prior to the determination of the adsorption isotherm, the samples (approximately 0.45 g) were outgassed for 12 h at 25°C. The samples are in powder format and load as they are. The compressed sample was compressed by the FT-IR sample compressor.

X-Ray diffraction

All the samples for XRD were determined by powder X-ray diffraction (XRD) analyses with X-ray diffractometer (Cu Kα, λ=1.5418Å) and generated at 30 kV, 30 mA. Scattering angles were ranged from 1° to 50°.

Results and Discussion

Melt process is considered the most efficient and flexible way for the polymer/organoclay nanocomposites fabrication and the organoclay dispersion i.e. exfoliation is determined by several aspects such as the extruder and screw configurations and residence times during the process. The shortcoming of this process is the degradation of the organoclay which is mainly affected by the process temperature. The melt temperature for the polymer matrix can be as high as several hundred Celsius while the tolerant temperature for the organoclay surfactants normally much lower than that value. And thus the degradation happens. Though the whole degradation is much complicated in the real exfoliation melt process, here we used the commercial organoclay 93A as an example to simplify this process by heating the organoclay in the furnace at different temperature to module the melt exfoliation process and then characterise the surface area and gallery distance to see the mechanic of clay degradation. The 93A samples that heated under different temperatures for 2 minutes is showed in Figure 1 and from the picture we can see that organoclay colour changed from white to grey from the range of room temperature 25°C to 350°C (Figures 1a-1f) and the colour become darker when the temperature increase. The 93A sample treated with 600°C for 2 minutes became black (Figure 1g). And this colour change indicates the degradation occurred during this heating process.

Weight difference of the samples during this thermal heating process was also recorded and the result is showed in Figure 2. From the curve we can clearly see that the sample resident weight remained from 96% to 70% when heated from 200°C to 350°C and at the temperature 600°C only 61% left. And these results indicate that during this thermal degradation process, the surfactant in the organoclay was released which contributed to the weight loss of the clay samples. Though in the test condition, the situations that the organoclay undergo are different from the real melt extrusion where the organoclay is in the melt polymer and lack of oxygen environment, the thermal treatments are similar as far as the effects on the organoclay are concerned.

As we know 93A organoclay is made from the cations exchange and the inter-gallery surfactant is quaternary alkyl ammonium salts and the microstructure of the clay is showed in Figure 3a and we hypothesize there are four possibilities for the degradation of organoclay 93A which is showed in the Figure 3b (1-4). In the first situation (Figure 3b-1), the clay platelet breaks down into smaller ones while the gallery distance keeps the same. This process is associated with weight loss and surface area increase significantly; in the second situation (Figure 3b-2), the clay platelets change the arrangements and the inter platelet distances and angels changes, and there is slightly changes in surface area and the no change in gallery distance; in the third situation (Figure 3b-3), the platelet half break and the damaged layers form inter-platelet pores and this associated with the weight loss and surface area increase, while gallery distance keep the same; in the forth situation (Figure 3b-4), the platelets break down into smaller ones and the deformation occur at the edge part of the layers which is associated with weight loss. In order to further confirm these possibilities, the surface area and pore size distribution were calculated through nitrogen absorption/desorption and the gallery distances (d-space, [001]) were monitored by X-ray diffraction.

Nitrogen absorption/desorption tests were carried out for the above mentioned samples. And isotherm curves showed that all the samples have low BET surface areas which mean that the samples have the similar low microspore surface area (Table 1) but have large portion of macrospore and mesopore characteristic. And the results show these samples don't have too much different in the BET surface area which range from lower 8.860 to higher 14.169 m^2/g, as the N_2 cannot get into the galleries of the clay [15-18], this surface area is mainly contributed by the external platelet macrospores/mesopores. These results indicate during the degradation process the clay did not break down into smaller platelets.

In order to further characterise the heat treated samples of 93A, X-ray diffraction experiments were carried out and results are showed in Figure 4. From results we can see that the [001] reflection appeared at 2θ=3.8 for all the samples (200°C-350°C) and did not change in all these samples compared with the original 93A at 25°C (Figure 4), but for the sample treated at 600°C, the [001] reflection disappeared. All other peaks kept the same in all these samples include the 600°C treated. These results showed that during the degradation process, the gallery distance does not change when the heat temperature is lower than 350°C (for 2 minutes). However, if the temperature increased to 600°C, the organoclay is no longer in stacking order which characterises layered platelets and this structure is suppressed and replace by the edge to edge or edge to face assemblies (no 001 reflection). These results indicate that upon the thermal treatment, the interlayered surfactant is transformed into oxide pillars which stabilize the gallery spacing between the adjacent clay layers [16].

SEM observation in Figure 5 showed that the clay sample at room temperature has the typical flask layered structure (Figure 5A) but when the samples treated with high temperature from 250°C, the

Figure 1: The images of 93A organoclay heated at different temperature 25°C (a); 150°C (b); 200°C (c); 250°C (d); 300°C (e); 350°C (f) and 600°C (g).

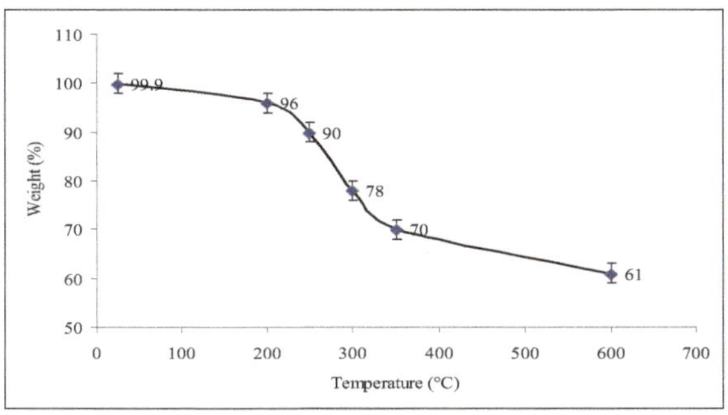

Figure 2: The sample resident weight curve in the thermal heating process.

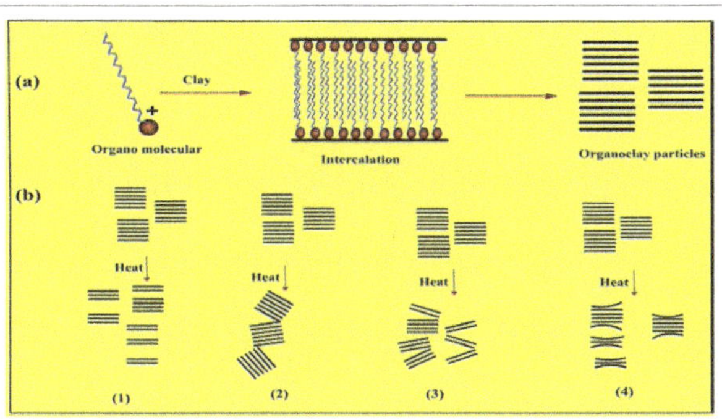

Figure 3: The illustration of organoclay microstructure (a) and the possibilities of the degradation process (b)

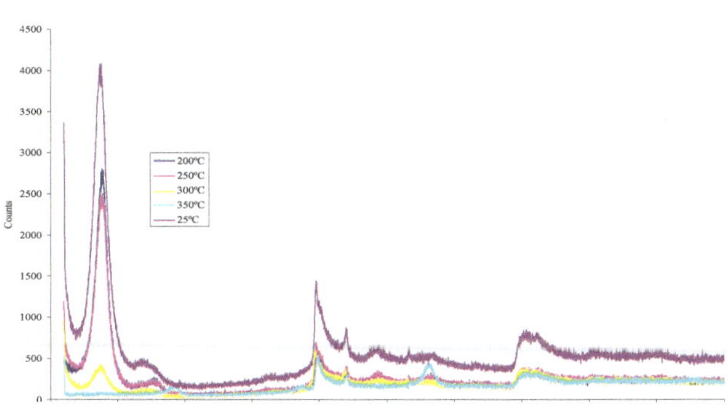

Figure 3: XRD spectra of the 93A organoclay 25°C (a); 200°C (b); 250°C (c); 300°C (d); 350°C (e) and 600°C (f).

structure began to change and the crashed surface morphology was observed (Figure 5B); and when the temperature continue to increase to 350°C and 600°C (Figures 5C and 5D), the clay particles completely lose their layered structure. SEM morphology showed that during the thermal degradation, the organic molecular inside the galleries is first undergoing the degradation and the surfactant molecules are the key issue when the orgnoclays are treated at high temperatures.

From the above analysis, it is clearly indicate that in these 4 hypothesis, only the second and the forth situation are possible. And one reasonable explanation for the degradation during the thermal treatment process is that at the relatively lower temperature, due to the silicate clay skeletons thermal resistant ability, the surfactant (which intercalated in the galleries) at the edge part of the galleries is first to degrade while the far deep into the galleries surfactant molecules can still hold in the position and this process is showed in the Figure 6. To this point, the higher heating temperature just accelerates the vapour of the surfactants in the degradation process.

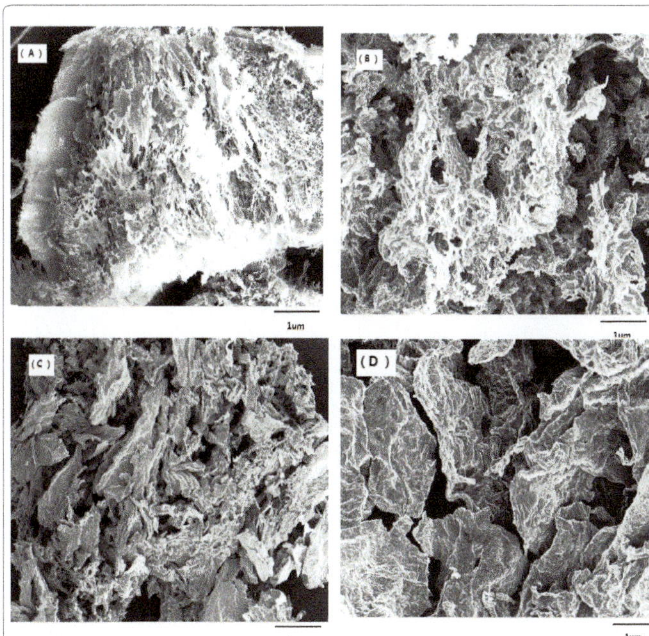

Figure 5: Typical morphology of clay samples in SEM. (A) 25°C; (B) 250°C; (C) 350°C; (D) 600°C.

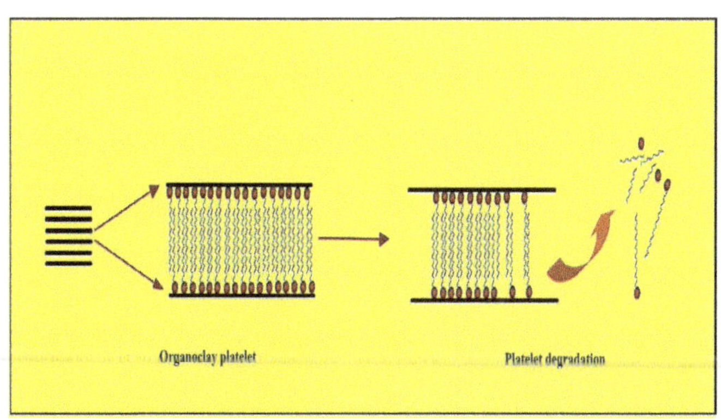

Figure 6: The illustration of organoclay platelet degradation process

Temperature (°C)	25	200	250	300	350	600
S_{BET} (m²/g)	10.402	9.455	14.169	8.860	9.198	

Table 1: The BET surface area (S_{BET}) of the samples at different temperature

Conclusions

The organoclay degradation is not always associated with the surface area and interlayer distance changes. The nitrogen absorption/desorption surface area is mainly contributed by the inter-particle pores. The surface area and the gallery distance do not change during the degradation process. And the degradation process starts at the edge part of galleries in the organoclay platelets which is associated with the weight loss. This study might be useful for the guide of designing the target organoclay for the different purposes. And also it might be useful for the reference of the thermal based polymer process regarding the degradations.

Acknowledgement

The Project was sponsored by the Scientific Research Initial funding for the advanced talent of Janghan University, Hubei Province Innovative Young Research Team in Universities (T201318) and Hubei Provincial Key Natural Science Foundations (2014CFA098).

1. **References**

2. Khudyakov IV, Zopf RD, Turro NJ (2009) Polyurethane Nanocomposites. Designed Monomers and Polymers 12: 279-290.
3. Ray SS, Okamoto M (2003) Polymer/layered silicate nanocomposites: a review from preparation to processing. Progress in Polymer Science 28: 1539-1641.
4. Pavlidou S, Papaspyrides CD (2008) A review on polymer–layered silicate nanocomposites. Progress in Polymer Science 33: 1119-1198.
5. LeBaron PC, Wang Z, Pinnavaia TJ (1999) Polymer-layered silicate nanocomposites: an overview. Applied Clay Science 15: 11-29.
6. Paul DR, Robeson LM (2008) Polymer nanotechnology: nanocomposites. Polymer 49: 3187-3204.
7. Luo JJ, Daniel IM (2003) Characterization and modeling of mechanical behavior of polymer/clay nanocomposites. Composites Science and Technology 63: S1607-S1616.
8. Nigmatullin R, Gao F, Konovalova V (2008) Polymer-layered silicate nanocomposites in the design of antimicrobial materials. Journal of Materials Science 43: 5728-5733.
9. Cui L, Khramov DM, Bielawski CW, Hunter DL, Yoon PJ, et al. (2008) Effect of organoclay purity and degradation on nanocomposite performance, Part 1: Surfactant degradation. Polymer 49: 3751-3761.
10. Ratinac KR, Gilbert RG, Ye L, Jones AS, Ringer SP (2006) The effects of processing and organoclay properties on the structure of poly(methyl methacrylate)/clay nanocomposites. Polymer 47: 6337-6361.
11. Ryu JG, Park SW, Kim H, Lee JW (2004) Power ultrasound effects for in situ compatibilization of polymer–clay nanocomposites. Materials Science and Engineering: C 24: 285-288.
12. Zampori L, Stampino PG, Dotelli G (2009) Adsorption of nitrobenzene and orthochlorophenol on dimethyl ditallowyl montmorillonite: A microstructural and thermodynamic study. Applied Clay Science 42: 605-610.
13. Chavarria F, Paul DR (2006) Morphology and properties of thermoplastic polyurethane nanocomposites: Effect of organoclay structure. Polymer 47: 7760-7773.
14. Chow WS, Abu Bakar A, Mohd Ishak ZA, Karger-Kocsis J, Ishiaku US (2005) Effect of maleic anhydride-grafted ethylene–propylene rubber on the mechanical, rheological and morphological properties of organoclay reinforced polyamide 6/polypropylene nanocomposites. European Polymer Journal 41: 687-696.
15. Xu B, Zheng Q, Song Y, Shangguan Y (2006) Calculating barrier properties of polymer/clay nanocomposites: Effects of clay layers. Polymer 47: 2904-2910.
16. Nguetnkam JP, Kamgac R, Villiéras F, Ekodeck GE, Razafitianamaharavo A, et al. (2005) Assessment of the surface areas of silica and clay in acid-leached clay materials using concepts of adsorption on heterogeneous surfaces. Journal of Colloid and Interface Science 289: 104-115.
17. Remy MJ, Vieira Coelho AC, Poncelet G (1996) Surface area and microporosity of 1.8 nm pillared clays from the nitrogen adsorption isotherm. Microporous Materials 7: 287-297.
18. Lantenois S, Nedellec Y, Prélot B, Zajac J, Muller F, et al. (2007) Thermodynamic assessment of the variation of the surface areas of two synthetic swelling clays during adsorption of water. Journal of Colloid and Interface Science 316: 1003-1011.
19. Koo J, Park S, Satija S, Tikhonov A, Sokolov JC, et al. (2008) Characterization of Langmuir–Blodgett organoclay films using X-ray reflectivity and atomic force microscopy. Journal of Colloid and Interface Science 318: 103-109.
20. Mohino F, Martin AB, Salerno P, Bahamonde A, Mendioroz S (2005) High surface area monoliths based on pillared clay materials as carriers for catalytic processes. Applied Clay Science 29: 125-136.
21. Kawi S, Yao YZ (1999) Silica bonded K10 montmorillonite (SBM): a high surface area catalytic clay material. Microporous and Mesoporous Materials 28: 25-34.
22. Kawi S (1999) A high-surface-area silica–clay composite material. Materials Letters 38: 351-355.

Conversion of Langbeinite and Kieserite in Schoenite With Mirabilite and Sylvite in Water and Schoenite Solution

Mariia Artus[1]* and Ivan Kostiv[2]

[1]*Precarpathian National University named after V. Stephanyk 57, Shevchenko str., 76025 Ivano-Frankivsk, Ukraine*
[2]*State Enterprise Scientific-Research Gallurgi Institute 5a, Fabrychna str., 77300 Kalush, Ukraine*

Abstract

The hydration of soluble Langbeinite and kieserite was investigated in the presence of potassium chloride and sodium sulfate. The addition of sodium sulfate affects water activity less than in the case of clean water. Therefore, we have conducted a research by adding water instead of schoenite solution. According to the results, hydration of Langbeinite by adding sodium sulfate in the first 240 hours occurs more intensively than without sodium sulfate with formation of a readily soluble schoenite and halite.

Keywords: Langbeinite; Kieserite; Conversion ; Sodium sulfate; Schoenite solution

Introduction

The polymineral potassium-magnesium ores contain both readily soluble minerals: kainite ($KCl{\cdot}MgSO_4{\cdot}3H_2O$), sylvite (KCl), halite (NaCl) as well as hardly soluble: langbeinite ($K_2SO_4{\cdot}2MgSO_4$), polyhalite ($K_2SO_4{\cdot}MgSO_4{\cdot}2CaSO_4{\cdot}2H_2O$), kieserite ($MgSO_4{\cdot}H_2O$). During their hot dissolution the readily soluble potassium minerals dissolve almost completely, and hardly soluble - only partially. So, during ore processing of Carpathians (Ukraine) the langbeinite dissolves only by 22-23% [1], kieserite - still less, and polyhalite is practically insoluble. Low degree of dissolution of sulfate mineral - langbeinite leads to saturation of the solution at the stage of ore dissolution by chloride salts. The mixture of chloride salts and schoenite crystallizes from the obtained solution at vacuum-crystallization, therefore, we received a low-quality potassium magnesium containing admixtures of chloride to 25% and an extract of potassium from the ore in potassium magnesium was no more than 50-60%.To increase the degree of dissolution of langbeinite from polymineral potassium-magnesium ore, the technology of complete dissolution has been developed [2], which consisted in additional dissolution of halite from solid residue with water with obtaining the halite solution as a raw material for the production of table salt, the next hot langbeinite dissolution, obtaining langbeinite solution and residue of polyhalite, by drying of which the low-grade potassium-magnesium sulfate fertilizer is obtained. However, the specified technology was complex, multistage, energy intensive and required significant capital costs. The technology of processing [3] of polymineral potassium-magnesium ore from previous hydration of hardly soluble langbeinite in the readily soluble schoenite has been developed by the reaction:

$$K_2SO_4{\cdot}2MgSO_4 + 13H_2O = K_2SO_4{\cdot}MgSO_4{\cdot}6H_2O + MgSO_4{\cdot}7H_2O \quad (1)$$

As a result of the processing of the hydrated ore of Carpathian, extract of potassium from ore in potassium-magnesium increased to 82%. The technology of processing of polymineral potash ores from the previous conversion of hardly soluble langbeinite with reversible carnallite in readily soluble kainite [4] has been developed by the reaction equation:

$$K_2SO_4{\cdot}2MgSO_4 + KCl{\cdot}MgCl_2{\cdot}6H_2O + nH_2O = 3[KCl{\cdot}MgSO_4{\cdot}3H_2O] + (n-3)H_2O \quad (2)$$

It is also known the technology that includes the dissolution of readily soluble minerals of polymineral ore, including halite, by means of water or dilute solutions with subsequent separation of the undissolved residue of hardly soluble langbeinite, kieserite and polyhalite, its drying with obtaining potassium-magnesium concentrate, containing K_2O - 20,4%, MgO – 16,0% and Cl - less than 1%. The solution after the stage of dissolution of ore is illuminated, evaporated, separated at first sodium chloride and then schoenite [5].

During processing of kieserite hartzalts of Germany, containing 17-28% of kieserite, crushed ore is subjected to hot dissolution with the subsequent crystallization from the received solution of potassium chloride. Insoluble halite-kieserite residue is filtered and washed from halite. Kieserite residue is filtered and given on the more hotly dissolution by reversible epsomite solution with obtaining a concentrated solution of magnesium sulfate. This solution is cooled for crystallization of epsomite. The latter is filtered and given on the conversion of potassium chloride to receive potassium sulfate. The concentration of magnesium chloride increases in the process of conversion in reverse solutions, therefore, the part of the solution is evaporated with crystallization of kainite, and excess magnesium chloride solution is discarded [6].

Langbeinite ores of Carlsbad field (USA) [7], contain not only langbeinite, but also halite and sylvite. The breed, crushed to a size of 3.3 mm, is classified according to the class of 1.6 mm. The large class is washed with cold water, enriched on the hydrocyclones, filtered on centrifuges and dried. The fraction of 1.6 mm is washed and dried. The dried concentrate, containing 96-98% of langbeinite, is classified with obtaining a product that contains 22% of K_2O, 18% of MgO and no more than 2.5% of chlorine. Small fraction of dried langbeinite is used for the conversion of chloride potassium and by receiving potassium sulphate. The excess schoenite solution, received in the conversion, is evaporated to highlight potash salts, and the concentrated magnesium chloride solution is discarded. In all technologies through conversion of potassium chloride in schoenite, the excess magnesium chloride solution has been formed that should be evaporated to highlight salts

***Corresponding author:** Mariia Artus, Precarpathian National University named after V. Stephanyk 57, Shevchenko str., 76025 Ivano-Frankivsk, Ukraine

of potassium, magnesium, sodium and sulfates that considerably increased capital costs. The final solution of chloride magnesium with a concentration of 25-32% had limited application and it was discarded.Conversion of hardly soluble and readily soluble minerals of polymineral potassium-magnesium ore with natural mirabilite is accompanied by formation of readily soluble schoenite and sodium of chloride without the formation of double salts with sodium sulfate by reaction equations:

Sylvite:

$2KCl + Na_2SO_4{\cdot}10H_2O + MgSO_{4(p.)} = K_2SO_4{\cdot}MgSO_4{\cdot}6H_2O + 2NaCl + 4H_2O$ (3)

Kainite:

$2[KCl{\cdot}MgSO_4{\cdot}3H_2O] + Na_2SO_4{\cdot}10H_2O = K_2SO_4{\cdot}MgSO_4{\cdot}6H_2O + NaCl + MgSO_4{\cdot}7H_2O + 3H_2O$ (4)

Carnallite:

$2KCl{\cdot}MgCl_2{\cdot}6H_2O + 2Na_2SO_4{\cdot}10H_2O = K_2SO_4{\cdot}MgSO_4{\cdot}6H_2O + 4NaCl + 10H_2O$ (5)

Hydration of hardly soluble langbeinite and kieserite in the presence of potassium chloride and sodium sulfate occurs by the reaction equation:

$K_2SO_4{\cdot}2MgSO_4 + Na_2SO_4{\cdot}10H_2O + 2KCl + 2H_2O = 2[K_2SO_4{\cdot}MgSO_4{\cdot}6H_2O] + 2NaCl$ (6)

$MgSO_4{\cdot}H_2O + 2KCl + Na_2SO_4 + 5H_2O = K_2SO_4{\cdot}MgSO_4{\cdot}6H_2O + 2NaCl$ (7)

Addition of mirabilite in the process makes it possible to regulate the ratio between $MgSO_4$ and $MgC1_2$ in the dissolving medium and set its optimum value at the stage of dissolution of convertible ore.

Experimental section

The monomineralic samples of natural langbeinite of the Dombrowskyi career of Kalush-Golyn fields of potassium salts were selected for the laboratory researched. Samples were crushed to size of 5 mm, the fraction of 3-5 mm was sifted on wire sieve, it was washed out from impurities with cold tap water, residual surface moisture was removed by initially 50% -, then 98% alcohol and dried at temperature of 50°C.

The chemical composition of the washed out and dried mineral, mas. %: K^+ - 18,31, Mg^{2+}- 11,35, Ca^{2+}- 0,11, Na^+ - 0,49, Cl^- - 0,66, SO_4^{2-} -67,74, H_2O - 1,35. In terms of mineral composition it included, mas. %: langbeinite - 97,79; halite - 1,17; polyhalite - 1,13.

Kieserite was taken from samples of core material, received from exploratory drilling on the areas of occurrence of zechstein breeds. Core material was crushed to the size of less than 5 mm and washed from impurities of readily soluble minerals by means of limited amount of cold water, the residual moisture was removed by initially 50% then 98% ethanol. The chemical composition of the washed out and dried mineral, mas. %: K^+ - 4,37, Mg^{2+}- 6,08, Na^+ - 21,45, Cl^- - 39,49, SO_4^{2-} - 20,86, H_2O - 7,70. In terms of mineral composition it included, mas. %: kieserite - 30,09 halite - 54,71; karnalite - 9,32; sylvite - 5,89.

1020.4 g of washed Langbeinite, 352.2 g of waterless sodium sulfate, 369.7 g of potassium chloride and 267.8 g of water were taken for laboratory research. The amount of water was 50% from the stoichiometric on equation 6 of a condition of formation of the wet mixture. The mixture without sodium sulfate was also prepared. Mixures were thoroughly stirred, filled up in tight polyethylene packs and placed in desiccator in an air thermostat, installed at the temperature of experiment. The experiments on hydration of Langbeinite with addition of waterless sodium sulfate and without it were conducted separately, a predetermined amount of water in which was added as a schoenite solution of structure, mas. %: K^+ - 2,50, Mg^{2+}- 3,19, Na^+ - 3,91 , Cl^- -14,19, SO_4^{2-} - 4,62, H_2O - 71,59.

The experiments with a sample of washed out kieserite were carried out separately. 360.4 g of washed out natural kieserite, 287.9 g of potassium chloride, 369.9 g of sodium sulfate and 257.8 g of water were taken according to the reaction equation (7). Water was taken 100% of stoichiometry, on the equation of reaction (7), while the mixture remained damp and solution from it did not drain.The researches were conducted at temperature of 40 °C, based on the exothermicity of the reactions of conversion and data given earlier [8].

The samples were taken in 1, 2, 3, 5, 7, 10, 15 and 20 days. For this purpose the contents of polyethylene pack was thoroughly mixed and 100 g was withdrawn. Then the sample was placed into a double-walled temperature-controlled reactor equipped with a stirrer and filled by 400 ml of distilled water. The dilution was carried out at 20 ± 0.5°C. The solution samples were withdrawn every 10 and 15 min and filtered using a Buchner funnel fixed on the Bunsen flask into the test tube which was previously weighted. The contents of the test tube were analyzed using well-known methods: methodtetraphenylborate weight method (fo K^+ions), flame photometric method (for Na^+ions), complexometric method (for $Mg^{2+}$and Ca^{2+} ions), mercurimetry (for Cl^-ions) and weight method (for SO_4^{2-} ions). The blank experiment was also carried out with the initial mixture before the reaction. The conversion degree of langbeinite was calculated using the results of liquid phases chemical analysis and Eq [9].

$$\alpha = \frac{(m_l \cdot C_{Mg^{2+}}) - (m_{l(b)} \cdot C_{Mg^{2+}(b)})}{m_{La} \cdot C_{Mg^{2+}La}} \cdot 100\% \quad (8)$$

where m_l - weight of liquid phase after dilution, g; $m_{l(b)}$ - weight of liquid phase after dilution in blank experiment, g; $C_{Mg^{2+}}$ - cncentration of Mg^{2+} ions in the sample, %; $C_{Mg^{2+}(b)}$ - concentration of Mg^{2+} ions in the sample of the blank experiment, %; m_{La} - weight of langbeinite in 100 g of the mixture withdrawn for the experiment, g; $C_{Mg^{2+}La}$ - content of Mg^{2+} ions ion the natural langbeinite, %.

The weight of liquid phase after dilution was calculated according to the formula:

$$m_1 = \frac{400 + m_{H_2omix}}{C_{H_2ol}} \quad (9)$$

where m_{H_2omix} - weight of water in 100 g of the mixture withdrawn for dilution, g; C_{H_2ol} - concentration of water in the sample after dilution, %.

The experimental results are represented in (Figure 1).

Results and Discussion

As it is seen from the obtained data, Langbeinite hydration by water occurs intensively in the absence of sodium sulphate (curve 1). During the first 168 hours the degree of conversion increases to 39.4% and further practically does not change. In the presence of sodium sulfate the Langbeinite hydration slows down and for the same period makes 25.3%. The slower process of hydration in the presence of Langbeinite waterless sodium sulfate can be explained by decreased activity of water in solution, which in addition to potassium and magnesium sulfates has sodium ions. In addition, sodium sulfate in the mixture is in crystalline form, therefore, the part of water is spent fot its crystalline hydrate formation. So, for the Langbeinite hydration there is a smaller amount of free water.

Langbeinite is in mix with readily soluble minerals in polymineral ore. Therefore, the intercrystalline saline solution is formed, when it is mixed with water, and Langbeinite hydration occurs in saline solution. Addition of sodium sulfate affects the water activity less than in the case of pure water. Therefore, the research was conducted by adding the schoenite solution instead of water. According to the results, the hydration of Langbeinite by adding sodium sulfate during the first 240 h proceeds more intensively (curve 4) than without sodium sulfate (curve 3). This can be explained by the positive influence of sodium chloride on the Langbeinite dissolution process that is known from the literature [10,11]. Reacting mixture firmly graspes and becomes dry after 10 days. Moisture does not remain to continue hydration reaction. The mixture of sodium sulfate remains damp and the reaction continues throughout the research period.

Hydration of kieserite (Figure 1, curve 5) proceeds during the first 240 h and its degree reaches 73.0%. Reacting mixture hardens and becomes almost dry. Water, which had to be spent for the course of the reaction (7) contacted in crystalline sodium sulfate and due to lack of reaction medium the part of kieserite remained not reacted. After this its value remains practically constant for 480 h. So, we can expect that the hardly soluble kieserite of polymineral ore will convert to readily soluble shenity with a sufficient amount of water. We have executed X-ray analysis of the products of Langbeinite hydration with sodium sulfate in the schoenite solution medium. Radiograph is shown in (Figure 2), which shows that the Langbeinite hydration products are leonite ($K_2SO_4{\cdot}MgSO_4{\cdot}4H_2O$) and halite (NaCl) [12,13].

So, polymineral ores that contain hardly soluble minerals of Langbeinite and kieserite in the process of mixing with sodium sulfate in the presence of sylvite and saline solution converte into the readily soluble schoenite. The readily soluble minerals such as sylvite, kainite and carnallite also convert in these circumstances into schoenite. Convertible ore, which consists of conversion products - readily soluble schoenite and halite - are easily processed into schoenite and potassium sulphate by halurhiynym method. This makes it possible to increase the extraction of potassium and magnesium in targeted products at

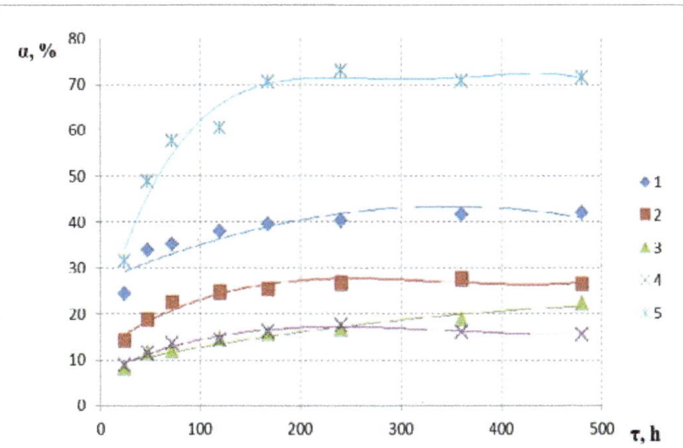

Figure 1: The dependence of the degree of hydration (α) of langbeinite by water (1) from the time (τ), water in the presence of sodium sulfate (2), langbeinite with sodium sulfate (4) and without it (3), kieserite hydration with sodium sulfate in scheonite solution (5).

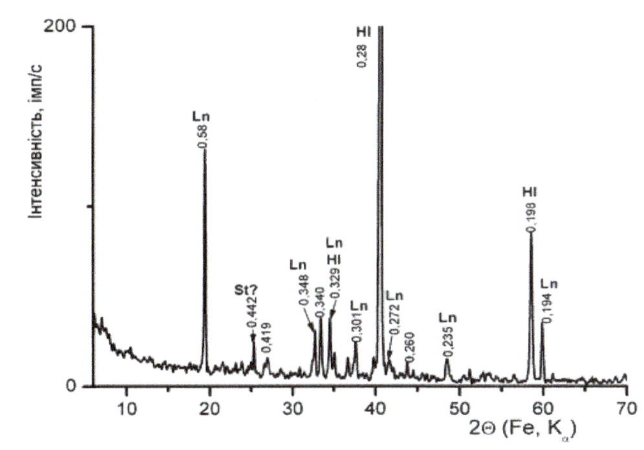

Figure 2: Radiograph of Langbeinite product hydration with sodium sulfate in schoenite solution medium.

reducing losses in comparison with known methods and to simplify processing technology of polymineral potassium-magnesium ore in magnesium and potassium sulphate.

References

1. DV Hrebeniuk, IM Okrepky (1975) Refinement of the degree of langbeinite dissolution in polymineral ore processing at KHMK, Proc VNII 74: 75-78.
2. Akselrud GA (1978) Kinetic method of the separation of salt mixtures. The Journal of Applied Chemistry 10: 2213-2218.
3. Hrebeniuk DV (1995) Technological schemes of polymineral potash ore processing Carpathian. Chem Industry Ukraine 2: 41-45.
4. Kostiv IYu(2004) Investigation of the kinetics of the conversion process of natural langbeinite with carnallite .The Journal of Applied Chemistry.
5. Lothar R,Ulrich R,Martin M,Uwe H,Wolfgang F(1991) Second look at suspect terranes in southern Mexico.8: 1233-1236.
6. Karpeets M V(2008) Kinetics of process of conversion of a langbeinite in schoenite in saturated solution. The Journal of Applied Chemistry 5: 705-708.
7. Processing of natural salts and brines (1985) Reference book. Under the editorship of ID Sokolov L: Chemistry -208.
8. New Mexico Geological Society Guidebook, 44th Field Conference, Carlsbad Region, New Mexico and West Texas, 1993 economic Geology Of The Carlsbad Potash District, New Mexico James M. Barker And George S. Austin
9. Artus MI (2013) The reaction of conversion of natural Langbeinite surface with sodium sulfate and potassium chloride in schoenite. FHTT1: 136-136.
10. Artus MI(2014) Kinetics of Langbeinite Conversion into Schoenite in the Presence of Mirabilite, Sylvine and Water. Chemistry and Chemical Technology 4: 455 - 459
11. Kostiv IYu (2011) Polymineral potash ore of Carpathians. Processing with Langbeinite hydration and chloride potassium mineral conversion with mirabilite in schoenite . Chemical Industry of Ukraine 5: 17-25.
12. Artus MI (2014) Chlorine-free schoenite processing of polymineral potash ore by conversion of potassium chloride and potassium-magnesium minerals with sodium sulfate. Chemical Industry of Ukraine 2: 53-57.
13. Reference book on solubility of salt systems (1975) T.2. L.: Chemistry 1064.

Application of the Thermodynamic Model to the Couple Water vapor-Raw Clay

Jhider N* and Bagane M

Applied Thermodynamics and Environment National engineering Research Unit, School of Gabes, Gabes, Tunisia

Abstract

This work is based on data from adsorption isotherms and isosteric enthalpies of El Hamma raw clay at different temperatures (30, 40, 50 and 60°C). These previously obtained data were used here to show the variation of the thermodynamic properties of water adsorbed on the surface of the clay: Gibbs energy, enthalpy and entropy. Then we present a model of solid solution based on a polynomial expression of degree 7. This model was applied to calculate the excess quantities: excess Gibbs energy of mixing, entropy term and Gibbs energy of mixing. The variations of free energy, enthalpy and entropy of adsorbed water show fluctuations that characterize three areas. For low water contents the entropy of the adsorbed water is greater than that of liquid water. As the amount of adsorbed water increases the entropy gradually approaches the normal value. Close to saturation, the model shows that the term of excess energy of mixing is always positive, although the water remains bound to the solid. Since the term of enthalpy of mixing is negative this means that the solution is stable.

Keywords: Adsorption; Hydration; Thermodynamic properties; Solid solution

Introduction

The study of water transfers in soil has interested scientists and engineers in different fields. Indeed, the amount and condition of the water in the soil can significantly affect its agronomic, hydraulic and geological. If knowledge of the two extreme states of the soil, saturated or dry, seems well under way, an unsaturated state, particularly at low water contents, merits further investigation.

Application techniques of adsorption are found in many industrial processes. Systems using solid adsorbents such as activated carbons, activated aluminas and clays play an important role in the process of separation, purification, drying, dehumidification, air handling and water treatment. In the case of the solid adsorption, the mode of association between the molecules of water vapor and the adsorbent is purely physical. The adsorption does not change the volume of the solid, it means a variation of the adsorbed mass and physical and thermodynamic properties of couple's adsorbent/adsorbate.

El Hamma raw clay was chosen as the adsorbent material in this study. The object of this work was to investigate the equilibrium and thermodynamic parameters of the adsorption of water vapor on raw clay surface and determine the adsorption.

A thermodynamic model can be implemented using known thermodynamic properties of pure poles and writing that all the adsorbent properties are expressed by varying the activities of these poles. This work focuses on the thermodynamic modeling of defining a model with parameters that have a physical meaning that it is possible to determine from experimental data. For solid solutions the model used is that based on polynomial expression. This model takes into account the short range order in the solution via the interaction between first and second neighbors.

We propose in this study a model in terms of ideal solid solution between the surface layer of El Hamma raw clay and different amounts of water adsorbed.

Characterization of Natural Clay

The adsorbent used in this study is a raw clay material from Jebel Haidoudi. Its physical and chemical characteristics were studied by Srasra [1] and confirmed by Ben Zina [2].

The results of physicochemical analysis are listed in table 1. The chemical composition of El Hamma raw clay is obtained by using various techniques, as classical analysis, atomic emission spectroscopy (ICP) and X-ray diffraction, and categorized in table 2. Its specific area was found about 60 m^2/g using BET method.

The results of this analysis and the swelling index have shown that this raw clay is an illite-smectite system with the presence of iron, calcium and very little of kaolinite.

Formula	$[Al_{1,94}Fe_{0,7}Mg_{0,5}][Si_{7,5}Al_{0,5}]O_{20}(OH)_4Ca_{1,56}Na_{0,32}K_{0,154}$
Molecular weight	791,711 g
Density	2,75 g/cm^3
BET specific area (Nitrogen)	60 m^2/g
Methylene Blue specific area	520 m^2/g
Total pore volume	143,9.10^{-03} cm^3/g
Average particle diameter	140 μm
Elemental mass of grain	40.10^{-6} g

Table 1: Textural characterization of raw clay.

i	SiO_2	Al_2O_3	Fe_2O_3	CaO	MgO	K_2O	Na_2O	P_2O_5	L.I*
%	50,8	18,8	8,5	2,3	1,5	1,4	2,2	1,9	14,0

*L.I: loss on ignition.

Table 2: Chemical composition of raw clay.

***Corresponding author:** Jhider N, Applied Thermodynamics and Environment National engineering Research Unit, School of Gabes (ENIG) Omar Ibn Elkhatab street, Zrig 6029 Gabes, Tunisia

The swelling index values (between 35 to 75%) were determined by addition of 2 g of raw clay to 100 ml of distilled water. After 24 hours the volume of sediment was read. The results obtained were relative values in order to compare the behavior of different samples.

On The basis of 20 structural oxygen plus two OH, the structural formula of El Hamma raw clay was calculated

using Ross and Hendricks technique [3,4]: $[Al_{1,94}Fe_{0,7}Mg_{0,5}][Si_{7,5}Al_{0,5}]O_{20}(OH)_4Ca_{1,56}Na_{0,32}K_{0,154}$. The cation exchange capacity [1] of calculated material was found between 29 to 71 meq/100 g.

Materials and Methods

The simplest way to characterize the adsorption at a surface is to calculate the number of adsorbed molecules as a function of applied pressure. The adsorption isotherms displays information of how much material will adsorb for a given set of state variables (pressure P and temperature T when using gases as adsorptive, concentration c and temperature T when dealing with adsorption from solutions). The amount of adsorbed material can be given as surface coverage, or by the volume taken up by the sum of particles adsorbed in the first monolayer.

The experimental determination of the adsorption isotherm can be done by several methods: volumetric, gravimetric and chromatography. In our case the volumetric method is difficult to use, and the chromatographic method, as we have seen [5,6] allows only the part of the adsorption isotherm located at low relative pressures p/p^0. To accomplish this work, we opted for a dynamic gravimetric method, which is relatively slow but allows for complete isotherms.

Gravimetric adsorption is a method to evaluate directly the mass of adsorbate by making successive weighing. This technique is based on measuring changes in mass of the porous sample when it is contacted with the adorable gas. The first systems based on this principle have been developed by McBain and Bakr [7] and were equipped with silica springs at the end of which is placed a crucible containing the porous adsorbent.

Nowadays, the weighing are performed either on principles based on oscillators or from displacement measurement using electromagnet. Among the principles based on oscillators, we can cite the microbalance quartz crystal. It involves placing the adsorbent, usually in liquid form, onto a quartz crystal.

With the magnetic suspension balance [8-10] gravimetric measurements in a wide temperature and pressure range and under aggressive media can be done. Such a magnetic suspension balance consist of a balance for recording the measurement values, a suspension coupling which carries the weight of the sample, a sensor for the position measurement of the levitation part and a control unit to control the suspension coupling.

In this work, we have chosen a gravimetric system and we used an analytical balance (type Sartorius: 1409, sensitivity 10^{-4}). Our installation is composed of an air compressor, 5 thermostatic baths, a vacuum pump and a main tube through which derive four smaller diameter tubes each emerging into an adsorption chamber where the temperature is fixed as desired. This experimental device can lead four parallel adsorption experiments at four different temperatures (30, 40, 50 and 60 °C).

The main tube is covered with electric heating resistors and then wrapped with insulation in order to avoid heat loss and therefore the risk of condensation. The power of the heating resistors and the temperature of the water thermostatic baths are adjusted so that the temperature of the gas inside each adsorption chamber has the value set in advance (with precision of one degree).

Samples that have previously spent more than 48 hours in an oven at T=60°C are first cooled under vacuum for 24 hours and then weighed and finally introduced very quickly within the confines of adsorption. The two bubblers are used to generate the air saturated with water vapor at the temperature of the thermostatic bath. Empty columns are designed to hold any water droplets entrained by the air flow. Finally, the column filled with silica gel is used to remove air derived from the compressor from his residual moisture.

The amount of adsorbed water vapor is measured from the change in weight of the samples by means of an analytical balance. The balance is considered established when the change in mass of the solid is less than 1%.

Thermodynamic Properties of Adsorbed Water

To understand the adsorption properties of adsorbed water along the adsorptions isotherms into raw clay surface, we have to study the energy of the clay-water interface. So, we have used the thermodynamic approach which is considered the most widely used approach.

Thermodynamic data relating to adsorbed water reflect the evolutions of water-clay system during adsorption process. Thermodynamic properties are estimated using the experimental results obtained from adsorption isotherms and isosteric enthalpy, over a range of temperature.

The thermodynamic values which reflect the changes of the water system from the adsorbed state to the free liquid state are:

Gibbs differential of adsorption

Free energy is the total energy required to transfer water molecules from vapor state to a solid surface or vice-versa. The value of the Gibbs differential adsorption is obtained from the adsorption isotherm (Figure 1).

At each value of the water activity corresponds a value to the Gibbs

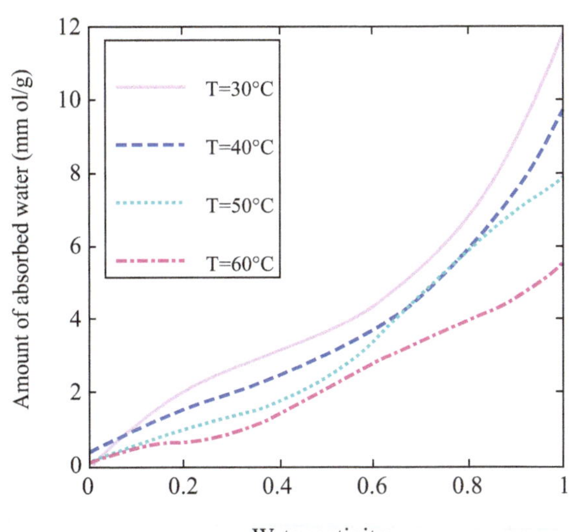

Figure 1: Sorption isotherms on raw clay at T=30°C, T=40°C, T=50°C and T=60°C.

energy of adsorption calculated for one mole of water corresponding to the energy of the adsorption reaction of dN moles of water to the surface clay in which are already adsorbed N moles of water.

The Gibbs energy (kJ/mmol) can be calculated as follows:

$$dG_R = dN \times RT \times \ln(\frac{p}{p_0}) \quad (1)$$

$$\Delta G_{R,N} = RT \times \ln(\frac{p}{p_0}) \quad (2)$$

Differential enthalpy of adsorption

The differential enthalpy of adsorption reaction is calculated from the curve of variation of the isosteric heat as a function of N recovery rate of raw clay. For a given value p/p^0, N moles of water are adsorbed on the raw clay. A small change in enthalpy of the hydration reaction, when measured dN moles of water added to the clay surface already covered by N moles of water is obtained as follows:

$$d\Delta H_{R,N} = -\Delta H_{is}(N + dN) + \Delta H_{is}(N) \quad (3)$$

For one mole of adsorbed water, we have:

$$\Delta H_{R,N} = -\frac{d(\Delta H_{is})}{dN} \quad (4)$$

Differential entropy of adsorption

Differential entropy of adsorption is determined by Tardy et al. [11] from the following equation:

$$\Delta S_{R,N} = \frac{1}{T}(\Delta H_{R,N} - \Delta G_{R,N}) \quad (5)$$

According to tardy et al. [11] the average free energy of all the water adsorbed to the stage p/p^0, is calculated as follows:

$$\Delta G^*_{R,N} = \frac{1}{N}\int_0^N RT \times \ln\left(\frac{p}{p^0}\right) dN \quad (6)$$

Gibbs adsorption average

To avoid errors in the evaluation of the mean energy of adsorption, Trolard et al. made from the integration of the first known experimental point by asking:

$$N' = N - N_1 \quad (7)$$

Where, N_1 represent the first amount of adsorbed water. Equation (7) becomes:

$$\Delta G^*_{R,N} = \frac{1}{N'}\int_{N_1}^N RT \times \ln\left(\frac{p}{p^0}\right) dN \quad (8)$$

Average enthalpy adsorption

The average enthalpy of adsorption is defined, while putting into consideration the assumption made in the preceding paragraph by the following expression:

$$\Delta H^*_{R,N} = \frac{1}{N'}\int_{N_1}^N \Delta H_{R,N} dN \quad (9)$$

Average entropy of adsorption

The average entropy of adsorption of one mole of water is the difference between the average free energy and enthalpy average, writes:

$$\Delta S^*_{R,N} = \frac{1}{T}(\Delta H^*_{R,N} - \Delta G^*_{R,N}) \quad (10)$$

Modeling and calculation of mixing energy

Solid solution model

Saxena et al. has introduced the concept of a solid solution, and the associated mathematical models to try to predict the behavior of these phases depending on environmental conditions. We can also see those of Ganguly and Saxena [12], Glynn and Reardon [13], Geiger [14].

Several models have been developed [15]. Most of these empirical models include for example: Quasi-chemical notions of order, solid solution ionic and molecular and models based on a polynomial expression of the excess enthalpy of mixing. The model on which our study is based is described below.

Models based on a polynomial expression of the excess enthalpy of mixing

The expression of the energy of mixing between i poles in a solid solution is decomposed into two terms: entropy function (TE) and excess enthalpy of mixing. The first of these terms results from the ability of the constituents of the solid solution to mix freely in this phase. It reduces, in fact, an entropic term type:

$$TE = RT\sum_i X_i \ln X_i \quad (11)$$

X_i represent the mole fraction of a macroscopic component i.

The difference between the Gibbs energy of a real solution and that of an ideal solution is known as the excess Gibbs energy of mixing of the solution. Thus, we have:

$$\Delta G_M^{ex} = RT\sum_i X_i \ln \gamma_i = RT(X_1 \ln \gamma_1 + X_2 \ln \gamma_2) \quad (12)$$

γi which represents the activity coefficient of the pure dipoles in the solid solution.

$$\Delta G_M = TE + \Delta G_M^{ex} \quad (13)$$

Depending on the model of solid solution can be expressed in different ways. One of these classical forms is to approximate the excess Gibbs energy of mixing by a polynomial function. For a binary solid solution between a pure dipole 1 and 2, Guggenheim suggested that the excess energy of mixing can be represented by the following polynomial expression [15]:

$$\Delta G_M^{ex} = X_1X_2\left[A_0 + A_1(X_1 - X_2) + A_2(X_1 - X_2)^2 + ...\right] \quad (14)$$

$$\Delta G_M^{ex} = X_1X_2\sum_{j=0}^{J} A_j(X_1 - X_2)^j \quad (15)$$

The coefficients A_j (i = 0, 1, 2) are constants at fixed P and T. The simplest models of this class are obtained when all coefficients are zero except A_0 which represents the interaction energy between species 1 and 2.

If the coefficients A_1, A_2, ... are different from 0, this amounts to consider that the 1-2 interaction depends on the chemical nature of the first neighbors if $A_1 \neq 0$ and the nature of the first and second neighbors if $A_1 \neq 0$ and $A_2 \neq 0$. In this case, equation (15) shows that the shape of the Gibbs energy of mixing function is asymmetric with respect to the composition of the solution, so this type of model is solid solution called asymmetric.

Powell and Ganguly [16,15] showed that in the case of a regular solution, we have:

$$RT \ln \gamma_1 = A_0X_2^2 = A_0(1 - X_1)^2 \quad (16)$$

and: $$RT \ln \gamma_2 = A_0X_1^2 \quad (17)$$

By combining equation (16) and (17), we obtained:

$$RT(X_1 \ln \gamma_1 + X_2 \ln \gamma_2) = A_0 X_1 X_2 \quad (18)$$

Thus, the major difficulty in the definition of a thermodynamic model describing the behavior of a solution is how to express properties excess.

Mixing energy: Implicated at the adsorption isotherm

Averaging the data of adsorption isotherms, Trolard et al. propose to compute a function that best describes these isotherms, the shape of the expression (19) and a polynomial of degree 7 has proved adequate for the description of ln (p/p⁰) according to X_{H2O}.

$$\ln\left(\frac{p}{p^0}\right) = \ln X_{H_2O} + \frac{1}{RT}\left[\left(1-X_{H_2O}\right)^2 \sum_{j=0}^{J=7} A_j \left(2X_{H_2O}-1\right)^j\right] + \frac{1}{RT}\left[2X_{H_2O}\left(1-X_{H_2O}\right)^2 \sum_{j=0}^{J=7} jA_j \left(2X_{H_2O}-1\right)^{j-1}\right] \quad (19)$$

The first-degree polynomial model representing this function is given by the following equation:

$$Y_{the} = A_0 + A_1X_1 + A_2X_2 + A_3X_3 + A_4X_4 + A_5X_5 + A_6X_6 + A_7X_7 \quad (20)$$

Experimental expression is giving by:

$$Y_{\exp} = \frac{RT}{(1-X_{H_2O})^2}\left[\ln\left[\frac{p}{p^0}\right] - \ln X_{H_2O}\right] \quad (21)$$

With: $X_j = (2X_{H_2O}-1)^{j-1}\left[2(j+1)X_{H_2O}-1\right]$ (22)

Where X_{H2O} represents the mole fraction of oxygen bound to the water relative to the total number of oxygen atoms involved in the adsorption reaction of N moles of water, is:

$$X_{H_2O} = \frac{N_O(H_2O)}{N_O(H_2O) + N_O(rawclay)} \quad (23)$$

The number of oxygen atoms of water is given by:

$$N_O(H_2O) = N.Av \quad (24)$$

The number of oxygen atoms on the surface of raw clay $N_O(rawclay)$, is a fixed number:

$$N_O(rawclay) = 2{,}5848.10^{22} \quad (25)$$

Results and Discussion

The determination of the thermodynamic properties of the adsorbed water at the surface of raw clay is carried out using the experimental data of adsorption isotherms obtained at different temperatures. The numerical values of these functions at temperature T=30°C, are compiled in table 3.

Adsorption isotherms

Figure 1 illustrates adsorption isotherms of adsorbed water vapor in the El Hamma raw clay surface at different temperature (30, 40, 50 and 60°C). These results are obtained by Kalfat et al. [17] using gravimetric method. The amount of adsorbed water vapor is measured from the change of mass of the samples using an analytical balance. Equilibrium is considered established when the change in mass of the solid is less than 1%.

The general shape of these isotherms corresponds to type II of Brunauer and classification. Whereas, when the temperature increases the shape of the isotherm approaches type III rather than type II. This phenomenon can be explained by the swelling of the clay under the influence of moisture.

By comparing the adsorption isotherms at different temperatures over the range of water activity from 0 to 1, it was noted that the experimental curves obtained show that the ability of raw clay to adsorb water is greater when the temperature decreases, which confirms the exothermic character of adsorption process. For a water activity p/p⁰=0.95, the amount of adsorbed water at 30°C is about 10 mmol/g of clay, whereas for a higher temperature T=60°C is about 5 mmol/g of clay.

Isosteric enthalpies

The molar enthalpy of adsorption isosteric enthalpy is often called a fundamental factor for the calculation of energy balances, noting that it is related to the adsorption isotherms by the Clapeyron equation:

$$\Delta H_{is} = RT\left[\frac{\partial \ln p}{\partial T}\right]_N \quad (26$$

Usually the isotherms are from a plot the content of adsorbed water related to the activity and repotted as ln (p) related to (-1/T) for various values of the converge.

Figure 2 show that the isosteric enthalpy of adsorption, in absolute terms, has been decrease with increasing water content. We also note that the isosteric enthalpy is greatest for very low values of N, and it decreases rapidly and tends to zero as far as N grows. This implies that the formation of the first monolayer of water vapor is more energetic than the upper layers and the water was adsorbed in the most accessible places on the outer surface of the solid. As the water content increased swelling of the adsorbent and thus, new high-energy sites are open for

p/p⁰	N	n	e	N'	ΔGR,N	ΔH_{is}	ΔHR,N	ΔSR,N	$\Delta G^*_{R,N}$	$\Delta H^*_{R,N}$	$\Delta S^*_{R,N}$
0.06	0.52	0.28	0.85	0.0	-7.091	-5.1680	-791.300	-2586.87	-	-	-
0.16	1.96	1.07	3.21	1.4	-4.619	-4.0285	-739.149	-2422.99	-5.855	-765.22	-2504.93
0.20	2.09	1.14	3.43	1.6	-4.056	-4.0075	-316.520	-1030.72	-5.729	-745.57	-2440.50
0.25	2.36	1.29	3.87	1.8	-3.494	-3.9019	-432.287	-1414.46	-5.442	-691.10	-2261.79
0.29	2.48	1.36	4.07	2.0	-3.120	-3.8389	-324.215	-1059.46	-5.312	-671.95	-2199.04
0.33	2.88	1.57	4.72	2.4	-2.794	-3.7333	-238.883	-778.79	-4.913	-605.78	-1982.08
0.40	3.01	1.64	4.93	2.5	-2.309	-3.7123	-283.846	-928.70	-4.789	-587.80	-1923.17
0.46	3.40	1.86	5.57	2.9	-1.957	-3.5857	-324.215	-1063.03	-4.430	-549.37	-1797.60
0.56	3.53	1.93	5.79	3.0	-1.461	-3.5437	-116.062	-378.03	-4.312	-535.15	-1751.08
0.61	4.31	2.36	7.07	3.8	-1.246	-3.4801	-25.658	-80.53	-3.703	-439.60	-1437.89
0.70	6.01	3.28	9.85	5.5	-0.899	-3.4801	0.000	2.97	-2.889	-307.45	-1004.65
0.79	7.06	3.86	11.57	6.5	-0.594	-3.4801	0.000	1.96	-2.545	-258.09	-842.96
0.84	7.58	4.14	12.43	7.1	-0.439	-3.4801	0.000	1.45	-2.395	-239.08	-780.74
0.89	8.11	4.43	13.30	7.6	-0.294	-3.4801	0.000	0.97	-2.254	-222.38	-726.14
0.94	10.30	5.63	16.89	9.8	-0.156	-3.4801	0.000	0.51	-1.799	-172.59	-563.37

Table 3: Thermodynamic properties of adsorbed water on the surface of raw clay at T =30°C.

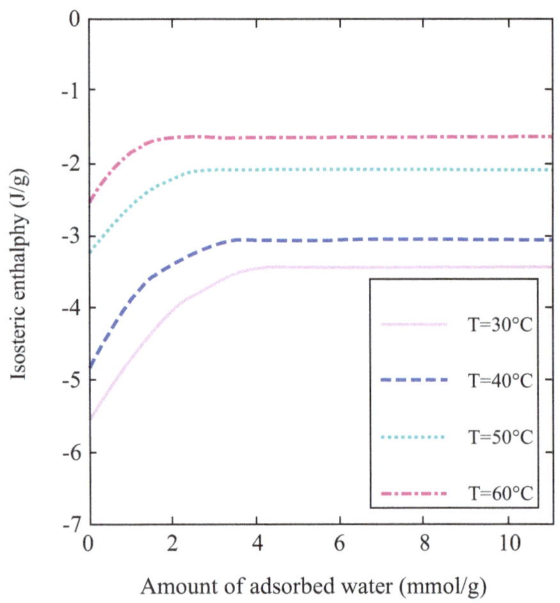

Figure 2: Isosteric enthalpies of raw clay at T=30°C, T=40°C, T=50°C and T=60°C.

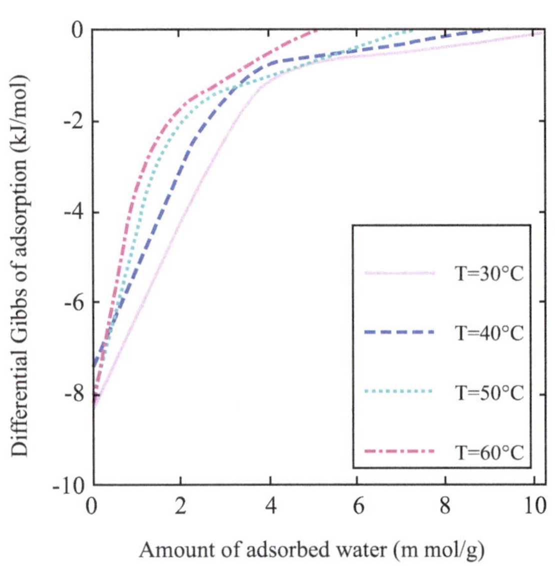

Figure 3: Differential Gibbs of adsorption at T=30°C, T=40°C, T=50°C and T=60°C.

water: this causes an increase in the isosteric enthalpy of adsorption with decreasing water content.

Differential Gibbs of adsorption

The differential Gibbs of adsorption of adsorbed water on raw clay surface calculated from Eq (3) is presented in Figure 3. We can observe that the free energy decreases as the water content increases. Above 8.5 mmol/g of raw clay the Gibbs differential adsorption in the different temperature is similar, and approaches zero. This indicates that as water content increases, the adsorbed water has properties similar to those of bulk water and the influence of the clay surface does not extend beyond a few molecular layers of adsorbed water. The negative values of Gibbs differential adsorption confirm the feasibility of the adsorption process.

Differential adsorption enthalpy

The differential enthalpy of adsorbed water on raw clay obtained from water adsorption isotherms at different temperatures is shown in figure 4. The variation of differential enthalpy with water content showed that when the amount of water increase the differential enthalpy decrease and the trend seemed to become asymptotic as a water content of 2 mmol/g was approached. The differential enthalpy is negative; this implies that water molecules lose energy upon being transferred from bulk water into the adsorbed state and indicate the exothermic nature of the adsorption process. This also confirms the possibility of physical adsorption as with increase in temperature of system. Above 5mmol/g of raw clay the evolution of the function is very rapid and gets very close to zero.

Differential entropy function

The variation of differential entropy function is given in figure 5. This differential entropy decreases continuously as the amount of adsorbed water increases and thereafter, remained constant. Figure 5 showed that the entropy decreases with increasing temperature and which is very close to zero for a water content greater than 5 mmol/g of raw clay. The negative values of differential entropy of adsorption correspond to decrease of freedom of the adsorbed water.

Gibbs adsorption average

Figure 6 shows the variation of Gibbs adsorption average as related to the adsorbed water content at different temperatures. The increase in the amount of water adsorbed on the surface of the clay leads to a decrease in the average free energy of adsorption. While the increase in temperature causes an increase in the enthalpy. For a temperature T=30°C and a water content of 2.8 mmol/g of clay Gibbs adsorption average is about -5 kJ/mol, while for a higher temperature T=60°C and for the same water content the Gibbs adsorption average is about -3 kJ/mol. From the figure 4 we noticed that there is a very rapid evolution of the function that gets very close to zero for water content greater

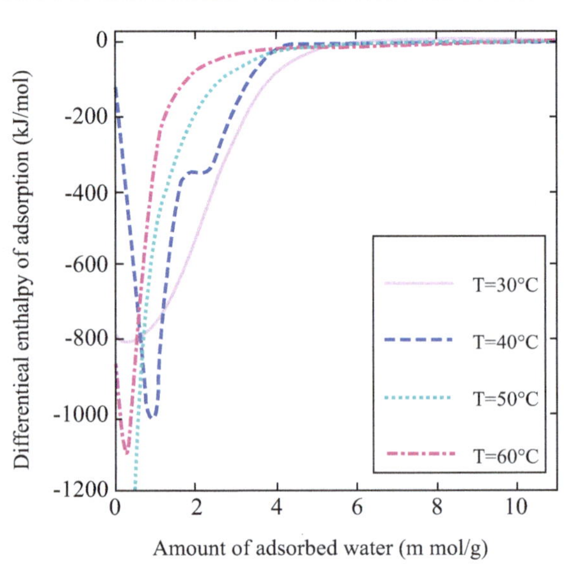

Figure 4: Differential adsorption enthalpy at T=30°C, T=40°C, T=50°C and T=60°C.

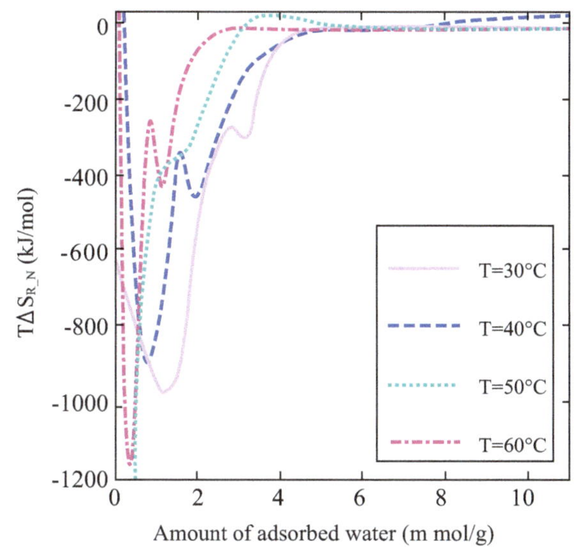

Figure 5: Differential entropy function at T=30°C, T=40°C, T=50°C and T=60°C.

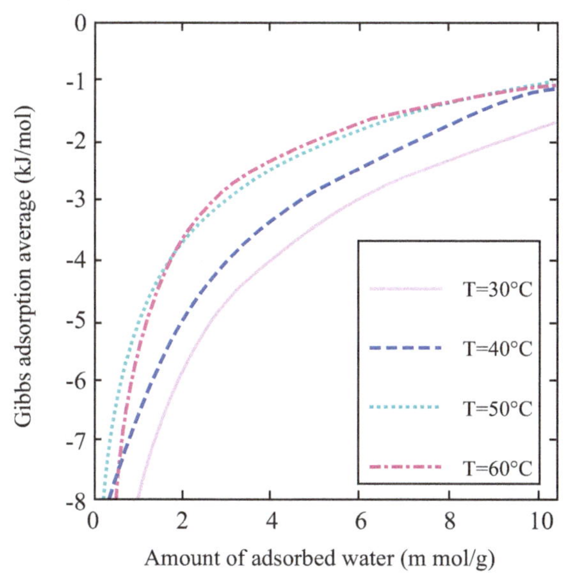

Figure 6: Gibbs adsorption average at T=30°C, T=40°C, T=50°C and T=60°C.

than 2 mmol/g of clay. The negative values of average Gibbs adsorption average indicate that the hydrated raw clay is stable.

Average adsorption enthalpy

The results of this integration, illustrated on the figure 7, show that the average adsorption enthalpy has a shape similar to that of the differential enthalpy of adsorption. Indeed, the increase in the amount of water adsorbed on the surface of the studied adsorbent causes a decrease of the enthalpy and that any increase in temperature causes a decrease of the function. We have also noted a very rapid evolution of the function that gets very close to zero for water content greater than 6 mmol/g of clay.

Average entropy function

Figure 8 present the changes of average entropy function. If we consider integrating the change of entropy of adsorbed water, it is clear that the average entropy decreases continuously when the amount of adsorbed water increases. It is also clear that the strong interaction with adsorbed water molecules at a high level of water content resulting in lower entropy.

Coefficients of the polynomial

The coefficients of the regression equation of first order of raw clay at different temperatures are summarized in table 4.

Thermodynamic functions modeled by using the polynomial equation

The determination of thermodynamic functions of water adsorbed

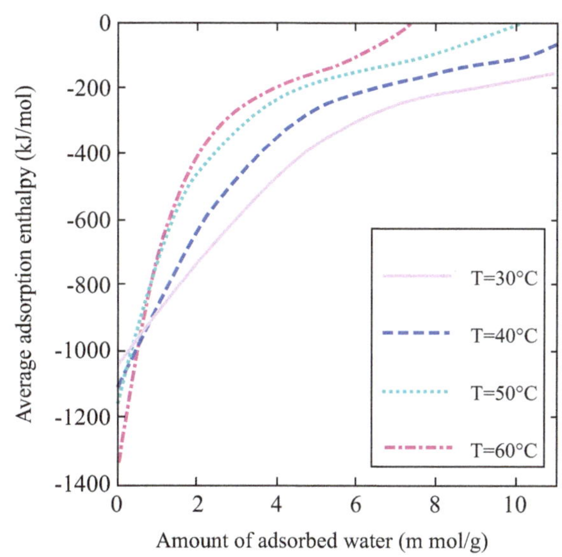

Figure 7: Average adsorption enthalpy at T=30°C, T=40°C, T=50°C and T=60°C.

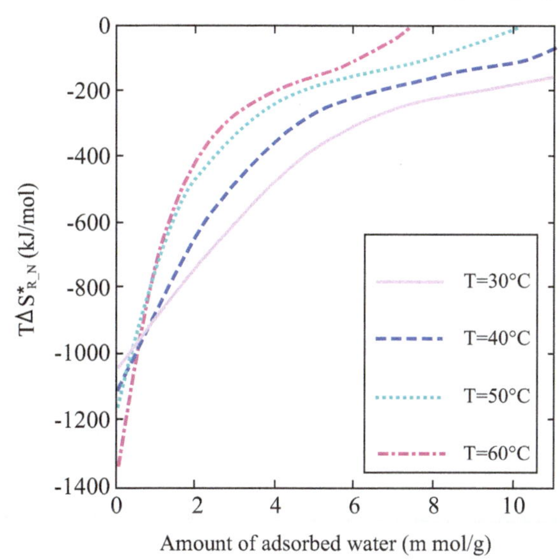

Figure 8: Average entropy function at T=30°C, T=40°C, T=50°C and T=60°C.

	T=30°C	T=40°C	T=50°C	T=60°C
A_0	-324738.4	-606940.27	685955.22	-969697.62
A_1	-1929226.5	-3159694.2	2861343.1	-4007621.1
A_2	-3921027.8	-4853810.7	2726160.3	-5272665.8
A_3	-2569524.4	565890.18	-4196964.5	-2406942.1
A_4	1127999	7537255.9	-9973877.8	-3117353.5
A_5	1236270.4	4622348.9	-6039924.7	-7950928.6
A_6	-916203.12	-1589693.4	-160087.31	-6996808.8
A_7	-777951.17	-1550488	647780.06	-1996296.7

Table 4: Coefficient of the polynomial equation.

p/p^0	N	XH2O	ΔG_M^{ex}	T.E	ΔG_M
0.06	0.52	0.012	60	-163.5	-103.5
0.16	1.96	0.0437	165.3	-452.3	-287
0.2	2.09	0.0464	175.3	-473.6	-298.2
0.25	2.36	0.0521	197.1	-516	-318.9
0.29	2.48	0.0546	207.2	-534.1	-326.9
0.33	2.88	0.0629	242.4	-591.9	-349.5
0.4	3.01	0.0655	254.2	-609.8	-355.6
0.46	3.4	0.0734	290.8	-661.3	-370.5
0.56	3.53	0.076	303.2	-677.7	-374.5
0.61	4.31	0.0913	378.1	-769.9	-391.8
0.7	6.01	0.1228	529.8	-939	-409.2
0.79	7.06	0.1413	611.8	-1026.5	-414.6
0.84	7.58	0.1501	649.9	-1065.9	-416
0.89	8.11	0.1589	687.4	-1103.7	-416.3
0.94	10.3	0.1936	829.7	-1238.4	-408.6

Table 5: Thermodynamic functions modeled by using the polynomial.

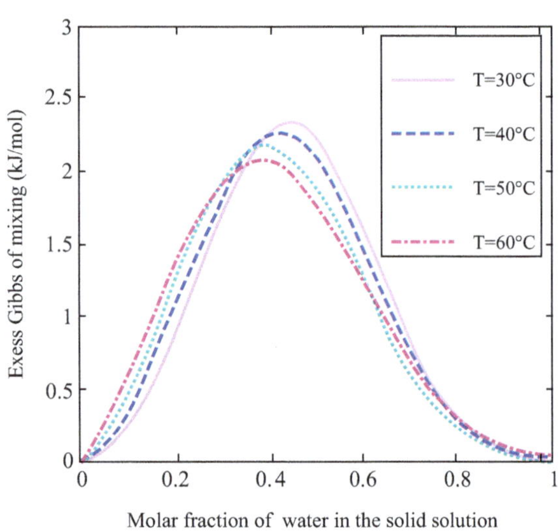

Figure 9: Excess Gibbs of mixing at T=30°C, T=40°C, T=50°C and T=60°C.

on the surface of raw clay is performed using the experimental data of adsorption isotherms obtained at different temperatures. The numerical values of these functions at a temperature T=30°C, are grouped in table 5.

Excess Gibbs energy of mixing: Figure 9 show the variation of the excess Gibbs energy of mixing at a range of temperature going from 30°C to 60°C for the raw clay of El Hamma depending on the composition of adsorbed water. Positive values of the excess Gibbs energy of mixing as shown by the figure 9 indicate a strong interaction between adsorbate/adsorbent and a high stability of the solid solution.

The excess Gibbs energy of mixing is symmetrical and has a maximum for a composition of adsorbed water by 40%, approximately. For a temperature of 50°C, we obtain an excess Gibbs energy of mixing of about 2 kJ/mol at x = 0.4. The figure 9 also shows that changes in the excess free energy of mixing are very close to that of a simple solution.

Entropy Function: Figure 10 show the changes in entropy Function of adsorbed water on El Hamma raw clay surface related to molar fraction of water in the solid solution. The values of the entropy of mixing calculated by means of the relation (12) are all negative as shown in table 5. The entropy of mixing decreases while increasing the temperature, this decrease is due to connections of water molecules to the adsorbent and their rearrangement leads to the appearance of order in their mutual arrangement. Entropy of mixing, as in the case of the excess Gibbs energy of mixing, has a minimum value for a given composition. For a temperature T=50°C, the mixing entropy minimum is -2.3 kJ/mol for El Hamma raw clay.

Gibbs energy of mixing: The results obtained from modeling

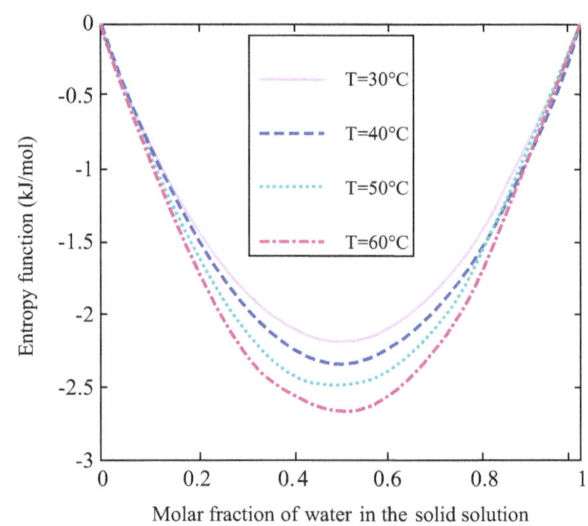

Figure 10: Entropy of mixing at T=30°C, T=40°C, T=50°C and T=60°C.

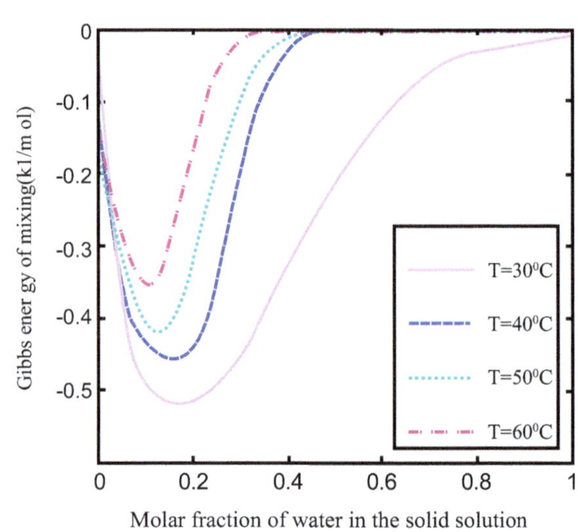

Figure 11: Gibbs of mixing at T=30°C, T=40°C, T=50°C and T=60°C.

equation (14) for El Hamma raw clay are shown in figure 11. We note from the result of modeling of the studied solid solution that he has a negative value of Gibbs energy of mixing which proves the system stability. The free energy of mixing for a given composition of water, increases with temperature indicating a decrease in the stability and the molecules are pressed against each other.

Conclusion

In this study, we conducted a comprehensive analysis of the adsorption isotherms of water vapor adsorbed on the surface of El Hamma raw clay in a range of temperature from 30°C to 60°C. The adsorption isotherm of water vapor on clay corresponds to type II of BDDT classification.

The isosteric enthalpy is maximum for very small values of N (adsorbed water content), which implies that the formation of the first layer of water vapor is more energetic than the upper layers and the water was adsorbed in the most accessible places on the external surface of the solid.

Also, this work is devoted to the calculation of thermodynamic properties of water adsorbed on the surface of the studied adsorbent, namely, the differential and average Gibbs energy, enthalpy and entropy. For low pressures, the differential and average enthalpies are greater than 1000 kJ/mol. To high pressures, the differential and average enthalpies are decreasing functions and tend to the enthalpy of liquid water. The entropy of adsorbed water is lower than that of liquid water which is the fact that the adsorbed molecules are ordered. When the amount of adsorbed water on the solid solution increases the thermodynamics functions are close to those of liquid water, on one hand, on the other hand, we see that the entropy of adsorbed water tends rapidly to that of liquid water, which corresponds to the fact that water molecules are ordered in a state.

The last part was devoted to the thermodynamic modeling of solid solutions. The expression of Saxena allowed us to estimate the excess energy of mixing and enthalpy of mixing of the solid solution under consideration. The modeling results show that the free energy of mixing for a given composition of water increases with temperature indicating a decrease in stability, a strong interaction between water vapor and the adsorbent, the rearrangement of water molecules leads to the appearance of order in their arrangement and the increase in temperature leads to a decrease in the stability of solid solutions.

Acknowledgments

The author gratefully acknowledges the Research Unit of Applied Thermodynamics in the National Engineering School of Gabès ENIG for providing support of a project entitled: application of the thermodynamic model to the couple Water vapor-El Hamma raw clay.

References

1. Srasra E, Kebir AN, Ayedi F, Bergaya F, Van DH (1988) Mineralogical identification of a bentonite clay deposit located near gabes (Tunisie). Journal de la société chimique de tunisie 2: 37-45.
2. Ben Zina M (1987) Mineralogical characterization, physicochemical properties and application of the clay deposit Haidoudi, PhD Thesis, Faculty of Science of Tunis.
3. Bagané M, Guiza S (2000) Elimination des colorants des effluents de l'industrie textile par adsorption. Ann Chim fr 3: 615-625.
4. Weaver CE (1989) Clays, Muds and Shales Developments in sedimentology. Elsevier.
5. Gannouni A, Bagané M, Bellagi A (1994) Sulfur dioxide adsorption isotherms on activated clay. Annale de chimie 19: 27-37.
6. Bagané M, Gannouni A (1995) Study of water vapor adsorption at high temperatures on activated alumina. Annale de chimie 20: 72-80.
7. Mc Bain JW, Bakr AM (1926) A new sorption balance. J Am Chem Soc 48: 690-695.
8. Dreisbach F, Staudt R, Keller JU (1999) High pressure adsorption data of methane, nitrogen, carbon dioxide and their binary and ternary mixtures on activated carbon. Adsorption 5: 215-22.
9. Dreisbach F, Lösch HW (2000) Magnetic suspension balance for simultaneous measurement of a sample and the density of the measuring fluid. Journal of Thermal Analysis and Calorimetry 62: 515-521.
10. Dreisbach F, Lösch HW, Harting P (2002) Highest Pressure Adsorption Equilibria Data: Measurement with Magnetic Suspension Balance and Analysis with a New Adsorbent/Adsorbate-Volume. Adsorption 8: 95-109.
11. Tardy Y, Fritz B (1980) An ideal solid solution model for calculating solubility of clay minerals. Clay minerals 16: 361-373.
12. Ganguly J, Saxena SK (1987) Mixtures and mineral reactions. Springer-Verlag 291.
13. Glynn PD, Reardon EJ (1990) Solid-solution aqueous-solution equilibria: thermodynamic theory and representation. American Journal of Science 290: 164-201.
14. Geiger CA (2002) Solid solution in Silicate and Oxide systems. Mineralogy Magazine 66: 1107-1108.
15. Ganguly J (2001) Thermodynamic modeling of solid solution, EMU Notes in Mineralogy 3: 37-69.
16. Powell R (1987) Darken's quadratic formulism and the thermodynamics of minerals. American Mineralogist 72: 1-11.
17. Kalfat (1995) Adsorption isotherms of water vapor on natural clay, Final Project study. Ecole National d'Ingénieur de Gabès.

Dearomatization of Normal Paraffin by Synthesized Nax Zeolite

Hossein Faghihian[1]*and Leila Riazi[1,2]

[1]*Department of Chemistry, Shahreza Branch, Islamic Azad University, Iran*
[2]*R&D Department, Iranian Chemical Industries Investment Company, Isfahan, Iran*

Abstract

Aromatic compounds produced in linear alkyl benzene (LAB) production process decrease the yield and quality of the product and enhance the deactivation rate of the catalysts. Therefore, the de-aromatization of different streams in LAB production complex is inevitable. In this research, the recycled paraffin from alkylation to dehydrogenation unit was de-aromatized by synthesized NaX zeolite. The effect of different parameters such as temperature, amount of the adsorbent, contact time and initial aromatic concentration was studied and optimized. The Langmuir and Freundlich adsorption models were applied to describe the equilibrium isotherms. It was concluded that the Langmuir model agreed very well with the experimental data. The calculated thermodynamic parameters of the adsorption showed that the adsorption process was spontaneous and exothermic. The reaction rate was estimated by the pseudo-second order kinetic model. To examine the re-generation ability of the adsorbent a fixed bed column was designed and the breakthrough curves were prepared for three successive regeneration cycles. The results indicated that the adsorbent capacity remained unchanged after three regeneration cycles.

Keywords: Zeolite; Chemical process; Petroleum; Adsorption; Separations; Dearomatization

Introduction

Linear alkylbenzene produced by C_{10}-C_{13} linear olefins, constitutes the active ingredients of many household detergents [1]. In the Universal Oil Products (UOP) design, the linear alkylbenzene process consists of pre-fraction, hydro-treating, extraction, dehydrogenation and alkylation units [2]. In the dehydrogenation unit, due to some side reactions such as dehydrocyclization, aromatization, cracking and isomerization processes non-linear by-products are formed that undergoes additional alkylation processes to form aromatic by-products such as alkylated benzenes, poly alkylbenzenes, naphthalenes and biphenyls [3,4]. These aromatic by-products are accumulated on the surface of the catalyst increasing its deactivation rate and decreasing dehydrogenation, alkylation selectivity and sulfonability of LAB. On the other hand, the sulfonated alkyl benzene will be colorful due to facile electron transition in the conjugated double bonds of polyaromatic by-products [5]. Among the method used for de-aromatization, adsorption is an effective method [6]. Different activated carbon has been studied for adsorption of phenanthrene as a model aromatic compound [7]. Susu [8] developed a mathematical model to predict the effluent concentration and breakthrough profiles of aromatic compounds in kerosene deodorization by commercial clays. Molecular sieve zeolites as low price and resistant materials are increasingly used in commercial separations. A controlling factor in whether molecules can be adsorbed in the cavities of the adsorbent is the pore-opening of the channels. Additionally, adsorption capacity can be enhanced by electrostatic interactions between the sorbent and the cations held within the pores of the adsorbent [9]. Adsorption of aromatic compounds occurred as a result of interaction between π-electron of the aromatic ring with the cation or hydroxyl group of the zeolites. Adsorptive separation of aromatics from kerosene cut by NaX has been reported [10]. In a recent study, adsorption of liquid aromatic/alkane mixtures on NaY zeolite has been investigated [11]. Many patents relevant to the separation of aromatic by-products in linear alkylbenzene process for either improving the quality of linear alkylbenzene or for economic aspect have been registered [12, 13].

In this paper, we report the application of synthetic zeolite NaX for aromatics adsorption from recycled paraffin. The experimental breakthrough data in a fixed bed adsorption column are given and the regeneration ability of the adsorbent has been reported.

Material and Methods

Synthesis of NaX zeolite

To prepare spherical type NaX zeolite, aluminum and silicon gels were prepared separately. Typically 7.2 g sodium metasilicate pentahydrate (Fluka, purum, ≥ 97%) was dissolved in 48.0 g of de-ionized water. After addition of 8.0 g of triethanolamine (TEA, Merck, 99%) under stirring, the solution was filtered through a 0.2 μm membrane filter (solution A). Solution B was prepared by dissolving 2.9 g sodium aluminate (Riedel–de Hae¨n, technical grade, 54 wt.% Al_2O_3, 41 wt.% Na_2O, rest: water) in 48.0 g of deionized water. After adding 8.0 g of TEA under stirring, the resulting solution was filtered through a 0.2 μm membrane filter. Solution A was mixed with solution B in a polypropylene bottle. The mixture was thoroughly shaken for 5 min. Zeolite crystallization occurred within 8 day at 85°C [14]. The spherical type zeolite adsorbent was prepared by thoroughly mixing of 20 wt.% kaolin clay with the synthesized NaX zeolite powder. The mixture was converted to paste and to spherical shaped by a granulator with size of 1mm × 2 mm. The products were air-dried for 24 h, and then calcined at 600-650°C for 2 h to convert kaolin into metakaolin. The thus-obtained solid product was treated with sodium hydroxide solution (1.0-1.2 M) at 90 °C for 6-8 h to convert the kaolin binder into zeolite. The treated products were separated, washed and dried [15]. The adsorbent was characterizes by FT-IR, XRD and XRF techniques. The FT-IR spectrum of the adsorbent was recorded by an Impact-

***Corresponding author:** Hossein Faghihian, Department of Chemistry, Shahreza Branch, Islamic Azad University, 86145-311, Iran

400D FT-IR instrument (Nicolet Co.) in region of 4000-400 cm^{-1} using KBr wafer technique. The crystalline structure of the adsorbent was examined through X-ray diffraction method (Bruker D8ADVANCE diffractometer using CuK αradiation). The chemical composition of the adsorbent was obtained with PANalytical Magix instrument. The samples for de-aromatization were prepared from the recycled C_{10}-C_{13} normal paraffin stream, with 3.5wt.% aromatic content from Iranian Chemical Industries Investment Company. A Hach Brand DR-5000 spectrophotometer was used for determination of aromatic concentration according to the procedure suggested by the UOP (Test method 495).

Adsorption studies

Batch experiments were designed and conducted at room temperature in order to determine the adsorption capacity, isotherm data and the effect of operating parameters. Before each experiment the synthetic zeolite was activated at 350°C for 2 hour and then cooled to room temperature in desiccators. Known amount of the activated sample was transferred to a beaker containing 50 ml of the recycled normal paraffin. The mixture was shaken for equilibration time and then filtered. The aromatics concentration was measured in the liquid phase and the removal percentage was calculated according to the equation 1.

Removal (%) = $((C_0-C_e)\times100)/C_0$ (1)

where C_0 and C_e are respectively the initial and equilibrium concentrations of aromatic compounds (mg/L).

Effete of adsorbent dose: To evaluate the effect of adsorbent dose, 50 ml of the sample was mixed with varying amounts of adsorbents in a 250 ml Teflon beaker. The mixture was shaken for 24 hour at 25°C. The solid phase was separated by filtration and the aromatic content of the filtrate were measured.

Kinetic Study: In order to evaluate the requisite time for maximal removal of aromatics, 10 g of the activated NaX zeolite was exposed to 50 ml of the recycled normal paraffin. The mixture was shaken at 25°C and the samples were taken at different interval time until equilibrium was attained. The solid phase was separated by filtration and the aromatic content of the samples was measured.

Effect of initial concentration: 50 ml of normal paraffin mixture with different aromatic content (0.01-5 wt.%) was mixed with 10 g of zeolite NaX and shaken for equilibrium time at 25°C. The solids were separated by filtration and the aromatic content of the filtrate was measured.

Effect of temperature: To study of the effect of the temperature, a rotary evaporator was used in which 50 ml of the recycled normal paraffin was mixed with 10 g of the activated NaX zeolite at different temperature from 25°C to 145°C. After equilibration, each sample was filtered and the aromatic content was measured in the filtrate.

Regeneration studies

To study the regeneration ability of the adsorbent, a pilot scale fixed bed column was designed and constructed (80 cm length and 2.54 cm inner diameter). All required auxiliary systems including, electrical furnace, injection pump, feed pre-heater, and collecting vessels were predicted. The column was packed with 265 g of the adsorbent and placed inside the electrical furnace (Excition model). Electrical heating was supplied to the column to control the bed temperature (Figure 1). Before each experiment, the adsorbent was activated in situ by electrically heating furnace at 350°C for two hours and then allowed to cool to room temperature at dry nitrogen atmosphere (5 ml/min). Once the adsorption process was started, samples were collected from the bottom of the bed at known time intervals. The concentration of aromatic in inlet and the outlet were determined in order to construct breakthrough curves and to calculate the adsorbent capacity. When the adsorbent was saturated, the regeneration cycle was carried out under dry nitrogen flow at 500°C for 15h. The adsorption-regeneration cycles were repeated three times, and adsorption capacity of the regenerated adsorbent was calculated for each cycle.

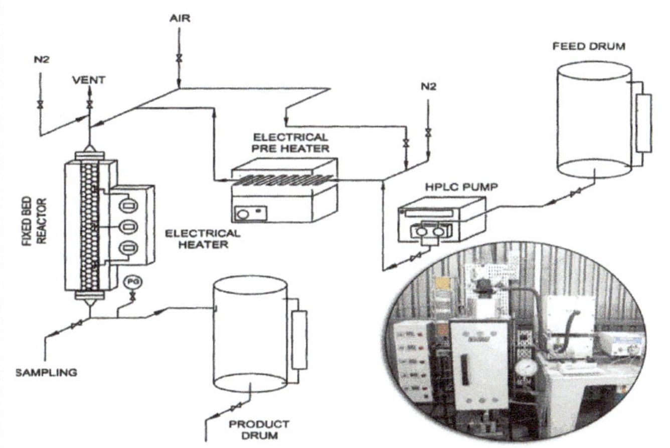

Figure 1: Schematic diyagram of fixed bed reactor for regeneration study.

Results and Discussion

Adsorbent characterization

In the FT-IR spectrum of the adsorbent (Figure 2), the band at 3400-3600 and 1630-1660 cm^{-1} were attributed respectively to OH stretching mode and bending mode of OH group of the water molecules. The two most intense bands of zeolites appeared at 980-1230 cm^{-1} and 420-500 cm^{-1} were clearly observed in the spectra. The first bands at 1630-1660 cm^{-1} was assigned to the asymmetric stretching mode of the T-O and the second one was related to bending mode of the T-O. The absorbance band at 700-800 cm^{-1} was assigned to symmetric stretching mode of the T-O. The mid-infrared region of the spectrum contained the fundamental framework vibration of Si(Al)O_4 groupings. These vibrations related to the external linkage of the TO_4 units in the structures that are sensitive to structural vibrations. All these observations confirmed that the adsorbent has an alumina silicate structure [16]. In the XRD pattern of the synthesized zeolite the line positions and the relative intensities were similar to those of the references, confirming that the synthesized adsorbent was pure NaX zeolite (Figure 3). The chemical composition of the sample obtained by XRF method agreed well with the results of the other workers (Table 1) [17].

Adsorption Studies

Effete of adsorbent dose: The removal efficiency of the aromatics was increased with increasing the adsorbent dosage, and remained constant at adsorbent dosage of 16 g (Figure 4). To economize the procedure, 10 g of adsorbent was considered as optimal adsorbent dosage.

Effect of contact time: By increasing the contact time from 0 to 195 min, the removal efficiency was first increased and then followed

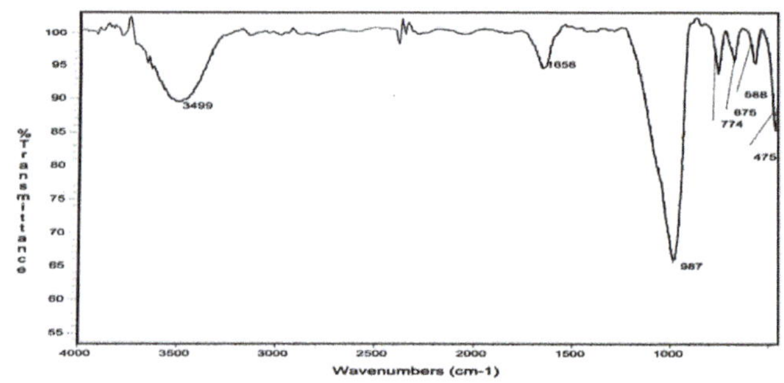

Figure 2: FT-IR spectrum of the synthetic zeolite NaX.

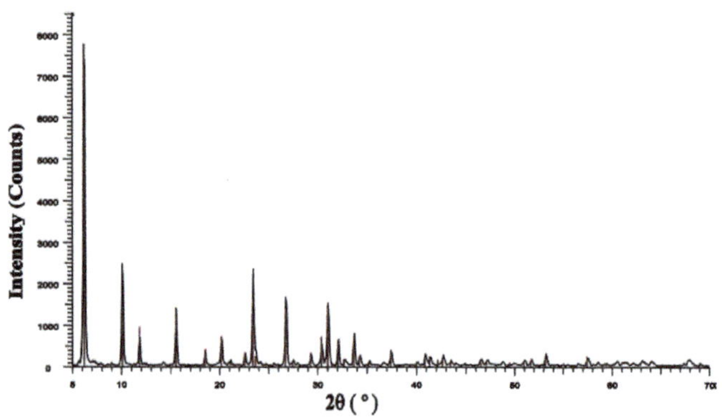

Figure 3: XRD pattern of synthetic zeolite NaX.

Constituent	Value (wt.%)
SiO_2	37.17
Al_2O_3	25.98
Na_2O	14.73
MgO	1.57
CaO	1.48
K_2O	0.27
Fe_2O_3	0.71
P_2O_5	0.11
LOI	17.8

Table 1: XRF analysis of synthetic zeolite NaX.

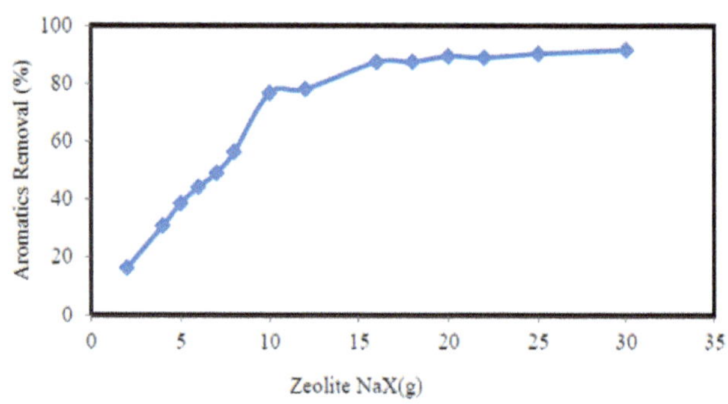

Figure 4: Effect of adsorbent dosage on the removal of aromatics (Init.Conc= 3.5%w, T= 25°C).

by constant adsorption (Figure 5). The results also revealed that the uptake of aromatics at initial stages was fast, with reduction in uptake rate being closer to equilibrium time. The rapid initial uptake is attributed to the adsorption of the aromatics on the surface sites. When the aromatics adsorption at the exterior sites reached the saturation level, the aromatics begins to enter the pores of the adsorbents and the adsorption rate decreased due to an increase in the diffusion resistance as the molecules have to travel further into the interior pores. The interior surfaces were active and showed affinity towards aromatics molecules, but the adsorption on the interior sites needed more contact time and the aromatics uptake after 3 hours was controlled by micro pore diffusion.

Effect of initial concentration: The influence of initial concentration on the removal efficiency is illustrated in Figure 6. The percentage removal decreased as the initial concentration of aromatics increased. At lower initial concentrations, sufficient adsorption sites are available for adsorption. As the concentration increased, the numbers of aromatic compounds is higher than adsorption sites; however the fractional adsorption was independent of initial concentration.

Effect of temperature: The removal of aromatics decreased with the increase of temperature (Figure 7). This is related to this fact that higher temperature is unfavourable to exothermic reaction once equilibrium is attained. The optimal temperature was found to be 25°C.

Equilibrium modeling

The Langmuir and Freundlich models as the most frequently employed models for adsorption processes were used to evaluate the experimental data. Langmuir model assumes uniform energies of adsorption onto the surface with no transmigration of the adsorbate in the plane of the surface. The linear form of Langmuir isotherm is given by the following equation [18]:

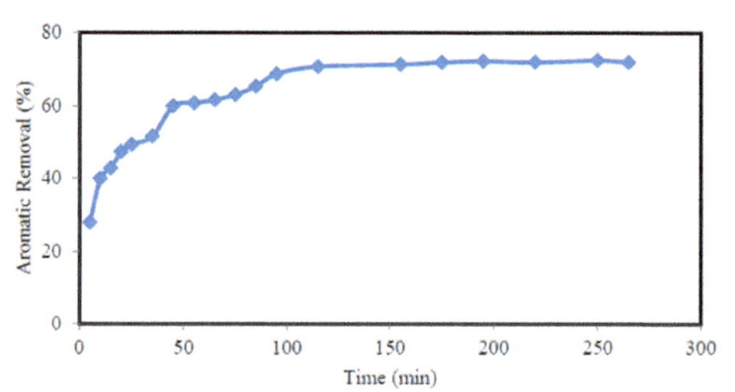

Figure 5: Effect of contact time on the removal of aromatics (Init.Conc= 3.5%w, T= 25°C).

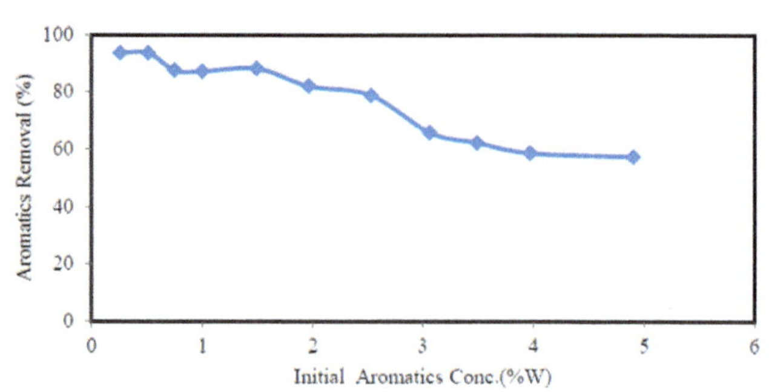

Figure 6: Effect of contact time on the removal of aromatics (Init.Conc= 3.5%w, T= 25°C).

$$1/q_e = 1/q_m + 1/q_m C_e \quad (2)$$

Where q_e is the amount of adsorbed species at equilibrium (mg/g), C_e is the equilibrium concentration of the adsorbate (mg/L), and q_m and b are the Langmuir constants related to maximum adsorption capacity and energy of adsorption, respectively. When $1/q_e$ was plotted against $1/C_e$, a straight line with slope $1/bq_m$ indicated that the adsorption followed Langmuir isotherm (Figure 8). The Langmuir constants q_m and b were determined from the slope and the intercept of the plot and are shown in Table 2. The adsorption data was also analyzed by the Freundlich model [18]. The logarithmic form of the Freundlich model is given by the equation 3.

$$Log\, q_e = \log K_f + 1/n \log C_e \quad (3)$$

where K_f and n are Freundlich constants related to adsorption capacity and intensity respectively. When log q_e was plotted against $\log C_e$, a straight line with slope 1/n was obtained indicating that the adsorption followed Freundlich isotherm (Figure 9). The Freundlich constants K_f and n were determined from the slope and intercept of the plot and were presented in Table 2. The results showed that Langmuir isotherm fitted well with the experimental data, indicating the homogeneous distribution of active sites of the adsorbent.

Thermodynamic parameters

The equilibrium data was used to calculate the thermodynamic parameters, ΔH°, ΔS° and ΔG°. The equilibrium constant was defined by the following equation [19]:

$$K_d = \frac{C_i - C_e}{C_e} \times \frac{V}{m} \quad (4)$$

C_i and C_e are respectively initial and equilibrium concentration (mg/L), V is volume of the solution (ml) and m is the adsorbent weight (g). ΔH°, ΔS° were calculated from the slope and intercept of linear plot of 1/T versus $\ln K_d$, according to equation (5), where T is the solution temperature (K) and R is the gas constant and is equal to 8.314 J/mol K (Figure 10).

$$\ln K_d = -\left(\frac{\Delta H^\circ}{RT}\right) + \left(\frac{\Delta S^\circ}{R}\right) \quad (5)$$

Values of the Gibbs free energy change for the adsorption process was obtained from equation 6:

$$\Delta G^\circ = \Delta H^\circ - T\Delta S^\circ \quad (6)$$

The values of thermodynamic parameters, ΔH°, ΔS° and ΔG° are summarized in Table 3. The calculated adsorption enthalpy were negative, the indication of an exothermic process and their magnitudes (<40 kJ/mol) showed a physical adsorption process. The negative value of ΔG° indicated the spontaneous nature of the adsorption process. Finally, the negative values of the adsorption entropy were consistent

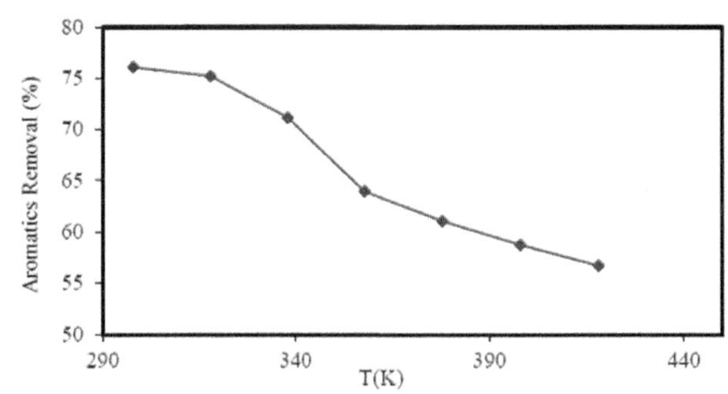

Figure 7: Effect of temperature on the removal of aromatics (Init.Conc= 3.5%w, T= 25°C, Cont.time = 195 minutes).

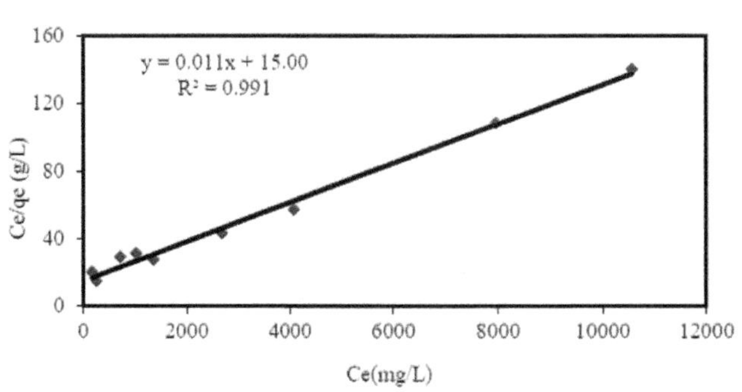

Figure 8: Langmuir isotherm for adsorption of aromatics by synthetic zeolite NaX.

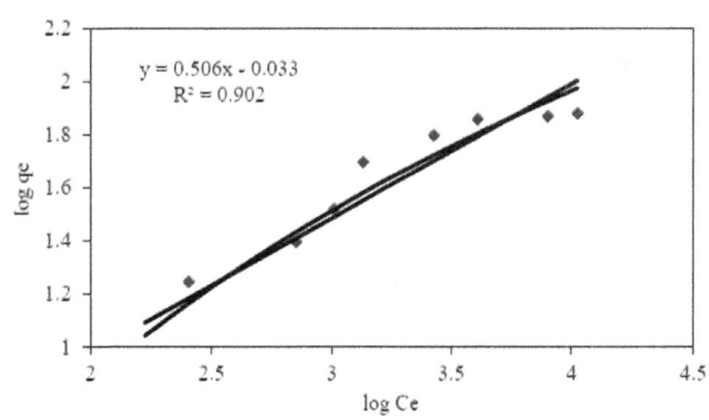

Figure 9: Freundlich isotherm for adsorption of aromatics by synthetic zeolite NaX.

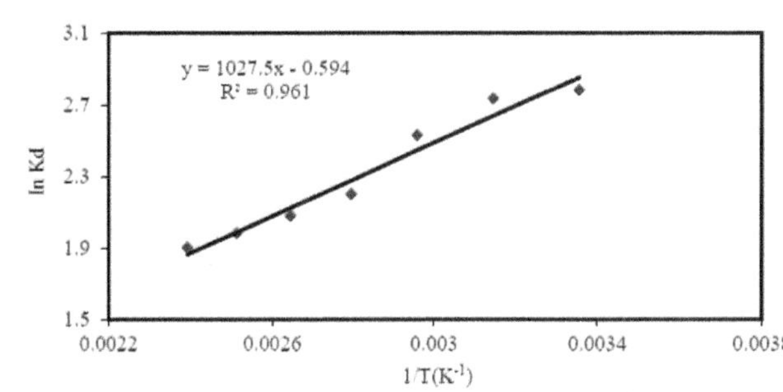

Figure 10: Equilibrium constant as a function of temperature.

Freundlich isotherm			Langmuir isotherm		
R^2	n	K_f((mg/g)(1/mg)$^{1/n}$)	R^2	b(L.mg^{-1})	q_{max}(mg.g^{-1})
0.902	1.98	0.93	0.991	7.3×10^{-4}	90.9

Table 2: Isotherm parameters for removal of aromatics by-products.

ΔH° (KJ/mol)	ΔS° (J/Kmol)	ΔG°(KJ/mol)						
		298°K	318°K	338°K	358°K	378°K	398°K	418°K
-8.54	-4.94	-7.07	-6.97	-6.87	-6.77	-6.67	-6.57	-6.47

Table 3: Thermodynamic parameters of the reaction.

with restricted mobility of the adsorbed molecules of aromatic compounds as compared with those molecules in the bulk.

Kinetic of adsorption

In order to investigate the mechanism of sorption process, various kinetic models were examined. The rate constant for the apparent adsorption of aromatics was obtained from the pseudo second-order equation [20]:

$$\frac{t}{qt} = \frac{1}{k_2 qe^2} + \left(\frac{1}{qe}\right)t \qquad (7)$$

Where q_t and q_e are the amount of adsorbate at time t(s) and equilibrium time (mg/g), and k_2 the second order rate constant (g/mg.min). If the equilibrium sorption capacity and the rate constant are known, then the adsorption capacity at any time can be calculated.

When t/q_t was plotted against t, a straight line was fitted to the data (Figure 11), the obtained correlation factor (R^2= 0.9965) suggested that aromatic adsorption by NaX zeolite followed pseudo second order kinetics.

Adsorbents regeneration

One of the challenging problems on adsorptive process is the regeneration of the adsorbents. Generally, there are two techniques for regeneration of the adsorbents, the thermal treating and the solvent elution. Zeolites based adsorbents are usually regenerated by solvent washing at ambient temperature but effective removal of the solvent is difficult. Therefore solvent washing is suitable for certain applications involving heat sensitive materials (i.e. active carbon). In this study, the thermal treatment was adopted and the adsorbent was regenerated by flowing nitrogen at 500°C for 15 h. The breakthrough curve of the regeneration step was constructed (Figure 12). It was concluded that after three regeneration cycles, the adsorption capacity remained unchanged. The XRD pattern of the regenerated adsorbent showed that the structure of zeolite was not affected by thermal treatment (Figure 13).

Conclusion

Characterization of the synthesized NaX zeolite by XRD, XRF and FT-IR confirmed that the pure phase of zeolite had been prepared. Experimental data fitted well with Langmuir model. It was also concluded that the aromatics adsorption was an spontaneous and exothermic process. The negative value of the standard entropy change (ΔS°) indicated that randomness decreased with the uptake of the aromatics. The high correlation coefficient value (R2=0.9965) indicated that the adsorption followed pseudo second-order model. The adsorbent was throughly regenerated with flowing nitrogen at 500°C and the initial adsorption capacity was kept after three sucsecive regenerations. Since the dearomatization increases the dehydrogenation selectivity, alkylation selectivity, and sulfonability of LAB, the suggested method of this research can be used for dearomatization of different streams in petrochemical industries.

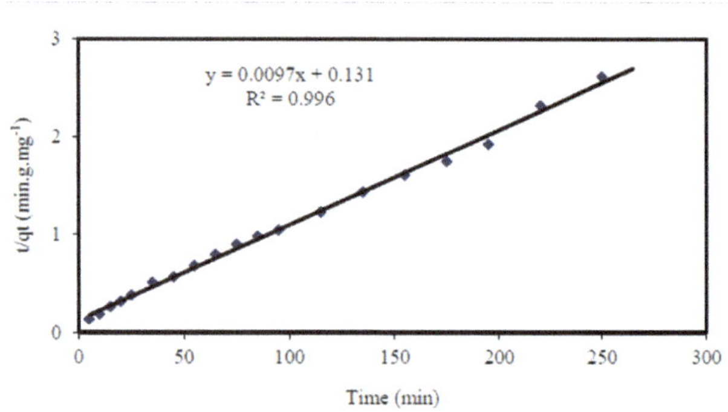

Figure 11: Pseudo second order kinetic plot for aromatics adsorption (Init.Conc= 3.5%w, 12 g synthetic zeolite NaX).

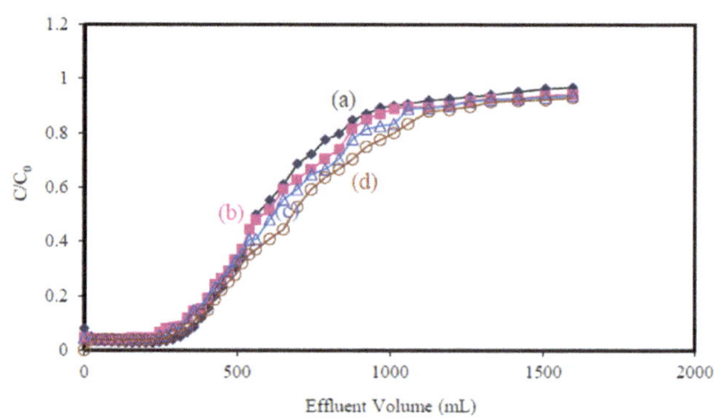

Figure 12: Breackthrough curve for three regeneration cycle (Init. Conc= 3.5%w, T= 25°C).

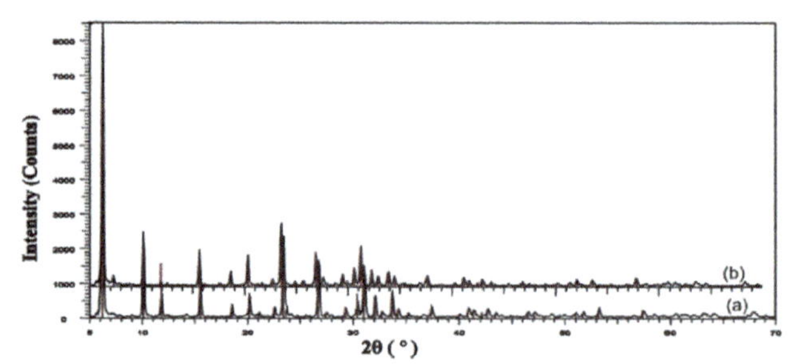

Figure 13: XRD pattern of synthetic zeolite NaX (a) before adsorption, (b) after regeneration.

Acknowledgements

Financial support of the Iran Chemical Industries Investment Company is greatly acknowledged.

References

1. Almeida JLG, Dufaux M, Bentaarit Y, Naccache C (1994) Linear Alkylbenzene. Jaocs 71: 675-694.
2. Kocal JA, Vora BV, Imai T (2001) Production Of Linear Alkylbenzenes. Appl Catal A 221: 295-301.
3. Kocal JA (1994) Increasing catalyst life and improving product linearity in the alkylation of aromatics with linear olefins. U.S. Patent No. 5334793.
4. Kocal JA, Korous DJ (1994) Alkylaromatic process with removal of aromatic by-products. U.S.Patent No.5276231.
5. Goncalvez De, Almeida G, Tejero J (2009) Method for purifying alkyl aromatic compounds. U.S.Patent No.0308728.
6. Yang RT (1973) Adsorbents: Fundamentals and Applications, Wiley interscience.
7. Murillo R, Garcia T, Aylon E, Callen M, Navarro M, et.al (2004) Adsorption of phenanthrene on activated carbons: Breakthrough curve modeling. J Carbon 42: 2009-2017.
8. Susu A (2000) Mathematical modeling of fixed bed adsorption of aromatics

and sulphur compounds in kerosene deodorisation. Chem Eng Process 39: 485-497.

9. Breck DW (1973) Zeolite Molecular Sieves, Krieger publishing company, Malabar, Florida.

10. Dorogochinskii A, Frid M, Borisova LV, Breshchenko V, Shutova TL (2004) Thorough dearomatization of kerosene cut on zeolite. Chem Technol Fuels Oils 9: 577-579.

11. Fathizadeh M, Nikazar M (2009) Adsorption of Aromatic fromAlkane/ Aromatic Mixtures by NaY Zeolite. J Chem Eng Jpn 42: 241-247.

12. Plee D (1998) Method for the purification of a paraffin cut. U.S Patent No.5731488.

13. Al-Zaid K, Owaysi F, Akashah S, Eltekov YA (1989) Sorption of aromatic compounds from solutions by zeolite 13X. Studies in Surface Science and Catalysis 49: 945-952.

14. Berger Ch, Glaser R, Rakoczy R, Weitkamp J (2005) The synthesis of large crystals of zeolite Y re-visited. Microporous and Mesoporous Materials 83: 333-344.

15. Zhao Zh, Cui X, Ma J, Li R (2007) Adsorption of carbon dioxide on alkali-modified zeolite 13X adsorbents. Int J greenh gas con 1: 355-359.

16. Keka O, Narayan C, Amar Nath S (2004) Zeolite from fly ash:synthesis and characterization. Bull Mater Sci 27: 555-564.

17. Zheng Hong, Han Lijie, Mu Hongwe (2008) Adsorption characteristics of ammonium ion by zeolite 13X. J Hazard Mater 158: 577-584.

18. Daneshvar N, Aber S, Khani A, Rasoulifard MH (2007) Investigation of adsorption kinetics and isotherms of imidacloprid as a pollutant from aqueous solution by adsorption onto industrial granular activated carbon. Int J Food Agric Environ 5: 425-429.

19. Guo L, Sun CM, Li GY, Liu CP, Ji CN (2009) Thermodynamic and kinetics of Zn(II) adsorption on crosslinked starch phosphates. J Hazard Mater 161: 510-515.

20. Bhattacharyya KG, Sen Gupta S (2006) Kaolin, montmorillonite and their modified derivatives as adsorbents for removal of Cu from aqueous solution. Sep Purif Technol 50: 388-397.

The Effect of Rapid Expansion of Supercritical Solution (RESS) Parameter on Sub-Micron Ibuprofen Particle Forming

Phung Kim Le[1]* and Kien Anh Le[2]

[1]*Department of Chemical Engineering, University of Technology, Ho Chi Minh City, Vietnam*
[2]*Institue of Tropical and Environmental, District10, Ho Chi Minh City, Vietnam*

Abstract

Rapid expansion of supercritical solution (RESS) is the most common method of pharmaceutical particle forming approaches using supercritical fluids. The RESS method is a technology which produces a small solid product with a very narrow particle size distribution, organic solvent-free particles. This process is also simple and easy to control the operating parameters in comparison with other methods based on supercritical techniques. In this study, Ibuprofen, anon steroidal anti-inflammatory drugs (NSAID), has been micronized by RESS; the size and morphology effects of three different RESS parameters including extraction temperatures, extraction pressures and expansion nozzle temperatures have been investigated and optimized by response surface method. The particle size distribution has been measured by Light diffraction scattering (LDS) method. SEM has been used to analyze the surface structure, DSC and FTIR for thermal and chemical structure analysis. The obtained results are average particle sizes below 1 µm decreased significantly compared to the initial particle size and the optimal operating conditions are identified, enabled the pilot-scale studies and industrial operation.

Keywords: Ibuprofen; Rapid expansion of supercritical solution (RESS); Response surface methodology; Supercritical fluid

Introduction

In the pharmaceutical industry, the large numbers of drugs are insoluble or poorly soluble in water, so it leads to a low bioavailability which is one of the mainproblems in processing new applications [1]. One of solutions is the reduction in particle size (increase in surface area), which leads to an improvement of the dissolution behaviour. There were several conventional techniques utilized for particle size reduction such as grinding, milling, spray-drying and high-pressure homogenization. However, the disadvantages of these techniques are thermal and chemical degradations of products, large amounts of solvent use, high energy requirements and broad particle size distributions (PSDs). New methods which can overcome these drawbacks are supercritical fluid technologies and The Rapid Expansion of Supercritical Solution (RESS) is the most popular among them [2]. RESS produces fine particles with narrow PSDs while being safe with environment and human health [3]. Although all the parameters affect particle size, nozzle and its characterization have a critical role in RESS process. Hezave AZ [4] and F. Esmaeilzadeh [5] studied the effect of nozzle diameter on particle size of creatine monohydrate and diclofenac. Kayrak [1] carried out a survey of the influence of spraying distance from nozzle to the collection surface on the size of ibuprofen particles. Z. Huang, [6] N. Yildiz [7] , A. Keshavars [8] researched the effect of extraction temperature on the average particle sizes of aspirin, salicylic acid raloxifene, cholesterol and lidocaine. Ibuprofen particle size was studied at different nozzle temperature by Z. Fatemeh [9]. A. Z. Hezave [4], F. Esmaeilzadeh [5], H. R. Satvati [10]; N. Yildiz [7] researched the effect of extraction pressure on sizes of creatine monohydrate, diclofenac, cholesterol and salicylic acid respectively.

Design of experiment (DOE) is an established tool in the pharmaceutical industry for the investigation of complex processes and to study a number of variables and their interactions. Several studies have investigated RESS parameters using an experimental design method. Despite the large number of reports, experimental design has not been used to evaluate the influence of variables in the case of Ibuprofen particle forming. The aim of this study was to evaluate the effect of process parameters on particle characteristics using the Rapid expansion of supercritical solution method, and compare the effect of these parameters on the size distribution. The particle mean size were the response functions (empirical models) of the three factors including extraction temperature, extraction pressure and nozzle temperature that have an effect on particle properties. The methodology applied in this work leads to a mathematical model which describes the effects of process variables on the studied response, and therefore the response behaviour can be predicted over the whole experimental domain.

Experimental

Material

Ibuprofen 99.7% is supplied by Aldrich Sigma in white powder form. CO_2 (>99.9%) originates from Singapore. Nonionic surfactant, polyoxyethylene sorbitan monooleate (Tween 80), originates from Aldrich Sigma.

Experimental Apparatus

The gaseous CO_2 from a cylinder capsule was cooled to make liquid CO_2. The liquid CO_2 was pressurized to the extraction pressure by a high pressure pump and then fed into an extraction vessel (100 ml), which contained dry powder ibuprofen. The extraction vessel was placed the temperature- controlled chamber of supercritical fluid extractor and the paddle to support the material dissolution. The RESS system was operated in 120 minutes with 100 mg ibuprofen for each experiment. Supercritical solution was expanded rapidly through a nozzle, where it

***Corresponding author:** Phung Kim Le, Faculty of Chemical Engineering, Ho Chi Minh City University of Technology, Ho Chi Minh City, Vietnam

was heated to avoid clogging, into ambient environment at atmospheric pressure to obtain small particle. CO_2 returned to gaseous form quickly and separated from the product.

Experimental design

The study of the influence of operating parameters on particle properties was based on an experimental design. Three variables were evaluated in term of their effect on the particle characteristics. These variables were: (Z_1) extraction temperature of extraction chamber, (Z_2) extraction pressure and (Z_3) nozzle temperature of the expansion chamber. The response functions were chosen is the mean size of particle (Y). Regression analysis of the data was carried out within a statistical design package ('Design-Expert' version 7.0.0, Stat Ease, Inc,) using a quadratic model with interactions:

$$Y = b_0 + \sum_i b_i X_i + \sum_j b_j X_i X_j \quad (1)$$

With i,j=1, 2, 3

Where Y is the estimated response, Xi is the scaled independent process variable (−1=low level, 0=central level and +1=high level) and the coefficients b_0, b_i, b_{ii}, b_{ij} characterise respectively the constant, the linear and quadratic effects of the variable X_i and the interactions between X_i and X_j. To define these coefficients, it is required a star point at two level in every variable X_i (+α=1.414 and -α=-1.414).

The complete model is based on the simultaneous variation of three factors at two levels implying the running of 18 trials including 4 ones at center points. Each point of experiment was repeated three times and the average was used for the statistic. The ranges of investigated parameters are extraction temperature (35-55°C), extraction pressure (90 -150 bar) and nozzle temperature (80-100°C)

Analysis

Differential scanning calorimetric (DSC) supplied information about physical properties (melting point and enthalpy of fusion) of unprocessed and RESS-processed ibuprofen. The sample (~1,5mg per run) was heated in an aluminum standard pan under a nitrogen gas flow of 20ml/min with heating rate of 5°C/min.

Light diffraction scattering (LDS) was used to determine the particle size and size distribution of the raw and processed particles. Solvent was the combine of ibuprofen saturated solution in distilled water and Tween 80 0.1% v/v, being sonicated for 30 minutes before performing the analysis.

Scanning Electron Microscope (SEM) was used to analyze the size and morphology of particles. The sample was sonicated for 3 minutes in the same solvent of LDS, and then few droplets were added on the SEM stub.

The FT-IR spectra of unprocessed and processed ibuprofen were taken to obtain information on the change of chemical structure after RESS processing.

Results and Discussion

DSC

Similar DSC curves were observed for the unprocessed material and the ibuprofen particles produced by RESS. On the basic of the DSC measurements, it was confirmed that the processed ibuprofen is still in a crystalline state. However the melting point of the product (74.28°C) was lower compared with the unprocessed (75.67°C). The result indicated a slight reduction in the degree of crystallinity of ibuprofen after processing with RESS. These data correspond with the results published by Hezave [11] and Mansouri [12] (Figures1 and 2).

FT-IR

It can be seen from Figure 3 that FT-IR spectra between unprocessed and processed ibuprofen do not show any significant differences. The assignment of bands were as follows: 1558.87 va 1507.33 cm^{-1} (C=C bond in benzene ring), 3090.2– 3018.75 cm^{-1} (=C-H bond in benzene ring), 1117 cm^{-1} (-O- stretching vibrations), 1379.52 cm^{-1} (-CH_3 stretching vibrations), 1720.6 cm^{-1} (C=O stretching vibrations), 2869.68–2921.68 cm^{-1} and 1364,19 cm^{-1}(-CH- stretching vibrations), 1461.27 cm^{-1}(-CH_2 stretching vibrations).

SEM and LDS

The particle size of the original ibuprofen was around 70 μm as

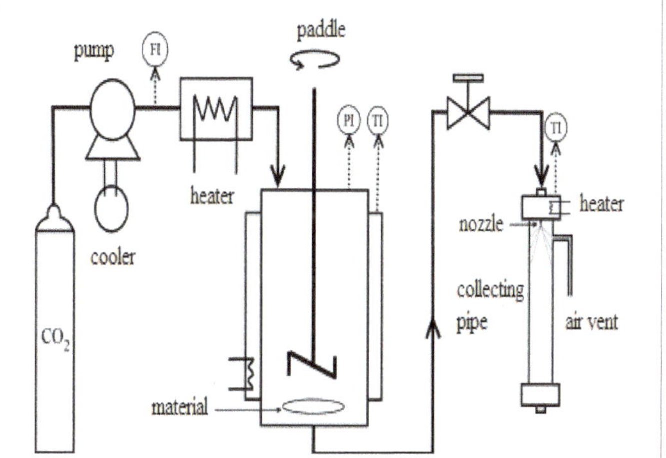

Figure 1: Schematic diagram of the RESS apparatus.

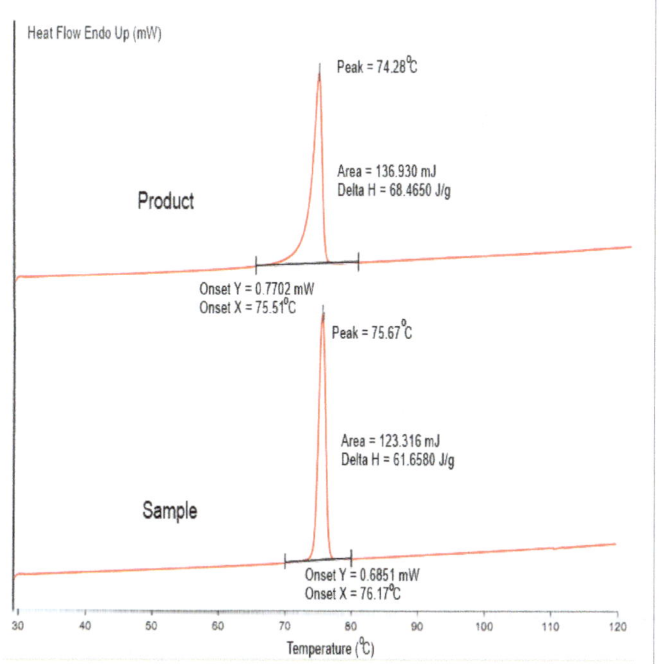

Figure 2: DSC analysis of ibuprofen before and after RESS processing.

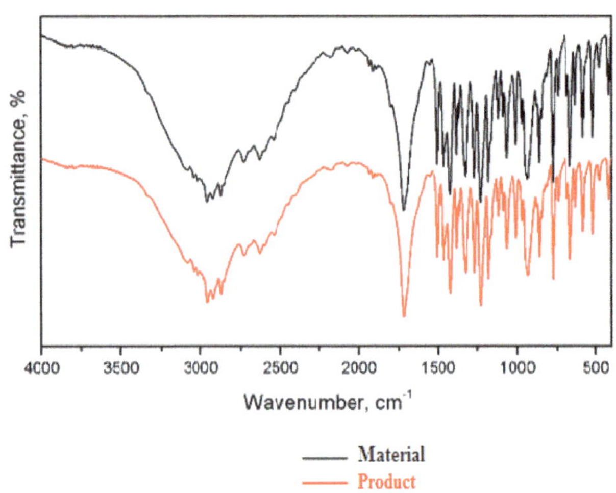

Figure 3: Comparison of ibuprofen FTIR spectra before and after RESS processing.

shown in Figure 4. After RESS processing, the particle size reduced significantly although the product was agglomerated. The SEM images revealed that the degree of aggregation was reduced after sonication. The similar results were found in researches of Kayrak [1] and Charoenchaitrakool [13].

The differences in size of ibuprofen before and after RESS process were also shown by LDS. The obtained results proved that the particle size of products was very small with the narrow PSDs, which was much smaller than that of the original sample.

Experimental design

To evaluate the influence of the considered variables on the measured responses, a mathematical model (Eq. (1)) was postulated with coefficients to be estimated. The he average particle sizes of products were the subject of a mathematical modelling (Figure 5).

The average particle size of products was measured by LDS and shown in Table 1.

The regression equation of objective function, which was mean particle size of ibuprofen, was integrated as following equation:

$$y = 5.55 + 0.59x_2 + 0.54x_3 - 0.88x_1^2 - 1.23x_2^2 - 1.39x_3^2$$

with y: average particle size, x_1: extraction temperature, x_2: extraction pressure, x_3: nozzle temperature

The effect of extraction temperature

According to the regression and the diagram (Figure 6), the mean particle size was seen to increase as the temperature increase from 35°C; reach the highest point at 45°C and then decrease till 55°C. The result might be explained as follows.

Increasing the extraction temperature led to a decrease inthe density of CO_2 and a concurrent increase in the solute'ssublimation pressure. The decrease of the solvent densitycaused a decrease of the solvent strength, which was the reason for the lower supersaturation and nucleation rate. Therefore, the particles might preferably grow to become larger. On the other hand, aconcurrent increase in the solute's vapor pressure was responsiblefor an increase in the ibuprofen solubility, which led to smaller particles. The net effect ofthese two competing factors resulted in the slight influence of extraction temperature. This result is suiTable with research of Z. Huang[6].

The Effect of extraction pressure

Extraction pressure had a considerable effect on ibuprofen particle size. The mean particle size increased dramatically as the pressure increased from 90 bar; reached a peak at 120 bar and then decreased slightly till 150°C. The possible explanation might be that the rise extraction pressure led to the growth of ibuprofen solubility, which resulted in the increase of higher supersaturations in the fluid upon expansion. Perhaps at high aspirin concentrations, the particle growth might be dominant. Therefore large particles could be obtained. Similar results havealso been reportedfor lidocaine by J.-T. Kim [14]. However, when pressure climbed to a certain level, the nucleation process was the deciding factor. So, the particle might become to decrease in size. Similar results can be referred in A. Z. Hezave [4] , A. Z. Hezave [5], H. R. Satvati [7] and N. Yildiz [10] researches for creatine monohydrate, diclofenac, cholesterol and salicylic acid respectively

The effect of nozzle temperature

In all runs, the nozzle temperature was kept higher than the

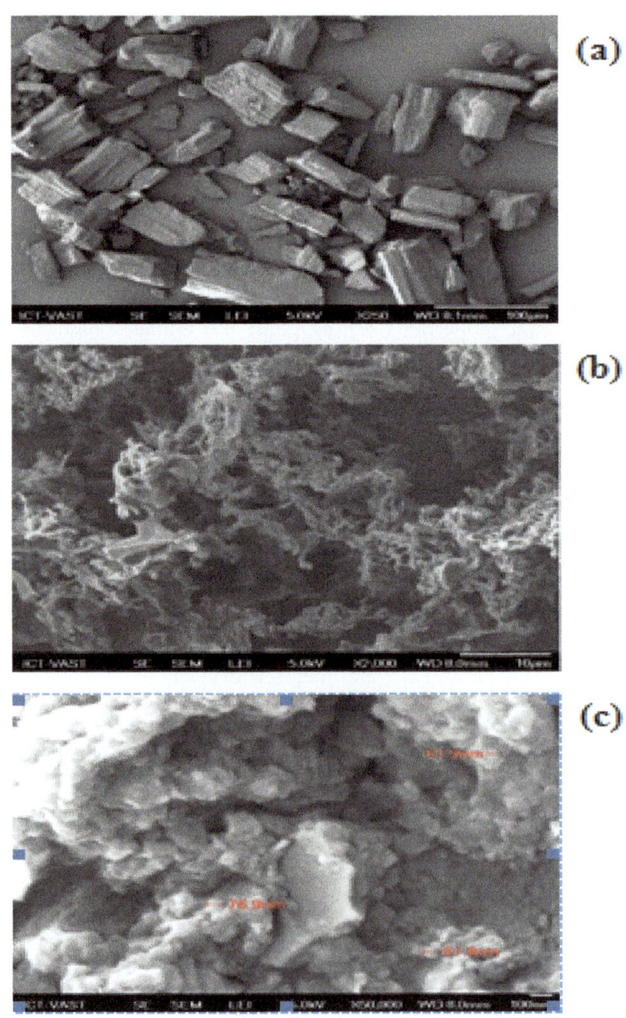

Figure 4: Typical SEM of (a) original and (b) Run 15 without sonication (c) Run 15 with 3 mins of sonication in solvent.

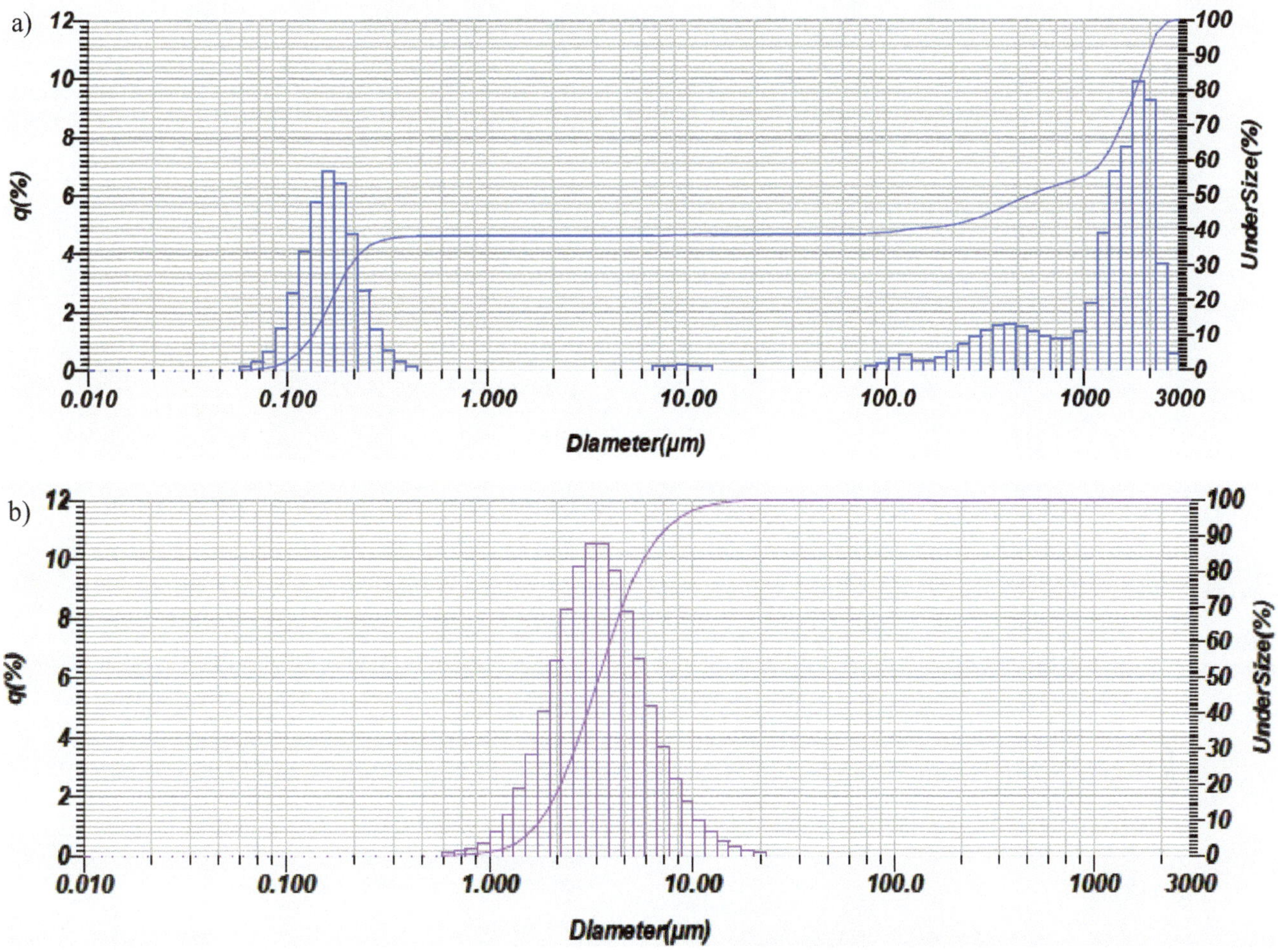

Figure 5: LDS analysis of ibuprofen before (a) and after RESS processing (b).

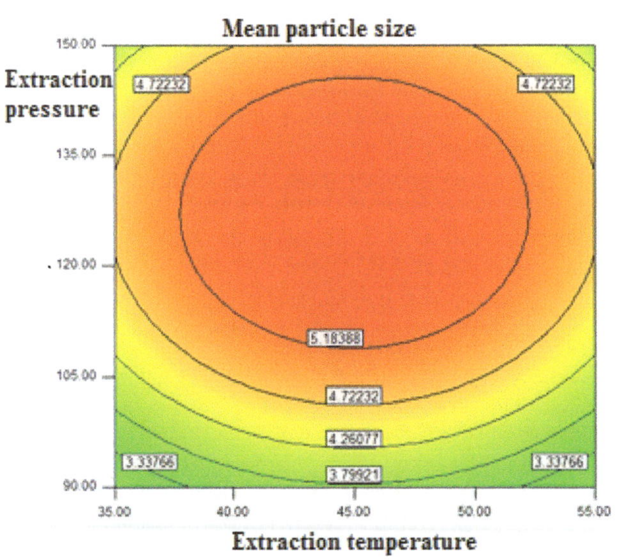

Figure 6: The effects of extraction temperature and extraction pressure on particle size at nozzle temperature 90 °C.

extraction temperature to prevent the early precipitationof ibuprofen solute along the expansion line.The pattern of extraction pressure was similar for nozzle temperature. The mean particle size increased as the nozzle temperature rose significantly in the range of 80°C-90°C. When the temperature continued to climb from 90°C to 100°C, the particle size of product reduced gradually. The reason for this phenomenon might be the high supercritical solution concentration andnumerous nucleus collisions that led to large clustersformation.However, when the nozzle temperature reached to 90°C, the nucleation rate might increase, so the particle size declined. This result is suiTable with research of Zahibi F.[9].

The optimization of Ibuprofen particle formation by RESS process

The optimum condition to obtain the smallest particle diameter is shown in Table 2. There were 2 operating conditions that were able to form the particles below 1 μm. However, the first condition which had extraction temperature 35°C was preferable to select in industry. It was also suiTable with the actual experiment.

Conclusions

The particle size and particle size distribution of ibuprofen was investigated by RESS process. The obtained average particle diameters of product were from 0.7222 μm to 5.8077 μm, which were much

Run	Extraction temperature (°C)	Extraction pressure (bar)	Nozzle temperature (°C)	Mean particle size (μm)
1	35	90	80	0.7222
2	55	90	80	2.1030
3	35	150	80	1.5955
4	55	150	80	1.2511
5	35	90	100	1.3137
6	55	90	100	2.4398
7	35	150	100	3.0495
8	55	150	100	3.4618
9	30.86	120	90	4.6505
10	59.14	120	90	3.14055
11	45	77.58	90	1.6840
12	45	162.42	90	4.7346
13	45	120	75.86	2.2324
14	45	120	104.14	3.5306
15	45	120	90	5.5656
16	45	120	90	4.9153
17	45	120	90	5.76767
18	45	120	90	5.72292

Table 1: Experimental conditions and the results of the mean particles size of the particles produced by RESS process.

Extraction temperature (°C)	Extraction pressure (bar)	Nozzle temperature (°C)	Mean particle size (μm)
35	90	80	0.922
55	90	80	0.922

Table 2: Results of the optimization.

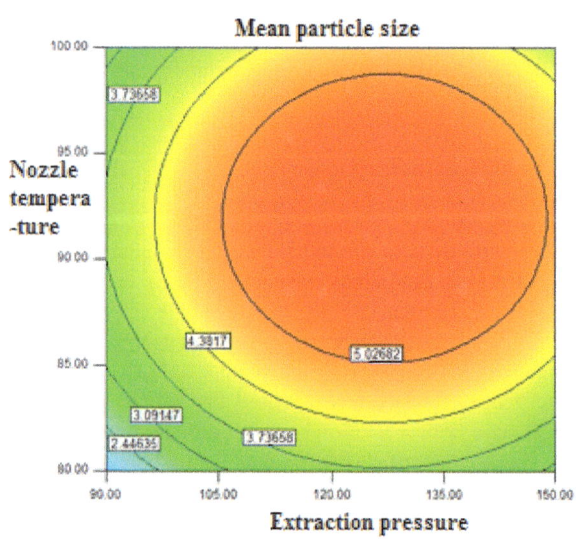

Figure 7: The effects of extraction pressure and nozzle temperature on particle size at extraction temperature 45°C.

smaller than that of material. The processed ibuprofen also had narrow PSDs compared with unprocessed one. The original and processed ibuprofen was analyzed by DSC and FTIR. No significant change was defined in structure of ibuprofen during the process. The effects of process parameters such as extraction temperature, extraction pressure and nozzle temperature were researched on ibuprofen particle size. Response surface methodology (RSM) was successfully applied for optimization of particle size design parameters by RESS. The optimum values of the variables (extraction temperature 35°C, extraction pressure 90 bar and extraction temperature 35°C) were determined inorder to obtain the smallest median particle diameter. The extraction pressure and nozzle temperature had significantly effects; however, extraction temperature affects slightly on particle size. The mathematical model then can be used for predicting the behaviour which describes the effects of process variables on the products (Figures 7 and 8).

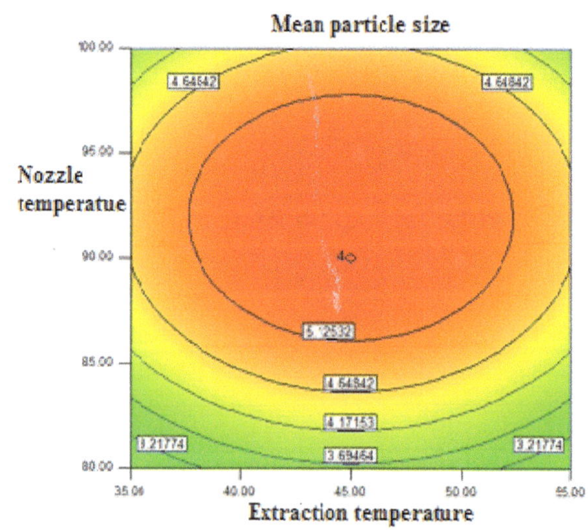

Figure 8: The effects of extraction temperature and nozzle temperature on particle size at extraction pressure 120 bars.

Acknowledgements

This research is funded by Vietnam National University Ho Chi Minh City (VNU-HCM) under grant number B2013-20-04.

References

1. Defne Kayrak UA, Oner Hortacsu (2003) Micronization of Ibuprofen by RESS. The Journal of Supercritical Fluids 26: 17-31.
2. Masoomeh PSF, Aliraza VARN (2012) Production of ultrafine drug particles through rapid expansion of supercritical solution; a statiscal approach. Powder Technology 225: 21-26.
3. Gupta BR (2006) Nanoparticle Technology for Drug Delivery. Drug and The Phamaceutical sciences ed. Taylor & Francis Group: New York 51-82.
4. Hezave AZ, Esmaeilzadeh F (2010) Micronization of creatine monohydrate via Rapid Expansion of Supercritical Solution (RESS). The Journal of Supercritical Fluids 55: 316-324.
5. Hezave AZ, F Esmaeilzadeh (2011) The effects of RESS parameters on the diclofenac particle size. Advanced Powder Technology 22: 587-595.
6. Huang Z (2005) Formation of ultrafine aspirin particles through rapid expansion of supercritical solutions (RESS). Powder Technology 160: 127-134.
7. Nuray Yildiz, Se T, Onur D oker, Ayla C (2007) Micronization of salicylic acid and taxol (paclitaxel) by rapid expansion of supercritical fluids (RESS). The Journal of Supercritical Fluids 41: 450-451.
8. Keshavarz A (2013) Preparation and characterization of raloxifene nanoparticles using Rapid Expansion of Supercritical Solution (RESS). The Journal of Supercritical Fluids 63: 169-178.
9. Fatemeh Zabihi, MA, Ali Akbar Sakordi (2011) Drug Nano-Particles Formation by Supercritical Rapid Expansion Method; Operational Condition Effects Investigation. Iran J Chem Chem Eng 30: 7-15.
10. Satvati HR , Lotfollahi MN (2011) Effects of extraction temperature, extraction pressure and nozzle diameter on micronization of cholesterol by RESS process. Powder Technology 210: 109-114.
11. Ali Zeinolabeedini Hezave FE (2010) Micronization of drug particles via RESS

process. The Journal of Supercritical Fluids 52: 84-98.

12. Mansour Mansouri HRP, Vida Vosoughi (2011) Preparation and Characterization of Ibuprofen Nanoparticles by using Solvent/ Antisolvent Precipitation. The Open Conference Proceedings Journal 2: 88-94.

13. Charoenchaitrakool M, FD, Foster NR (2000) Micronization by Rapid Ezpansion of Supercritical Solutions to Enhance the Dissolution Rates of Poorly Water-Soluble. Ind Eng Chem Res 39: 4794-4802.

14. Kim JT, H LK, Ju CS (2010) Micronization and characterization of drug substances by RESS with supercritical CO2. Korean Journal of Chemical Engineering 1139-1144.

Simulation of a Vacuum Cooling System for the Phosphoric Slurry: "Flash Cooler"

Abdennebi N[1]* and Hajji N[2]

[1]National School of Engineers of Gabes (ENIG), University of Gabes, University City, Gabes, Tunisia
[2]Laboratory of Energy and Environment, ENIG, University City, Gabes, Tunisia

Abstract

The present study concerns the phosphoric slurry cooling using a vacuum system "flash cooler". This system has a dual purpose, the first is to ensure the desired cooling of the phosphoric slurry and the second and the mainly purpose is to reduce the fluorine content in the exhaust gas outlet of the phosphate attack units. A simulation was made through the Aspen Plus software features of this vacuum system which is composed of a flash cooler to cool the slurry, a spray tower to clean the fluorinated gas and recovered it to a fluosilicic acid solution and a barometric condenser to create the vacuum all over the system. The determination of the different operating parameters of each device in order to minimize the amount of fluorine released into the atmosphere and maximizes his recovery as a solution of fluosilicic acid, were investigated. The obtained results confirmed the good performance of this system and showed that it is possible to cool the slurry to the desired phosphoric slurry temperature and get a good discount rate of fluorine released into the atmosphere. This pollutant is recovered as fluosilicic acid solution to 20% by mass.

Keywords: Fluorine removal; Flash cooler; Absorption; Fluosilicic acid; Simulation

Introduction

Emissions from the phosphoric acid process in the Chemical Tunisian Group include fluorides gaseous as hydrogen fluoride (HF) and silicon tetra fluoride (SiF_4). The gases are released during the attack on the phosphate rock. These gases, if they are released into the atmosphere unchecked, can have a negative impact on the environment. The reduction of these pollutants is in absorption towers operated under vacuum. In these towers, fluorinated gases are absorbed and recovered into a solution of fluosilicic acid at 20% by mass. This solution can be neutralized or it can serve as feedstock for the production of aluminum fluorine, hydrofluoric acid and cryolite later. Note that the temperature of gas outlet from the phosphoric unit is at 70°C, the maximum rates allowed fluoride in the effluent is determined by the Tunisian standard norm to 10 mg/Nm^3[1].

Vacuum system reducing fluorine content

The system used for cooling the phosphoric slurry and for reducing the amount of fluoride released during the phosphate attack is shown in Figure 1.

This system includes a "flash cooler" which is the mainly equipment to cool the phosphorus slurry. A pump is installed at the exit of the flash to recycle the cooled slurry in the attack tank. The steam derived from the flash cooler full of fluoride is washed in a column using a spray solution of fluosilicic acid as an absorption solution. The purpose of this column is to absorb the fluoride found as a form of HF and SiF_4 and recovered it into a fluosilicic acid solution at 20% by mass. The washed gas derived from the spray tower then passes into a barometric condenser which is responsible for creating the vacuum in all the system adopting a sea water drop.

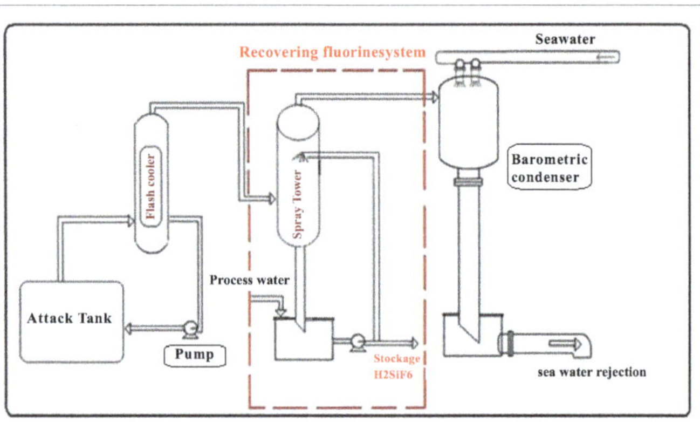

Figure 1: Flash cooler system.

Process simulation

The main objective of the process simulation is to find the optimal operating conditions:

- To maximize the recovery of fluorine,
- To ensure cooling of 3-4 degrees Celsius of the slurry at most,
- To obtain a solution of fluosilicic acid at 15-20 wt%.

To simulate this vacuum system, we used the software features over Aspen Plus. The different models of unit operations used are presented in Table 1, while interconnections between these different models are shown in the block diagram (Figure 2).

Note that the unit system used is the "METCBAR" which expresses the temperature in °C, the pressure in bar and the mass flow in kg/h.

Operating parameters of the vacuum system

The simulation was made on the basis of an estimated composition

***Corresponding author:** Abdennebi. N, National School of Engineers of Gabes (ENIG), University of Gabes, University City, 6029, Gabes, Tunisia

of the phosphoric slurry [2,3]. The different parameters of the system to get the desired aims are listed in Table 2. The flash and the barometric condenser are assumed adiabatic.

The principles reactions considered in this system are:

$H^+ + F^- \leftrightarrow HF$ (1)

$SiO_2 + 6F^- + 2H_2O \leftrightarrow SiF_6^{--} + 4OH^-$ (2)

$SiO_2 + 4HF \leftrightarrow SiF_4 + 2H_2O$ (3)

Results and Interpretation

It was found that the three degrees of cooling of the phosphoric slurry is provided by a pressure of 240 mmHg at the flash. The

Equipments	Aspen Plus model
Flash cooler	Flash 2
Scrubber	Absorber (RADFRAC)
Mixer	Fsplit
Barometric condenser	Flash 2
Pump	Pump

Table 1: Different models of equipments used.

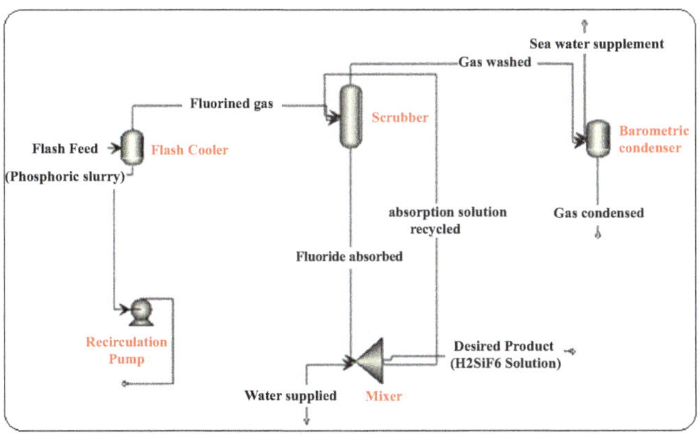

Figure 2: Block diagram of the vacuum cooling system.

Equipments	Operating parameter
Flash cooler	Pressure (bar) 0,32
Scrubber	Number of stage 10
Mixer	Rate of H_2SiF_6 (wt %) 15-20
Barometric condenser	Pressure (bar) 0,28

Table 2: Operating parameters.

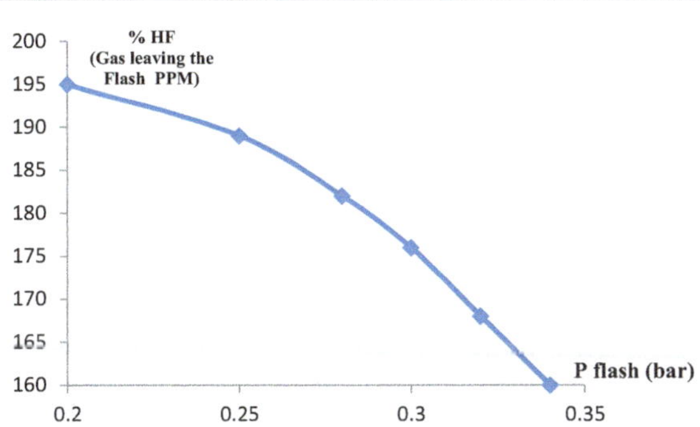

Figure 3: Pressure influence on HF rate in the fluorinated gas.

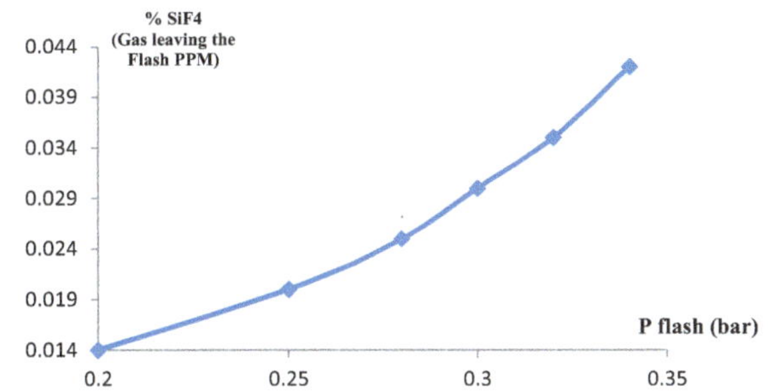

Figure 4: Pressure influence on SiF_4 rate in the fluorinated gas.

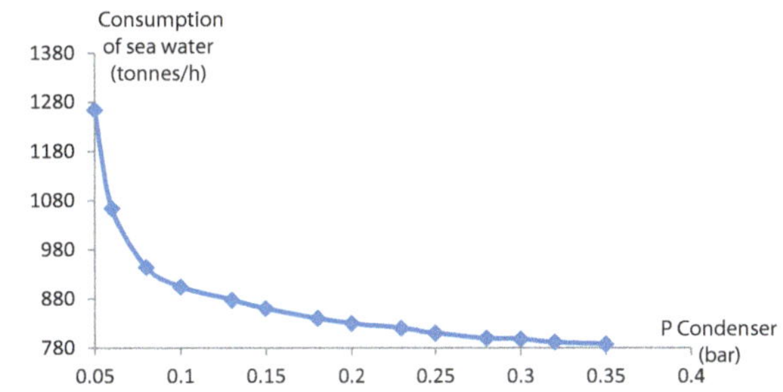

Figure 5: Pressure influence on the consumption of sea water in the barometric condenser.

	Flash Feed	Fluorinated Gas	Washed Gas	Desired Product (H_2SiF_6)	Condensed gas released at sea water
Temperature (°C)	79	76	109,5	45,3	34,2
Pressure (bar)	1	0,32	1	1,02	0,28
Flow (Tons/h)	7250	41	100,6	112	900,6
% HF	8,00E-05	168 PPM	75 PPB	traces	58 PPB
% SiF_4	0,0002	0,035	657 PPB	820 PPM	Traces
% SiF_6^{--}	0,044	0	0	20,8	Traces

Table 3: Results of the simulation.

simulation results showed that the pressure in the flash cooler has a much influence on the composition of the gas leaving it. Note that the rate of HF in the gas leaving decreases with the pressure of the flash while the rate of SiF_4 increases (Figure 3 and Figure 4). Figure 5 shows the effect of barometric condenser pressure on the consumption of sea water that provides a total condensation. It is clear that a greater flow of sea water is necessary to achieve a higher vacuum.

In addition, Table 3 recapitulate the different results obtained after this simulation. We can conclude that this vacuum system has almost eliminate the fluorine rate from the gas atmospheric emission and recovered it as a fluoscilicic acid solution 20% by mass. It's clear that also the condensed gas released at the sea water contain some traces of fluorine, which confirm a good performance of this system.

Conclusion

In this work, we have elaborated a simulation of a vacuum cooling system of phosphoric slurry by the functionality of Aspen Plus features.

This system comported a flash cooler, a spray tower responsible for the absorption of fluorine, a barometric condenser for reaching the vacuum on the whole of the system. This simulation has approved that we can assure the three degrees of cooling by a pressure of 0, 32 bars in the flash and offers a good abatement of fluorine which is recovered into a fluosilicic acid solution at 20 % by mass.

References

1. President of the Republic (2010) Decree No. 2010-2519 of 28 September 2010 laying down the limits to the source of air pollutants from stationary sources.
2. Paul M, Mathias P, Mike M (1998) Simulation of Phosphoric Acid Production by the Dihydrate Process. Aspen Technology Inc.USA.
3. Paul M, Mathias P, Chau-Chyun C, Marten W (2000) Modeling the Complex Chemical Reactions and Mass Transfer in a Phosphoric Acid Reactor. KEMWorks Technology.

Physico-Chemical Properties of Before and After Anaerobic Digestion of *Jatropha* Seed Cake and Mixed With Pure Cow Dung

Nisha Yadav[1], Rajdeo Kumar[2*], Dr. (Mrs.) Laxmi Rawat[1,2] and Sarita Gupta[3]

[1,2]Forest Ecology and Environment Division, Forest Research Institute, Dehradun, India
[3]CSWCRTI, Dehradun, India

Abstract

In this study, the biogas production potential of *Jatropha* seed cake (JSC) was assessed in the laboratory. The physico-chemical analysis data Total solids (TS%), Volatile solids (VS%), Organic carbon (OC%), Total nitrogen (N%), Total phosphorus (P%), Total potassium (K%), Moisture content (MC%) of the cakes confirmed that JSC is feasible of anaerobic digesters. By blending, JSC with cow dung in different ratios, the feedstock of the biogas digesters was prepared. From the nine treatments to analyze the changes in physico-chemical parameters of the fresh and digested slurry of moisture content is highest after digestion Treatment control-2, 15.8% and lowest before digestion (12.2%). But before digestion, Treatment control-2, TS is 98.2% and VS is 93.12% higher than after digestion. The organic nutrients content Treatment control-2 is C%, 27.37%, N% 2.67 and K % 1.98 of BD is higher than AD but P% 1.078% thus the phosphorous content had increased during the digestion process.

Keywords: *Jatropha* seed cake; Pure cow dung; Anaerobic digestion; Total solid; Volatile solid; Organic nutrient content.

Introduction

Jatropha curcas is a perennial shrub to small evergreen tree of up to 6m height, adapted to all kinds of soil and does not demand any special nutritive regime [1]. It has been introduced in Africa and Asia and is now cultivated worldwide. In India, it was believed to be introduced by the Portuguese settlers during 16th century. It is a multipurpose, deciduous, small tree reported to be cultivated in drier regions of central and western parts of India. Recently, it has also been introduced in the northern and southern states of India. The plant is widely distributed and fits easily into agriculture system in the form of hedges, windbreak and erosion barrier or as a source of fire wood [2].

In recent years there are growing concerns about the utility of *Jatropha curcas* as a source of biofuel plantations in order to generate alternate source of energy. *Jatropha* seed cake is used as a fertilizer but the literature and data regarding the efficacy of improving crop yield or conditioning soil is limited. *Jatropha curcas*, commonly known as physic nut, belong to the euphorbiaceous family. It is a hardy plant, thrives on degraded land and requires limited amount of nutrients and water. The seeds contain about 300-350 gkg^{-1} oil, which can be used as a fuel directly or, in its trans esterified form, as a substitute for diesel. Large scale planting of *jatropha* has taken place or is planned in India, china, Madagascar, Myanmar and many other developing countries, with the aim of using the oil as biodiesel [3] in these countries the oil is produced from whole seeds using a screw press. The seed has a hard, black outer shell and kernel range from 350 to 400 $gm.kg^{-1}$ and from 600 to 650 $gm.kg^{-1}$ respectively.

The seed cake left as a by-product after oil extraction by screw press can contain as much as 500-600 $gm.kg^{-1}$ indigestible shells. The generation of biogas from these cakes would be a best solution for its efficient utilization. Anaerobic digestion has been considered as waste-to-energy technology, and is widely used in the treatment of different organic wastes for example: organic fraction of municipal solid waste, sewage sludge, food waste, animal manure, etc. Anaerobic treatment comprises of decomposition of organic material in the absence of free oxygen and production of methane, carbon dioxide, ammonia and traces of other gases and organic acids of low molecular weight. Anaerobic digestion provides an alternative option for energy recovery and waste treatment.

Material and Methods

Statistics

The included Mean, Standard Error analysis was done using SPSS package and (Minitab 13 for windows software package) where used to analyze the Physico-chemical parameters VS, TS of feedstock and digestate, result of *jatropha* seed cake.

Experimental apparatus setup

Nine treatments in three replications were taken for the study. The treatment, T1 contained *Jatropha* seed cake (JSC) alone and the subsequent treatments consisted of cow dung (CD) and *Jatropha* seed cake in 1:1,1:2, 1:3, 1:4, 1:5, 2:1 and 3:1 for T2, T3, T4, T5, T6, T7 and T8 respectively. T9 with cow dung alone was taken as the control. The experiment was conducted during 84 days constant temperature around 35 ± 1°C for optimization of the biogas yield.

Hydraulic retention period

The optimum period for the economical gas production in batch fermentation depends largely on pattern of daily gas production and the pattern of changes in the calorific value of gas produced i.e. CH_4 content. The study period in the experiment was 84 days.

***Corresponding author:** Rajdeo Kumar, Forest Ecology & Environment Division, Forest Research Institute, Dehradun-248006 (Uttarakhand), India

Inoculum

The inoculum for the anaerobic digestion of non-edible oil seed cakes and cow dung will be prepared by composting the fresh cow dung under anaerobic condition for a period of five days in ambient temperature. The fresh cow dung contains active micro-organisms. A pinch of urea will be added in different treatments to supplement the nutrients so that it enhances the growth and multiplication of consortium of micro-organisms.

Anaerobic batch digestion assembly

Laboratory studies will be conducted by using 2L capacity conical flasks, and graduated cylinders, as batch digesters for carrying out the anaerobic digestion. The feed stocks will be diluted with the requisite amount of water so as to make 7-10% total solid concentration. The gas generated will be collected in inverted calibrated gas collection jar filled with water which will be partly immersed in water bath.

Measuring yield and quality of biogas

The gas production is measured by water displacement method whereas the quality, which is the percentage of methane from the biogas, is estimated by the displacement of sodium hydroxide, with a process held one next to the other. The gas volume produced in the anaerobic reactor was captured in a bottle filled with water, which was kept under pressure. When a gas bubble entered the bottle with water, the gas replaced the water, which was then forced out of the bottle into an empty bottle. The volume of water in the measuring cylinder thus resembled the gas production in the reactor. The displaced water is collected and then using a measuring cylinder its volume would be calculated. The displaced water indicates the total volume of biogas produced. Note that the gas coming out of the digester is stored in the displaced bottle. Adding back the displaced water to the displacement bottle would push out the biogas stored before; and passing it through 5% NaOH solution. The CO_2 from the biogas would be retained in the solution whereas the methane would displace its equivalent volume of NaOH. Collecting the displaced solution and measuring its volume using a measuring cylinder would give the volume of methane from the produced biogas [4]. Hence, it would be possible to estimate the percentage of methane in the biogas, using the following simple equation.

$$\text{CH4\%} = \frac{displaced\ NaOH}{displaced\ water} \times 100$$

Quantitative analysis of biogas yield

Cumulative yields of biogas (expressed in litres/kg of dry matter) from JSC and its admixtures with cow dung and the trends in biogas production are shown in Table 1.

The rate of biogas production from JSC was observed to be rapid as compared to cow dung during the second week of fermentation, the rate of production in the mixtures of cow dung and JSC increased substantially and biogas yield was faster than that from treatment with cow dung alone. JSC alone produced biogas of 76.5 L/kg dry matter (dm) within 14 days and there after its production attained steady pace. The trend pointed out a gradual increase to a maximum between 14 and 21 days, and then the gas production is maintained at a high level to the end of the experiment [5]. The total period taken for digestion was 84 days, beyond which the gas yield were more or less stopped as observed for all the treatments except CD:JSC in 2:1. But the treatment, CD: JSC in 2:1 ratio showed a steady increase in biogas generation throughout the study period and yielded the highest amount of biogas i.e. 500.15 L/kg dry matter followed by CD: JSC in 3:1, generating 420.7 L/kg dm. All the treatments except treatment with *Jatropha seed* cake alone have shown biogas potential higher than CD and produced 263, 256.34, 237.82, 273.54 and 276.94 L/kg dm of biogas by CD:JSC in 1:3,1:2,1:1,1:4 and 1:5 respectively. Compared to 184.34 L/kg from cow dung. This higher yield may be due to the synergistic action of micro-organisms from the co-digestion process.

Sample preparation and feed solution

Jatropha curcas seed was collected from Dehradun. The seed cake was stored in a plastic bag at constant temperature around 35 ± 1°C and was blended prior to experiment. After blending the seed cake were mixed with tap water at ratios of JSC alone, 1:1, 1:2, 1:3, 1:4, 1:5, and 2:1, 3:1, CD alone by weight to create slurry.

Estimation of Volatile Solids, Total Solids, Moisture Content, Organic Carbon, Total Nitrogen, Total Potassium, Total Phosphorus

Total solid (TS)

The total solid content of the *Jatropha* seed cake (JSC) and cow dung (CD) can be determined by the procedure given by APHA [6]. A 10 g of freshly collected sample has to be taken in a disc and placed inside a hot air oven at 105°C for one hour. Then it is taken out and cooled to room temperature in desiccators and weighed. The percent of total solids is computed using the formula.

$$\%\ \text{total solids} = \frac{C-A}{B-A} \times 100\ \%$$

Where,

A= Mass of empty clean and oven dried silica crucible, g

B= Mass of silica crucible + sample, g

C= Mass of silica crucible + sample after oven drying, g

Volatile solids (VS)

The volatile solid content of sample can be determined by adopting the following procedure furnished by APHA [6]. A knows mass of residue obtained from the determined of the total solids is placed in a silica crucible and ignited in a muffle furnace at 550 ± 500°C for 30 minutes. Then the crucible is taken out of the muffle furnace, partially cooled in air, kept in desiccators for few minutes and then weighed. Volatile solids is computed using the formula given below,

$$\%\ \text{VS} = \frac{B-C}{B-A} \times 100\ \%$$

Where,

A= Mass of empty clean and oven dried silica crucible, g

B= Mass of silica crucible + sample, g

C= Mass of silica crucible + sample after ignition, g

Moisture content

The moisture content will be determined using AOAC [7] method, about 15g of sample will be weighed in converted dish previously dried at 98-100°C, cooled in desiccator and weighed soon after reaching room temperature. Cover will be loosened and heated at 98-100°C to constant weight. At the end of drying, cover will be immediately tightened on dish, transferred to desiccator and weighed soon after reaching room temperature.

$$\text{Moisture\%} = \frac{\Delta SB - \Delta SA}{weight\ of\ sample} \times 100$$

Period (weeks)	TI JSC	T2 (1:1)	T3 (1:2)	T4 (1:3)	T5 (1:4)	T6 (1:5)	T7 (2:1)	T8 (3:1)	T9 CD
I	22.04	25.33	27.18	13.21	16.60	24.02	20.91	34.93	6.54
II	76.5	27.31	33.65	15.22	18.81	25.81	23.38	36.38	11.18
III	68.78	30.13	39.87	17.63	20.74	29.74	25.62	37.16	14.55
IV		33.65	38.65	18.26	24.62	34.03	26.05	38.32	18.48
V		31.39	31.23	20.25	27.11	37.96	28.14	38.48	21.59
VI		31.11	27.42	23.84	28.59	39.83	35.29	38.56	25.94
VII		30.34	28.06	33.43	32.79	41.74	38.12	38.63	28.32
VIII		28.56	30.28	36.46	34.54	43.81	45.32	38.78	28.52
IX				41.98	36.63		52.75	39.83	29.2
X				42.71	33.11		59.23	39.87	
XI							69.36	39.76	
XII							75.98		
TOTAL	**167.32**	**237.82**	**256.34**	**263**	**273.54**	**276.94**	**500.15**	**420.7**	**184.34**
Methane content (%)	**56.8**	**57.2**	**57.7**	**56.2**	**58.3**	**58.5**	**61**	**58.7**	**54**

Table 1: Periodic And Total Of Biogas And Methane Content (%) Obtained From Jatropha Seed Cake And Admixtures With Cow Dung (L/Kg At Dry Weight).

ΔSB = weight of dish and sample before drying.

ΔSA = weight of dish and sample after drying.

Carbon content

0.1g seed cake powder into a 500ml conical flask. Add 10ml of 1N potassium dichromate solution. Add 20 ml sulphuric acid and mix by gentle rotation for 1 minute taking care to avoid throwing seed cake powder up onto the sides of the flask. Let stand for 30minutes and dilute to 200 ml with distilled water. Add 10ml phosphoric acid, 0.2 g ammonium fluoride and 10 drops diphenylamine indicator. Titrate with 0.5 N ferrous ammonium sulphate solutions until the colour changes from dull green to a turbid blue. Add the titrating solution drop by drop until the end point is reached when the colour shifts to a brilliant green and prepare and titrate a blank in the same manner.

% Organic Matter=10[1(S/B)] X0.67

S=sample titration

B=blank titration

Nitrogen content

Nitrogen content will be determined using (Kjeldahl method, 1950) half a gram powdered sample will be digested with concentrated H_2SO_4 in presence of digestion mixture ($CuSO_4$ + K_2SO_4 + selenium powder in 20:10:1) till the digest give clear green colour. The digested sample will be further diluted carefully with distilled water to known volume. Then a known amount of aliquot will be transferred to distillation unit (Micro Kjeldahl apparatus) and liberated ammonia will get trapped in boric acid containing mixed indicator. Later, it will be titrated against standard H2SO4 and the amount of ammonia liberated indicator.

Potassium content

The potassium content of the samples will be estimated using the standard procedure as suggested by flame photometer method [8] which is explained below. For 10 ml of dried sample, 50 ml of 0.01 N ammonium acetate will be added. The sample gets leached with an additional 50 ml of ammonium acetate and made up the volume to 100 ml. The standard of potassium is fed to the flame photometer and the readings were noted down using K-filter. The procedure was repeated thrice and the average values are reported.

Phosphorus content

Plant and seed samples were digested with diacid mixture of HNO_3 and $HClO_4$ in the ratio of 9:4 and the extract was made to a definite volume. Total phosphorous was determined by vanado molybdate phosphoric acid yellow colour method at 730 nm [8].

Feedstock compositions of cow dung mixed with *jatropha* seed cake with different ratio were prepared as shown in Table 2.

Results and Discussion

In this study analysis of the different ratio of treatment composition of JSC and with CD showed that Table 3. The physico-chemical properties of the feedstock composition of the substrates before and after digestion in terms of moisture content, VS%, and TS% etc. Were analyzed and presented above Table 3.

Total solids (TS)

Different ratio of seed cake wastes were diluted and total solids of all treatments seed cake and cow dung which were held at for example one treatment TC-2, 98.2% showed a decrease after the retention period of 84 days which were found to be only 93.67%. Thus there was a 40% decrease in the TS% and thus indicated that the continuous feeding for digestion process is very small efficient in removing the total solids. Total solids are very important as far as the process is concerned. According to Abdulkarim et al. [9] at some point in the increase of the percent TS, no further rise in the volume of the biogas is obtained. Balsam [10] and Zennaki et al. found that the optimum solid content obtained for biogas production is in the range 7- 9%. Furthermore, Baserja [11] reported that the process was unstable below a total solids level of 7% (of manure) while a level of 10% caused an overloading of the digester. Itodo and Awulu [12] showed that slurries of higher TS concentrations were more acidic than that of lower TS concentrations.

Volatile solids (VS)

Volatile solids reduction occurred in the slurry during the digestion. The VS fed on an average basis into the digester were found to be all treatment for example one treatment like TC-2, 93.12%. The VS found after the digestion process was found to be 83.31%. Thus there was about 12% decrease in volatile solids. Volatile solids are the solids that are lost on ignition of the dry solids at 55°C centigrade. VS are responsible for biogas production. Since seed cake waste contains 93.12% of volatile solids thus it has a great potential of biogas production and can be used easily and potentially as a raw material for biogas production.

Label Treatment composition	Replication	Amount of seed cake(g)	Amount of cow dung(g)	Amount of water added(ml)	Total volume of slurry (ml)
TC-1 seed cake alone	R1	108.108	0	891.892	1000
	R2	108.108	0	891.892	1000
	R3	108.108	0	891.892	1000
TC-2 1:1	R1	54.05	225.733	720.22	1000
	R2	54.05	225.733	720.22	1000
	R3	54.05	225.733	720.22	1000
TC-3 1:2	R1	72.432	148.98	778.58	1000
	R2	72.432	148.98	778.58	1000
	R3	72.432	148.98	778.58	1000
TC-4 1:3	R1	81.081	112.86	806.06	1000
	R2	81.081	112.86	806.06	1000
	R3	81.081	112.86	806.06	1000
TC-5 1:4	R1	86.486	90.293	823.221	1000
	R2	86.486	90.293	823.221	1000
	R3	86.486	90.293	823.221	1000
TC-6 1:5	R1	90.05	74.94	835.006	1000
	R2	90.05	74.94	835.006	1000
	R3	90.05	74.94	835.006	1000
TC-7 2:1	R1	35.67	302.48	661.846	1000
	R2	35.67	302.48	661.846	1000
	R3	35.67	302.48	661.846	1000
TC-8 3:1	R1	81.081	112.86	806.059	1000
	R2	81.081	112.86	806.059	1000
	R3	81.081	112.86	806.059	1000
TC-9 Cow dung alone	R1	0	451.467	548.53	1000
	R2	0	451.467	548.53	1000
	R3	0	451.467	548.53	1000

Table 2: Feedstock composition of cow dung and jatropha seed cake and taken in 9 treatments with 3 replications.

Treatment Composition (TC)	Moisture content (%)		Total Solids (%)		Volatile Solids (%)	
	BD	AD	BD	AD	BD	AD
TC-1 JSC alone	4	7.5	92.5	88.21	93.77	82.13
TC-2 1:1	12.2	15.8	98.2	93.67	93.12	83.31
TC-3 1:2	11.7	14.7	97.5	92.99	92.88	82.11
TC-4 1:3	10.3	13.5	96.28	91.52	91.55	81.1
TC-5 1:4	8.8	11.3	94.56	90.21	90.22	79.22
TC-6 1:5	7.4	10.2	92.33	88.12	89.71	77.90
TC-7 2:1	5.1	8.6	90.18	84.01	88.95	76.66
TC-8 3:1	5.9	8.3	90.11	85.21	88.62	74.32
TC-9 CD alone	59.31	77.95	22.15	15.44	92.66	76.61

Table 3: Moisture content, Volatile solid and total solid content of treatment before and after anaerobic digestion.

Moisture Content (MC)

The moisture content of the digestate increased from all treatment for example one treatment TC-2, 12.2% to 15.8% thus there was an increase (Table 3). According to Sadaka and Engler [13], water content is one of very important parameter affecting anaerobic digestion of solid wastes. There are two main reasons i.e., (a) Water make possible the movement and growth of bacteria facilitating the dissolution and transport of nutrient and (b) water reduces the limitation of mass transfer of non-homogenous or particulate substrate. In general, the moisture content of the digestate increased with increase in the amount of VS and TS reduction. The moisture content to be maintained for the degradation depends upon the type of the waste. In the present study final moisture content was at 15.8%.

Organic carbon content (OC%)

Organic carbon content reduction occurred in the slurry during the digestion. The OC fed on an average basis into the digester were found to be all treatment for example one treatment like TC-2, 27.37%. The OC found after the digestion process was found to be 24.78%. Percent carbon content in the feedstock was found to 24.80 in cow dung slurry and 48.80 in the JSC waste slurry. The reduction of C by anaerobic processes was therefore probably limited to the production of organic acids, H_2 and CO_2 by facultative bacteria [14]. Thus continuous process of feeding is also efficient in removing carbon content. The carbon compounds are converted to the CH_4 and other gases like CO_2. According to Richard and Wilkie [15,16], anaerobic bacteria do not or very slowly degrade lignin and some other hydrocarbons. In other word, the higher lignin content will lower biodegradability of waste. As food waste was used, it contains enough

Treatment composition	OC%		P%		N%		K%	
	BD	AD	BD	AD	BD	AD	BD	AD
TC-1(JSC alone)	48.80	46.21	2.09	3.77	4.5	3.85	3.22	2.65
TC-2(1:1)	27.37	24.78	0.066	1.078	1.890	1.24	0.86	0.78
TC-3(1:2)	28.64	26.05	0.063	1.068	2.520	1.87	0.90	0.87
TC-4(1:3)	30.97	28.38	0.060	1.043	2.310	1.56	0.87	0.75
TC-5(1:4)	31.22	28.63	0.058	0.081	2.390	1.61	0.90	0.88
TC-6(1:5)	31.58	28.99	0.056	0.077	2.260	1.56	0.89	0.82
TC-7(2:1)	22.44	23.85	0.057	0.070	2.030	1.38	0.81	0.79
TC-8(3:1)	26.72	24.13	0.061	0.067	2.013	1.413	0.87	0.81
TC-9(CD alone)	24.80	22.21	0.064	0.075	1.92	1.61	1.12	0.78
SE ± (m)	1.329	1.328	0.108	0.200	0.134	0.129	0.125	0.099

Table 4: Organic nutrient content of various treatments before and after anaerobic digestion.

carbohydrates and less cellulose and lignin content, thus the removal of carbon content was fast.

Macronutrients content

In this study the amount of macronutrients content like K, N and P% which were present at the start of the digestion P is0.066%, K is 0.86% and N is 1.890% of the waste. The macronutrients content were again analysed at the end of the study period (84 days) and it were found to be P is 1.078%, K is 0.78% and N is 1.24%. Thus the macronutrients content like phosphorus content had increased, potassium content had decreased and nitrogen content had decreased during the digestion process. Increase and decrease in the quality of macronutrient were recorded in the slurry; (Table 4). Phosphorous, Nitrogen and Potassium a macronutrient is very important as far as plant growth is considered. It is present in every living cell. It is very important to enhance process stability and maintain a stable operation for anaerobic digestion of municipal solid waste [17].

Conclusion

The supplementation of JSC to cow dung up to 75 per cent can be advantageously used for the economic biogas generation from a total solid concentration of 8-10 per cent. This shows that the mixing of these two materials in different proportions enhanced the gas production by utilizing more of the complex substrates than when used alone. The effect may be due to the synergistic action of a variety of cellulolytic and hydrolytic bacterial species in the breakdown of raw material [18]. The decline in gas production in treatment with JSC alone may be due to high lignin content. It is worth mentioning that slurry made of 2:1cow dung and JSC was found the best among the various treatments in biogas generation. The admixtures of cow dung and JSC is a viable option for the energy production and for efficient management of *jatropha* seed cake which was otherwise creates environmental problems.

Acknowledgment

Authors are thankful to the Director FRI for his kind permission to publish this research article. We also acknowledged the help of Mr. Rajendra kumar for arrangement of all necessary chemicals, reagents etc.

References

1. Patil V, Singh K (1991) Oil gloom to oil book – *Jatropha curcas*. Agroforestry Federation, Maharashtra and National Wastelands Development Mission. MoEF 81.
2. Paramathma M, Parthiban KT, Neelakantan KS (2004) A Book on "Jatropha curcas". Published by FC&RI, Mettupalayam.
3. Francis G, Edinger R, Becker K (2005) A concept for simultaneous wasteland reclamation, fuel production and socio-economic development in degraded areas in India: need, potential and perspectives of Jatropha plantations. Nat Resources Forum 29: 12-24.
4. Veeken A, Hamerlers B (1999) Effect of temperature on the hydrolysis rate of selected bio waste components. Bioresour Technol 63: 249-255.
5. Wong MH, Cheung YH (1988) Anaerobic digestion of pig manure with different agro industrial wastes. Biol Wastes 28: 143-155.
6. APHA (1995) Standard Methods for examination of water and waste water 19th Ed. American Public Health Association, Washington DC, USA.
7. AOAC (1995) Official methods of Analysis. Association of Analytical chemist's 14th edn Alinton, Virginia 22: 20-9.
8. Jackson ML (1967) Soil Chemical Analysis (Second India Print). Prentice Hall of India. Pvt. Ltd, New Delhi.
9. Abdulkarim BI, Evuti AM (2010) Effect of Buffer (NaHCO3) and Waste Type in High Solid Thermophilic Anaerobic Digestion. International Journal of ChemTech Research Coden (USA) 2: 980-984.
10. Balsam J (2002) Anaerobic digestion of animal wastes: Factors to consider. ATTRA-national sustainable agriculture information service. United States Department of Agriculture's, USA.
11. Baserja U (1984) Biogas production from cowdung: Influence of time and fresh liquid manure. Swiss- Bio Tech 2: 19-24.
12. Itodo IN, Awulu JO (1999) Effects of total solids concentrations of poultry, cattle, and piggery waste slurries on biogas yield. Transactions of the ASAE 42: 1853-1855.
13. Sadaka SS, Engler CR (2003) Effect of initial total solids on composting of raw manure with biogas recovery. Compost Sci and Utilization 11: 361-369.
14. Hobson PN, Bousfield S, Summers R (1981) Methane production from agricultural and domestic wastes. Applied Science Publishers Ltd., London, UK.
15. Richard T (1996) The effect of lignin on biodegradability. Cornell Composting, Cornel Waste Management Institute.
16. Wilkie AC (2005) Anaerobic Digestion of Dairy Manure: Design and Process Consideration. In: Dairy Manure Management: Treatments, Handling, and Community Relations, Natural Resource, Agriculture, and Engineering Service, Cornell University, USA.
17. Kayhanian M, Rich D (1995) Pilot scale high solids thermophilic anaerobic digestion of municipal solid waste with an emphasis on nutrient requirements. Biomass & Bioenergy 8: 433-444.
18. Hills DJ, Roberts DW (1981) Anaerobic digestion of dairy manure and field crops residues. Agric Wastes 3: 179-189.

Modeling and Output Feedback Distributed Control for an Absorption Packed Column

Selatnia M*, Illoul R and Boucherit MS

Laboratoire de commande des processus Ecole Nationale Polytechnique d'Alger, BP 182 Avenue Hassen Badi, El Harrach, Algeria

Abstract

This work consists of modeling, simulating and finally multiple models control of an industrial absorption packed column designated to remove CO_2, from natural gas. The multiple models approach is an elegant way of turning nonlinear problems into linear ones. We used in this paper the output feedback distributed control (ODC) coupled with local linearization of the model of the absorption packed column. We compared the results with those obtained with the traditional PID control and the results are satisfactory.

Keywords: Multiple models; PDC control; Methyldiethanolamine (MDEA) Absorption packed column; LMI; Lyapunov function.

Introduction

The absorption packed column is a physicochemical separation unit largely used in the chemistry industry. It consists of a tube where we send gas mixtures in order to separate one or more compounds from the principal mixture. It is largely used for the separation of acid gases (CO_2, H_2S) from natural gas.

The model presented in this paper is a dynamic model of the absorption packed column and consists of a set of nonlinear partial differential equations; it is elaborated starting from considerations on CO_2 and MDEA mass balance in gas and liquid phases and considers also the energy balance [1,2]. We finally obtain a nonlinear distributed parameters system.

Few studies were carried out on modelling and controlling the absorption column. Crosby and Durbin [3] studied the performance of a state controller. Roffel [4] developed a sub-optimal output controller with state inequality constraint. Darwish and Fantin [5] used a decentralized command with pole placement. Petrovsky [6] developed a multivariable PI regulator. Najim [7] developed a self-adjusting regulator in the case of CO_2 absorption by a diethanolamine solution and also multilevel learning control [8]. It took again the problem later on with predictive control [9].

Figure 1: The absorption packed column of In Salah Gas (ISG).

Few studies have also been published concerning the modelling and simulation of CO_2 absorption by aqueous solutions of MEA or MDEA on pilot and industrial columns [10-12].

For the model developed in our study, it seemed interesting for us to use the multiple models approach for the command of the absorption packed column because it enables us to obtain good performances for complex dynamics processes. We develop in first stage the PID regulation to compare the performances of the classical techniques with the performances of the Takagi-Sugeno multiple model approach.

Modeling and Open Loop Simulation of the Industrial Absorption Packed Column

The absorption packed column presented here is located at Khrechba and is part of the In Salah Gaz project, it removes CO_2 from natural gas by using an aqueous solution of methyldiethanolamine (MDEA) as a washing liquid, It is a packed type column measuring 8 meter height and 4 meter in diameter with Pall rings to improve the surface of contact between phases. For a better elimination of CO_2 from the natural gas, the liquid flow (water+MDEA) is counter-current with gas flow. The working pressure and temperature is respectively 71.5 bar at 55°C [2].

At contact between liquid and gas phase occurs on the surface of the Pall rings, CO_2 passes from the gas phase to the liquid phase; this diffusion is accelerated by chemical reaction of CO_2 with the MDEA in the liquid phase. The liquid flow (water+MDEA) and the CO_2 concentration in the gas mixture are respectively selected as control variable and output variable (Figure 1).

***Corresponding author:** Selatnia M, Laboratoire de commande des processus Ecole Nationale Polytechnique d'Alger, BP 182 Avenue Hassen Badi, El Harrach, Algeria

Model equations

In order to simplify the model, the following assumptions are done [2,13]:

- There is no resistance in gas phase
- The reaction between CO_2 and MDEA is fast (Ha>5)
- Axial dispersion is negligible in the gas phase and the liquid phase
- The MDEA does not pass in gas phase

The mass balance on an elementary section dz of the column for CO_2 in the gas phase is written [2,13,14]:

Quantity of aqueous solution at input Z=quantity of aqueous solution at the output (z+dz)+quantity of aqueous solution transferred from the liquid phase to the gas phase+accumulation.

This gives:

$$(GC_{Ag})_z=(GC_{Ag})_{z+dz}+\varphi S\,dz+S\frac{dC_{Ag}}{dt}dz \quad (1)$$

Where G (m^3/s) is the volumic gas flow, ϕ the CO_2 flow transferred from the gas phase to the liquid phase, S the section of the column and C_{Ag} (mol/m^3) the CO_2 concentration in the gas phase. Given $U_G=G/S$ (m/s) the gas flow velocity, we obtain then:

$$U_g\frac{dC_{Ag}}{dz}+\varphi=-\frac{dC_{Ag}}{dt} \quad (2)$$

The chemical reaction between CO_2 and the MDEA is [10-12]:

$$CO_2+R_1R_2NCH_3+H_2O \rightarrow R_1R_2NCH_4^{+}+HCO_3^{-} \quad (3)$$

The reaction rate rA has the following form [14,15]:

$$r_A=kC_{AL}\,C_{BL} \quad (4)$$

Where k is the constant for reaction rate [14,15]:

$$k=2{,}9610^5\exp\left(-\frac{5332.8}{T}\right) \quad (5)$$

C_{AL} is the CO_2 concentration in the liquid phase and C_{BL} the MEA concentration in the liquid phase. The mass balance for CO_2 in the liquid phase gives finally:

$$\varphi=[kC_{AL}\,C_{BL}] \quad (6)$$

Which means that the totality of CO_2 transferred to the liquid phase reacts with the MDEA.

The mass balance for the MDEA in the liquid phase gives:

$$(LC_{Bl})_z=(LC_{Bl})_{z+dz}-[k\,C_{Al}\,C_{Bl}]\,S\,dz-S\frac{dC_{Bl}}{dt}dz \quad (7)$$

Where L is the volume liquid flow. By taking account of (5) and noting by $U_L=L/S$ (m/s) the mean liquid flow velocity, we obtain:

$$U_l\frac{dC_{Bl}}{dz}-\varphi=-\frac{dC_{Bl}}{dt} \quad (8)$$

Our absorption packed column is finally described by the following set of partial derivative equations:

$$\begin{cases} U_g\dfrac{dC_{Ag}}{dz}+\varphi=-\dfrac{dC_{Ag}}{dt} \\ U_L\dfrac{dC_{BL}}{dz}-\varphi=-\dfrac{dC_{BL}}{dt} \end{cases} \quad (9)$$

The procedure to compute flow ϕ is given in [2] according to [14-16].

We have finally to consider the boundary conditions which for gas phase are the CO_2 concentration at the column bottom or input concentration C_{AGe} and for liquid phase the MEA concentration at the column top or input concentration C_{BLe}.

$$\begin{cases} C_{Ag}\big|_{z=0}=C_{Age} \\ C_{BL}\big|_{z=h}=C_{BLe} \end{cases} \quad (10)$$

Chemical reactions within the industrial column induces a strong heat emission and the appearance of a temperature gradient throughout the column; the temperature variation is approximately 5°C between the input and the output of the column, which leads us to establish an energy balance in order to describe the temperature changes which affects the various concentrations along the column [16]:

$$\begin{cases} U_g\dfrac{\partial T_g}{\partial z}+\dfrac{a.h_{g/l}(T_l-T_g)}{\left[\sum_i cp^g_i C^g_i\right]}=\dfrac{\partial T_g}{\partial t} \\ -U_L\dfrac{\partial T_l}{\partial z}+\dfrac{1}{\sum_i cp^l_i C^l_i}\left[\Delta H_r r_A-a.h_{g/l}(T_l-T_g)\right]=\dfrac{\partial T_l}{\partial t} \end{cases} \quad (11)$$

With:

C_i^g: concentration in gas phase at the interface (mol/m^3)

C_i^l: concentration in liquid phase at the interface (mol/m^3)

cp_i^g: Specific heat in the gas phase at the interface (J/mol.K)

$h_{g/l}$: coefficient of heat transfer (convection) ($J/m^2.K.s$)

T_l: Liquid temperature (K)

T_g: Gas temperature (K)

ΔH_r: enthalpy of the reaction (J/mol)

cp_i^l: specific heat in the liquid phase at the interface (J/mol.K)

We finally takes into account the boundary conditions for the temperature which are the temperatures for gas and the liquid at the column input.

$$\begin{cases} T_g\big|_{z=0}=T_{ge},\ \dfrac{\partial T_l}{\partial z}\Big|_{z=0}=0 \\ T_l\big|_{z=h}=T_{le},\ \dfrac{\partial T_g}{\partial z}\Big|_{z=h}=0 \end{cases} \quad (12)$$

Model validation

A test was carried out on our industrial absorption column to compare the output CO_2 concentration given by the model with the real one and this for a step input variation of 10 t/h. the data were collected on a horizon of 6800 seconds. The results are grouped in Figure 2 where we represent respectively, the flows of MDEA and gas at the output and then the concentrations of CO_2 at the column output either experimental or given by the model [2]. We note that the model dynamics of the CO_2 concentration at the column output agree with the experimental results.

Open loop Simulation of the industrial column

By considering the equations (2) and (8), the dynamic model of the absorption column is that of a nonlinear, distributed parameters system. The results from open loop simulations are presented in Figures 3 and 4.

Simulations show that the system is stable. It presents a dead time

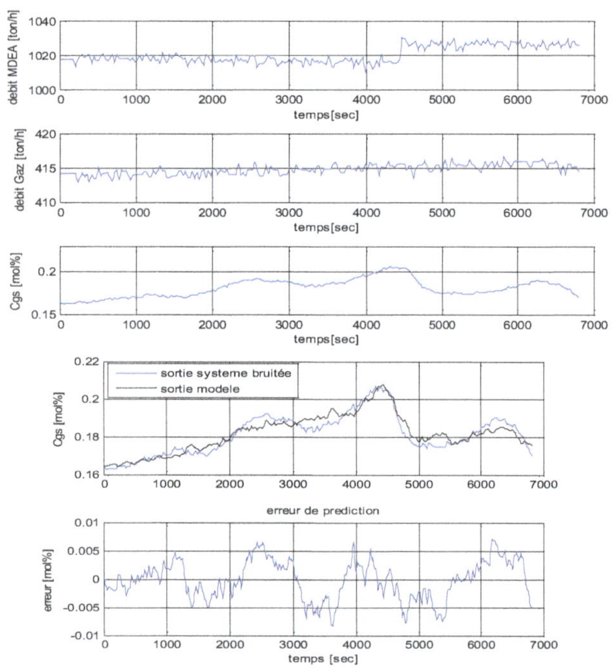

Figure 2: Output CO_2 Concentrations from the system and the model.

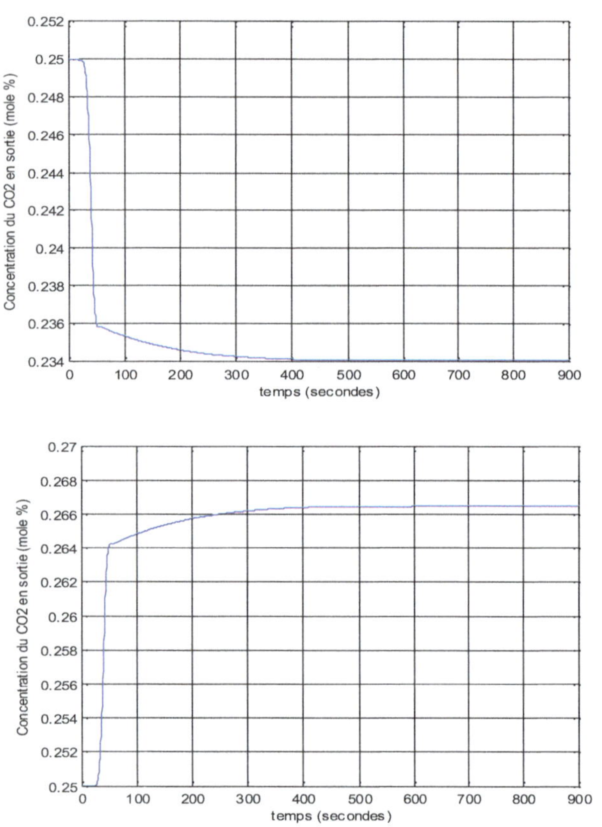

Figure 3: Output CO_2 Concentration for an input step disturbance of ± 5% on gas concentration.

in response to a step input disturbance on the CO_2 concentration due to the gas propagation along the absorption column.

PID Regulation of the Industrial Absorption Packed Column

We apply a PID regulation to the dynamic model of our absorption packed column. We chose a sampled control with a sampling period of 10 seconds. The reference for the input CO_2 concentration is s 0.25 mole%, which corresponds to a concentration of 7.05 mole CO_2/m^3. The parameters of the PID regulator were optimized using trial and error.

The simulation results are satisfactory; the PID regulator cancels the permanent error and ensures a quick response due to the derivative action. The regulation shows a net asymmetrical behaviour between responses to positive and negative step input disturbances due to the strong non linearity of the relationship between the input and the output.

Multiple Models Control of the Industrial Absorption Packed Column

Introduction

The multiple models are a nonlinear modeling technique which allows achieving a good compromise between precision and model complexity. In the light of the numerous work related to it in recent years [17-20], it arises a great interest, especially in applications dealing with simulation and control. It can also be seen as a particular fuzzy modeling technique [21,22], corresponding to a Takagi Sugeno (TS) approach [23]. A TS model is a composed of a finite number of linear models

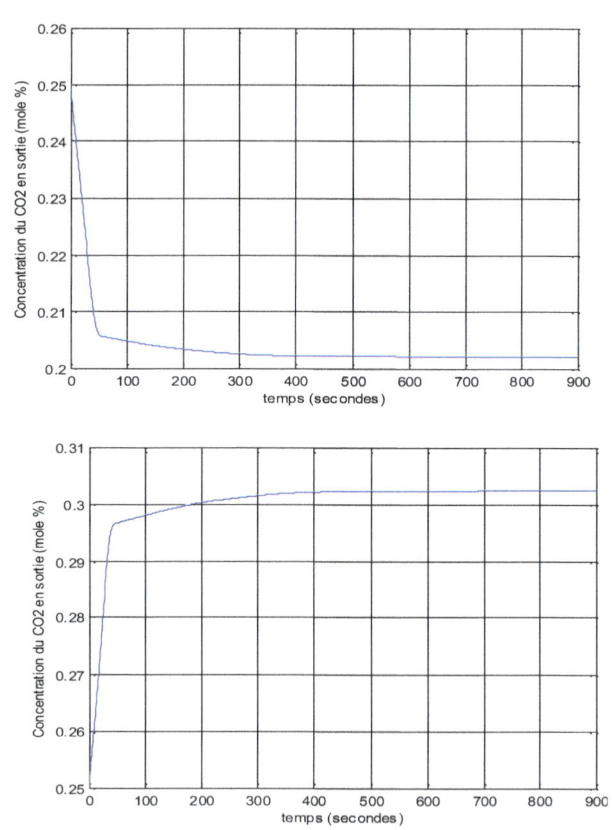

Figure 4: Output CO_2 Concentration for an input step disturbance of ± 10% on gas flow.

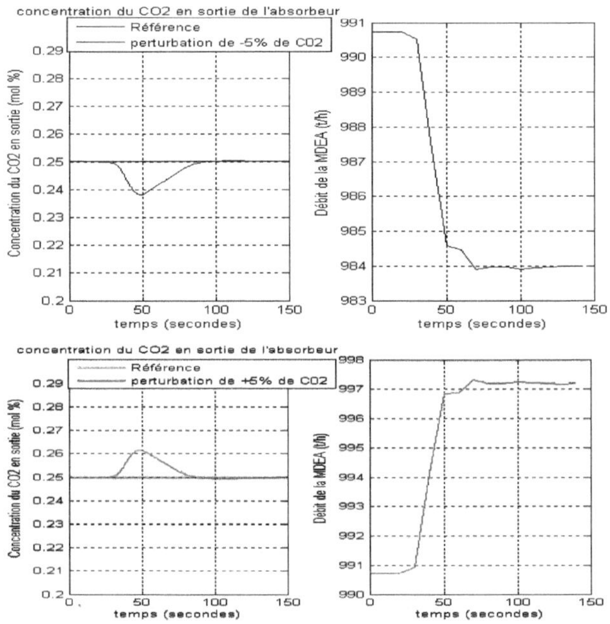

Figure 5: Output CO_2 Concentration and washing liquid flow for an input step disturbance of ± 5% on gas concentration.

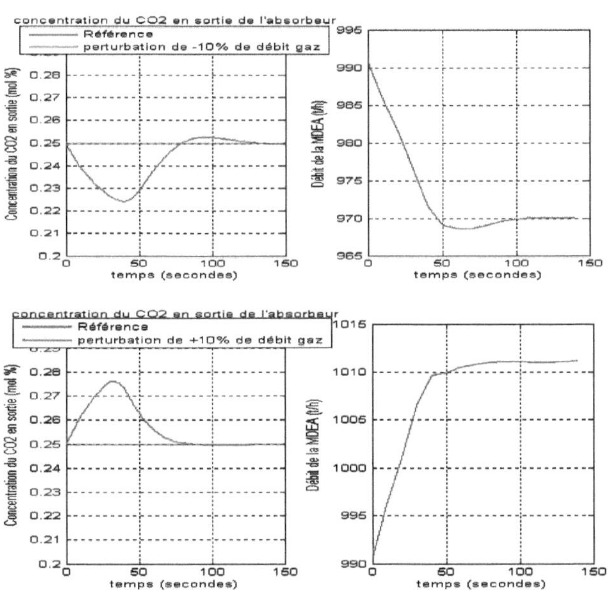

Figure 6: Output CO_2 Concentration and washing liquid flow for an input step disturbance of ± 10% on gas flow

connected with nonlinear functions called membership functions and verifying the convex mapping property (i.e. they are non-negative and their sum is equal to 1). It allows us to solve various problems of control, observation and diagnosis for nonlinear systems with linear techniques.

The approach associated with multiple models in control is known as the Parallel Distributed Compensation (PDC) [24]. This method is based on a set of linear controllers designed for each linear model, and stability of the overall closed loop is guaranteed via a Lyapunov function common to all the linear models.

In this paper, we identify the absorption packed column as a multiple model of the TS type and proposes a control based on State feedback Distributed Control (PDC) (Figures 5 and 6).

Problem formulation

The Multiple model approach: The multiple models have three basic structures: Coupled states (TS), uncoupled states [25] and hierarchical structure. The coupled states Structure (TS) is the most popular in the analysis and synthesis of the multiple models. it is written in the following form:

$$\begin{cases} \dot{x}(t) = \sum_{i=1}^{M} \mu_i(z(t))(A_i x(t) + B_i u(t)) \\ y(t) = \sum_{i=1}^{M} \mu_i(z(t)) C_i x(t) \end{cases} \tag{13}$$

$x(t) \in R^n$ being the state vector, $u(t) \in R^m$ the input vector, $y(t) \in R^p$ the output vector $z(t) \in R$ is the decision variable or premises and the matrices $A_i \in R^{n\times n}$, $B_i \in R^{n\times m}$ et $C_i \in R^{p\times n}$, $\forall$ i=1,2........, M are constant and supposed to be known.

The activation function or membership function μ_i (z (t)) determines the degree of activation of the ith local model. It allows a progressive passage from this model to the other close local models. These functions can depend of the measurable variables of the system (the input and output signals) or of the non-measurable variables of the system (the states). They can be of triangular or Gaussian form and satisfy the properties of convex mapping:

$$\begin{cases} \sum_{i=1}^{M} \mu_i(z(t)) = 1, \forall t \\ 0 \le \mu_i(z(t)) \le 1, \forall i = 1, ..., M, \forall t \end{cases} \tag{14}$$

The multiple models can be viewed as universal approximations since any nonlinear system can be approximated by a multiple models representation with sufficient accuracy and this simply by increasing the number of sub-models. In practice, a reduced number of sub-models can be sufficient to obtain a satisfactory approximation, and we can use the tools of linear systems analysis to achieve this goal.

There are three approaches largely used in the literature allowing us to obtain a TS model: Identification, transformation by nonlinear sectors [26], or linearization. This last one is used in this work (Figure 7).

Using convex analysis for regulator synthesis, the multiple models allows us to obtain control laws by the simultaneous resolution of a finite number of Linear Matrix Inequalities (LMI). In this case, the number of LMI inequalities is polynomial with respect to the number

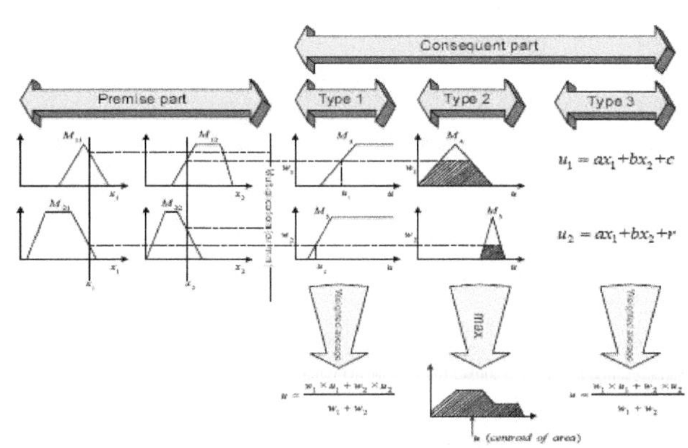

Figure 7: PDC control for TS multiple models.

of local models. Thus, it is advisable to minimize the number of local models to limit the conservatism of the method.

In the case of TS multiple models, this technique of regulator synthesis corresponds to the PDC method, It supposes that all the linear sub-models are at least stabilizable. Subsequently, they will also be supposed commendable.

Given the TS model given by equation (13), a control law resulting from PDC synthesis will thus be the combination of linear control laws for each sub-model, given by:

$$u(t) = -\sum_{i=1}^{M} \mu_i(z(t)) F_i x(t) \quad (15)$$

Applying this control law to the TS multiple model, we obtain in closed loop:

$$\begin{cases} \dot{x}(t) = (A_z - B_z F_z) x(t) \\ y(t) = C_z x(t) \end{cases} \quad (16)$$

Or, in a more explicit way:

$$\left\{ \dot{x}(t) = \sum_{i=1}^{M} \sum_{j=1}^{M} \mu_i(z(t)) \mu_j(z(t)) (A_i - B_i F_j) x(t) \right. \quad (17)$$

The stability conditions for the closed loop system amounts to find a control gain F_j such that the derivative of the candidate Lyapunov function associated with the system is negative. Stabilizing the system thus amounts to solve the following problem:

Find a positive definite matrix P and F_i matrices, i=1. , M such that:

$$(A_z - B_z F_z)^T P + P(A_z - B_z F_z) < 0 \quad (18)$$

We notice that this inequality is nonlinear with respect to P and F_i. By using the congruence of the symmetrical full row matrix:

$$X = P^{-1} \quad (19)$$

We get:

$$XA_z^T + A_z X - XF_z^T B_z^T - B_z F_z X < 0 \quad (20)$$

By using the bijective variable change

$$M_i = F_i X,\ i=1,\ldots,M, \quad (21)$$

The problem becomes LMI in variables X and M_i.

$$\gamma_{ij} = XA_i^T - M_j^T B_i^T + A_i X - B_i M_j < 0 \quad (22)$$

We finally get:

$$\sum_{i=1}^{M} \sum_{j=1}^{M} \mu_i(z(t)) \mu_j(z(t)) \gamma_{ij} \quad (23)$$

And we can express the following result:

Theorem 4.1 [24] Given a continuous TS model, the PDC control law (14) and the γ_{ij}, if it exists a positive definite matrix X and M_i matrices, such that (21) is satisfied for all *i,j*=1,..., M, then the closed loop is overall asymptotically stable. Moreover, if the problem has a solution, the gains of the PDC control are given by:

$$F_i = M_i X^{-1} \quad (24)$$

And the PDC control is:

$$u(t) = -\sum_{i=1}^{M} \mu_i(z(t)) F_i x(t) \quad (25)$$

If $F_i = F,\ \forall i=1,\ldots,M$, then we define a linear control law. In practice, to determine the matrix P and the control gain F_i, we have to solve (21) for all *i,j*=1,..., M. In the particular case where the multiple models verify the positive co linearity of the input matrices, that is:

$$B_i = B,\ \forall i \in I_n \quad (26)$$

The closed loop multiple models system of (16) is rewritten without the crossing terms $B_i F_j$:

$$\dot{x}(t) = \sum_{i=1}^{M} \mu_i(z(t)) (A_i - B_i F_i) x(t) \quad (27)$$

The stability conditions of theorem 4.1 reduce then to the stability of the dominant models: P > 0,

$$(A_i - B_i K_i) P + P(A_i - B_i K_i) < 0,\ \forall i=1,\ldots,M \quad (28)$$

Substituting Bi by B, the control law leads to similar conditions.

Multiple models identification

The structural identification of a multiple models representation consists in the determination of the local models structures and the operation zones (or validity zones) for each local model [25]. The local models can be of various structures but in general we use simple structures, such as linear models.

The identification leads to a family of functions parameterized by the parameters vector $\underline{\theta}_i$ defining the structure of the ith local model, and the parameters vector $\underline{\beta}_i$ characterizing the zone of validity of this local model. The parametric estimate consists in determining for each local model *i* the parameters vector: $\underline{\Theta}_i = \left[\underline{\theta}_i^T \ \ \underline{\beta}_i^T \right]^T$

The parametric estimation (also called training) is done on the basis of minimization of a functional binding the inputs and outputs system to the characteristics parameters of the model.

$$J_G = \frac{1}{2} \sum_{k=1}^{N} (y_s(t) - \hat{y}(t))^2 = \frac{1}{2} \sum_{k=1}^{N} \left[y_s(t) - \sum_{i=1}^{M} \omega_i(\zeta(t), \beta_i) f_i(\varphi(t), \theta_i) \right]^2 \quad (29)$$

In order to simplify the model, we chose linear sub-models of ARX type (auto regressive with exogenous inputs) with 3 inputs (system MISO):

- Liquid Flow U_l [ton/h]
- Gaz flow *Ug* [ton/h]
- Input CO_2 concentration C_{ge} [mol%]

The chosen local models are second order ones, they are written in the following form:

$$C_{gsi}(t+1) = A_{i1} C_{gs}(t) + A_{i2} C_{gs}(t-1) + B_i U_l(t) + C_i U_g(t) + D_i C_{ge}(t) + P_i + e_i(t+1) \quad (30)$$

With:

$i=1,\ldots,M$: local model indices

$C_{gsi}(t+1)$:*output* (concentration of CO_2) of local model i

$e_i(t+1)$: Gaussian white noise

$A_{i1}, A_{i2}, B_i, C_i, D_i, P_i$: parameters of local model i

We chose Gaussian membership functions with the decision variable $\xi(t) = C_{gs}(t)$ being the system output at time t.

The expression of the membership function is:

$$\omega_i\left(C_{gs}(t), [c_i, \sigma]\right) = \frac{exp\left(-\frac{(C_{gs}(t) - c_i)^2}{2\sigma^2} \right)}{\sum_{i=1}^{M} exp\left(-\frac{(C_{gs}(t) - c_i)^2}{2\sigma^2} \right)} \quad (31)$$

Where c_i is the mean. And σ the standard deviation for the Gaussian membership function

The global identification diagram is shown in Figure 8.

Training data (Figure 9)

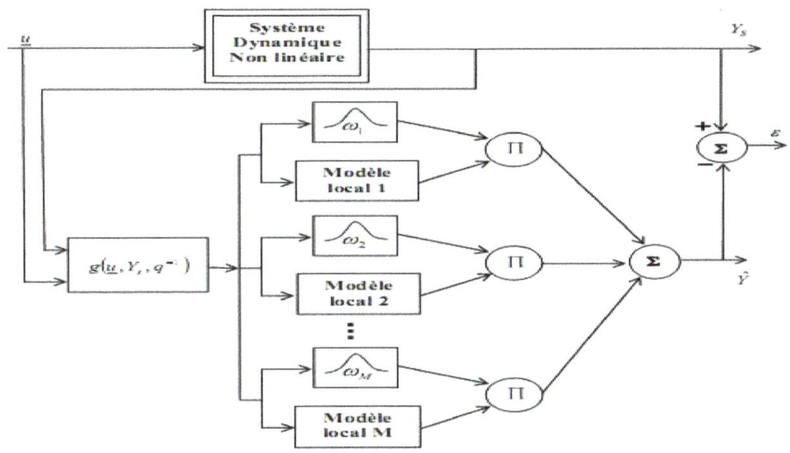

Figure 8: Structure of the multiple models identification.

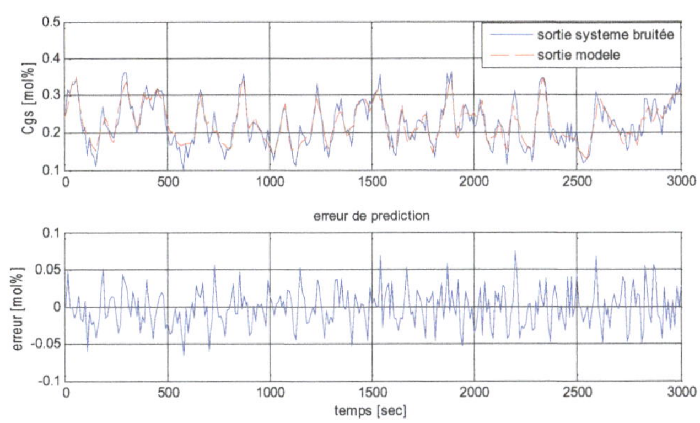

Figure 9: Training with the ARX structure.

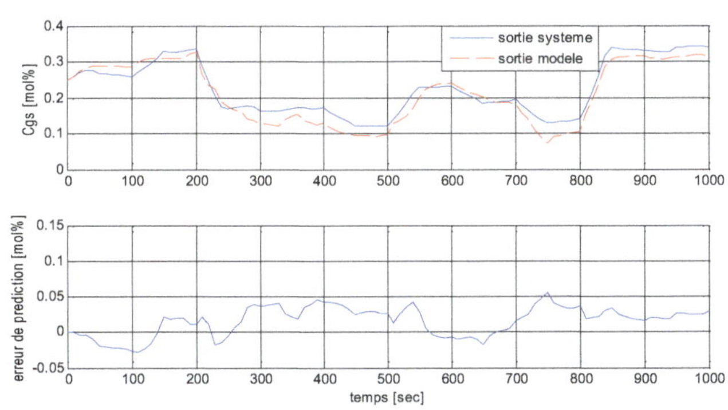

Figure 10: Validating with the ARX structure.

Validating data (Figure 10)

Implementation of the method and simulation results

The decision variable z(t) is in our case the state vector *x(t)* output gas concentration of the industrial column C_{gs},

$x(t)=[C_{gs}(t)\ C_{gs}(t-1)]^T$. A multiple models with four sub-models can easily be obtained in the form:

$$\dot{x}(t) = \sum_{i=1}^{4} \mu_i(x(t))\,(A_i - B_iF_i)\,x(t) \tag{31}$$

With the memberships functions $\mu_i(x(t))$ either triangular or Gaussian. The parameter of the four models is:

$$A_1 = \begin{pmatrix} -0.4436 \\ 0.5744 \end{pmatrix} A_2 = \begin{pmatrix} -0.4772 \\ -0.2732 \end{pmatrix} A_3 = \begin{pmatrix} -1.8937 \\ 0.3093 \end{pmatrix} A_4 = \begin{pmatrix} 0.5339 \\ -0.2304 \end{pmatrix}$$
$$B_1 = (-0.0006)\, B_2 = (-0.0006)\, B_3 = (-0.0006)\, B_4 = (-0.0006) \tag{32}$$
$$C_1 = C_2 = C_3 = C_4 = \begin{pmatrix} 1 \\ 0 \end{pmatrix}$$

And the State Feedback Distributed Control (PDC) is:

$$U_L(t) = -\sum_{i=1}^{4} \mu_i\,(C_{gs}(t))\,F_i\,x(t) + U_{Lin} \tag{33}$$

U_L is the liquid flow velocity, and U_{Lin} the mean flow velocity corresponding to the chosen operating point of the column.

The F_i feedback gains are determined by resolution of *LMI* in order to ensure good regulation performances, less than a 0.25% variation of CO_2 around the operating points.

The selected gain is finally $F_{i=}$[980 985 995 1002]t/h

Closed loop simulation:

Case (a): triangular membership functions

The column simulation shows us the output CO_2 concentration for step input disturbances on either the gas flow or the CO_2 concentration. The evolutions of the output CO_2 concentration and the corresponding control are on Figures 11-13

Case (b): Gaussian membership functions

The evolutions of the output CO_2 concentration and the corresponding control are on Figures 14-16

Discussions

The oscillations of the output CO_2 concentration for the industrial column are mainly due to fuzzy control, but the operating point of is quickly reached in less than 50 seconds . We note on all the curves (5, 6 and 11-16) that disturbance is always rejected with both PID and PDC regulation. But comparison of the peak values as well as the oscillations show us that PDC control acts more quickly than PID control.

Conclusion

In this paper, a TS multiple models obtained by linearization was

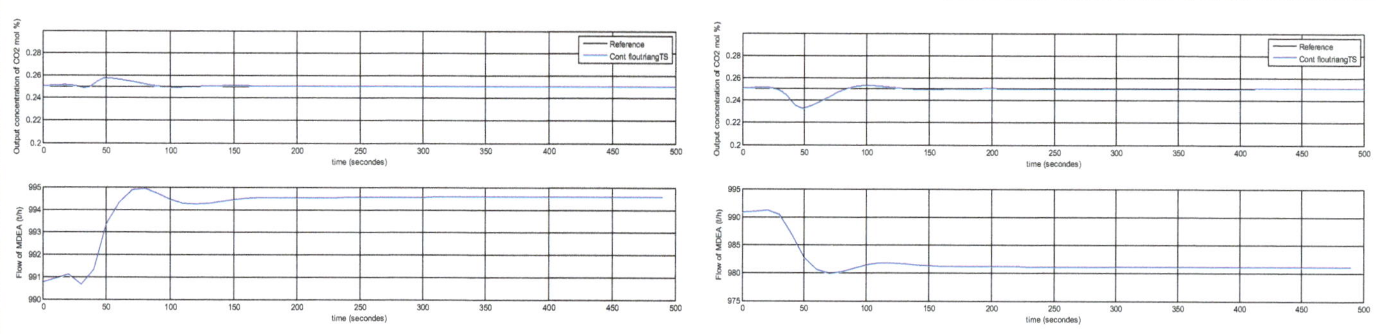

Figure 11: Output CO_2 Concentration and washing liquid flow for a step input disturbance of ± 5% on gas concentration.

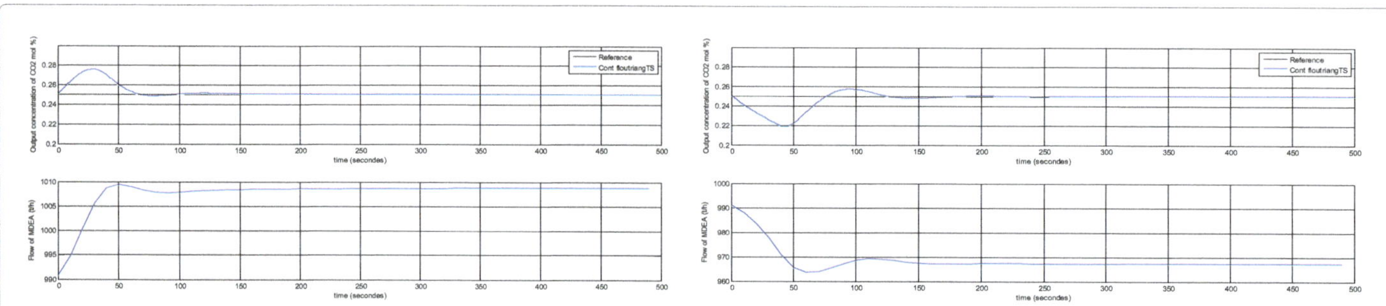

Figure 12: Output CO_2 Concentration and washing liquid flow for a step input disturbance of ± 10% on gas flow.

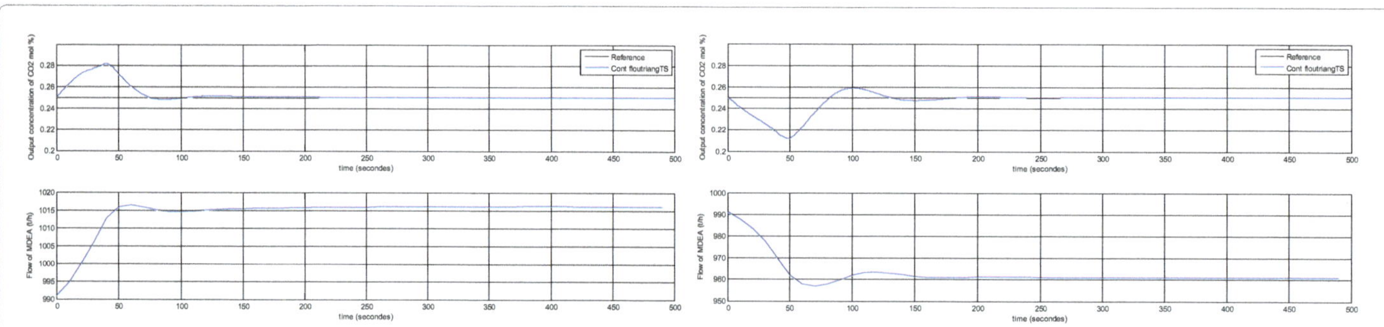

Figure 13: Output CO_2 Concentration and washing liquid flow for a step input disturbance of ± 5% on gas concentration and ± 10% on gas flow.

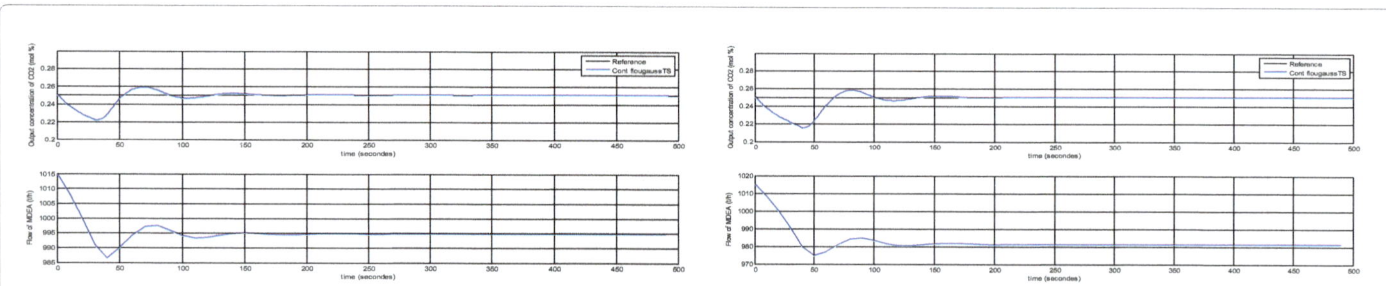

Figure 14: Output CO_2 Concentration and washing liquid flow for a step input disturbance of ± 5% on gas concentration.

used for the regulation of an industrial absorption packed column used for gas washing .We take a reduced number of sub models, four, to ease identification. The results obtained with state distributed feedback (PDC) are better than those by classical PID regulation, for a method which is not significantly complicated. Further investigations will be undertaken in the use of multiple models in control, multiple observers and multiple models in systems diagnosis.

References

1. Selatnia A, Seddari S (2006) Simulation of CO_2 absorption in aqueous solution of monoethanolamine (MEA) on a pilot absorption column. Distillation and absorption conference, IChemE, London, England.
2. Zidane I, Zebiri H (2010) Identification multimodèle et commande par logique floue d'une colonne d'absorption industrielle. Projet de fin d'études, ENP, Alger.
3. Crosby JE, Durbin LD (1976), "Design and application of state variable feed-

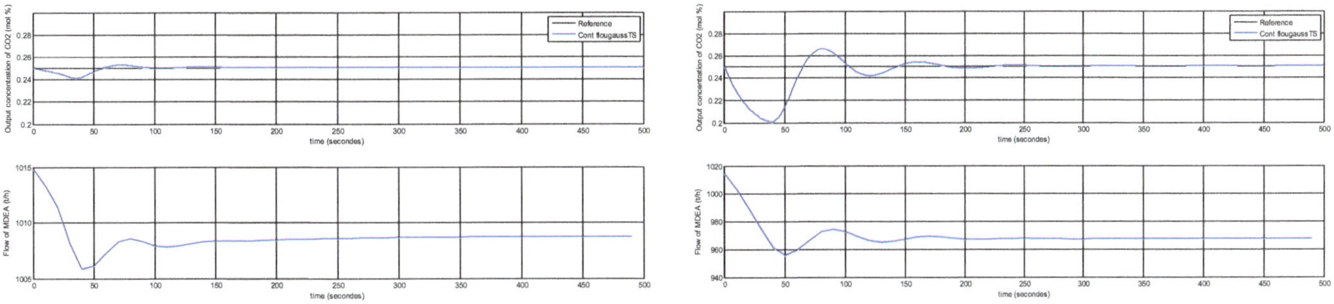

Figure 15: Output CO_2 Concentration and washing liquid flow for a step input disturbance of ± 10% on gas flow.

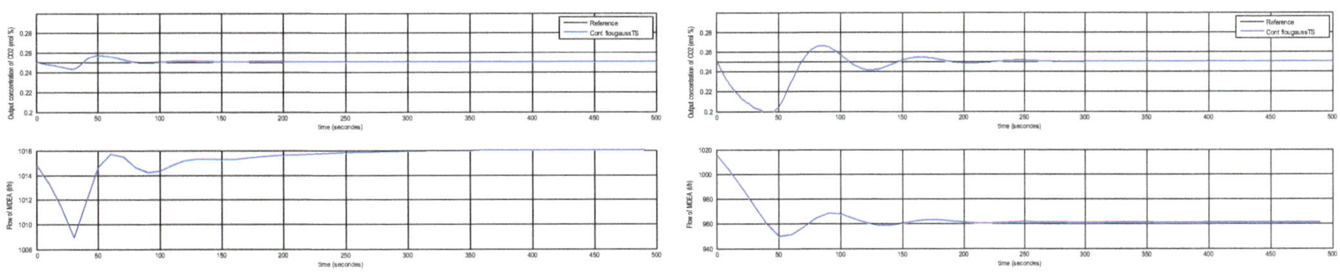

Figure 16: Output CO_2 Concentration and washing liquid flow for a step input disturbance of ± 5% on gas concentration and ± 10% on gas flow.

back controller for a packed trickle bed column with carbon dioxide absorption", in Joint Automatic Control Conference,USA.

4. Roffel B (1976) Control of processes with a state variable inequality constraint. Automatica 17: 19-28.
5. Darwish M, Fantin J (1980) Stabilization and control of absorber tower chemical process. Third IFAC/IFIP/IFORS Conference on System Approach for Development, Rabat.
6. Petrovski D (1980) A design scheme for multivariable output constrained PI regulators with applications to absorption column. Automatic Control Theory Applications 8: 61-69.
7. Najim K (1991) "Modelling and self-adjusting control of an absorption column", International Newspaper of Adaptive Control and Processing Signal, 5: 335-345.
8. Najim K, Chtourou M (1991) "Multilevel learning control of an absorption column", Optimal control applications & Methods, 12: 189-195.
9. Najim K, Ruiz V (1995) "Long range predictive control of an absorption packed column", Applied Mathematical Modelling Journal, 19: 39-45.
10. Idem R, Wilson M, Tontiwachwuthikul P, Chakma A, Veawab A, et al. (2006) Pilot Plant Studies of the CO_2 Capture Performances of Aqueous MEA and Mixed MEA/MDEA Solvents at the University of Regina CO_2 Capture Technology Development Plant and the Boundary Dam CO_2 Capture Demonstration Plant. Ind Eng Chem Res 45: 241-242.
11. Tontiwatchwuthikul P, Meisen A, Lim CJ (1992) "CO_2 Absorption by NaOH, monoethanolamine and 2-amino-2-methyl-1-propanol solutions in a packed column", Chemical Engineering Science, 47: 381-391.
12. Pintola T, Tontiwatchwuthikul P, Meisen A (1993) Simulation of pilot plant and industrial CO2 -MEA absorbers. Gas separation and purification 7: 47-52.
13. Seddari S (2004) "Simulation d'une colonne à garnissage d'absorption du CO_2 par une solution aqueuse de monoéthanolamine à 25°C et 40°C", Thèse de Magister, ENP, Alger.
14. Leye LD, Froment GF (1986) "Rigorous simulation and design of columns for gas absorption and chemical reactions packed columns", Comp. Chem. Eng, 10: 493-504.
15. Trambouze P, Landegheim HV, Wauquier JP (1984) Les réacteurs chimiques, conception/calcul/mise en œuvre, Paris, Technip.
16. Bedelbayev A, Greer T, Lie B (2008) "Model based control of absorption tower for CO_2 capturing", Telemark University College, Porsgrunn, Norvège.
17. CO_2 absorber CO_2 control. Data Sheet ISG. Ain Salah Gas.
18. Murray-Smith R, Johansen TA (1997) Multiple model Approaches to modelling and control ,Taylor & Francis,London.
19. Orjuela R, Marx B, Ragot J, Maquin D (2008) Conception d'observateurs robustes pour des systèmes non linéaires incertains : une stratégie multimodèle,Conférence Internationale Francophone d'Automatique,CIFA.
20. Ichallal D, Marx B, Ragot J, Maquin D (2008) Diagnostic des systems non linéaires par une approche multimodèle,Workshop Surveillance, Sûreté et sécurité des grands systèmes,3 SGS 08,Troyes,France.
21. Takagi T, Sugeno M (1985) Fuzzy identification of systems and its application to modelling and control,IEEE Trans. On systems Man and Cybernetic, 15: 116-132.
22. Wang HO, Tanaka K, Griffin MF (1996) An approach to fuzzy control of nonlinear systems: stability & design issues,IEEE Trans. On Fuzzy Systems, 4: 14-23.
23. Guerra TM, Vermeiren L (2004) LMI- based relaxed nonquadratic stabilization conditiond for nonlinear systems in the Takagi-Sugeno's form, Automatica, 40: 823-829.
24. Wang HO, Tanaka K, Griffin MF (1995) Parellel Distributed Compensation of nonlinear systems by Takagi-Sugeno fuzzy model. Proc of the Fuzzy-IEEE/IFES.
25. Orjuela R, Maquin D, Ragot J (2006) Identification des systèmes non linéaire par une approche multimodèle à états decouples. Journées Identification et Modélisation Expérimentale JIME, Poitiers, France.
26. Ichalal D (2009) Estimation et diagnostic des systèmes nonlinéaire décrits par des modèles de Takagi-Sugeno. Centre de recherche en automatique de Nancy, INPL, France.

An Alternative Gas Chromatography Setting for Geochemical Analysis

Rodrigues LF*, Goudinho FS, Laroque DO, Lourega LV, Heemann R and Ketzer JMM

Pontifical Catholic University of Rio Grande do Sul-PUCRS, Center of Excellence in Research and Innovation in Petroleum, Mineral Resources and Carbon Storage-CEPAC, Brazil

Abstract

Gas chromatography is one of the most important techniques used to characterize gas composition of geochemical samples. This work presents a new gas chromatography setting to characterize gaseous composition using three detectors: a Flame Ionization Detector (FID), a Thermal Conductivity Detector (TCD) and a Flame Photometric Detector (FPD). The new gas chromatography setting developed is able to quantify hydrocarbons gases (C_1-C_5) and non-hydrocarbons (H_2S, CO, CO_2, N_2, O_2 and H_2) in only one injection and in the same analysis. This alternative gas chromatography is able to reduce time of analysis and save sample that can be used for other kind of analyses.

Keywords: Gas chromatography; Geochemical analysis; Gases detections

Introduction

Gas Chromatography (GC) technique is widely used in geochemical studies [1-3]. One of its extensive applications is the determination of gaseous hydrocarbons (C_1-C_5) and non- hydrocarbons gases (N_2, CO_2, H_2, O_2) in marine sediments. The knowledge of composition and concentration of these gaseous is used to constrain the origin of gases [4].

Gas chromatography is applied to conventional petroleum exploration to determine the quantity of light hydrocarbon gases (methane, ethane, ethylene, propylene, propane, butane, iso-butane, butylene, and pentane) in marine sediments which are proxies for deeper petroleum reservoirs. Differences in concentration of the gases above may reflect origin of hydrocarbons and their upward migration to seabed [5]. The gas composition in combination with stable isotopic composition ($\delta^{13}C$-CH_4) can be used to distinguish thermogenic from biogenic gases [6].

Other GC application is for the exploration of unconventional gas reservoir such as gas shale as gas hydrates [7]. Thermal maturity provides an indication of the maximum palaeo temperature reached by a source rock, which determines the type and quantity of the hydrocarbon product from kerogen. Therefore, gas shale is characterized by widespread hydrocarbons gases purity content. Furthermore, it is possible to distingue thermogenic, biogenic or combined biogenic/thermogenic origin using gas composition and additional geochemical parameters [8,9].

Composition of trapped gas in gas hydrate can also be determined using gas chromatography techniques [10]. Methane is the most common gas in gas hydrate, but others gases like CO_2 and H_2S also occur [11,12]. The gas composition of hydrates will determine its crystalline structure and stability [13].

This work presents an alternative gas chromatography setting to characterize gaseous composition using an innovative device containing three detectors: (1) Flame Ionization Detector (FID), (2) Thermal Conductivity Detector (TCD) and (3) Flame Photometric Detector (FPD). In this paper we will present the results of standard gases used to methodology development of the new gas chromatography setting which is able to quantify hydrocarbons gases (C_1-C_5) and non-hydrocarbons (H_2S, CO, CO_2, N_2, O_2 and H_2) in only one injection and in the same analysis.

Equipment descriptions

The analytical methodology was developed using a Shimadzu Gas Chromatograph Model GC-2014. Helium was used as carrier gas at a 30 ml/min flow rate. This device is equipped with FID detector use to quantify principally light hydrocarbons (C_1-C_5), CO and CO_2. Hydrogen and air, both with 30 ml/min flow rate, are used as combustion gas in the FID detector at 200°C. The methanation reaction of CO and CO_2 occur using methanator (MTN-1) with excess of hydrogen. The methane produced is detected in the FID. For CO, CH_4, CO_2, N_2, O_2 and H_2 detection it is used a TCD detector at 200°C. The reference gas and the capillary makeup flow rates are 30 ml/min. A FPD detector at 200°C is used to detect H_2S. Hydrogen and air were used as detector gases at 30 ml/min flow rate.

The gas chromatograph is equipped with a Carboxen capillary column (30 m length, 0.53 mm ID), two Hayesep Q columns (2 m length, 3.0 mm ID) and a Molecular Sieve column (3 m length, 3.0 mm ID). The GC conditions are as follows: starting temperature of the oven of 80°C (hold for 3.5 min) ramp of 8°C min^{-1} to 170°C (hold 10.2 min).

The injection system is equipped with three different loops. In a single injection, the gas sample is distributed in the loops and to the specific detectors. The FID detector has a 100 μL loop, while the TCD and FPD detectors have loop volumes of 500 and 250 μL, respectively. The properties of the columns and volumes of loops used in the development of the methodology can be seen in table 1.

The gases required for operation of the equipment are helium (99.999% purity), hydrogen (99.999% purity) and nitrogen (99.999% purity). Mixture of helium and methane (C_1), ethane, ethylene, propane, propylene, butane, butylenes and pentane (C_2-C_5) is used as standards. The methane concentration is 1,000 ppm (v/v) while C_2-

***Corresponding author:** Rodrigues LF, Pontifical Catholic University of Rio Grande do Sul-PUCRS, Center of Excellence in Research and Innovation in Petroleum, Mineral Resources and Carbon Storage-CEPAC, Brazil

Detector	Loop (µL)	Column	Length (m)	Inside diameter (mm)
FPD	250	Haysep Q	2	3
TCD	500	Haysep Q Molecular Sieve	2 3	3 3
FID	100	Carboxen 1006	30	0.53

Table 1: Properties of the columns and loops used in the development of the methodology.

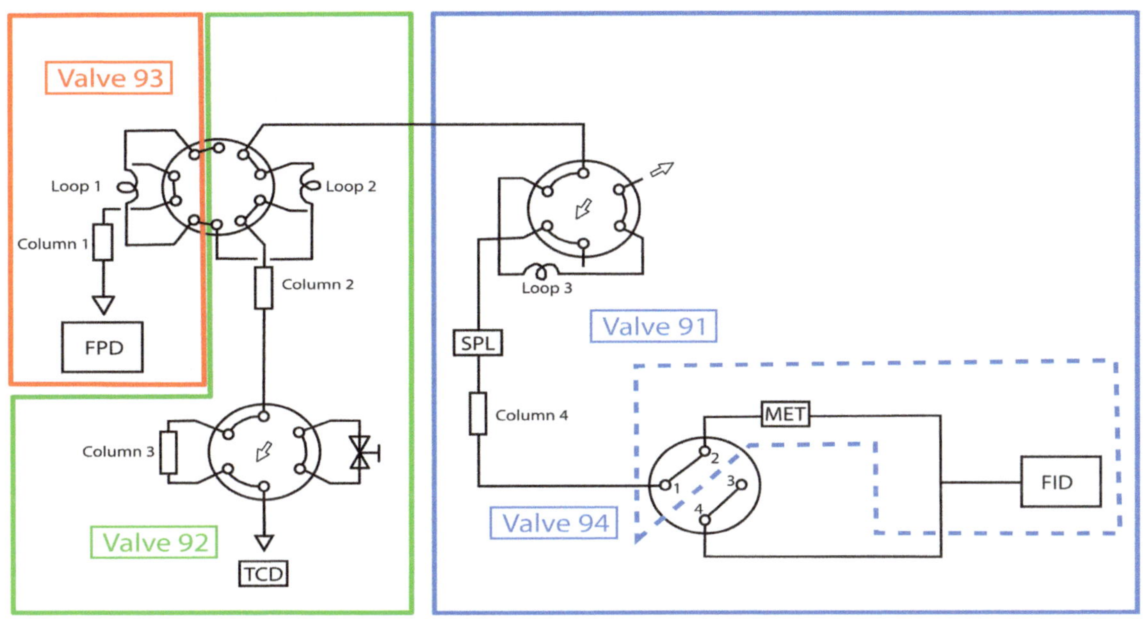

Figure 1: Configuration of equipment developed.

Flow Rate (mL min^{-1})	Heating Rate (°C min^{-1})	Temperature (°C)	Time (min)
30	-	80	3.5
	8	170	10.2

Table 2: Gas chromatography temperature program.

C_5 concentrations are 250 ppm (v/v). Other standard gases used are a mixture of helium and CO, CO_2, O_2, N_2 and H_2 with 2,500 ppm (v/v) of concentration. Finally, 10 ppm of H_2S in helium also was used in this work. All gas standards were provided by White Martins Company.

The gas chromatograph configuration was designed to detect C_1-C_5, CO, CO_2, O_2, N_2 and H_2 in only one injection. Injection of gases is made using syringes of BD PrecisionGlide™ coupled of three-way tap rotation and BD PrecisionGlide™ hypodermic needle.

Figure 1shows four rectangles with different colors. Each color represents a possible pathway which the gas sample will be directed until its detection. The load position is shown in figure 1, where the gases are filling the sampling loops.

After injection (A), the gas sample is divided in three loops (B, C and D) with different volumes. The gas sample will fill the loop B with 250 µL, loop C with 500 µL and loop D with 100 µL and will be carried to the FPD, TCD and FID detectors, respectively, by helium. The gases pathway after injection is described below:

Way 1 - Red rectangle

After starting time, the valve 93 allows 250 µL of the sample to be carried until the Haysep Q (column 1). After gases separation the H_2S is detected in the FPD (Figure 1).

Way 2 - Green rectangle

The valve 93 also allows the sample to the loop with 500 µL and, therefore, the gas helium carries the sample to Q Hayesep column (column 2) and after that to the molecular sieve packed column (column 3). After gas detection in TCD the valve 92 allows a back flush flux to discard the undesirable compounds. These CO, CO_2, O_2, N_2 and H_2 will not be detected in TCD (Figure 1).

Way 3 - Blue rectangle

After starting time, the valve 91 allows 100 µL of the sample to be carried to the Carboxen 1006 (column 4). There are two main directions controlled by valve 94: to the FID detector (where C_1-C_5 will be detected by FID detector) or to the methanator (dashed blue rectangle) (where after methanation, CO and CO_2 will be detected by FID detector).

Results and Discussions

Detection of C_1-C_5, CO_2 and CO by FID detector

The oven's temperature was maintained constant during the other optimization. Table 2 shows the gas chromatography temperature program during this work.

Initially, it was injected the standard containing a mixture of hydrocarbon (C_1-C_5) to adjust their signal behavior and retention time. The separation and identification of hydrocarbons is made with the capillary column (Carboxen 1006) and the FID detector. For this purpose, the valve 91 is allows the injection of 100 µL in the column, and the valve 94 avoids deviation of the sample to the methanator. In the beginning, it was chosen an isotherm temperature (100°C) but in

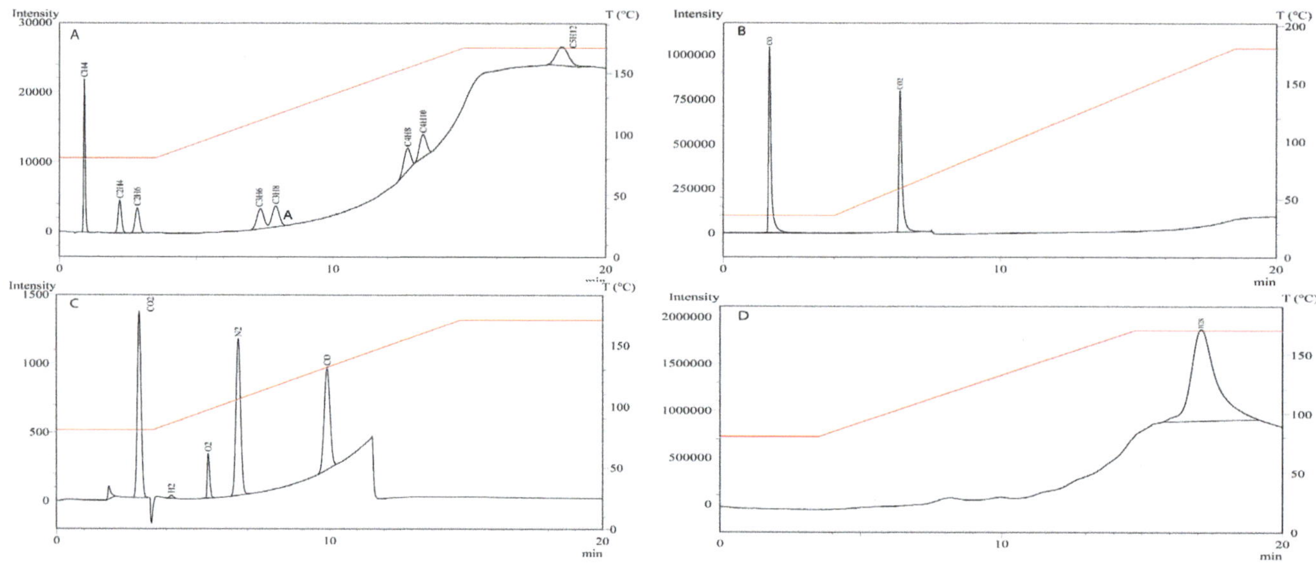

Figure 2: A) Detection of 1000 ppm (v/v) of C_1 and 250 ppm (v/v) of C_2-C_5 by FID; B) Detection of 10,000 ppm (v/v) of CO and CO_2 by FID; C) Detection of 10,000 ppm (v/v) CO_2, H_2, O_2, N_2 and CO by TCD; D) Detection of 10 ppm (v/v) of H_2S by FPD.

this temperature, butane, butylene and pentane had their retention time above 50 min therefore, the time of analysis would stay infeasible. It was chosen a temperature program with two ramps to reduce the retention time for these compounds. The initial temperature was fixed at 35°C for three minutes and after methane (0.9 min), ethylene (2.2 min) and ethane (2.8 min) separation, the temperature is increased to 100°C using a heating rate of 50°C min^{-1}. The temperature must remain at 100°C for 10 min and during this period, propane and propylene were separated in 7.4 and 7.9 min, respectively. Finally, using a heating rate of 25°C min^{-1}, the temperature is increased until 250°C, hold for 12 min to separate butylene (12.7 min), butane (13.3 min) and pentane (18.4 min). The chromatogram for C_1-C_5 is shown in Figure 2A.

After the C_1-C_5 analysis, CO and CO_2 are analysed with the injection of 100 μL of standard containing 25,000 ppm (v/v) of CO and CO_2. The separation and identification of these gases is made with a capillary column (Carboxen 1006) and the FID detector, respectively. Carbon dioxide and carbon monoxide CO and CO_2 must be reduced to methane in the methanator before their detection by FID detector. This procedure is achieved by closing the valve 94, allowing the CO and CO_2 passing through the methanator.

The main purpose of this optimization is to analyse hydrocarbon (C_1-C_5), CO_2 and CO in the same run. Methane will be first detect in the FID detector and after its signal integration, CO is the next compound (retention time in 1.7 min). Because ethylene has the retention time (2.2 min) similar to CO (1.7 min), the valve 94 must be kept closed. This valve will be kept closed until CO_2 is detected by the FID detector. This procedure will allow ethylene and ethane with 2.2 min and 2.8 min respectively, get in the methanator because CO_2 has a retention time of 5.3 min. After CO_2 signal integration, the valve 94 will actuate again and C_3-C_5 will go direct to the FID detector. Figure 2B shows the chromatogram of CO (1.7 min) and CO_2 (6.4 min).

Detection of CO, CO_2, N_2, O_2 and H_2 by TCD detector

For detection of CO, CO_2, N_2, O_2 and H_2 in TCD detector it was used a standard containing 25,000 ppm (v/v) of CO, CO_2, N_2, O_2 and H_2. The valve 93 directs the sample to the loop with 500 μL and helium carries the sample to Q Hayesep column (column 2). In this column occurs the first separation of the compounds. While the hydrocarbons C_2-C_5 have a high affinity with the stationary phase, the compounds with Low Molecular Weight (LMW) such as CO, CH_4, CO_2, N_2, O_2, H_2 do not have great affinity and cross the column earlier than C_2-C_5. The purpose of this column is only to separate the hydrocarbons C_2-C_5 to the LMW. After the last LMW compound detection in TCD, the valve 92 is closed. This valve will act as a back flush flux in order to discard the retained fraction (C_2-C_5) in column 2 and therefore, only CO, CO_2, N_2, O_2 and H_2 will reach the TCD detector. Figure 2C shows the results of LMW gases detection in the TCD: CO_2 (3.0 min) is the first compound that is detected in TCD followed by H_2 (4.2 min), O_2 (5.4 min), N_2 (6.7 min) and CO (9.9 min). Although is possible to detect C_2-C_5 in TCD it was chosen to discard these compounds because their detection take more than 50. Furthermore, C_2-C_5 are in low concentration, therefore, it is necessary to use one specific detector for this purpose.

Detection of H_2S by FPD detector

When the valve 93 is open, 250 μL of the sample is also carried to the Haysep Q (column 1). A standard containing 10 ppm (v/v) of H_2S is injected. In our first trial, no signal after followed injections was observed. How the H_2S is a reactive gas, probably it reacted with the tubing material (stainless steel). Therefore, it was changed the tubing and loop to sulfinert material to avoid some possible reaction. After that, it was possible to detect H_2S how can be viewed on the retention time 17.05 min (Figure 2D).

Limits of Detections (LODs)

Limits of Detection (LODs) of each gas are shown in Table 3. Methane, carbon dioxide and carbon monoxide can be analyzed using TCD and FID detectors. This is advantage because if the samples have low concentration, it will be necessary to use FID detector due its lower LODs and higher sensitivity. Carbon dioxide and carbon monoxide are determined by FID only due the methanator actuation. Even TCD is known as universal detector, in this gas chromatography there are no

Compounds	Limit of Detection ($ppm_{v/v}$)		
	FID	TCD	FPD
Methane	3.0	250	ns
Ethene	0.5	n.a.	ns
Ethane	0.5	n.a.	ns
Propene	0.5	n.a.	ns
Propane	0.5	n.a.	ns
Butene	0.5	n.a.	ns
Buthane	0.5	n.a.	ns
Pentane	0.5	n.a.	ns
Carbon dioxide	3.0	290	ns
Carbon monoxide	1.0	670	ns
Hydrogen	ns	3000	ns
Oxygene	ns	800	ns
Nitrogen	ns	700	ns
Sulfur dioxide	ns	n.a.	0.001

ns - nonspecific detector
n.a. - not applicable for this methodology

Table 3: Limits of detection of gases.

values of LODs for C_2-C_5 in TCD detector because the configuration was performed to detect these gases only in FID detector. Sulfur dioxide is only detected in the selective FPD detector and the LOD was 0.001 $ppm_{v/v}$. FID detector is nonspecific for H_2S compound and there is no value of LOD for H_2S in TCD detector because our standard gas contained 10 $ppm_{v/v}$ which is insufficient concentration to get some signal in this detector.

Conclusion

The alternative gas chromatograph developed and presented here was able to quantify hydrocarbons (C_1-C_5) and non-hydrocarbons (H_2S, CO, CO_2, N_2, O_2 and H_2) gases in only one injection in the same equipment. Our methodology is very useful in situations where gas samples are scarce for multiple injections such as in head space mud samples from marine sediments.

Our work demonstrated that it is possible to detect low gas content ($ppm_{v/v}$) and high concentration ($\%_{v/v}$) using FID detector for light hydrocarbons (C_1-C_5). Samples contain high concentration ($\%_{v/v}$) of CO_2 and CO can be detected by the TCD detector, whereas samples containing low concentrations ($ppm_{v/v}$), need to be directed to the methanator redirected to FID detector.

Other advantage of this device is the H_2S detection by FPD at the same time with other gases, which normally has to be done in separate equipment. Finally, our new equipment setting will extend the gases characterization reducing the time analysis and saving samples that can be used in other analyses such as stable isotopes. Using this alternative gas chromatography, many types of geochemical samples can be characterized with this device.

Acknowledgement

We would like to thank Petróleo Brasileiro S.A. for financial supporting, Pontifical Catholic University of Rio Grande do Sul and Shimadzu Corporation and the Brasilian National Council for Scientific and Technological Development (CNPq).

References

1. Galimov EM, Kodina LA, Stepanets OV, Korobeinik GS (2006) Biogeochemistry of the Russian Arctic. Kara Sea: Research results under the SIRRO project, 1995–2003. Geochemistry International 11: 1053-1104.
2. Nemirovskaya IA, Brekhovskikh VF, Kazmiruk TN (2009) Origin of hydrocarbons in bottom sediments of the Ivankovo Reservoir. Water Resources 36: 337-344.
3. Korobeinik GS, Tokarev VG, Waisman TI (2002) Geochemistry of Hydrocarbon Gases in the Kara Sea Sediments. Polar and Marine Research 419: 158-164.
4. Safronova NS, Grishantseva ES, Korobeinik GS (2013) Hydrocarbon gases (C_1–C_5) and organic matter in bottom sediments of the Ivankovo Reservoir on the Volga River. Water Resources 40: 285-296.
5. Vice DH, Halleck PM (1999) The effects of soil environment on the ability of surface geochemical surveys to detect underlying hydrocarbon traps. Journal of Geochemical Exploration 66: 457-468.
6. Whiticar MJ (1996) Stable isotope geochemistry of coals, humic kerogens and related natural gases. International Journal of Coal Geology 32: 191-215.
7. Curtis JB (2002) Fractured shale-gas systems. AAPG Bulletin 86: 1921-1938.
8. Jarvie DM, Hill RJ, Ruble TE, Pollastro RM (2007) Unconventional shale-gas systems: The Mississippian Barnett Shale of north-central Texas as one model for thermogenic shale-gas assessment. American Association of Petroleum Geologists Bulletin 91: 475-499.
9. Jenkins CD, Boyer CM (2008) Coal-bed and shale-gas reservoirs. Journal of Petroleum Technology 60: 92-99.
10. Schicks JM, Ziemann MA, Lu H, Ripmeester JA (2010) Raman spectroscopic investigations on natural samples from the Integrated Ocean Drilling Program (IODP) Expedition 311: indications for heterogeneous compositions in hydrate crystals. Spectrochim Acta A Mol Biomol Spectrosc 77: 973-977.
11. Lu H, Seo YT, Lee JW, Moudrakovski I, Ripmeester JA, et al. (2007) Complex gas hydrate from the Cascadia margin. Nature 445: 303-306.
12. Bourry C, Chazallon B, Charlou JL, Donval JP, Ruffine L, et al. (2009) Free gas and gas hydrates from the Sea of Marmara, Turkey: Chemical and structural characterization. Chemical Geology 264: 197-206.
13. Kida M, Hachikubo A, Sakagami H, Minami H, Krylov A, et al. (2009) Natural gas hydrates with locally different cage occupancies and hydration numbers in Lake Baikal. Geochemistry, Geophysics, Geosystems 10: 1-8.

Detection of Pathogenic *Escherichia coli* (*E.coli*) Using Robust Silver and Gold Nanoparticles

Boken J[1]*, Dalela S[2], Sharma CK[3] and Kumar D[4]

[1]*Department of Physics, Banasthali Vidyapith, Rajasthan, India*
[2]*Department of Pure and Applied Physics, University of Kota, Kota, Rajasthan, India*
[3]*Assistant Professor, Department of Bioscience & Biotechnology, Banasthali University, Rajasthan, India*
[4]*Department of Chemistry, Banasthali Vidyapith, Rajasthan, India*

Abstract

The paper deals with differently synthesized robust metal nanoparticles using chemical synthesis approach for detection of a new pathogenic strain *E. coli* in water sample. The reactions have been completed using aqueous metal salt solution and trisodium citrate as a capping agent to obtain nanoparticles. For different concentration of sodium borohydride, surface plasmon resonance peaks are obtained at 390 nm to 402 nm for silver nanoparticles and 518 nm to 524 nm for gold nanoparticles. It is to be shifted from 395 nm to 408 nm and 520 to 527 nm for gold and silver nanoparticles after the interaction with *E. coli*. These nanoparticles have also been synthesized at 80°C, exhibited strong surface plasmon resonance peaks respectively at 433 nm and 529 nm. Remarkable shift can be noticed in surface plasma resonance peaks for small amount of *E. coli* added in the solution. The nanoparticles and their interaction against microorganisms were characterized using UV-Vis spectroscopy and Transmission Electron Microscopy (TEM).

Keywords: *E. coli* strain of JB-26; Metal nanoparticles; Silver and gold nanoparticles; Surface plasmon resonance

Introduction

Escherichia coli (*E. coli*) are members of a large group of bacterial germs that inhabit the intestinal microorganism of humans and other warm blooded animals. The presence of *E. coli* in water is strong indication of sewage or animal waste contamination and as few as 10 cells can cause serious human illness and even death. That's why presence of *E. coli* in foodstuffs and drinking water is a chronic worldwide problem. The world wide food production industry is worth about billion and the demand for the biosensors to detect waterborne pathogens and pollutants in food stuffs is growing day by day [1]. Recently, metal nanoparticles as antimicrobial properties have great interest because they have unique optical and electrical properties and perform excellent prospects on biological and chemical sensing [2]. It is estimated that due to bacterial cause thousands of food borne illness and hundreds of hospitalized and deaths each years. It is very important to monitor pathogenic *E. coli* strain and stop their growth by simple methods [3]. Mostly, water quantity standards were set at 103 coliforms per 100 mL and then health goal for total coliforms at zero for drinking water [4]. In addition, human population growth and climate change are expected to increase in the number of species and concentrations of waterborne pathogens in surface water [5].

Bionanotechnology has emerged up as integration between biotechnology and nanotechnology for developing environment friendly and biosynthetic technology for preparation of nanomaterials [6]. The importance of nanomaterials for science and technology is highly increased in past years. The metallic nanoparticles have become more attractive because of their fascinating properties such as optical, physicochemical, electronic and photonic etc. properties due to their high surface area to volume ratio. Mostly the properties are highly affected by their size, morphology and size distribution of nanoparticles. Nanoparticles can be synthesized by growing, shaping or assembling the materials by physical, chemical and biological methods.

Nowadays, silver is currently used to control bacterial growth in a variety of applications. In fact, it is well known that silver ions and silver based compounds are highly toxic to microorganism which shows strong biocidal effects on as many species of bacteria including *E. coli* [7]. The effect of silver ions on bacteria can be observed by the structural and morphological changes. For the last decades, silver ions and silver salts have been used as antimicrobial agents in various fields because of their growth inhibitory capacity against microorganisms. Silver nanoparticles can be synthesized by various methods such as wet chemical method, electrochemical [8], photochemical [9], laser ablations [10], γ- radiations method [11] etc. Silver nanoparticles can be synthesized through reduction of silver nitrate by sodium borohydride along with the stabilizing agents in aqueous solution. Stabilizing agent is used for protecting the growth of nanoparticles by aggregation [12]. Silver nanoparticles were prepared by chemical reduction method in which silver nitrate is reduced by sodium borohydride in aqueous medium at room temperature [13]. Although various approaches have been developed to synthesize stable and monodispersed particles but chemical route method has been found to be most suitable as it is easy, versatile and economical. By controlling the reaction parameters such as temperature, pH, stabilizing agents, reactant concentration, oxidizing or reducing agents etc, variety of particles of different shape and size can be produced. Now a day using wet chemical method, it is possible to produce not only spherical metal nanoparticles and nanoshells but also nanodisks, multipods, triangular nanoprisms etc. When spherical nanoparticles are transformed into one of these shapes the surface plasmon resonance are strongly affected [14].

***Corresponding author:** Boken J, Department of Physics, Banasthali University, Banasthali, Rajasthan-304022, India

Gold nanoparticles have been employed in multiple application involving biological systems, it has remarkable binding properties which makes it attractive for attaching ligands to enhance various bimolecular interactions [15]. It has been well known that inorganic nanoparticles can act as antifungal and antibacterial agents and thus have the ability to interact with microorganisms [16]. Due to its various shapes and sizes, it has been very difficult to predict the positive and negative effects and its mode of action in environment and within living microorganisms. Small size of nanoparticles can modify the physiochemical properties of materials which can lead to adverse biological effects on living cells [17]. In general, the main aspect of the nanoparticles is that the small size of nanoparticles provides for a large surface area for the particle and hence increases the effect. The size of the particles also increase the penetration potential of the silver particles hence again aiding in better utilization of the metal properties [18].

In our present study, simple methods for the detection of *E. coli* and its interaction with gold and silver nanoparticles have been presented. To understand the interaction of *E. coli* with silver and gold nanoparticles various analytical techniques have been used to discuss surface morphology and optical properties of the solution.

Materials and Methods

Materials

Silver nitrate ($AgNO_3$), trisodium citrate ($Na_3C_6H_5O_7$), sodium borohydride ($NaBH_4$) and tetrachloroauric acid ($HAuCl_4$) of Sigma Aldrich make have been used for synthesizing silver and gold nanoparticles. All the solutions have been freshly prepared for the synthesis of nanoparticles especially $NaBH_4$ aqueous solution that was ice bathed before use. Distilled water passed through Millipore system that has been used for all the solution preparation and throughout the experiments. All glassware has been first rinsed with aqua regia solution and then thoroughly with distilled water which is followed by Millipore water.

For identification of *E. coli* stain present in the ground water of near Banasthali area, water sample is sent to IMTECH, Chandigarh, INDIA and finally we have obtained pure culture of *E. coli* for sensing purposes.

Synthesis of robust metal nanoparticles at room temperature

For the synthesis of silver and gold nanoparticles sodium borohydride and trisodium citrate have been used as reducing and capping agent, respectively. $AgNO_3$ (0.2 mM in 10 mL), trisodium citrate (2 mM in 10 mL) have been used as a metal precursor and stabilizing agent, respectively. $NaBH_4$ solution (0.1 M) in 0.1 mL to 1 mL of water (to obtain different samples) has been added drop wise to the above silver nitrate solution and further stirred at 15 min. As soon as the freshly prepared $NaBH_4$ added in the solution, the colour of solution turned light yellow to golden yellow. As we increased the concentrations of $NaBH_4$ solution then the color of silver nanoparticles have been changed. All the reaction has been carried out at room temperature.

Similarly, gold nanoparticles have been synthesized by adding trisodium citrate (10 mM in 10 mL) in $HAuCl_4$ (2 mM in 10 mL) solution and further stirred for 15 min. After 15 min NaBH4 (1 mM) solution in 1 mL to 10 mL of water has been added drop wise in the reaction mixture to obtain gold nanoparticles. The final solution turned from light ruby red to dark ruby red in colour with increased concentration of $NaBH_4$.

Synthesis of robust metal nanoparticles at 80°C

In this method, $AgNO_3$ (2 mM in 50 mL) which has been heated at 80°C for 15 min then added trisodium citrate (20 mM in 50 mL) to metal precursor solution and resulting mixtures refluxed at 80°C for 30 min. After 15 min, initially clear solution of $AgNO_3$ turned pale yellow then brownish yellow in colour indicating the formation of silver nanoparticles.

Similarly, the synthesis of gold nanoparticles have been performed using $HAuCl_4$ (2 mM in 50 mL) heated at 80°C for 15 min then added trisodium citrate (10 mM in 50 mL) to this solution and finally refluxed at 80°C for 30 min. After 15 min, initially clear solution of $HAuCl_4$ turned light red to ruby red colour indicating the formation of gold nanoparticles.

Characterization techniques

The particle size and surface morphology of the nanoparticles have been examined using TEM of TECNAI make G20 operated at an accelerating voltage of 220 keV. All the samples for TEM have been deposited on carbon coated copper grid by placing the drops of diluted samples of gold and silver. Optical absorption measurements have been performed using a Perkin-Elmer Lambda 750 UV-VIS-NIR Spectrophotometer with pre-aligned tungsten, halogen and deuterium sources. The resolution of the spectrophotometer is 0.17-5.00 nm for UV-VIS and 0.20-20.00 nm for NIR.

Results and Discussions

Silver and gold nanoparticles have been synthesized by wet chemical method using trisodium citrate as well as sodium borohydride at different reaction temperatures. Trisodium citrate merely serves as a capping agent because it is weak reducing agent and cannot perform reduction at room temperature. It is well known from the literature that trisodium citrate merely serves as a capping agent because it is weak reducing agent and cannot perform reduction at the room temperature. $NaBH_4$ being a strong reducing agent, reaction takes place almost instantly and very small particles are obtained [19]. Reduction using sodium borohydride is done at room temperature while trisodium citrate needs higher temperature ~80°C for the reduction. The sizes and shapes of the nanoparticles were tuned, selecting the reducing and capping agents.

A simple mechanism for the formation of silver nanoparticles using a metal precursor which is given below:

$$AgNO_3 \rightarrow Ag^+ + NO^-_3$$

$$Na_3C_6H_5O_7 \rightarrow 3Na+ + C6H5O_7^{3-}$$

$$3AgNO_3 + Na_3C_6H_5O_7 \rightarrow 3Ag^+ + C_6H_5O^-_7 + 3NaNO_3$$

Similarly, the chemical reactions of the precursor leading to the simply formation of gold nanoparticles which is given below:

$$HAuCl_4 \rightarrow Au^{3+} + 4Cl^- + H^+$$

$$Na_3C_6H_5O_7 \rightarrow 3Na^+ + C_6H_5O_7^{3-}$$

$$HAuCl_4 + Na_3C_6H_5O_7 \rightarrow 3Au^+ + C_6H_5O_7^- + 6NaCl + 3Cl_2$$

Simple mechanism of gold and silver nanoparticles formation is clear from the above equations. When $AgNO_3$ is dissolved in water and it breaks in Ag^+ and NO_3^- ions due to its less hydration energy whereas tri sodium citrate breaks into Na^+ and $C_6H_5O_7^{3-}$. These silver ions get reduced by $NaBH_4$ by accepting one electron and forms Ag^0 atom. When the number of Ag^0 atoms in the solution approaches to saturation then nucleation starts. After the nucleation, particles grow due to the surface. It may be attributed that some unreduced Ag^+ ions remain on the surface of the particles which attracts negatively charged citrate ions towards them. These negatively charged citrate ions from a layer around the nucleation site and act a capping agent for controlling the shape of the particles. It can be concluded that the formation of double layer is responsible for controlling the size of particles. At lower concentration of $NaBH_4$ formation of silver nuclei will be low so silver nuclei can be combed with other silver ions to form large particles in the solution mixture. On the other hand, at the higher concentration of $NaBH_4$, the rate of formation of silver nuclei will be higher and obtain less nuclei growth which lead to the smaller particles. Similarly this mechanism is valid for gold nanoparticles [20].

Optical properties of silver and gold nanoparticles and its correlation with $NaBH_4$ Concentration

All the spectra show prominent surface plasmon resonance peaks, characteristics of spherical nanoparticles at 529 nm for gold and 433 nm for silver nanoparticles as shown in Figure 1(a) & 1 (b). In silver nanoparticles, the absorption spectrum is showing a hump at 322 nm along with the intense peak at 433 nm. The presence of hump at lower wavelength may owe its presence either due to origin of higher order multipole modes or due to incomplete reduction of silver ions.

Wine red colour of colloidal solution of gold nanoparticles is due to strong absorption of gold nanoparticles. Peak is arising due to absorption of energy via the collective oscillations of free (dipolar plasmon) and interband transitions (HOCO to LOCO). Surface plasmon resonance gets coupled when the gap between the particles decreases.

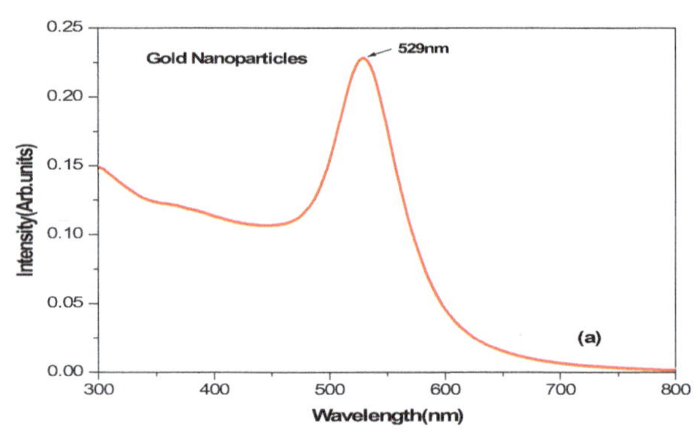

Figure 1a: UV-Vis spectrum of gold nanoparticles synthesized with trisodium citrate at 80°C.

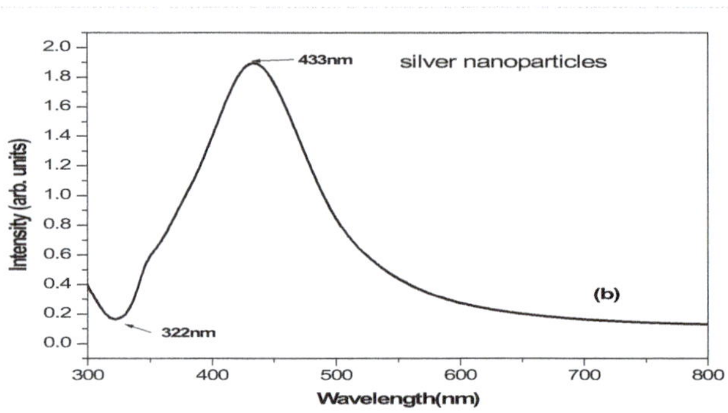

Figure 1b: UV-Vis spectrum of silver nanoparticles synthesized with trisodium citrate at 80°C.

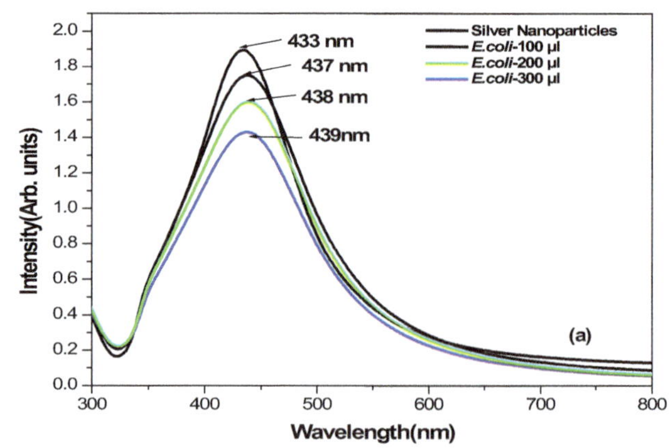

Figure 2a: UV-Vis spectra of silver nanoparticles with interaction of *E. coli* at different concentration e.g. 100 µL, 200 µL, 300 µL synthesized at 80°C.

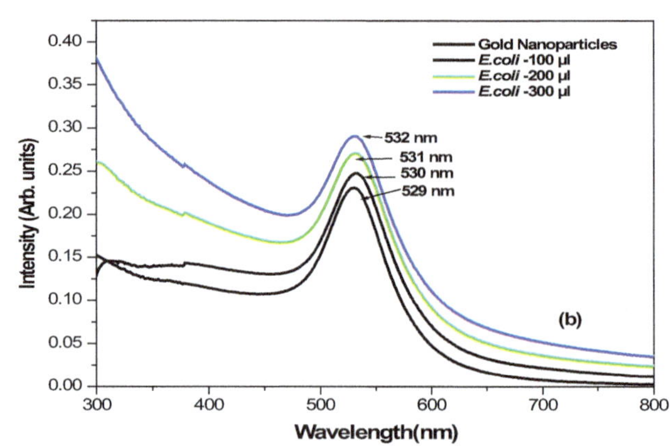

Figure 2b: UV-Vis spectra of gold nanoparticles with interaction of *E. coli* at different concentration e.g. 100 µL, 200 µL, 300 µL synthesized at 80°C.

The optical absorption spectra have been also recorded for the solution of gold and silver nanoparticles to study the interaction between *E.coli* and metal nanoparticles.

Similarly, red shit is also occurred for the gold nanoparticles as shown in Figure 2b. As the size of the particles increases then the light can be longer polarized the nanoparticles homogeneously and higher modes at lower energy dominate. This causes a red shift, broadening and with less amplitude of the surface plasmon band. Therefore, small gold nanoparticles aggregates will make their surface plasmon combine and results in the color change from red to purple.

Hence we can say that the surface plasmon resonance plays an important role in the determination of optical absorption spectra of metal nanoparticles which shifts to a longer wavelength with increasing the particles size. More and more particles begin to aggregate due to which HUCO-LUCO gap between the particles decreases. Less energy is requires to excite the electron which leads to red shift in surface plasmon resonance wavelength. So red shift and broadening were due to both surface plasmon coupling and aggregation between closely spaced nanocomposites.

If the naked clusters are in contact with one another aggregation occurs, particle size increases and quantum properties are lost. All the HOCO's merge as do all LOCO's to from band structure.

However, TEM image of silver nanoparticles show the size of particles are less than 10 nm as can be seen from Figure 3(a & b). For this size of particles, higher order of plasmon modes cannot be expected. In Figure 4, a TEM images represent the size of gold nanoparticles before and after the interaction of *E. coli* in the solution. Here we can see that the particles size have been increased after the interaction of *E. coli.*

All the spectra show prominent surface plasmon resonance peaks, characteristics of spherical nanoparticles in the 518 nm to 524 nm ranges for gold nanoparticles and 390 nm to 402 nm for silver nanoparticles, respectively. $NaBH_4$ solution has been varying from 1 mL to 10 mL for gold nanoparticles and 0.1 mL to 1 mL for silver nanoparticles to examine the effect of $NaBH_4$ on the resulting particle size. We clearly observed a tendency to form a smaller particle size with increasing concentration of $NaBH_4$ in the reaction which can be concluded from a continuous red shift in the absorption spectra with an increasing $NaBH_4$ concentration. The size of these nanoparticles can be tuned by the $NaBH_4$ concentration within certain range.

The absorption spectrum in Figure 5(a) show the prominent surface plasmon resonance peaks, characteristics of nanoparticles in 518 nm to 524 nm ranges for gold nanoparticles with NaBH4 (different concentration of water e.g. 3 mL, 5 mL and 10 mL) and for the same amount of NaBH4, peaks shifted from 520 nm to 529 nm after the interaction of *E. coli* as shown in Figure 5(b).

We also obtained the absorption spectra for silver nanoparticles at 390 nm without and with interaction of *E. coli,* the peak shifted at 395 nm as shown in Figure 6(a). Similarly for another concentration

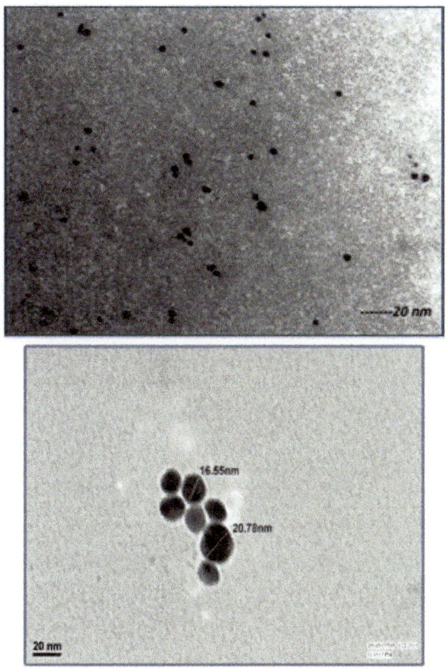

Figure 3: (a) TEM images of silver nanoparticles, (b) after interaction with *E. coli.*

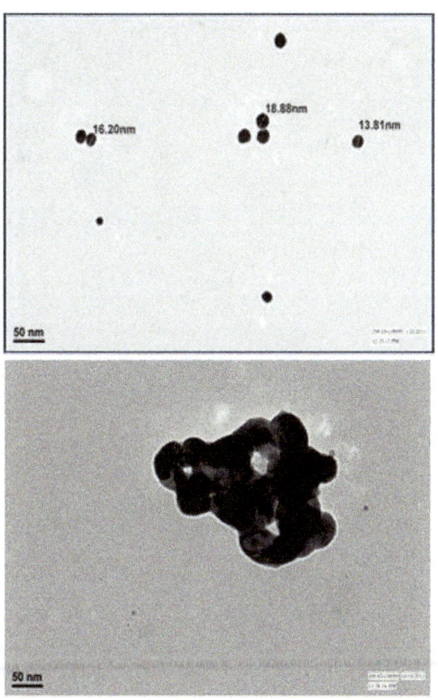

Figure 4: (a) TEM images of gold nanoparticles, (b) after interaction with *E. coli*

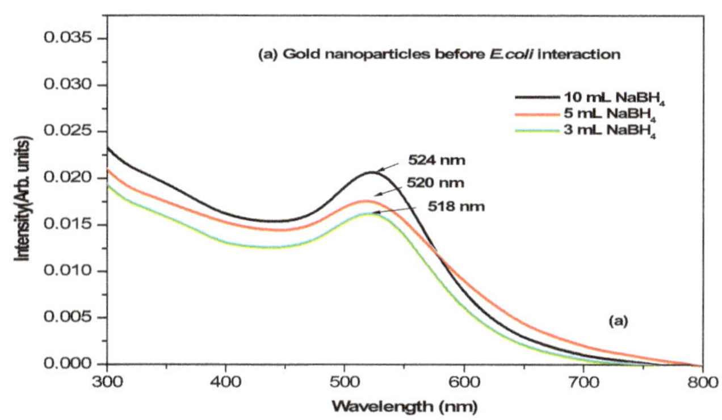

Figure 5a: UV-Vis spectra of gold nanoparticles before *E. coli* interaction of 0.01M $NaBH_4$.

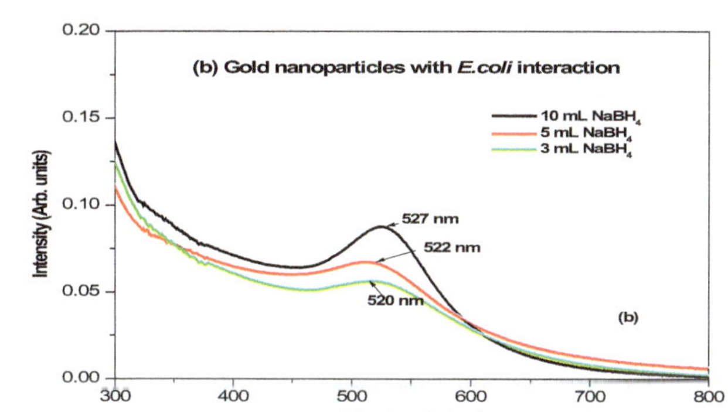

Figure 5b: UV-Vis spectra of gold nanoparticles after interaction with *E. coli* of 0.01M $NaBH_4$.

of $NaBH_4$ we obtained in Figure 6(b), the absorption peak at 402 nm for silver nanoparticles and after the interaction of *E. coli* the peak also shifted at 408 nm. This shows as we increased the concentration of the reducing agent then absorption peaks of bacteria will be increased in the solution, hence the red shift will occurs.

One more interesting result we would like to quote that after interaction with *E. coli* bacteria the surface plasmon resonance peaks shifted at 513 nm to 529 nm for gold nanoparticles, which may be ascribed due to particle size increase with interaction of *E. coli* with nanoparticles and it can be expected that a large number of *E. coli* bacteria coagulate around silver and gold nanoparticles and cluster formation can be expected. To confirm the effect of silver and gold nanoparticles, a comparative study of silver and gold nanoparticles activity against *E. coli* bacteria were performed. In case of silver nanoparticles, surface plasmon resonance peaks shifted at 395 nm to 408 nm after interaction with *E. coli*. TEM images of silver and gold nanoparticles show with and without interaction of *E. coli* in the solution as shown in Figure 7 & 8. In Figure 3(b), 4(b) & 8(b) represent the TEM images of nanoparticles which show the cluster formation after the interaction of *E. coli* bacteria, it may be due the size of the nanoparticles has large surface area to come in contact with the bacterial cell. The smaller nanoparticles interact with bacteria and produce electronic effects which enhance the reactivity of nanoparticles. Hence we can say due to this effect we can be used to determine the presence of the *E. coli* bacteria in the solution.

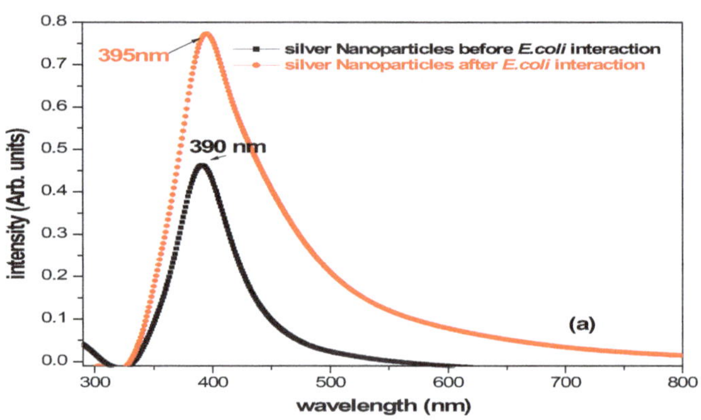

Figure 6a: UV-Vis spectra of silver nanoparticles before and after interaction with *E. coli* of 0.1 M $NaBH_4$.

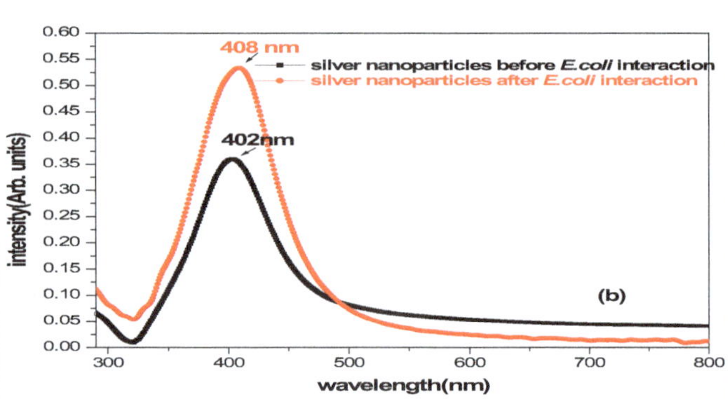

Figure 6b: UV-Vis spectra of silver nanoparticles before and after interaction with *E. coli* of 0.1 M $NaBH_4$.

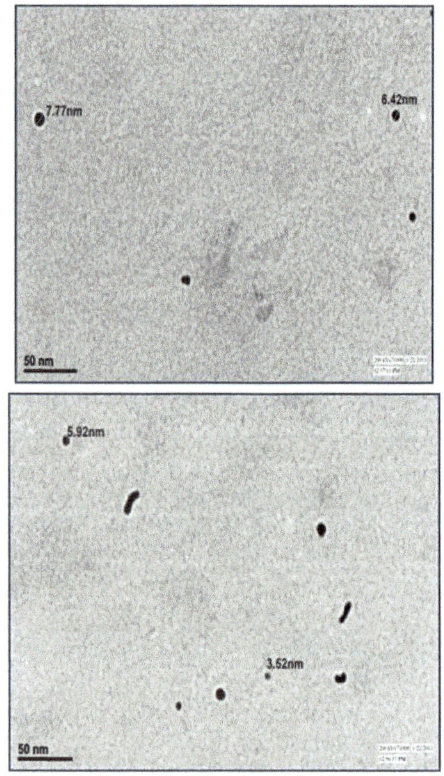

Figure 7: (a) TEM images of silver nanoparticles before interaction of *E. coli*, (b) after interaction of *E. coli* with 0.1 M $NaBH_4$.

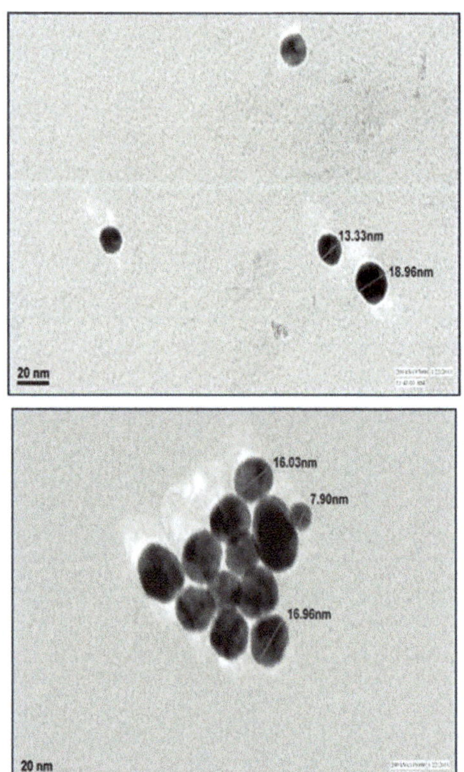

Figure 8: (a) TEM images of gold nanoparticles before interaction of *E. coli*, (b) after interaction of *E. coli* with 0.01 M $NaBH_4$.

Conclusions

Gold and silver nanoparticles were synthesized by wet chemical method using tri sodium citrate and sodium borohydride. As we increase the concentration $NaBH_4$ than a continuous red shift in the absorption spectra is obtained. The size of gold and silver nanoparticles could be tuned by the borohydride concentration within certain range. After the interaction of *E. coli*, the size of particles will increase with increasing the concentration of $NaBH_4$. The size of the nanoparticles implies that it has a large surface area to come in contact with the bacterial cells. Hence, it will have a higher percentage of the interaction than bigger particles. Thus we can say that the bacterial effect of silver nanoparticles and gold nanoparticles are size dependent and can be used to determine the presence of *E.coli* bacteria in the solution.

Acknowledgement

The one author (J. Boken) is thankful to Department of Science and Technology (DST), New Delhi for its financial assistance, grant no F.No.SR/WOS-A/PS-35/2010.We are also thankful to Professor Aditya Shastri, Vice Chancellor of Banasthali Vidyapith for kindly extending the facilities of "Banasthali Centre for Education and Research in Basic Sciences" sanctioned under CURIE programme of the Department of Science and Technology, New Delhi. The authors would like to thank AIIMS (All India Institute of Medical Sciences, New Delhi) using its TEM (Transmission Electron Microscopy) facility for analysis.

References

1. Singh AK, Senapati D, Wang S, Griffin J, Neely A et al. (2009) Gold nanorod based selective identification of Escherichia coli bacteria using two photon rayleigh scattering Spectroscopy. ACS Nano 3: 1906-1912.
2. Zhang X, Geng P, Liu H, Teng Y, Liu Y et al. (2009) Development of an electrochemical immunoassay for rapid detection of E.coli using stripping voltammetry based Cu@Au nanoparticles as antibody labels. Biosens Bioelectron 24: 2155-2159.
3. Rastogi SK, Rutledge VJ, Gibson C, Newcombe DA, Branen JR et al. (2011) Ag colloids and Ag clusters over EDAPTMS- coated silica nanoparticles: synthesis, characterization, and antibacterial activity against Escherichia coli. Nanomedicine: Nanotechnology, Biology and Medicine 7: 305-314.
4. Temur E, Boyaci IH, Tamer U, Unsal H, Aydogan N (2010) A highly sensitive detection platform based on Surface-enhanced raman scattering for Escherichia coli enumeration. Anal Bioanal Chem 397: 1595-1604.
5. Rule KL, Vikesland PJ (2009) Surface-enhanced resonance raman spectroscopy for the rapid detection of Cryptosporidium parvum and Giardia lamblia. Environ Sci Technol 43:1147-1152.
6. Rai M, Yadav A, Gade A (2009) Silver nanoparticles as a new generation of antimicrobials. Biotechnol Adv 27:76-83.
7. Kim JS, Kuk E, Yu KN, Kim JH, Park SJ et.al (2007) Antimicrobial effects of silver nanoparticles. Nanomedicine: Nanotechnology, Biology and Medicine 3: 95-101.
8. Vorobyova SA, Lesnikovich AI, Sobal NS (1999) Preparation of silver nanoparticles by interphase reduction. Colloids and Surfaces A: Physiochemical and Engineering Aspects 152: 375-379.
9. Xue C, Metraux GS, Millstone JE, Mirkin CA (2008) Mechanistic study of photo medicated triangular silver nanoprism growth. Journal of American Chemical Society 130: 8337-8344.
10. Tsuji T, Watansbe N, Tsuji M (2003) Laser induced morphology change of silver colloids: formation of nano-size wires. Appl Surf Sci 211: 189-193.
11. Choi SH, Zhang YP, Gopalan A, Lee KP, Kang HD (2005) Preparation of catalytically efficient precious metallic colloids by γ-irradiation and characterization. Colloids and Surfaces A: Physiochemical and Engineering Aspects 256: 165-170.
12. Lee SM, Song KC, Lee BS (2010) Antibacterial activity of silver nanoparticles prepared by a chemical reduction method. Korean J Chem Eng 27: 688-692.
13. Mazumdar H, Ahmed GU (2011) Synthesis of silver nanoparticles and its adverse effect on seed germinations in oryza sativa vigna radiate and brassica campestris. International Journal of Advanced Biotechnology and Research 2: 404-413.
14. Seydack M (2005) Nanoparticle lables in immunosensing using optical detection methods. Biosens Bioelectron 20: 2454-2469.
15. Williams DN, Ehrman SH, Pulliam HTR (2006) Evaluation of the microbial growth response to inorganic nanoparticles. Journal of Nanobiotechnology 4: 3.
16. Nam JM, Thaxton CS, Mirkin CA (2003) Nanoparticle based bio bar codes for the ultrasensitive detection of proteins. Science 301: 1884-1886.
17. Nel A, Xia T, Madler L, Li N (2006) Toxic potential of materials at the nanolevel. Science 311: 622-627.
18. Singh A, Shirolkar M, Lalla NP, Malek CK, Kulkarni SK (2009) Room temperature, water-based microreactor synthesis of gold and silver nanoparticles. Int J of Nanotechnology 6: 541-551.
19. Pradhu S, Poulose EK (2012) Silver nanoparticles: mechanism of antimicrobial action, synthesis, medical applications and toxicity effects. International Nano Letters 2:32.
20. Kalele S, Gosavi SW, Urban J, Kulkarni SK (2006) Nanoshell particles: synthesis, properties and applications. Current Science 91:1038-1052.

Biodiesel from Household/Restaurant Waste Cooking Oil (WCO)

Owolabi R U[1]*, Osiyemi N A[2], Amosa M K[3] and Ojewumi M E[1]

1Fountain University, Industrial & Environmental Chemistry Department, Osogbo Osun State, Nigeria
2 Ahmadu Bello University, Chemical Engineering Department, Zaria, Kaduna, Nigeria
3Covenant University, Chemical Engineering Department, Ota, Ogun State, Nigeria

Abstract

The use of waste cooking oil (WCO) from extended household and restaurants was considered as a suitable feedstock for biodiesel production in this paper. The paper also focuses to showcase the qualities of biodiesel produced from WCO compared to that of other virgin oil. The oil that was supposed to be disposed, at times indiscriminately was de-odoured and purified using appropriate solvents. The purified oil was characterized and used for biodiesel production. The physical and fuel properties such as density, viscosity, cloud point, pour point, cetane number etc were determined according to ASTM standards. The properties obtained were not only comparable with that of others but also within standard limits .

Keywords: WCO; Biodiesel; Cloud point; Pour point; Cetane number; ASTM

Abbreviations: WCO: Waste Cooking Oil; ASTM: American Society for Testing and Materials; API: American Petroleum Institute

Introduction

Biodiesel, a renewable fuel made by trans-esterification of vegetable oil with alcohol, is becoming more readily available for blending with conventional diesel fuel for use in transportation applications. Fossil fuels such as petroleum, coal and natural gas, which have been used to meet the energy needs of mankind are associated with negative environmental impacts such as global warming [1,2].Similarly, the fossil fuels accumulated over series of geological activities are irreversibly consumed at a rate more than million times faster than they were formed [3]. This has left us in a precarious position especially for petroleum products. The hike in price of petroleum and its products, both in national and international scenes is frequent for two reasons; the mounting demands and fast depletion of reserves [3] the duo of which call for alternative source of energy. Similarly in developing countries, the price paid for petrol, diesel and petroleum products now dominates over all other expenditures and forms a major part of country's import bill.Biofuel/Biodiesel made from natural oils and fats is being considered as a promising substitute for petrol diesel. Although biodiesel cannot entirely replace fossil fuel, the following reasons however further justify its development.

- It provides a market for excess production of vegetable oils.
- It decreases the dependence on imported petroleum.
- It does not contribute to global warming due to its closed carbon cycle.
- The exhaust emissions of carbon monoxide, unburned hydrocarbons and particulate emissions from biodiesel are lower than fossil fuel [4].

Though much has been done in the area of biodiesel production especially with the use of various feed stock; soya bean oil, corn oil, peanut oil, olive oil, cotton seed oil, safflower, rape seed oil/canola, line seed oil, sunflower, coconut oil, palm kernel oil, jathropha seed oil etc as further evidenced in the report of [5,6] but still, few research has been carried out on the use of WCO for biodiesel production.

The conversion of vegetable oil carried out by [7] under optimized conditions in presence of co-solvent (n-hexane) obtained with solid base catalyst Na/NaOH/γ-Al_2O_3 is comparable to that obtained with conventional homogeneous NaOH catalyst. Heterogeneous solid oxide catalysts of the type $(Al_2O_3)_4$ (SnO) and $(Al_2O_3)_4$ (ZnO), have been examined by [8] for the transesterification of vegetable oil using different chain length alcohols. The highest yield of 84% was observed with methanol. Also, [9] found that calcined Mg–Al hydrotalcite at 400°C was most effective catalyst for the transesterification of jatropha oil, obtaining conversion of jatropha oil of 96%.Sathivel ultimately feels that biodiesel production from waste salmon oil holds promise because the biofuel represents a use of a currently under-utilized resource—and salmon oil biodiesel burns more cleanly and efficiently than pure salmon oil [10]. Recently, report in the literature showed that satisfactory result was obtained when the mixture of canola oil and used cooking oil was used as feedstock for biodiesel production [11]. Biodiesel production with biocatalyst produced biodiesel with a very high degree of purity [12]. Different lipases, such as *Candida Antarctica* [13], *Pseudomonas cepacia* [14] and *Thermomyces lanuginosus* have been employed as biocatalysts in the production of biodiesel from vegetable oil.

Huge quantities of waste cooking oils are generated in the developed countries as contained in Table [15].Management of such oils poses a significant challenge because of their disposal problems and possible contamination of the water and land resources. Even though some of this waste cooking oil is used for soap production, a major part of it is discharged into the environment. The Energy Information Administration in the United States estimated that some 100 million gallons of waste cooking oil is produced per day in USA, where the

***Corresponding author:** Owolabi R.U, Fountain University, Industrial & Environmental Chemistry Department, Osogbo Osun State, Nigeria

average per capita waste cooking oil was reported to be 9 pounds [16]. Waste oil has the potential to become a new source of high-oleic oil and its full potential should be exploited beyond being a suitable way to reuse waste materials [17]. The total waste cooking oil produced in Canada could be approximately 135,000 tons/year. In the EU countries, the total waste cooking oil production was approximately 700,000-1,000,000 tons/yr [18]. UK produces over 200,000 tons of waste cooking oil per year [19]. As large amounts of waste cooking oils are illegally dumped into rivers and landfills, causing environmental pollution [20], the use of waste cooking oil to produce biodiesel as petro-diesel substitute offers significant advantages because of the reduction in environmental pollution. This work therefore focuses on the appreciation and workability qualitatively and quantitatively of transition from the use of virgin oil to spent cooking oil in biodiesel production under catalyzed conditions and the determination of the fuel properties of the waste oil produced biodiesel. In addition, further analysis was also carried out in this wok on the waste oil [15].

Methods and Materials

The vegetable oil after been used to fry large size fish several times for a considerable number of days in an extended household in Lagos metropolis of Nigeria (Latitude 6.5833 and Longitude3.75) was collected in a 2 litre bottle. The oil was de-odoured and purified.

Purification of spent oil

Accurately prepared 0.73% NaCl was poured in the spent oil contained in a separating funnel and shaked rigorously for about 15 minutes. After allowed to settle by gravity, the bottom sediments were removed. A mixture of Chloroform, Methanol and 0.58% NaCl in the ratio 5:48:47 was added to the partially purified oil in the separating funnel. Shaking took place for another 15 minutes followed by gravitational settling and later removal of the second sediments. The purified oil was poured in a beaker of wider surface area and placed in a water bath for any residue solvent to escape. The oil was kept for further analytical tests and the eventual trans-esterification reaction.

pH: The pH of the spent oil was determined using table pH meter.

Density: The density of the oil was determined using a density bottle at 25°C and this is the ratio of the mass of the oil to its volume.

Solvent Miscibility: Solvent miscibility of the oil was determined by physical observation of the uniform blending of oil samples in acid (HCL), alkali (NaOH), sodium hydrogen carbonate solution, ether and water at room temperature.

Congealing temperature: The congealing temperature of the oil was determined crudely by putting 20ml of the dried oil sample in a 100ml beaker and inserting a simple laboratory thermometer into the oil and putting it in a freezer. The oil sample was closely monitored as the temperature reduces. The temperature at which the initial fluid oil sample starts getting jelly was noted, and the temperature at which gelation took place within the oil was also noted and an average of the two gave approximate congealing temperature.

Further process details

Titration process: The essence of the titration process is to get the number of gram of NaOH that will be used per liter of oil in the transesterification process. This will give a rough guide on the amount of catalyst that will give an optimum yield. 1gram of NaOH was dissolved in 1 liter of distilled water (0.1%NaOH) solution. Phenolphthalein solution was used to get the end point. In a smaller beaker, 1ml of WCO oil is dissolved in 10ml of pure isopropyl alcohol. The beaker was warmed gently by standing it in some hot water, stirred until all the oil dissolves in the alcohol and the mixture turns clear.2 drops of phenolphthalein solution was added. Using a burette, 0.1% NaOH solution was added drop by drop to the oil alcohol phenolphthalein solution, stirring all the time, until the solution stays pink for 10 seconds. The number of ml of 0.1% NaOH solution used added to 5.0 will gives the number of NaOH to be used per liter of oil [21].

Heating of oil: 43g of the oil in a beaker was pre- heated to about 60°C using heating mantle with temperature range of 0-100°C to further reduce the reaction time.

Mixing of methanol and catalyst: Less than 1g of NaOH (Lye) as roughly described by the titration process was added to methanol (about 60% in excess) prior to methanolysis of the heated oil. Continuous agitation for 90 minutes using shaker was done to ensure complete dissolution of the lye catalyst.

Transesterification (biodiesel reaction): The methanol in excess was added to the pre- heated oil in a 250ml beaker serving as a batch reactor. The mixture was agitated in a shaker at 200rpm for 90 minutes and then left overnight for phase separation to take place due to gravity. A yield of about 30% was obtained; though it can be greatly improved

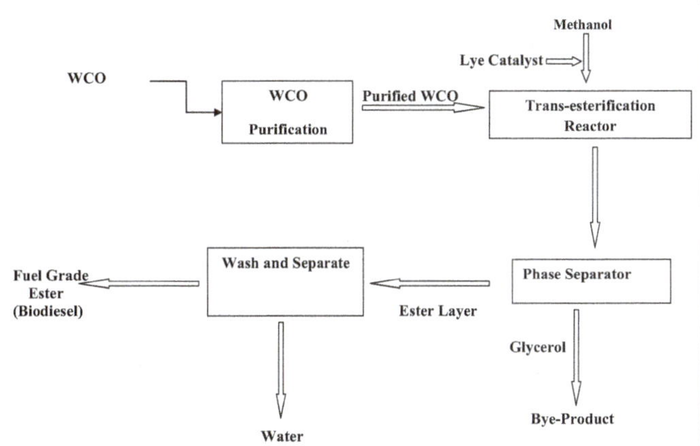

Figure 1: Biodiesel Manufacturing Flow Sheet From Waste Cooking Oil (WCO).

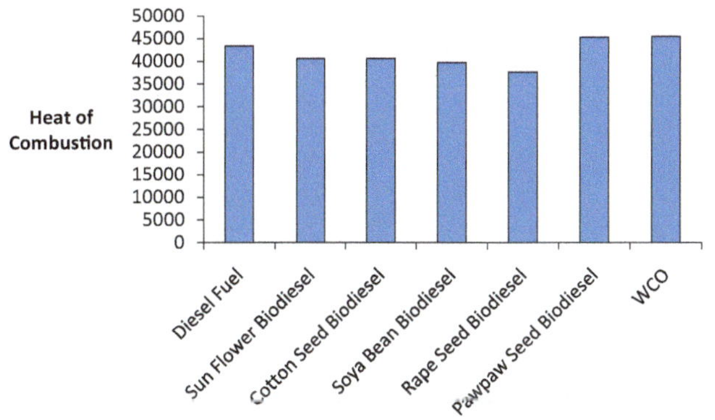

Figure 2: Comparison of Heat of Combustion of WCO biodiesel with other biodiesel fuel.

by tampering with the process thermo-kinetics conditions. This aspect is on-going and shall be separately published. The flow chart indicating the detailed experimental algorithm is stated in Figure 1 while Figure 2 & Figure 3 indicate the WCO undergoing purification and the produced biodiesel respectively.

Washing of fuel: This was done using a small quantity of water. (a misting of water over the fuel, draining water off the bottom).The purpose is to wash out the remnants of the catalyst and other impurities. The washing process was repeated for several times until the washed water became clear. The clean biodiesel was dried in an incubator for 48h. The final product was analyzed to determine its fuel properties.

Fuel properties of oil

Specific gravity, density & API (America Petroleum Institute) gravity: Density is the mass of unit volume of a material at a specific

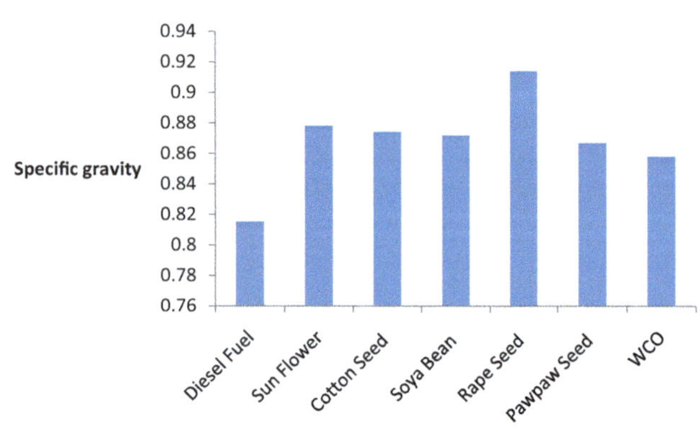

Figure 3: Comparison of Specific Gravity of WCO biodiesel with other biodiesel fuel.

Restaurant	Years in Business	Volume of Waste Oil Disposed Per Month In Gallons	Amount Paid For Disposal Per Month in Dollars
Miraku	1	117lbs	Free
Sandy's Place	1year 2 months	80	Free
Las Magarita	5	100	120
Moraghot	2.5	160lbs	Free
Boston Market	8	100	150
JP Gators	20	50lbs	Free
Napolatano	12	300lbs	<10
Calypso Bar &Grill	1.5	72-80	Free
El Toro	14	600lbs	40
Mildred's Cafe	8.5	30	Free
Chik-Fil-A	24	1052lbs	Free
Kotobuki	14	600lbs	Free
Ballyhoo Bar &Grill	3	3400lbs	511-550 per year.
Porter's	4	20	80
Clara's	1	420	150
Hot Wok	10 months	100	230
Mr.Han's Restaurant &Night Club	17	140	60
Miya Sushi	8	12	100
Timber Creek Steakhouse	12	560	130

Table 1: Selected Restaurants in the City of Gainesville (U.S.A) to Determine Supply Lines for Waste Cooking Oil.

Test	Result
PH	3.91
Specific Gravity	0.895
Congealing Temperature	11°C
Solvent Miscibility	Miscible with ether

Table 2: Results of the Waste oil characterization.

temperature. A more useful unit used by the petroleum industry is specific gravity, which is the ratio of the weight of a given volume of a material to the weight of the same volume of water measured at the same temperature. Specific gravity is used to calculate the mass of oils. The API (American Petroleum Institute) gravity is another way to express the relative masses of oils. The API gravity could be calculated mathematically using the equation:

$$^{\circ}API = \frac{141.5}{Sp.Gr60/^{\circ}60F} - 131.5.$$

A low API gravity indicates heavier oil, while a higher API gravity means a lighter

crude or product. Specific gravities of crude oils roughly range from 0.82 for lighter crudes to over 1.0 for heavier oil.

Flash point: This is the minimum temperature at which the vapour from oil sample will give a momentary flash on application of a standard flame under specific test conditions. This is used to predict the possible fire hazard during transportation, storage and handling.

Pour point, Cloud point: The pour point of a crude oil or product is the lowest temperature at which oil is observed to flow under the conditions of the test. Handling and transporting oils and heavy fuels is difficult at temperatures below their pour points.Often, chemical additives known as pour point depressants are used to improve the flow properties of the fuel. The temperature at which wax crystals begin to form on the surface of the biodiesel is the cloud point.

Aniline point, Diesel Index: Aniline Point is the minimum temperature at which equal volumes of anhydrous aniline and oil mix together. A low aniline point indicates low diesel index. Diesel index is a measure of ignition quality of a fuel. Aniline point can also predict the amount of carbon present in the oil as given by the equation

$$\%C = 1039.4n_D^{20} - 470.4d_{20} - 0.567AP(^{\circ}C) - 1104.42.$$

Where n_D^{20} = Refractive index at 20°C and d_{20} = Density at 20°C.

Cetane Number: This is a corollary of the Index Number.

Heat of Combustion: The amount of heat evolved when 1 mole of the biodiesel is burnt.

Viscosity: This is the resistance to flow of oil. Ease of starting depends on viscosity.

Results and Discussions

Conclusion and recommendation

Table 1 shows selected restaurants in some part of the world and the volume of WCO generated. This will determine the supply lines for waste cooking oil. The purified WCO was used for biodiesel production. Specific gravity, cetane number, diesel index, viscosity, pour point, cloud point and other fuel properties determined exhibited similar properties with those of virgin oil. The specific gravity and

Fuel Property	[1](Petroleum diesel)	[2](PKO Biodiesel	[3]Rapeseed Biodiesel	[4]Canola Biodiesel	[5]Jathropha Biodiesel	[6]Pawpaw Biodiesel	[7]WCO Biodiesel
Specific Gravity	0.853	0.883	0.876	0.878	0.880	0.867	0.858
Flash Point °C	74	167	124	177	170	135	150
Pour Point°C	–16	2	–2	–1		4.000	7
Cloud Point °C	–12	6	–10	–6		7.000	10
Aniline Point °C						5.000	4
API Gravity						31.710	33.42
Heat of Combustion Cal/g						10821.45	10854.06
Viscosity @ 40°C mm²/s	2.847	4.839	6.170	4.892	5.65cSt	1.650	1.89
Cetane Index					50	11.140	10.96
Diesel Index						1.585	1.337
pH						7.400	7.43

[1-4] Alamu et al, 2008; [5] Reddy et al. 2005 & [22]; [6,7] (Owolabi et al. 2011)

Table 3: Characterization of the WCO based biodiesel.

Fuel Type	Heat of Combustion kj/kg	Specific Gravity
Diesel Fuel	43350	0.815
Sun Flower Biodiesel	40579	0.878
Cotton Seed Biodiesel	40580	0.874
Soya Bean Biodiesel	39760	0.872
Rape Seed Biodiesel	37620	0.914
Pawpaw Seed Biodiesel	45342	0.867
WCO	45479	0.858

(Doysan, 1999; Paksoy, 1999; Recep Altim et al., 2000, Owolabi et al, 2011 (This Study)

Table 4: Heat of a Combustion and Specific gravity of other vegetable biodiesel fuel compared to WCO biodiesel fuel.

viscosity at 40°C of the WCO biodiesel was 0.858 and 1.89 respectively which are the closest to the petro-diesel of all the biodiesel from other feed stocks reported in literature. The 7.43 PH also indicate neither acidic nor basic fuel. The temperature at which the WCO biodiesel will give a momentary flash was 150°C which was close to that of biodiesel from PKO. The pour point, cloud point, aniline point and API were found to be 7°C, 10°C, 4°C and 33.42. The API gravity value of WCO biodiesel is greater than that of pawpaw seed oil biodiesel indicating that WCO is lighter.The pour and cloud point were not too far away from that of PKO while the low value of aniline point is an indication of likely low carbon % of the WCO biodiesel. The difference in heat of combustion of the WCO biodiesel compared to that of others especially pawpaw seed oil and petrol-diesel was not significant. The low value of cetane number will lead to higher ignition delay and higher rates of pressure rises. This tends to increase the combustion noise, decreases engine efficiency and power input. However, the WCO biodiesel can be improved to have the desired high cetane number with the use of certain additives.

Therefore used cooking oil can be recommended as the supplementary oil feedstock for biodiesel production. Production of biodiesel from waste cooking oils for diesel substitute is particularly important because of the decreasing trend of economical oil reserves, environmental problems caused due to fossil fuel use and the high price of petroleum products in the international market.

References

1. Munack A, Schroder O, Krahl J, Bunger J (2001) Comparison of relevant exhaust gas emissions from biodiesel and fossil diesel fuel. Agricultural Engineering International: the CIGR J Sci Res Dev 3(EE 01- 001)
2. Saravanan S, Nagarajan G, Rao GLN, Sampath S (2007) Feasibility study of crude rice bran oil as a diesel substitute in a DI-CI engine without modifications. Energy for Sustainable Development 11: 83- 92.
3. Bhaskara BK (1990) Modern Petroleum Refining Processes, (2ndedn).
4. Ayhau D (2009) Progress and Recent Trends In Biodiesel Fuels. Energy Convers Manage 50: 14-34.
5. Van JG, ShanksB, Pruszko R (2004) Biodiesel Production Technology. Department of Energy Office of Energy Efficiency and Renewable Energy, Midwest Research Institute,Battelle.
6. Koji Y, Atsushi U, Yuzuru S (2001) Influence of Physical and Chemical Properties of Biodiesel Fuel on Injection. Combustion and Exhaust Emission Characteristics in a DI-CI Engine,COMODIA,Nagoya.
7. Kim H-J, Kang B-S, Kim M-J, Park Ym, Kim D-K, et al. (2004) Transesterification of vegetable oil to biodiesel using heterogenous base catalyst. Catal Today 93: 315.
8. Xie W, Huang X (2006) Synthesis of biodiesel from soybean oil using heterogenous KF/ZnO catalyst. 107: 53-57
9. Om Tapanes Nc, Aranda Da, Carneiro Jw, Antunes Oa (2008) Extraction of Jatropha curcas oil glycerides: Theoretical and experimental studies of biodiesel reaction. Fuel.
10. Sathivel S (2006) Oil from Fish Processing Byproducts and Underutililized Fish as a Viable Renewable Resource for Biodiesel Production.
11. Issariyakul T (2007) Biodiesel production from mixtures of canola oil and use cooking oil. Chemical Engineering Journal.
12. Fukuda H, Kondo A, Noda H (2001) Biodiesel fuel production by transesterification of oils. J Biosci Bioeng. 92: 405-416.
13. Hernández-Martín E, Otero E.C (2008) Different enzyme requirements for the synthesis of biodiesel. Novozym 435 and Lipozyme TL IM Bioresour Technol 99: 277-286.
14. Kumari V, Shah S, Gupta MN (2007) Preparation of Biodiesel by Lipase-Catalyzed. Transesterification of High Free Fatty Acid Containing Oil from *Madhuca indica*, Energy Fuels 21: 368-372.
15. AneurinTJ (2003) A Model For Potential Cooperative Biodiesel Production In Gainesville, Alachua County, University of Florida,Florida.
16. Radich A (2006) Biodiesel performance, costs, and use. US Energy Information Administration
17. DaSilva N, Rivera E, Batistella CB (2008) Biodiesel production from vegetable oils: Operational strategies for large scale systems, Campinas, Brazil.
18. Kulkarni MG, Dalai AK (2006) Waste Cooking Oils an Economical Source for Biodiesel: A Review. Ind Eng Chem Res. 45: 2901-2913.
19. Carter D, Darby D, Halle J, Hunt P (2005) How To Make Biodiesel; Low-Impact Living Initiative, Redfield Community, Winslow, Bucks.
20. Yang HH, Chien SM, Lo MY, Lan JCW, Lu WC, et al. (2007) Effects of biodiesel on emissions of regulated air pollutants and polycyclic aromatic hydrocarbons under engine durability testing. Atmospheric Environment 41: 7232–7240.
21. Tiwari AK, Kumar A, Raheman H (2008) Biodiesel production from jatropha oil (*Jatropha curcas*) with high free fatty acids: an optimized process. Biomass and Bioenergy 569 – 575.
22. Sandip KH, AhindraN (2008) Utilization of Three Non-Edible Vegetable Oils for the Production of Biodiesel Catalysed by Enzyme. The Open Chemical Engineering Journal 2: 79-83.

Removal of Iron and Manganese in Water Samples Using Activated Carbon Derived from Local Agro-Residues

Magda A. Akl*, AbdElFatah M. Yousef and Sameh AbdElnasser
Chemistry department, Faculty of Science, Mansoura University, Mansoura, Egypt

Abstract

Three olive stones-derived activated carbons (ACOS) with different chemical characteristics, appropriate for the removal of iron and manganese in groundwater are prepared. The steam activated carbon is obtained from carbonized olive stones in the presence of nitrogen in the temperature range from 700 to 900°C and modified by HNO_3 and ammonium persulphate. The structure of the activated carbons was characterized by N_2 adsorption at 77 K, scan electron microscopy and FTIR. B.E.T and α-methods are used to deduce the effective surface areas. The parameters (such as initial pH, temperature, etc) affecting the adsorption capacity of ACOS toward iron and manganese cations removal from aqueous solutions are investigated using batch experiments. The study of kinetic models including pseudo first order and pseudo second-order are carried out. Langmuir adsorption isotherm is investigated. Equilibrium adsorption data fitted the Langmuir adsorption isotherm well with R^2>0.9908. The maximum adsorption capacities of ACOS for the removal of iron and manganese cations are calculated. The results obtained revealed that the sample activated by HNO_3 has the highest adsorption capacity followed by ammonium persulphate and steam activated samples. The mechanism of adsorption is proposed.

Keywords: Activated carbons; Adsorption; Iron; Manganese; Groundwater; HNO_3; Ammonium persulphate; Langmuir isotherm

Introduction

Ground water is the major source of drinking water in rural and semi-urban areas of Egypt. Both iron and manganese are common contaminants in groundwater, in concentrations exceeding 10 μg/L but they are rarely present in contaminations exceeding 1.0 g/L [1]. The World Health Organization (WHO) has set a guideline value of 0.3 mg/L of iron and 0.05 mg/L of manganese in drinking water [2], and many of the countries, including Egypt, have adopted this value in their national drinking water standards. In the Delta district of Egypt, where Dakahlia Governorate is located, a substantial part of groundwater used for potable uses shows contamination significantly exceeding the maximum contaminant level (MCL). The presence of high concentrations of iron and manganese in water causes economic and technological problems. On contact with air, they precipitate a dark sludge with influences the development of ferruginous and manganese bacteria on walls of pipes enhancing thus the corrosion of the pipes [3,4]. Manganese is also toxic to the brain resulting in neurological disorder similar to Parkinson's disease [5,6]. Although iron is an essential mineral for human, its presence in ground water above a certain limit makes the water un-usable for aesthetic considerations such as metallic taste, odour, staining of laundry and plumping fixtures.

A number of specialized processes have been developed for the removal of metals from water and waste discharges. These unit operations include: chemical precipitation [7,8], coagulation/flocculation [9,10], ion exchange/solvent extraction [11,12], cementation [13,14], complexation [15,16], electrochemical operation [17], biological operations [18], evaporation [19], filtration [20], membrane processes[21] and adsorption [22-26].

Adsorption of heavy metals on activated carbons (ACs) and modified ACs proved to be the most efficient technique in this application. It is recommended to use low cost precursors that are available in large amount, renewable and contain high carbon content and low inorganic content. These requirements are found in many agricultural wastes [25]. Olive stones are agricultural waste which has been widely used for the preparation of ACs [27,28].

The objective of the present investigation is the preparation of AC developed from olive stones and to modify the surface of these carbons by surface oxidation with concentrated nitric acid and ammonium persulfate. The textural properties and the chemistry of the surface of the carbons prepared were determined. The adsorption of Mn (II) and Fe (III) from their aqueous solution, at different conditions of solution volume, adsorbent mass, solution concentration, pH and temperature, kinetic and equilibrium adsorption were performed.

Materials and Methods

Materials

The chemicals used HCl, HNO_3, $(NH_4)2S_2O_8)$, NaOH and standard stock of iron ($FeCl_3$), manganese ($MnCl_2$) were of analytical grade. Stock solutions containing 1000 mg/l of $Fe_2(SO_4)_3 12H_2O$ and 1000 mg/l of $MnCl_2$ by dissolving the calculated amounts in doubly distilled water.

Preparation of the activated carbons

The activated carbon C was prepared by carbonizing clean dry crushed olive stones (0.4-0.6 mm) in nitrogen atmosphere at 450°C, followed by gasification with steam at 900°C to a burn off=29%. Details of preparation are reported elsewhere [25]. Modification of steam activated carbon "C" was carried out by oxidation with 65% nitric acid to get HNO_3-modified carbon "C1". Oxidation with ammonium persulphate of carbon C was also carried out to get persulphate-modified carbon "C2" Details of procedures are given elsewhere [28].

***Corresponding author:** Prof. Magda A. Akl, Chemistry Department, Faculty of Science, Mansoura University, Mansoura, Egypt

Characterization of the activated carbons

The adsorption of nitrogen at 77 K was studied using a conventional volumetric apparatus. Prior to adsorption, the carbon sample was degassed at 473 K under a reduced pressure of 10^{-5} Torr. The adsorption was followed until a relative pressure p/po=0.95 and the desorption was continued until the closure of the hysteresis loop.

Scanning electron microscopy of activated carbons was examined using the SEM (JEOL-JEM-100S) under the following conditions: The samples were prepared by sprinkling the powder lightly onto a double-sides adhesive tape which was mounted on SEM specimen stub. The edges of the double-sided tape were printed with gold. Micrographs were obtained in a secondary electron imaging mode using a potential difference of 25 KV. SEM observations were carried out at magnification of 400x. Representative sections of surfaces were recorded and developed on monochrome paper (IL Ford FP4).

A Fourier transform infrared (FTIR) spectrometer (Perkin-Elmer) model 1430 was used to detect the IR-observable functional groups on the carbon surface, in the wave number range 4000-400 cm^{-1}. The samples were first mixed with KBr and then pressed into pellets. The KBr pellet contains 0.5 wt% carbon.

Adsorption of iron and manganese

Batch experiments were carried out by shaking a stopper flasks containing specific concentration of Fe^{3+} or Mn^{2+} cations solution and specific amount of ACOS at 80 rpm for desirable time, temperature and initial pH. The suspensions were then filtered and the filtrates were analyzed using flame atomic absorption spectrophotometer (Perkin-elmer model 2380). The pH values of suspensions were adjusted with dilute HCl or NaOH solution. The experiments were carried out by varying the initial metal ion concentration, contact time, amount of the adsorbent, pH of the initial suspension.

Results and discussion

Characterization of activated carbons

Textural properties: The surface area and porosity of carbons are prominent factors in determining their adsorption capacities [29]. The textural properties of solids are conventionally determined from the adsorption of nitrogen at 77 K and the adsorption data are usually analyzed by the application of the BET equation [30].

The adsorption of nitrogen by the carbons investigated was found to be relatively rapid with the equilibrium attained within 25 min indicating that the adsorption is not controlled by activated diffusion encountered in ultrafine pores and meanwhile refers to the accessibility of the entire pore structure to the nitrogen molecules [24]. Figure 1 depicts the nitrogen adsorption isotherms measured at 77 K for the carbons C, C1 and C_2.The isotherms of carbon C is mainly type I according to the BDDT classification with no hysteresis loop [31], this isotherm is also characterized by very steep initial portion and gave the highest BET-C constant (Table 1).

The isotherms of carbon C1 and C2 show mixed characteristics of type I and type П, they are less steep and exhibited closed hysteresis loops. The nitrogen adsorption data fitted very well the BET equation. The linear BET plots are shown in figure 2. The surface areas SBET (m^2/g) were calculated and the total pore volume VT (ml/g) were read from the volume of nitrogen adsorbed at P/P0 = 0.95. Included also in table 1 are the BET-C constant and the mean pore radius r (nm), calculated from the relationship: $r=(2V_T(ml/g)*10^3)/S_{BE}T\ (m^2/g)$.

More knowledge on the texture of the carbons investigated could be obtained by considering the α-method [32] in analyzing the nitrogen adsorption results. The application of the α-method has been performed using the standard data reported [33].

This method is considered as an independent method for the determination of the specific surface area S_α (m^2/g) from nitrogen adsorption data. The method allows also the determination of some important textural parameters. These parameters are the area of micropores S_m^α and the area of non-micropores S_n^α. The volumes of micropores and of non-micropores V_m^α and V_n^α have been also determined. Figure 3, shows the α-plots of C, C_1 and C_2. The textural parameters determined by the α-method are listed also in table 1.

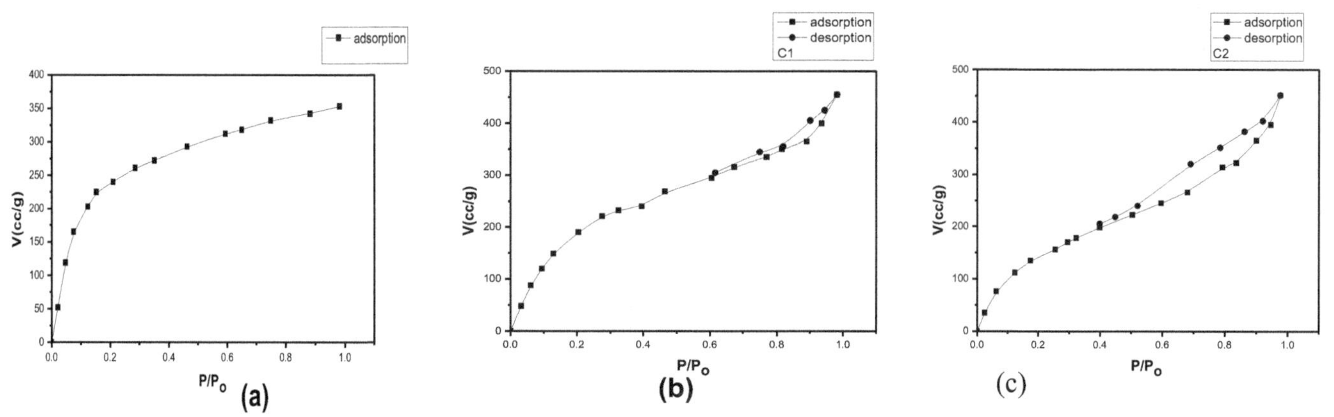

Figure 1: The nitrogen adsorption isotherms at 77 K of (a) C; (b) C1 and (c) C2.

Carbon	$S_{BET} m_2/g$	V_T m/g	r nm	BET-C constant	S_α m2/g	S^α_n m²/g	S_m^α m_2/g	V_n^α ml/g	V_m^α ml/g	V_m^α/VT %
C	831.3	0.547	1.32	25	812	136	676	0.136	0.401	73.3
C1	729.6	0.705	1.93	18.5	740	385	335	0.39	0.25	35.5
C2	576	0.699	2.43	11	525	265	260	0.599	0.1	14.3

Table 1: Textural characteristics of the carbons investigated.

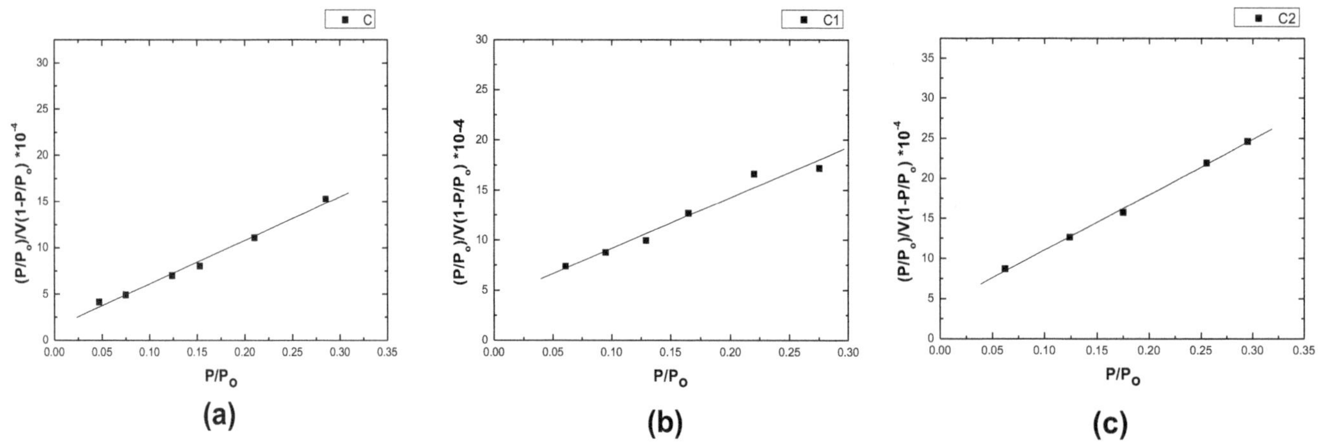

Figure 2: Linear BET plots of nitrogen adsorption at 77 K by (a) C; (b) C1 and (c) C2.

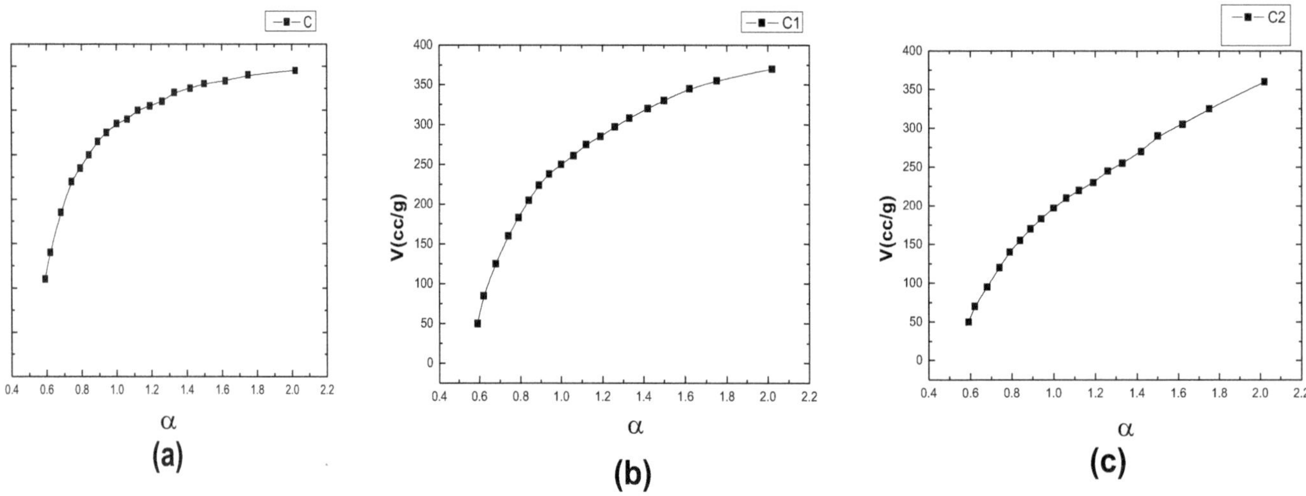

Figure 3: The α plots of nitrogen adsorption at 77 k of (a) C; (b) C1 and (c) C2.

Table 1 depicts: (i) comparable surface areas are obtained from the BET-method and the α-method. This may be taken as an evidence for the reliability of the measurements. (ii) The surface area of the un-oxidized activated carbon is higher than those of the oxidized derivatives. (iii) The total pore volume of the un-oxidized carbon C is lower than those of the oxidized carbons C1 and C2. This could be attributed to the pore widening brought about by oxidation. Evidently r of carbon C was calculated to be 1.32 nm, where as r of C1 was calculated to be 1.93 nm and r of C2 was found to be the largest , i.e=2.43 (nm). Included also in table 1 are the ratio V_m^{α}/V_T, the values obtained may be taken as a support for the pore widening due to oxidation of activated carbons. Thus, the ratio was 73.3% for C, 35.5% for C1 and dropped to 14.3%. (iv) The textural changes depend among other factors on the oxidizing agent used.

The scanning electron microscopy: The scanning electron microscopy images gave insight into the olive stone structure with respect to its shape. The dark areas are macro pores and the pale grey areas are due to the carbon matrix. Figure 4, represents the morphology of the resulting activated carbon prepared via physical (C) and chemical (using HNO_3, C1) activation methods, respectively. It can be clearly seen that physical and chemical activations show some changes in the surface of the particles after activation. The different pore structures of the activated carbon prepared from either physical (using nitrogen as activating agent) or chemical (using HNO_3) activation are observed, which depend upon different reaction mechanisms. Figure 4, depicts two pictures (a) for physically activated carbon (C) and (b) for HNO_3 activated carbon (C1). In this concern, the sample (a) activated by N_2 gas indicated that the carbon matrix are greater than the dark areas. This is referred to the nature of the micro pores resultant from the activated N_2 gas. Whereas, the dark areas are greater than the carbon matrix using HNO_3 as activator in chemical activation. This implies to the higher surface area of sample (b) due to mesopores as well as to micro-pores.

Also, this shows that the olive stone activated by HNO_3 have irregular carbon matrix. i.e. of different shapes and sizes. This was proved by nitrogen adsorption isotherms of activated olive stone which indicated that they were mainly micro pores only in physical activation and mixture from micro- and mesopores in chemical activation.

This may be explained by progressive changes in the surface of the cellular structure of the parent material with HNO_3 solution used as shown in figure 4. The cellulose units are hydrolyzed by the acid and thus the main components of inter cellular wall are broken down to smaller structure.

Thus it is apparent that the external surface of olive stone is quite rough, consisting of cavities, cracks, and irregular protrusions.

Chemistry of the carbon surface: The chemistry of the surface of a carbon is more important than its textural properties in determining its adsorption from aqueous solution particularly when the adsorption involves interaction with the surface functional groups via ion exchange and/or complex formation which is most probably the case in adsorption of metal ions on activated carbons.

The chemistry of the carbon surface is attributed to the existence on the surface of carbon-oxygen functional groups of acid or basic character. The FTIR spectra of samples C, C1 and C2 (before and after adsorption of iron and manganese) have been recorded. Because the surface C-O groups, particularly those of acid type are the most important sites for the adsorption of heavy metals from their aqueous solutions [34], concentration is directed to such functional groups.

The recorded IR spectra (before adsorption of iron and manganese) depict\the existence of absorption bands at 1610 cm^{-1}, 1365-1330 cm^{-1} and 1100-1081 cm^{-1} for C,C1 and C2. These bands are assigned respectively to symmetric COO- vibration [35], stretching C-O of OH groups and ether type structure and phenolic or aliphatic OH [36]. The band 1710-1705 cm^{-1} is shown only for the oxidized carbons C1 and C2, the 1710-1705 cm^{-1} band is assigned to stretching C=O vibration of lactone, quinine and carboxylic acid [37] which are all of acid type. Based on this, the existence, absence or shift in the location in the 1700-1750 cm^{-1} should be considered.

The FTIR spectra of C, C1 and C2 after the adsorption of Fe (III) showed the following changes: for carbon C, the disappearance of the 1610 cm^{-1} and the down ward shift of the 1340 cm^{-1} band to 1292 cm^{-1} and the appearance of a new band centered at 1207 cm^{-1} may account for Fe (III) interaction with the un-oxidized carbon C. A new band centered at 1750 cm^{-1} made its appearance upon Fe (III) uptake by this particular sample, taking in consideration that the FTIR spectra of carbon C, before iron adsorption did not show such band. This observation may be explained by the significant decrease in the pH of iron solution after equilibrium adsorption. It seems that Fe (III) can be adsorbed via ion exchange under certain circumstances. It has been reported on the carbon surface some C-O groups increased and new groups have appeared [38], possibly via a reaction such as

$$C+4Fe^{3+} + H_2O = CO_2 + 4Fe^{2+} + 4H^+.$$

For C1 and C2, the interaction with Fe (III) brought about a significant shift of the 1710 cm^{-1} band to 1740 cm^{-1} and the appearance of a new band centered at 1232 cm^{-1} together with the disappearance of the 1610 cm^{-1} and 1330 cm^{-1} bands.

The FTIR spectra of C, C1 and C2 after the adsorption of Mn (II) showed the following changes: for carbon C, the adsorption of Mn (II), the disappearance of 1340 cm^{-1} band and the shift of the 1610 cm^{-1} band to 1579 cm^{-1} may indicate the possible interaction of Mn (II) via complexation. For carbon C1, the interaction of Mn (II) with the surface was associated with the disappearance of the bands at 1610 cm^{-1} and 1330 cm^{-1} referring thus to the involvement of the corresponding groups in the interaction with Mn (II) via complexation. However the small shift of the 1710 cm^{-1} to 1705 cm^{-1} may not account for any mode of interaction. For C2, the interaction with Mn(II) proceeded also with the disappearance of 1610 cm^{-1} and 1365 cm^{-1} bands, together with the appearance of a band centered at 1207 cm^{-1} standing for =C-O symmetric stretching. These significant changes are related to the uptake of Mn (II) by carbon C2 via complex formation. It remains, here to exclude the uptake of Mn (II) via ion exchange process. The slight shift of the band at 1710 cm^{-1} upon Mn (II) interaction either with C1 or C2 may be taken as an evidence for this explanation. Evidence was shown by the negligible of the pH of Mn (II) solution prior to or after the adsorption.

The pH of the aqueous slurry of the carbon material provides a convenient indicator of the type and concentration of the chemical parameters of the ACs investigated. Table 2 reveals that (i) the surface pH of carbon C (8.5) indicate its surface basicity, i.e the basic functional groups on the surface of non-oxidized carbon are more dominating compared with those of the acid type. The same is also true, but to a less extent for C2 of surface pH=7.6. Steam activation at 950°C usually leads to the formation on the carbon surface of C-O groups of basic

carbon	Surface pH	pH zpc	Carboxyl (m.eq/g)	Lactonic (m.eq/g)	Phenolic (m.eq/g)	Carbonyl (m.eq/g)	Total (m.eq/g)
C	8.5	5.4	0.21	0.20	0.18	0.24	0.83
C1	5.2	4.4	1.00	0.89	1.02	1.55	4.45
C2	7.6	5.1	0.48	0.44	0.64	0.69	2.25

Table 2: Surface chemical parameters of the investigated carbons.

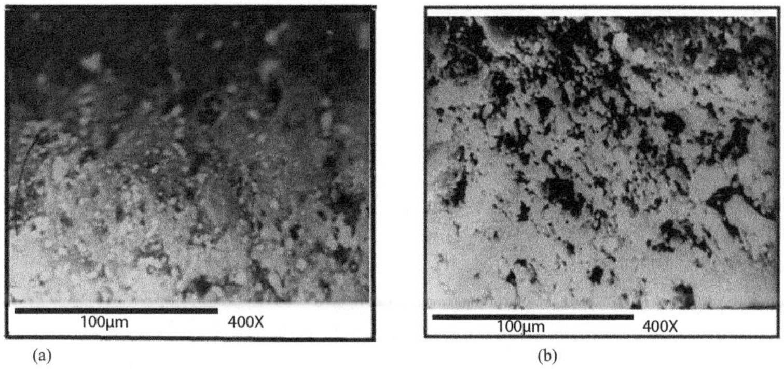

Figure 4: Scanning electron micrographs of (a) steam activated carbon and (b) HNO_3-oxidized carbon at magnification of 400 x.

character [38]. Oxidation of steam activated carbon with nitric acid creates C-O groups of acid character decreasing thus the surface pH. Treatment with persulphate (C2) decreased the surface pH from 8.5 to 7.6, yet the strength of the acid was not high enough to turn over the surface pH in the acid range. Oxidation of carbon C with 45% nitric acid (C1) brought about this turn-over. Thus the surface pH of C1 was determined to be 5.2.

The pH of zero point charge pH pzc of C, C1, C2 were determined to be 5.4, 4.4 and 5.1, respectively (column 3 of Table 2). It seems that pH zpc is always lower than the corresponding surface pH, but both follow the same trend.

The surface acidic groups could be determined in (m.eq/g) by the selective neutralization with a series of bases of varying strength. $NaHCO_3$, Na_2CO_3, NaOH and $NaOC_2H_5$. $NaHCO_3$ neutralizes carboxylic groups where as those neutralized by Na_2CO_3 but not by $NaHCO_3$ are lactones. The weak acid groups neutralized by NaOH but not by Na_2CO_3 were postulated as phenols.

The reaction of $NaOC_2H_5$ is not considered as a true neutralization reaction since it did not involve change by H^+ or Na^+ ions.

The groups reacting with $NaOC_2H_5$, but not with NaOH, were suggested to be carbonyl groups [39]. Table 2 indicates that the concentration of the acids groups considerably increase with treatment with nitric acid with this increase depending of the acid strength used in the oxidation of the steam-activated carbon C. Table 2 reveals that:

i. The total surface acidity of HNO_3-oxidized carbon "C1" is the highest where as the non-oxidized activated carbon "C" measured the lowest surface acidity. Evidently carbon "C2" modified by oxidation with persulphate exhibited intermediate total surface acidity.

ii. Oxidation with persulphate was found to be associated with more than two fold increase with respect to the concentration of carboxyl and lactonic groups and about three fold increase in the concentration of both phenolic and carbonyl groups.

iii. On the other hand, oxidation with nitric acid brought about 5-6 fold increase in the concentration of the different types of the surface acid groups, compared with the non-modified carbon.

Adsorption of Fe (III) and Mn (II)

Effect of contact time: The effect of contact time is an important parameter in determining the equilibrium time that should be allowed for the maximum uptake of an adsorptive by a given adsorbent. Meanwhile, the determination of the uptake from an adsorptive at different time intervals permits the construction of an adsorption kinetic curve from which the adsorption rate constant and the order of the adsorption process could be obtained. The kinetics of Fe (III) and Mn (II) uptakes were followed using solutions of initial pH=6 and initial concentration of the adsorption solution=5 ppm and adsorbent dosage=0.010 g.

The kinetic adsorption curves of Fe (III) and Mn (II) by carbon C1 at 25°C are shown in figure 5. The data obtained show that the metal ion uptake by the oxidized carbon C1 increased with increasing contact time, with equilibrium being established in less than one hour. This equilibrium time was considered, but more time was allowed particularly in determining the equilibrium adsorption isotherms to ensure the attainment of equilibrium conditions.

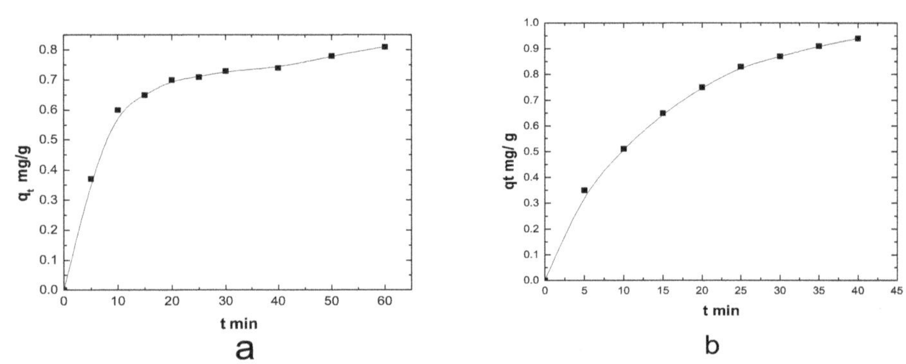

Figure 5: The kinetic adsorption curves of (a) Fe(Ш) and (b) Mn(П) by the activated carbon C1 at 25 °C.

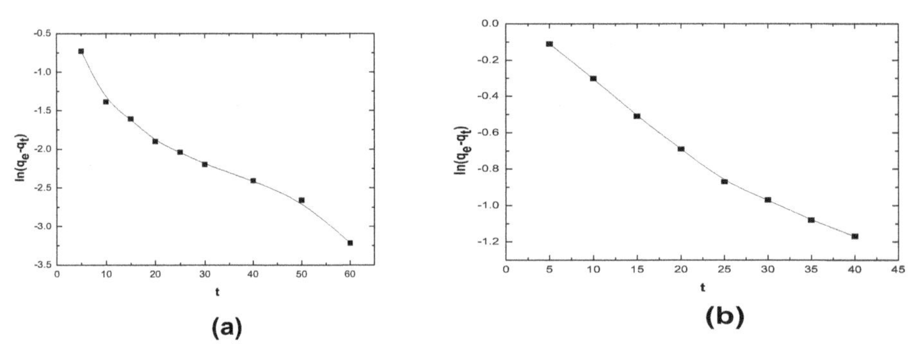

Figure 6: The sorption kinetics of Fe(Ш) and Mn(П) at initial concentration on oxidized carbon C1.

Usually the adsorption of heavy metal ions is analyzed using the Lagergren-first-order and pseudo-second order kinetic equations.

The Lagergren-first-order kinetic equation [40] is based on adsorbent capacity and may be written as

$$dq_t/dt = K_{1,ads}\,(q_e\text{-}q_t) \quad (1)$$

Where $K_{1,ads}$ (1/min) is the first-order sorption rate constant, q_e and q_t are the equilibrium uptake and the uptake at time t, respectively. Equation (1) can be integrated using the boundary conditions q_t=0 at t=0 to give

$$\ln (q_e\text{-}q_t)=\ln q_e\text{-}\ _{K1,ads}\ t \quad (2)$$

Thus if the adsorption kinetics obey the Lagergren-first-order kinetic equation, a plot of ln (qe-qt) versus t should result in a straight line. However, in many cases the Lagergren-first-order equation only fits the sorption kinetics over the first 15-20 min of the sorption process [41]. Figure 6 shows a fit of the sorption kinetics of Fe (Ш) and Mn (П) at initial concentration on oxidized carbon C1.

The pseudo-second-order kinetic equation [42] is, also based on the sorption capacity of the sorbent and is given by

$$dq_t/dt = K_{2,ads}(q_e\text{-}q_t)^2 \quad (3)$$

Where $K_{2,ads}$ is the rate constant for pseudo-second-order sorption . Integrating equation (3) with the boundary conditions q_t=0 at t=0 gives

$$t/q_t=t/q_e+1/K_{2,ads}\,q_e^{\,2} \quad (4)$$

Thus, for pseudo-second-order kinetics a plot of t/q_t versus t should result in a straight line, with the values of q_e and $K_{2,ads}$ being determined from the slope and intercept. Figure 7 shows a plot to test such pseudo-second-order kinetics. The values of sorption rate constant, the equilibrium uptake and the correlation coefficient are listed in table 3.

As can be seen from figure 7 and table 3, pseudo-second-order kinetics gave a good fit to the data for the sorption of Fe (Ш) and Mn (П) onto the HNO_3 oxidized carbon C1.

The kinetic data obtained were also fitted to the intra particle diffusion model presented by the equation [43]

$$q_t = K_d t^{0.5} \quad (5)$$

where K_d is the rate constant for intra particle diffusion (mg/g. min$^{1/2}$). The subsequent linear plots of uptake against the square root of the contact time obtained are depicted in figure 8. A good correlation

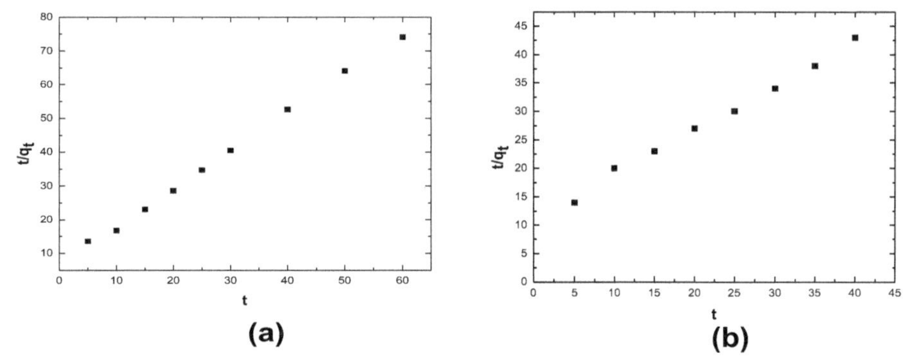

Figure 7: Linear plots of t versus t/q_t of a Fe(Ш) and b Mn(П) sorption at 25 °C on oxidized carbon C1.

Metal ion	Experimental		First-order model			Second-order model	
	$q_{e\,(exp)}$	$K_1(min^{-1})$	q_e(theor)(mg/g)	R^2	K^2(g/mgmin)	qe(theor)(mg/g)	R^2
Fe(Ш)	0.85	3.67*10^{-1}	0.273	0.81	2.08*10^{-1}	0.87	1
Mn(П)	1.25	3.5*10^{-2}	1.02	0.95	8.25*10^{-2}	1.21	1

Table 3: First-order and second-order constants for the adsorption of Fe (Ш) and Mn (П) by carbon C1.

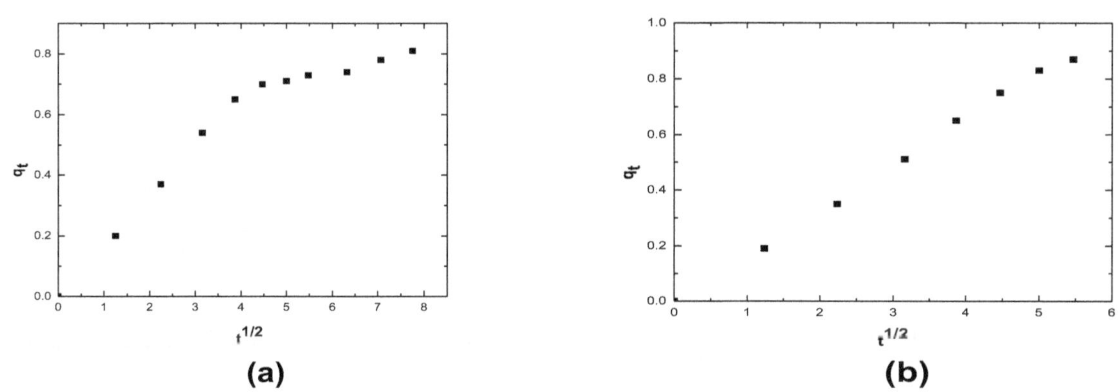

Figure 8: Linear plots of uptake against the square root of the contact time of (a) Fe(Ш) and (b) Mn (П).

coefficient for the experimental results with the linearized form of the model would indicate that the diffusion mechanisms are controlling factors in the sorption kinetics. Correlation coefficients were found to be very high, i.e $R^2=1$ for Fe (III) and Mn (П), respectively. The fact that the linear line passed through the origin and the linearity extended over a relatively wide range could indicate that intra particle diffusion is the only phenomenon occurring.

Effect of sorbent concentration: In this case, the effect of different amounts of sorbent on the removal of Fe (III) and Mn (П) was considered. Experiments were conducted using 50 ppm Fe (III) and 50 ppm Mn (П) at an initial pH 6 and different sorbent concentrations viz., 5,10,20,30 and 50 mg/L. The adsorption data for the residual concentration of Fe (III) and Mn (П) ions versus the amount of sorbent used are present in figure 9.

The results indicate that a greater amount of Mn^{2+} was removed from the solution with increasing amount of the sorbent in the system. This effect may be explained in terms of the availability of a greater surface area with a higher number of functional groups on the AC surface.

Effect of pH: The removal of heavy metal ions from aqueous solution is known to be strongly dependent on the pH value of the solution [44]. The solution pH affects the surface charge on the solid particles and the solubility of the metal ions. The effect of pH Fe (III) and Mn (П) adsorption by the HNO_3-oxidized carbon C1was studied over an initial pH range of 2-7. Figure 10 depict a plot for the uptake of Fe (III) and Mn (П), respectively as a function of initial pH of the solutions where the initial Fe (III) and Mn (П) concentrations were 5 ppm, adsorbent concentration 50 mg and adsorption solution volume 50 ml.

Effect of the solution volume: To determine the influence of the solution volume, the concentration of Fe (III) and Mn (П) was 5 ppm, the mass of the sorbent "carbon C1" was 50 mg, the initial pH 6. Figure 11 shows the metal ion uptake as a function of the volume of the adsorption solution. It is shown that for both metal ions, i.e Fe (III) and Mn (П), the metal ion uptake by carbon C1 decreased with the increase of the sorption solution volume. This may be explained by considering the increase of the thickness of the diffusion layer.

Equilibrium sorption of Fe (III) and Mn (П): The adsorption isotherms of Fe (III) and Mn (П) on ACs investigated are shown in figure 12. The isotherms are of Langmuir type, i.e type L with steep initial portion and with a plateau starting at relatively low concentration and covers a wide range of equilibrium concentrations.

The adsorption isotherms were analyzed using the linear form of the Langmuir equation, i.e., equation (6)

$$C_e/q_e = C_e/q_m + 1/K_{L.qm} \quad (6)$$

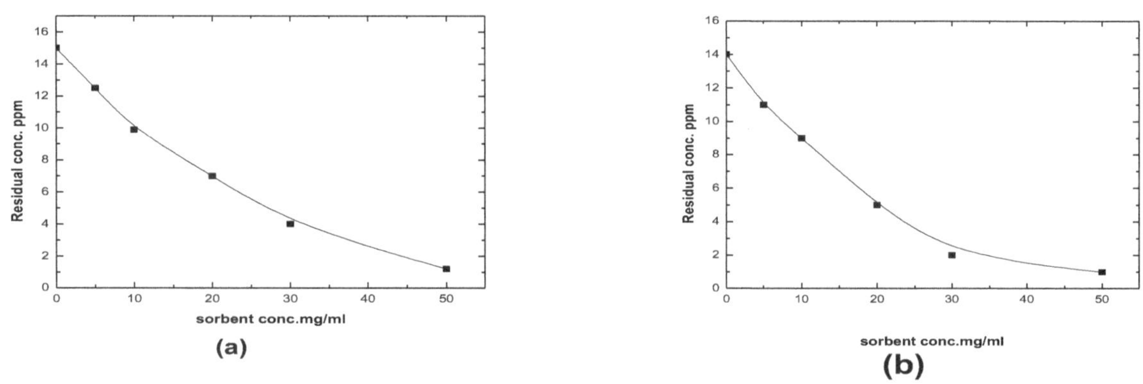

Figure 9: Influence of sorbent concentration "carbon C1" on the uptake of (a) Fe(Ш) and (b) Mn(П) from aqueous solution.

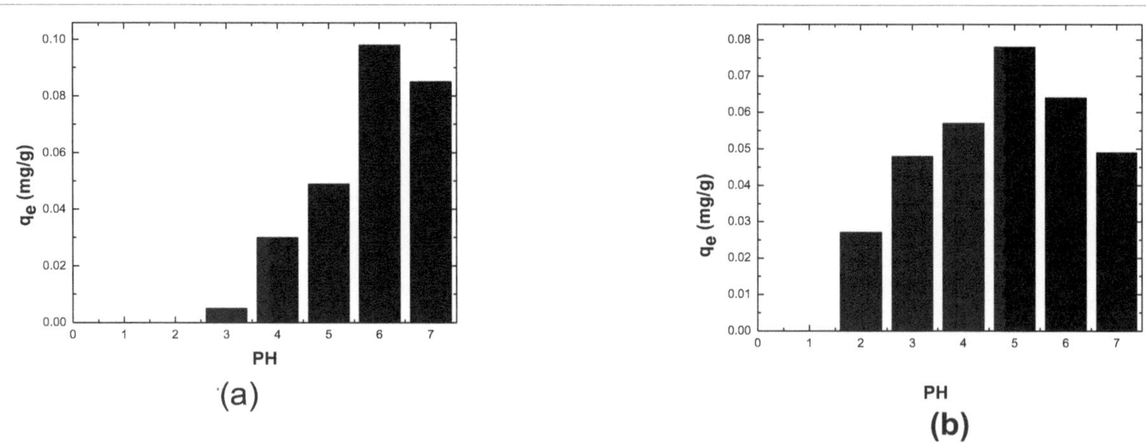

Figure 10: Uptake of (a) Fe(Ш) and (b) Mn(П) by "carbon C1" as a function of initial solution pH.

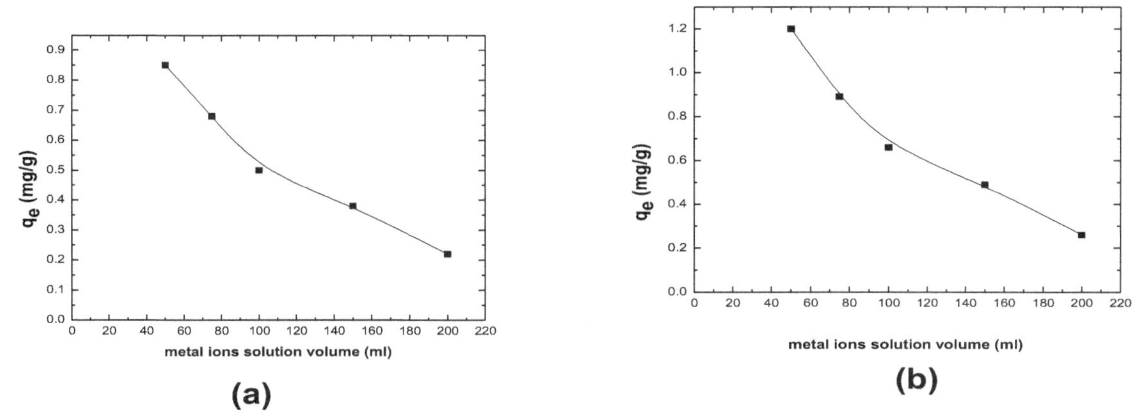

Figure 11: Influence of solution volume "carbon C1" on the uptake of (a) Fe(Ш) and (b) Mn(П) from aqueous solution.

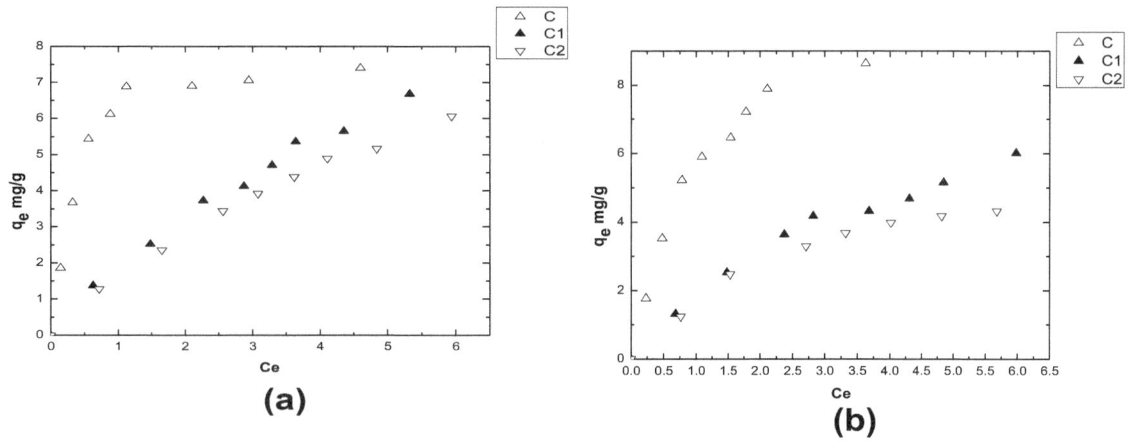

Figure 12: Equilibrium adsorption isotherms of (a) Fe(Ш) and (b) Mn(П) on C, C1 and C2.

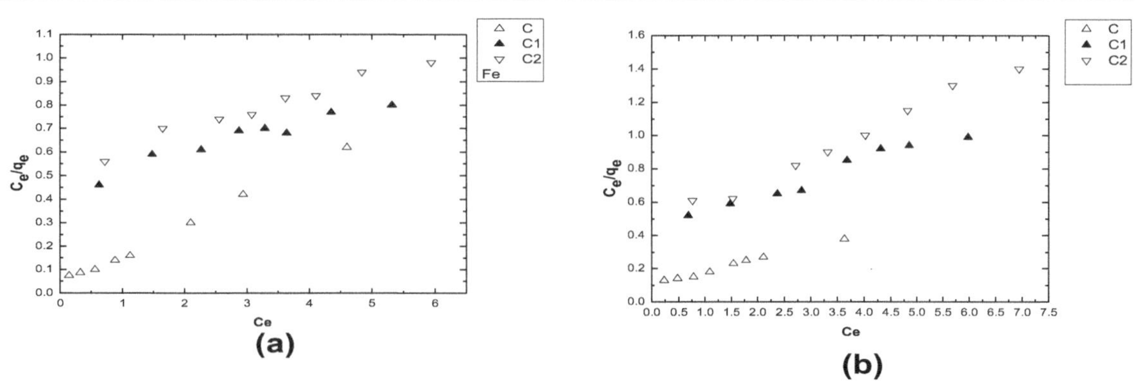

Figure 13: Linear Langmuir plots for equilibrium adsorption isotherms of Fe(Ш) and Mn(П) on C, C1 and C2.

Where q_e (ppm) is the amount adsorbed at equilibrium concentration C_e (ppm), qm (ppm) is the maximum sorption capacity, convention-ally considered as the mono layer capacity and K_L is the Langmuir adsorption constant (L/mg). The linear Langmuir plots of Fe (III) and Mn (П) adsorption on all the investigated carbons are shown in figure 13. Values of q_m and K_L are listed in table 4 together with the values of regression coefficients R^2 of each system. The high values of R^2 refer to the excellent fit of the results obtained to the Langmuir model.

Mechanism of adsorption

Modification of the carbon surface via oxidation with concentrated nitric acid introduces a variety of carbon oxygen functional groups of acidic nature. These groups dissociate in aqueous solutions and participate in the metal binding process. It is likely that other surface oxygen atoms (some of them may not be a component of protonogenic functional groups) may participate in metal complex formation. Thus,

Sample	Fe(Ш)		Mn(П)	
	q_m mg/g	K_L L/mg	q_m mg/g	K_L L/mg
C	8.06	2.07	9.52	1.62
C1	12.2	0.195	10	0.22
C2	10.64	0.192	6.66	0.38

Table 4: The values of q_m and K.

one would expect an ion exchange between Fe (III) and Mn (П) with the H^+ released from the carboxylic and phenolic surface groups and therefore these groups contribute significantly to the enhancement of metal ion uptake. Carbonyl and/or lactonic groups, on the other hand are more likely involved in a complex formation with metal ions like Fe (III) and Mn (П) [45].

Conclusion

The results obtained in this study clearly demonstrated the potential use of activated carbons derived from olive stones for the removal of Fe (III) and Mn (П) from aqueous solutions. The following conclusions can be drawn based on obtained results:

d. Nitric acid and persulphate-activated carbons possess polar surfaces due to the existence of a number of carbon-oxygen groups of different basicity.

e. The sorption of Fe (III) and Mn (П) ions onto chemically activated carbons is high and the mechanism of ion exchange is predominating.

f. The kinetic studies indicated that equilibrium in the adsorption of Fe (III) and Mn (П) on activated carbon was attained within one hour.

g. The optimum pH corresponding to the maximum adsorption was found to lie between 5.0 and 6.0.

h. The amount of Fe (III) and Mn (П) adsorbed increased with the increase of concentration

i. The extent of adsorption for Fe (III) and Mn (П) increased along with an increase of activated carbon dosage.

j. Pseudo-second-order kinetics gave a good fit to the data for the sorption of Fe (III) and Mn (П) into the HNO_3-activated carbon, and also the correlation coefficients for second-order model are higher than the first-order model for Fe (III) and Mn (П).

k. The high values of R2 in Langmuir model give an indication of favorable adsorption.

References

1. Giammanco S, Vatenza M, Pignato S, Giammanco G (1996) Mg, Mn, Fe and V concentrations in the ground water of Mount Etna (Sicity). Water Res 30: 378-386.
2. WHO (2003) Guidelines for Drinking Water Quality. World Health Organization, Germany.
3. Kontari E (1988) Groundwater, iron and manganese, an unwelcome trio. Water Eng Manag 25-26.
4. Komewska E, Lach J, Kacpzak M, Neeza E (2007) The removal of manganese, iron and ammonia nitrogen on impregnated activated carbon. Desalination 206: 251-258.
5. Takeda A (2003) Manganese action in brain function. Brain Res Rev 41: 79-87.
6. Donaldson J (1987) The physicopathlogic significance of manganese in brain. Neuro Toxicology 8: 451-462.
7. Bhattachayya D, Jumawan A, Sun G, Schwitzebel K (1980) Precipitation of sulphide: Bench Scale and Full scale experimental results. AICHE symposium series, Water 77: 31-42.
8. Linstedt KD, Houck CP, O'Connor JT (1971) Trace element removals in advanced waste water, treatment processes. J Water Pollut Control Fed 43: 1507-1513.
9. Daniels SL (1975) Removal of heavy metals by iron salts and polyelectrolytic flocculants. AICHE symposium series. Water 71: 265-271.
10. EPA (1987) Manual of treatment techniques for meeting the Interim primary drinking water regulations EPA600877005.
11. Knocke WR, Clevenger T, Ghosh MM, Novak JT (1978) Recovery of metals from electroplating waste waters. Proc. 33rd Purdue Industrial waste conf 33: 415-426.
12. McDonald CW, Bajwa RS (1977) Removal of toxic metal ions from metal finishing waste water by solvent extraction. Sep Sci 12: 435-445.
13. EPA (1979) Economic of waste water Treatment Alternatives for the Electroplating Industry. EPA 625/5-79-016.
14. Kim BM (1984) A membrane extraction process for selective recovery of metals AICHE Annual meeting, San Francisco, CA 26-30.
15. Muzzarelli R (1971) Application of polymers in marine ecology, Revue Int Oceanogr Med 21: 93-108.
16. Jellinek HHG, Sangel SP (1972) Complexation of metal ions with natural polyelectrolytes (removal and recovery of metal ions from polluted waters. Water Res 6: 305-314.
17. Dean JG, Bosqui FL, Lanoutte KH (1972) Removing heavy metals from waste water. Environ Sci Technol 6: 518-522.
18. Chang SY, Huang JC, Liu YC (1984) Heavy Metals Removal in affixed film biological system. Summer National AICHE meeting, Philadelphia PA 19-22.
19. Potterson JW, Minear RA (1975) Physical-chemical methods of heavy metal removal, in" Heavy Metals in Aquatic Environment. P.A. Krenkel, edn., Pergamon Press. Oxford, England 261-276.
20. De Mora SJ, Harrison RM (1983) The use of physical separation techniques in trace metal speciation studies. Water Res17: 723-733.
21. Cartwright PS (1981) Reverse osmosis and ultrafiltration in the plating shop. Plating Surf Fin 68: 40-45.
22. Youssef AM, El-wakil AM, El-Sharkawy WA, Farag AB, Toggan KH (1996) Adsorption of heavy metals on coal-based active carbons. Adsorpt Sci Technol 13: 115-124.
23. Youssef AM, Mostafa MR (1997) Removal of metal ions by modified activated carbon. J Appl Si 7: 1-15.
24. Youssefn AM, El-Nabarawy Th, Samra SE (2004) Sorption properties of chemically-Activated carbons. I. Sorption of Cd(11). Colloids and surfaces 235: 153-163.
25. Youssef AM, El-Nabarawy Th, El-shafey El (2006) Modified activated carbons from olive stones for removal of heavy metals. Carbon Science 6: 1-9.
26. Youssef AM, El-Khouly S, El-Nabarawy Th (2008) Removal of Pb(11) and Cd(11) from aqueous solution using activated carbons from pecan shells. Carbon Lett 9: 8-17.
27. Caturla F, Martin-Martinez JM, Molina-Sabio M, Rodriguez-Reinoso F, Torregrosa R (1988) Adsorption of substituted phenols on activated carbons. J Colloid Interface Sci 124: 528-534.
28. Alaya MN, Youssef AM, Karman M, Abdel-Al H (2006) Textural properties of activated carbons from wild cherry stones as determined by nitrogen and carbon dioxide Adsorption. carbon Sci 7: 9-18.
29. Youssef AM, Alaya MN, Nawar N (1994) Adsorption properties of Activated Carbons Obtained from Polymer wastes. Adsorp Sci Techol 11: 225.
30. Brunaur S, Emmett PH, Teller E (1938) Adsorption of Gases in Multimolecular Layers. J Am Chem Soc 60: 309-319.
31. Brunaur S, Deming LS, Deming WS, Teller E (1940) On a theory of the Vander Waals Adsorption of Gases. J Am Chem Soc 62: 1723-1732.
32. Sing KSW (1968) Chem Ind (London) 1520.

33. Sellez-Perez MJ, Martin-Martinez JM (1991) Classification of αs plots obtained from N2/77 K adsorption isotherms of activated. Fuel 70: 877-881.

34. Lodeirop, Barriada JL, Herrero R, Sastre de Vicente ME (2006) The marine macroalga cystoseira baccata as biosorbent for cadmium (П) and lead (П) removal: Kinetic and equilibrium studies. Environ Pollut 142: 264-273.

35. 35. Thrower PA (1989) Chemistry and physics of carbon. Marcel-Dekker, USA.

36. Karunakaran K, Thamilarasu P (2010) Removal of Fe (Ш) from aqueous solutions using Ricinus communis seed shell and polypyrrole coated Ricinus communis seed shell activated carbons. International Journal of Chem Tech Research 2: 26-35.

37. Shindo A, Izumino k (1994) Structural variation during pyrolysis of furfuryl alcohol and furfural-furfuryl alcohol resins. Carbon 32: 1233-1243.

38. Youssef AM, El-Khouly SM, El-Nabarawy Th (2008) Removal of Pb(II) and Cd(II) From Aqueous solution Using Oxidized Activated Carbons Developed From Pecan Shells. Carbon letters 9: 8-16.

39. Cox M, El-Shafey EI, Pichugin AA, Appeleton Q (1999) Preparation and characterization of a carbon adsorbent from flax shive by dehydration with sulfuric acid. Chem Technol Biotechnol 74: 1019-1029.

40. Gaid A, Kaoua F, Mederres N, Khodjsa M (1994) Surface mass transfer processes using activated date pits as adsorbent. Water SA 20: 273-278.

41. Asku Z (2001) Equilibrium and kinetic modeling of cadmium (II) biosorption by C. vulgaris in a batch system: effect of temperature. Sep Purif Technol 21: 285-294.

42. Ho YS, Mckay G (2000) The kinetics of sorption of divalent metal ions onto sphagnum moss peat. Water Res 34: 735-742.

43. Youssef AM, Mostfa MR (1992) Removal of copper ions by modified activated carbons. Indian J Technol 413.

44. Youssef AM, El-Nabarawy Th, SE Samra (2004) Sorption properties of chemically-Activated carbons. I. Sorption of Cd(11) ions. Collids and surfaces A 235: 153-163.

45. Jellinek HHG, Sangel SP (1972) Complexation of metal ions with natural polyelectrolytes (removal and recovery of metal ions from polluted waters). Water Res 63: 205.

Reprocessing the Natural Fibers to Procure Advanced Materials

Ashish Chauhan*, **Priyanka Chauhan and Balbir Kaith**

Department of Chemistry, Dr. B. R. Ambedkar National Institute of Technology, Jalandhar 144 011 (Pb) India

Abstract

Nature and natural resources has offered the mankind with precious gifts for survival and maintenance but with the ever increase in population, pollution and deforestation the sustenance of natural resources is in suspicion. A major part of the land on earth provides us with renewable biomass that mostly degrades with time. Graft copolymerization is a chemical technique to improve the properties of the waste biomass and modify its properties while sustaining the inherent trait. With the advancement of science and research many efforts have been made to modify and improve the synthesis, characterization and evaluation techniques for a better scientific and industrial achievements. The review explores a brief overview of this magnificent engineering, its types and factors that has helped to provides with outstanding results.

Introduction

The word 'polymer' was derived from the classical Greek word 'poly' meaning "many" and 'meres' meaning "parts". If simply stated, polymers are large molecules formed by the repetition of small and simple chemical units called monomers. Polymers may be natural, synthetic or semi-synthetic in origin. They are synthesized by the process called polymerization and if the combining monomers which may be two or more, are different then the resultant is called copolymer. Lord Todd, President of Royal Society of London, answered to the question, 'What do you think has been the chemistry's biggest contribution to science, to society?' "I am inclined to think that the development of polymerization is, perhaps, the biggest thing chemistry has done, where it has the biggest effect on everyday life. The world would be totally different place without artificial fibers, plastics and elastomers. Even in the field of electronics, what would you do without insulation? and there you get back to polymers again" [1].

Cellulose has been the most abundant form of terrestrial biomass. It finds applications in many spheres of modern industry. Cellulose, especially cotton linters, has immense use in the manufacture of nitro cellulose, used in smokeless gunpowder. Cellulose could be processed to make cellophane and rayon and other textile derivatives. It has also been used in the laboratory as a solid substrate for thin layer chromatography and a major constituent of paper. In the recent years, many attempts have been made to modify cellulose by graft copolymerization of vinyl monomers that help in the incorporation of desirable properties without drastically affecting its inherent traits. The formation of graft copolymers with long polymeric sequences of diverse chemical composition opens the ways to afford specialty polymeric materials [2,3].

Concept of Copolymerization

A polymer may be known as homopolymer if it has similar monomeric units, whereas, if two different monomers are the repeat units, the resultant product is called copolymer and the reaction may be termed as copolymerization. Reactivity of a particular monomer decides the sequence in which the different monomers attach themselves onto the trunk polymer. Common methods to induce grafting are radiation source initiation, chemical initiation and physical or mechanical treatment (thawing or beating). Grafting may proceed either through ionic or free radical reaction mechanism.

Since the probability for the production of newer monomers at low cost remains bleak, so the modification of the synthetic and natural polymer through graft copolymerization, by incorporating specific properties through comonomers has gained much scientific attention. Improvement in dyeing, printing, chemical resistance, water repellency, fiber strength, abrasion resistance, crease resistance and impact strength of the thermoplastics and rubber are some of the advantages of the graft copolymerization. Sequencing copolymer exhibit many properties of each constituent homo-polymer. Copolymers with substantially large polymer sequences of diverse chemical compositions are designed in such a manner that the desirable properties are retained while eliminating the rest. Mechanical properties of polymers such as tensile strength, extensibility and impact-strength exhibit a correlation with the percent grafting (Pg). On grafting crystal lattice of the polymer gets disrupted but the strength of the added material helps to reinforce the structure. However, if crystallinity does not get distorted on grafting, then continuous increase in strength could be obtained with the increase in Pg [4,5]. Copolymers can be classified into different types:

Random copolymer

There lies no definite sequence of distribution of monomer units on the polymer back-bone. The monomer units are randomly placed along the polymeric chain and could be exemplified as:

$$-A_1-A_2-A_1-A_1-A_1-A_2-A_1-A_2-A_1-A_2-A_2-A_1-A_1-A_1-A_1-$$

Alternate copolymer

In these copolymers, monomers are placed orderly forming a polymeric chain and having a perfect alternate arrangement of monomer units, for example, poly (styrene-alt-maleic anhydride)

***Corresponding author:** Ashish Chauhan, Department of Chemistry, Dr. B. R. Ambedkar National Institute of Technology, Jalandhar 144 011 (Pb) India

where styrene (A_1) and maleic anhydride (A_2) are copolymerized in a regular fashion to give an alternate sequence as shown below:

$$-A_1-A_2-A_1-A_2-A_1-A_2-A_1-A_2-A_1-A_2-A_1-A_2-A_1-A_2-$$

Block copolymers

They are composed of chemically dissimilar extended lengths or blocks or segments terminally connected to one another. They could be exemplified as:

$$-A_1-A_1-A_1-A_2-A_2-A_2-A_1-A_1-A_1-A_2-A_2-A_2-$$

Graft copolymers

Graft copolymers involve a preformed polymeric back-bone to which other polymeric chains of different chemical nature get attached at several points. Stannett defined graft copolymers as: "Graft copolymers consist of a polymer back-bone with lateral covalently linked side chains." Both the back-bone and the side chain polymers could be homopolymers or copolymers. Graft copolymers are made of molecules with one or more block of species having different configurationally features connected to main chain as side chains [6]. Alfrey and Bandel [7] were the first to synthesize the graft copolymers in 1950. They polymerized vinyl acetate (VA) in presence of styrene (Sty) and vinylidene chloride (ViCl). The resulting graft copolymer assumed its shape depending upon the structural format of the back-bone polymer. The generalized view of graft copolymer has been shown below, where the monomer (A_2) has been incorporated onto the back-bone polymer, of building units (A_1).

$$------A_1-A_1-A_1-A_1-A_1-A_1-A_1-----$$
$$|\quad\quad|\quad\quad|$$
$$(A_2)_n\,(A_2)_n\,(A_2)_n$$

Monomer (A_2) could be grafted as a unitary system or as a binary mixture with other monomers. Although, random and alternate copolymers show better properties than the constituent homopolymers yet they are not found suitable for their applications as specialty polymers. On the other hand, both block and graft copolymers have been found to show many technologically advanced characteristics.

Grafting Methods

Induction of monomeric polymerization, if carried-out by an active site generated along a polymer back-bone, it is also called as 'grafting from' technique that finally ends in a graft copolymer:

$$\wedge\wedge\wedge\wedge A_1 \rightarrow \underset{*}{\wedge\wedge\wedge\wedge A_1} \xrightarrow{A_2} \wedge\wedge\wedge\wedge\wedge\wedge \underset{\underset{A_2^*}{|}}{A_1}$$

Whereas, if there has been an attachment of the growing chain (A_2) onto polymer back bone (A_1), thereby, resulting in a graft copolymer, it may be named as 'grafting onto' technique:

$$\wedge\wedge\wedge\wedge\wedge\wedge A_1 \rightarrow \underset{*}{\wedge\wedge\wedge\wedge\wedge\wedge A_1}$$

$$A_2^* + nA_2 \rightarrow (A_2)^*_{n+1}$$

$$\underset{*}{\wedge\wedge\wedge\wedge\wedge\wedge A_1} + (A_2)^*_{n+1} \rightarrow \wedge\wedge\wedge\wedge \underset{\underset{(A_2)_{n+1}}{|}}{A_1}$$

Grafting could be achieved through free radical, anionic or cationic mechanism [1].

Synthesis of Graft Copolymers

Since last decades, numerous methods have been explored for the preparation of graft copolymers by conventional chemical techniques. The active sites are created on the polymeric back-bone for the synthesis of graft copolymers, either a free-radical or a chemical group that may get involved in an ionic polymerization or in a condensation process. Ionic polymerization has to be carried-out in presence of anhydrous medium or in presence of considerable quantity of alkali metal hydroxide. Usually, low molecular weight graft copolymers are obtained in ionic grafting, while in case of free radical grafting high molecular weight copolymers could be prepared [8-11]. A number of methods are used for the generation of active sites on the polymeric back-bone and are broadly classified as:

Physical methods

Active sites could be generated on the parent polymer by swelling it with suitable solvent that ruptures the bond leading to the formation of free radicals. Williams and Stannett [12] utilized freezing and thawing of polymer-monomer mixture to effect the grafting of vinyl monomers onto wool. Preparations of polymers with basic pendent moieties were carried-out by grafting dimethyl amino-ethylmethacrylate onto linear low density polyethylene (LDPE).

Chemical methods

Conventional radical initiators [13], benzoic peroxide (BPO), ceric ammonium nitrate (CAN) [14,15], potassium persulphate (KPS), Fenton's reagent (FAS-H_2O_2), potassium permanganate ($KMnO_4$), β-di ketones of transition metal alone and in conjugation with additives , Lewis acids, strong bases and metal carbonyl , have been successfully used for the chemical grafting similarly, Misra et al. [15-22] have successfully grafted a wide variety of monomers onto wool, cellulose, starch and gelatin by use of the radical initiators.

Physico-mechanical methods

In this method, the application of stress on parent back-bone caused segmental motions and molecular flow that leads to bond-scission and consequent formation of active sites on the back-bone. Free-radicals may also be generated by mastication and milling of back-bone polymers.

Radiation methods

The arena of graft copolymerization has been completely revolutionized by the radiation grafting of vinyl monomers onto the polymeric back-bone. When the electromagnetic radiations were passed through the reactants, there was a decrease in the intensity of the incident ray while the absorbed radiations created active sites on the polymer back-bone. Generally, UV and high-energy γ-radiations have been employed for generating the active sites where grafting of appropriate monomer takes place. In addition, low cost regulation over structural factors, such as the number and length of the grafted chain, has been done by careful selection of dose and dose rate [1].

Plasma radiation induced grafting

Plasma radiation induced grafting has gained much scientific attentions in the recent years. Plasma conditions attained through slow discharge offered the same possibilities as with ionizing radiation. Electron induced excitation, ionization and dissociation are some of the main processes. High energy accelerated electrons are the motivating agents that induce cleavage of chemical bonds in polymeric structure to form active sites.

Enzymatic grafting

In this method, an enzyme initiates the chemical or electro chemical grafting reaction [23], for example, phenol was converted to o-quinone by tyrosinase that underwent subsequent non-enzymatic reaction with chitosan. The enzymatic grafting on poly(dicarbazole-N-hydroxy-succinimide) film, thionine and toluidine blue have been inversely bound to the poly(dicarbazole) back-bone and grafting of polyphenol oxidase (PPO) on poly(dicarbazole) are some of the examples [1].

Factors affecting Graft Copolymerization

The percentage of grafting and grafting efficiency are usually affected by the nature of the monomer, concentration, reaction time, reaction temperature, reaction medium, additives and structure of the back-bone. Some of these factors affecting the graft yield have been discussed underneath [1,24]:

Role of inorganic ions on grafting

The free radical formation gets induced by the addition of some additives that increase the graft yield. In case of cellulose, Cu^{2+} ions help in complex formation. Only a small concentration of inorganic ions is needed to enhance the graft yield, whereas the higher concentration may adversely affect it. The effect of Cu^{2+}, Fe^{2+}, Fe^{3+} ions on radiation induced grafting in case of nylon was studied. The reactivity order observed was $Cu^{2+} > Fe^{2+} > Fe^{3+}$. Grafting efficiency of polar monomers was increased by facilitating monomer access on reaction site, by decreasing the crystallinity of cellulose and by the activation of the monomer through donor or acceptor complex formation with monomers after the addition of $ZnCl_2$ salt.

Different transition metals such as Zn, Fe, V, Co, Cr and Al have been used in the preparation of metal acetyl acetonates and other diketones. Metal acetyl acetonate could be used as an initiator for the vinyl polymerization. Metal chelates generally decompose on heating with the generation of free radicals that abstract hydrogen atoms from the polymeric back bone producing active sites.

Metal chelates are better initiators as compared to other redox systems. Since the reaction could be carried-out even at low temperatures, so the undesired reactions due to chain transfer, homo-polymerization could be avoided [1].

Effect of acid in graft copolymerization

The strength of the acid used in the process has a great impact on the magnitude of graft yield. In radiation induced grafting of cellulose, following changes are supposed to occur [1,25]:

$$\text{Cell-OH} + H^+ \rightarrow \underset{\text{(I)}}{\text{CellO}^+ H_2}$$

where, (I) could capture an electron to give (II).

$$\underset{\text{(I)}}{\text{CellO}^+ H_2} + e^- \rightarrow \underset{\text{(II)}}{(\text{Cell-OH})^*} + H^*$$

Structure (II) decomposes into different radical species that offer sites for the attachment of graft chains.

$$(\text{Cell-OH})^* \rightarrow \text{Cell-O}^* + H^*$$

Moreover, Cell and OH^* may also be formed depending on the dissociation point (II). H_2SO_4 has been reported to be the best acid amongst the series of acids to enhance the graft yield.

When Ceric Ammonium Nitrate (CAN) was used as a source of ceric ions, the presence of nitric acid has been found to play a significant role. In the absence of nitric acid no grafting was observed because of the large size ceric ions which were unable to form complex with the polymer back-bone. However, an increase in concentration of nitric acid resulted in the formation of complexes with functional groups on the polymer back-bone. The complex later decomposes to give free radicals on the back-bone where the grafting occurs [25].

Nature of the substrate

The rate of grafting has been found to increase with the increase in the surface area of the backbone. Therefore, the nature of the substrate as fiber, film or powder form plays an important role. It was reported that grafting of styrene onto cellulose acetate film decreased with increase in thickness of the film. Moreover, in case of cellulose, lignins adversely affect graft yield. Polymer analogous reactions of cellulose also affect grafting efficiency and affect the properties like swelling behavior, redox nature or the complex forming ability. Sometimes the groups introduced on the polymer back-bone decrease the graft yield due to increase in hydrophobicity, that blocks the diffusion of monomer and the initiator to the active sites.

Type of monomer and its concentration

Presence of aromatic rings, monomer polarity and the length of alkyl chain are among the structural factors that affect the graft yield and graft efficiency. Decrease in grafting efficiency has been reported as a function of steric and polar group. They had reported the decreasing reactivity order of various acrylates as methylacrylate (MA) > ethylacrylate (EA) > butylacrylates (BuA) > methylmethacrylate (MMA). Moreover, in case of simultaneous irradiation method, if monomer has high sensitivity to radiation than back-bone polymer, homo-polymerization is favored instead of graft copolymerization. A higher concentration of monomer may result in increased polymerization but at the same time graft copolymerization may not be the preferred process. In case of a redox reaction, graft copolymerization always starts from complex formation between monomer and the substrate that depends upon the concentration of monomer [1,24].

Effect of initiator concentration

Initiator concentration plays an important role to determine the graft yield. Generally at a molar ratio of 1:6 of FAS: H_2O_2 maximum grafting was observed. However, further increase in the molar ratio deteriorated the graft yield. This could be due to the fact that at higher concentration more Fe^{3+} ions are produced that served as chain terminator [26].

$$\text{Flax-(M)}^*_{n+1} + Fe^{3+} \rightarrow \text{Flax-(M)}_{n+1} + Fe^{2+}$$

The amount of initiator, monomer and the back bone polymer are the major factors responsible to determine the rate of grafting. There are various empirical relationships governing the dependence of grafting efficiency on the initiator concentration. It has been observed that once a certain initiator concentration is achieved, higher concentration of initiator do not further effects the graft yield [27].

Solubility of the initiator in the grafting medium shows a profound impact on the yield. Ideally, the initiator should be highly soluble in grafting medium in order to initiate the grafting through monomers [28,29].

Effect of reaction temperature and reaction time

The graft yield gets enhanced with the increase in reaction time and temperature. After reaching the optimum value, further increase in either, results in the decline of percentage grafting. It occurs due to the increase in various hydrogen-abstraction reactions and high viscosity of the reaction medium, due to homopolymerization. High viscosity of the medium creates hindrance in the path of free radicals to reach the active sites on the polymer back-bone. Moreover, a further increase in temperature, results in the dissolution of water soluble components of the fiber thereby, affecting the overall graft yield [30-32].

Effect of the pH

The pH of the reaction medium plays a key role during the grafting process.

Kaith et al. [27] observed maximum graft yield at pH 7.0 and further increase in pH resulted in decreased graft yield. It could be due to decrease in the concentration of the free radicals on back-bone polymers as *OH ions served as the chain terminator [27]:

$$\text{Flax-(M)}_{n+1}{}^{*} + \text{OH}^{*} \rightarrow \text{Flax-(M)}_{n+1}\text{-OH}$$

Effect of Grafting on the Properties of the Back-bone

The properties of the back-bone appreciably changes on grafting with vinyl monomers. As grafting involves covalent attachment of a monomer to a pre-formed polymeric back-bone, the physical nature and the chemical compositions play an important role in the process.

Grafting modifies the back-bone and makes it amorphous by the incorporation of monomer units. Grafting of hydrophilic monomers lead to increase in wettability, adhesion, dying and rate of release of oil stains. On the other hand, hydrophobic monomers affect the properties adversely. Grafting of acrylic acid (AA), methacrylic acid (MAA) and vinyl acetate (VA) onto the polymer back-bone increased water absorbance. Cellulose graft copolymers of styrene, acrylates and methyl methacrylate have been reported to be resistant to acids and imparted flame resistance to cellulosics [33-36].

Graft Copolymerization onto Lingo-Cellulosics

Since the chapter deals with the grafting of vinyl monomers onto cellulosic fibers, so it was pertinent to discuss in brief the chemistry of cellulose and review of literature related to the modification of cellulose through graft copolymerization.

Cellulose has been the most abundant form of terrestrial biomass. It finds applications in many spheres of modern industry. Cellulose, especially cotton linters, has immense use in the manufacture of nitro cellulose, used in smokeless gunpowder. Cellulose could be processed to make cellophane and rayon and other textile derivatives. It has also been used in the laboratory as a solid substrate for thin layer chromatography and a major constituent of paper. In the recent years, many attempts have been made to modify cellulose by graft copolymerization of vinyl monomers that help in the incorporation of desirable properties without drastically affecting its inherent traits. The formation of graft copolymers with long polymeric sequences of diverse chemical composition opens the ways to afford specialty polymeric materials [27,37,38].

So, we have seen above how the various researchers have utilized the low weight and high strength of natural fibers and polymers as backbone and graft copolymerized by various physico-chemical and radiation techniques using numerous monomers and characterization techniques to procure a number of modified graft copolymers. These graft copolymers are modified materials that can have desired physico-chemico-thermo-mechanical properties, depending upon the monomers and technique used. But, still a lot remains unexplored. Natural resources are the gift of the God to the mankind. But, with time these renewable resources and natural fiber will soon deplete. So, there is a great need to sustain and procure them for the future. They should be used economically and wisely. We should seek more fruitful means to explore the maximum potential and utilize the natural fiber for the development of science and technology [39-44].

References

1. Chauhan A (2009) Synthesis and Evaluation of Physico-Chemico- Mechanical properties of polymer matrix based Composites using Graft copolymers of *Hibiscus sabdariffa* as reinforcing agents.
2. Houtz R, Adkins H (1933) Catalysis of polymerization by ozonides (II). J Am Chem Soc 55: 1609-1617.
3. Flory P (1937) Kinetics of vinyl polymerization. J Am Chem Soc 59: 241.
4. Lodge TP (2003) Block Copolymers: Past Successes and Future Challenges, Macromol Chem Phys 204: 265-273.
5. Ham GE (1996) Encyclopedia of Polymer Science and Technology, Bikales, Wiley-Interscience, New York.
6. Alfrey T, Bandel D (1951) Paper Presented at 118th Am Chem Soc Meeting, Chicago.
7. Patterson GS, Hoffmann AS, Merill EW (1960) The Permeability of some graft copolymers of Polyethylene to gases and vapors. J Appl Polym Sci 4: 159-165.
8. Kaith BS, Singha AS, Chauhan A (2006) Physico-Chemico-Thermal transformation in waste biomass to novel regenerated *Hibiscus sabdarifffa*-g-poly (MA) copolymers, International Conference on Design of Biomaterials and XVII Annual Meeting of Society for Biomaterials and Artificial Organs, IIT, Kanpur.
9. Chauhan A, Kaith BS (2010) Screening the effect of 4-Vinyl Pyridine in *Hibiscus sabdariffa*-graft-poly(MA-co-4VP) for the utilization of the waste biomass to form novel regenerated graft copolymer, 2nd International Conference on Production and Industrial Engineering (CPIE 2010), Dr. B.R. Ambedkar NIT, Jalandhar.
10. Chauhan A, Kaith B (2011) Evaluation of the Morphological Transformations in Novel Graft Copolymers Obtained from *Hibiscus sabdariffa* Stem Fibre, Polymers from Renewable Resources 2: 49-67.
11. Chauhan A, Kaith B (2011) Recycling the Cellulosic Biomass to Competent Material. J Textile Science and Engineering 1: 1-5.
12. Williams JV, Stannett V (1970) Spontaneous grafting in presence of swelling agent. J Polym Sci B 8: 711-715.
13. Kaith BS, Kalia S (2008) Preparation and Characterization of Flax-g-copolymers, Advances in Polymer Science and Technology, IIT Delhi.
14. Chauhan GS, Bhatt S, Kaur I, Singha AS, Kaith BS (1999) Modification of Natural Polymers: graft copolymers of Methyl methacrylate onto Rayon Fibre Initiated by Ceric ions- A Study in the swelling and Thermal Properties. J Polym Mater 16: 245.

15. Chauhan GS, Bhatt SS, Kaur I, Kaith BS, Singha AS (2000) Evaluation of optimum grafting parameters and the effect of ceric ion initiated grafting of methyl methacrylate onto jute fibre on the kinetics of thermal degradation and swelling behavior. Polym Degrad Stab 69: 261-265.

16. Sharma BR, Kumar V, Sen PL (2003) Ce (IV)-ion initiated graft copolymerization of methyl methacrylate onto guar gum. J Macromol Sci Pure and Appl Chem 40: 49.

17. Misra BN, Dogra R, Mehta IK (1980) Grafting onto cellulose: Effect of complexing agents on Fenton's Reagent (Fe^{2+}– H_2O_2) initiated grafting of poly (ethyl acrylate). J Polym Sci Polym Chem Ed 18: 749-752.

18. Mondal IH (2003) Graft copolymerization of nitrile monomers onto sulfonated jute-cotton blended fabric. J Appl Polym Sci 87: 2262-2266.

19. Misra BN, Jassal JK, Dogra R (1981) Grafting onto cellulose: Graft copolymerization of methyl acrylate by use of metal helates as initiators. Journal of Macromolecular Science. Part A: Chemistry 16 : 1093-1106.

20. Sood DS, Rawat BR, Misra BN (1985) Grafting onto wool. XXVII. Graft copolymerization of mixed vinyl monomers by use of ceric amonium nitrate as redox Initiator. J Appl Polym Sci 130: 35.

21. Misra BN, Chandel PS, Dogra R (1978) Grafting onto wool, III. Graft copolymerization of poly(vinyl acetate) by use of fenton's reagent as initiator. J Polym Sci Polym Chem Ed 16: 1801-1805.

22. Misra BN, Sood DS, Sharma RK (1981) Grafting onto wool. XIV. Graft copolymerization of vinyl monomers by use of $Mn(AcAc)_3$ as an initiator. J Appl Polym Sci 26: 3797-3804.

23. Chen T, Kumar G, Harris MT, Smith PJ, Payne GF (2000) Enzymatic grafting of hexyloxyphenol onto chitosan to alter surface and rheological properties. Biotechnol Bioeng 70: 564-573.

24. Bhattacharya A, Misra BN (2004) Grafting: a versatile means to modify polymers Techniques, factors and applications. Prog Polym Sci 29: 767-814.

25. Kaur I, Misra BN (1997) Handbook of Engineering Polymeric Materials. New York.

26. Kaith BS, Singha AS, Kumar S, Misra BN (2005) FAS-H_2O_2 initiated Graft Co-polymerization of Methylmethacrylate onto Flax and evaluation of Some Physical and Chemical Properties. J Polym Mater 22: 425.

27. Kaith BS, Susheel Kalia (2007) Modification of Flax fiber through grafting with Binary Vinyl Monomer mixture and their Characterization, International Conference of Futuristic Trends in Textile. NIT, Jalandhar.

28. Chauhan A, Kaith B (2011) Physico-chemico-thermal Assessment of *H. sabdariffa*-graft-copolymers. Der Chemica Sinica 2: 30-40.

29. Chauhan A, Kaith B (2011) Evaluation of Dynamic Materials procured from waste Biomass, *ISRN Material Science*, ID 393489: 7.

30. Kaith BS, Chauhan A (2008) Synthesis, Characterization and Evaluation of the transformation in *Hibiscus sadarifffa*-g-poly (Butylacrylate), E-Journal of Chemistry 5: 980-986.

31. Chauhan A, Kaith B, Singha AS, Pathania D (2010) Induction of the morphological changes in *Hibiscus sabdariffa* on graft copolymerization with acrylonitrile and co-vinyl monomer in binary mixtures, Malaysian Polymer Journal 5: 140-150.

32. Chauhan A, Kaith B (2011) Synthesis, Characterization and Evaluation of Novel regenerated Hibiscus sabdariffa-graft-poly(Acrylonitrile-co-Vinyl Monomer), Malaysian Polymer Journal 6: 14-26.

33. Kaith BS, Chauhan A, Mishra BN (2008) Studying the morphological transformation in graft copolymers of Binary Mixture of Methyl acrylate and Acrylonitrile onto *Hibiscus sabdariffa* fiber by XRD and TGA/DTA, Journal of Polymer Materials 25: 69-76.

34. Kaith BS, Chauhan A, Singha AS, Pathania D (2009) Induction of the morphological changes in *Hibiscus sabdariffa* fiber on graft copolymerization with Binary vinyl monomer mixtures, Inter. Jour. Polym. Anal. and Characterization 14: 246-258.

35. Chauhan A, Kaith B (2010) Thermo-Chemical Evaluation of the Roselle Graft Copolymers, Polymer from Renewable Resources 1: 173-187.

36. Chauhan A, Kaith B (2011) Thermal and Chemical studies of *Hibiscus sabdariffa*-graft-(Vinyl monomers), International Journal of Polymeric Materials 60: 1-15.

37. Chauhan A, Kaith B (2011) Synthesis, Characterization and Evaluation of novel *Hibiscus sabdariffa*-g-poly(EA) copolymer from waste biomass, Journal of Applied Polymer Science (Accepted on 15 Jan, 2011, now is press).

38. Chauhan A, Kaith B (2011) Synthesis, Characterization and Chemical studies of *Hibiscus sabdariffa*-g-copolymers, Fibers and Polymers 12: 1-7.

39. Kaith BS, Singha AS, Chauhan A (2006) X-Ray Diffraction Studies and Thermogravimetric / Differential Thermal Analysis of Graft Co-polymers of Methylacrylate onto *Hibiscus sabdariffa* Fiber, Journal of Polymer Materials 23: 349-356.

40. Chauhan A, Singh B (2011) X-Ray Powder Diffraction Studies to Accredit the Morphological Transformations in *Hibiscus sabdariffa* Graft-copolymers, International Journal of Polymer Analysis and Characterization 16: 319-328.

41. Chauhan A, Kaith B (2011) Novel Materials Procured From Surface Modification of Biomass, Waste and Biomass Valorization 1-8: 12.

42. Chauhan A, Kaith B (2011) Transforming the waste bio-mass to novel grafted copolymer, World Journal of Engineering 8: 347-356.

43. Thakur VK, Singha AS, Thakur MK (2012) In-Air Graft copolymerization of Ethyl Acrylate onto Natural Cellulosics Polymer, International Journal of Polymer Analysis and Characterization 17: 48-60.

44. Singha AS, Rana RK (2010) Enhancement of hydrophobic character of lignocellulosic fiber through graft copolymerization. Adv Mater Letter 1: 156-163.

Effect of Immobilization of Cadmium Ions on the Hydration of Ordinary Portland Cement

Minocha AK and Manish Kumar Goyal*

Central Building Research Institute, Roorkee and NIPER, Mohali-Punjab, India

Abstract

The disposal of toxic wastes containing heavy metals may cause a peculiar environmental problem especially when these metals are present in water soluble form. In the present research varying concentration of cadmium ions have been studied for their effect on physical, chemical and engineering properties of 43 grade Ordinary Portland Cement like initial and final setting time, compressive strength, bulk density, leaching studies, microscopic studies etc. our Efforts have been made to establish a quantitative corelation between the different concentrations of cadmium ions used and the intensity of any change in above properties of the cement. Nitrate salt of cadmium has been taken in the concentration range 1000-5000 ppm and the results of these investigations are presented in this paper. The results show that Cd improves the Compressive strength as well as the bulk density of the mortar samples. The leaching studies shows that the concentration of cadmium is almost negligible in the leachate analysed by the Atomic Absorption Spectrophotometer and as the concentration increases in the mortar samples the leaching of cadmium ions decreased it shows all the added cadmium ions gets adsorbed on the surface of cement. Scanning electron microscopic studies shows that the C-S-H gel phases are denser than that of control samples and C-S-H phases are responsible for the improved compressive strength of cement samples.

Keywords: Heavy metals; Cement; Flyash; Setting time; Cadmium; Waste management

Introduction

Cadmium is generally present in a higher concentration in electroplating waste, paper industry waste, Battery waste etc. it is very toxic to human being and environment doses beyond the permissible limit causes Kidney damage, sterility on males, flu-like disorder and high blood pressure. Solidification/Stabilization (S/S) often uses cement and cement like material (and water if necessary) in the treatment of hazardous waste that are in liquid or sludge forms to produce a solid for land disposal in case of heavy metals containing wastes in which metal concentration are too low for economical recovery but high enough to represent a toxicity hazardous. The safe disposal of cadmium waste generating from electroplating, paper industry, paint industry etc. to protect human health and environment is an issue of concern. Cadmium is very interesting metal because of its toxic nature, high solubility and mobility characteristics. In the present work we explored the immobilization of cadmium ions with ordinary Portland cement. This paper also consist the leaching studies using Toxicity Characteristic leaching procedure recommended by USEPA-1311. In recent years there have been increased concern regarding the leaching of hazardous substances from land fill into local surrounding to assess the hazardous of heavy metal containing wastes prior to disposal. The leaching of heavy metal ions from cementitious waste has been extensively investigated by many researchers [1-4]. Solidification/Stabilization technology appears both cost effective and safe since the metals are converted into highly insoluble salts which donot leach into ground water at appreciable rates. Fixation is a technology used to transform potentially hazardous liquid or solid wastes into less hazardous or non hazardous solids before disposal in a landfill, thus preventing the waste from entering the environment [5-6]. The US Environmental Protection Agency also recognizes cementations.

Solidification/stabilization as "The Best Demonstrated Available Technology" (BDAT) for land disposal of most toxic element [7-10]. The solidification/stabilization (S/S) process would be the best practical technology to treat the Cadmium containing waste [11-20] and also found appropriate by other investigators in treating cadmium contaminated wastes [20-28]. Cullinane et al. [3] reported the solidification of hazardous waste. Vipulanandan [18] studied the effect of Chromium (VI) on the ordinary Portland cement. Deja and Ouki [10,28], reported the effect of different metal ions on activated slag and Portland cement. Halim et al. [24], Minocha et al. [25] reported the leaching effect and the effect on setting time of cadmium, chromium and other toxic heavy metal ions. Minocha et al. [26] reported the effect of copper metal ions on the ordinary Portland cement. Sadeghi et al. [29] reported Synthesis, characterization, and experimental investigation of surface activity of SERS substrates using neodymium oxide. Bahniman et al. [30] studied Novel design of combinational and sequential logical structures in quantum dot cellular automata. This paper presents the quantitative correlation between the varying concentrations of cadmium ions and the intensity of effect on physical, chemical and engineering properties of 43 grade Ordinary Portland Cement as there is a very less data reported in the literature about these variables and quantitative relationships are necessary in order to estimate the influence of the amount of metal ions on these properties. Therefore, attempts have been made to fill void in this data.

Materials and Methods

Materials

Nitrate salt of cadmium (Cd $(NO_3)_2.4H_2O$) from loba, India have been used for this study. The metal salt was used as obtained without

***Corresponding author:** Manish Kumar Goyal, Central Building Research Institute, Roorkee (India) and NIPER, Mohali-Punjab, India

any further purification. Double distilled water was used to prepare the metal ion solutions throughout the study. Commercial Ordinary Portland cement 43 grade was used. Fly ash is an industrial waste, a byproduct generated from the thermal power plants due to the combustion of the coal at 1350-1650 °C with about 20% excess air in the furnace. Flyash constitutes 80-85% of the total ash produced. Flyash is known as a pozzolanic material which has no cementing properties but in the presence of moisture; chemically react with calcium hydroxide at ordinary temperature to form compounds possessing cementitious properties. Benefit of flyash is the use of a zero-cost raw material. Chemical composition of the cement determined as per IS: 4032-1985 guidelines, presented in Table 1. The physical properties of the cement were tested according to IS: 4031-1996 and the results of these studies are given in Table 2.

Apparatus

Atomic Absorption Spectrophotometer (AAS), from Hitachi model no. Z-7000 has been used to determine the concentration of metal ions present in the leachate as well as in the waste. Hazardous waste filtration system from Millipore model no. YT-30142 HW was used to carry out toxicity characteristic leaching procedure as recommended by United State Environment Protection Agency (USEPA-1311) [7]. UV-Visible spectrophotometer (Aquamate) from Thermo Corporation was used for spectrophotometer studies. Vicat apparatus was used to determine the initial as well as final setting time of the cement. Compressive strength testing machine from central scientific instruments company was used to determine the compressive strength of mortar samples. Scanning electron microscope from LEO 438VP, UK was used for the microstructure visualization of the prepared cement samples.

Methods

Preparation of cement pastes and mortars: The effect of the addition of the varying concentrations of cadmium ions on the initial and final setting time of cement was determined according to IS: 4031-1996 guidelines and compared to the blank sample to know their effect on setting time of the cement. The samples were mixed by hand with a stirring rod. To determine compressive strength of blank mortar samples (cement: fly ash, 1:3) as well as metal ion containing samples were prepared. The mortar pastes were cast in 2.78" cubic iron molds. The samples were demoded next day and were kept dipped in water at constant temperature and humidity for curing. These samples were tested for bulk density, visual changes and compressive strength on 3, 7, 28, 60, 90 and 180 days of curing. Six samples were tested every time and the average values of these results were compared with blank sample values to assess the practical applicability of these compositions.

Constituent	Weight percentage
Silica	20.80
Aluminum oxide	4.40
Iron oxide	3.79
Calcium oxide	66.10
Magnesium oxide	3.30
Anhd. Sulfuric acid	3.00
Sodium oxide	0.20
Potassium oxide	0.70

Table 1: Chemical composition of the cement.

Physical parameter	Results
Loss on ignition	0.7%
Consistency	31.58%
Soundness	1.0 mm
Bulk density	1.421 g/cm^3
Initial setting time	175 min
Final setting time	300 min
Sodium oxide	0.20
Potassium oxide	0.70

Table 2: Physical properties of the cement.

Scanning electron microscopy: In this paper a technique called Scanning electron microscopy and digital image processing is used for obtaining two dimensional images of actual ordinary Portland cement as well as samples containing different concentration of cadmium ions. It is used for the identification of all major phases in OPC and the samples containing varying concentration of cadmium ions. SEM is generally used to know the fractured surface of cement and for the magnification of an image it can magnify an image upto three lakh times. The back scattered rays are used for imaging while secondary and X-rays are used for elemental analysis of the cement samples. Scanning electron microscopy is used for conducting materials. Scanning electron microscope from LEO 438VP, UK was used for the microstructure visualization of the prepared cement samples.

Leaching studies: Leaching experiments of cadmium ions from the solidified sample were conducted in accordance with the USEPA-1311-1996 for 28 days of curing. Metal containing samples were crushed to less than 2.4 mm in diameter was mixed with the distilled water at a liquid to solid (L/S) ratio of 10:1 and trumbled for 18 H at a speed of 30 rpm. The resulting solution (leachate) was filtered through 0.4 micro meter membrane filter to remove suspended solids and was then divided into two portions and the pH was measured. The samples were preserved below 4°C before metal ion analysis. Atomic Absorption Spectrophotometer (Hitachi Z-7000) was used for the determination of heavy metal ion leached out in the solution and the results are presented in Table 5. Leachate was analysed in triplicate and the mean values were reported to ensure the reproducibility of the data. The results were compared with those obtained for blank samples to know the actual enhancement in the concentration of the particular metal in the leachate.

Results and Discussion

Visual observation

Control mortar samples as well as the samples containing different concentrations of cadmium ions were dipped in water for curing and the samples were analysed for visual observations at different intervals of time. It was found that the surface of control samples appeared rough with comparison of samples containing cadmium ions. It was found that as the concentration increases of cadmium ions in the cement samples the surface becomes smoother than the samples containing lower concentration of cadmium ions.

Bulk density

The bulk density values of mortar samples containing different concentrations of cadmium ions are reported in Table 3. The results shows that the bulk density values for one day curing is 2.303 ± 0.001

Binder system	No. of days	Average bulk density (+ 0.001 g/cm3) Cadmium			
		Control	1000 ppm	2000 ppm	5000 ppm
Cement + fly ash	1	2.303	2.303	2.303	2.303
	28	2.303	2.345	2.380	2.395

Table 3: Average bulk density values of solidified products on 1 and 28 days of Curing.

g/cm^3 which is similar than that of control samples. But the result of the 28th day of curing is quite improved than that of control samples. The bulk density values for 1000, 2000, 5000 ppm cadmium containing samples are 2.345, 2.380, 2.395 ± 0.001 g/cm^3. The result shows that the bulk density values increased with the increasing concentrations of cadmium ions in the mortar samples (Figure 1).

Compressive strength

The result of the compressive strength of cadmium ions with cement based system are presented in Table 4. Four samples have been taken and the average values are reported for the compressive strength. The result of the compressive strength of cadmium containing samples shows that the addition of cadmium ions improved the strength of the mortar samples. The compressive strength values for the blank mortar samples are according to the IS: 8112-1989 for 43 grade ordinary Portland cement, should be at least 230 kg/cm^2, 330 kg/cm^2 and 430 kg/cm^2 for 3, 7 and 28 days of curing. The addition of 1000, 2000, 5000 ppm of cadmium shows improved compressive strength for all the concentration added. Figure 2 for 1000 ppm cadmium containing samples 3 day CS is 352.5, for 7 day is 382.5 and for 28 day it was observed 510.0 kg/cm^2 for the concentration 2000 ppm 3,7, 28 day CS is 380.0, 486.6, 556.0 kg/cm^2 for 5000 ppm the CS values are 380.0, 484.0, 620.0 kg/cm^2. The percent change in compressive strength for 1000 ppm is 108.9, 87.7, 104.9, 96.8, 105.3, 111.6 for 3, 7, 28, 60, 90, 180 days of testing. For 2000 ppm 117.5, 111.6, 114.4, 97.5, 105.2, 113.8 and for 5000 ppm it was found 117.5, 111.0, 128.0, 13.7, 117.8, 116.1 Figure 3 It is clear from Table 4 that there is a significant increase in compressive strength for the Cadmium containing samples in comparison with controls and the intensity of the effect is directly proportional to the concentration of cadmium added.

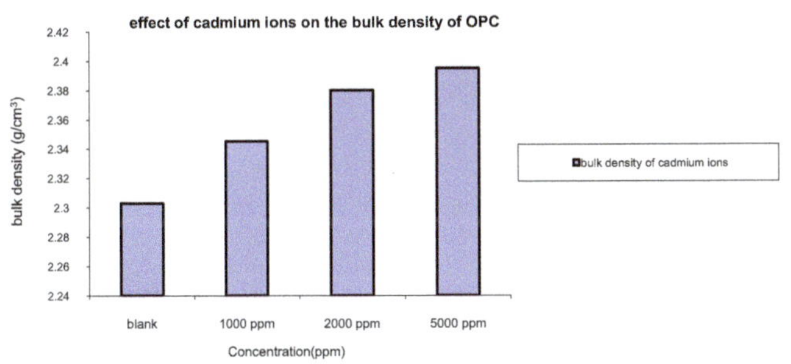

Figure 1: Bulk density of control and Cadmium on 28 day of curing.

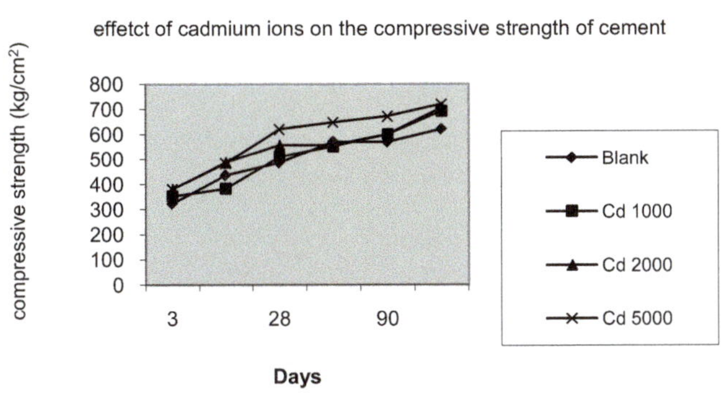

Figure 2: Effect of addition of cadmium ions on the compressive strength of cement.

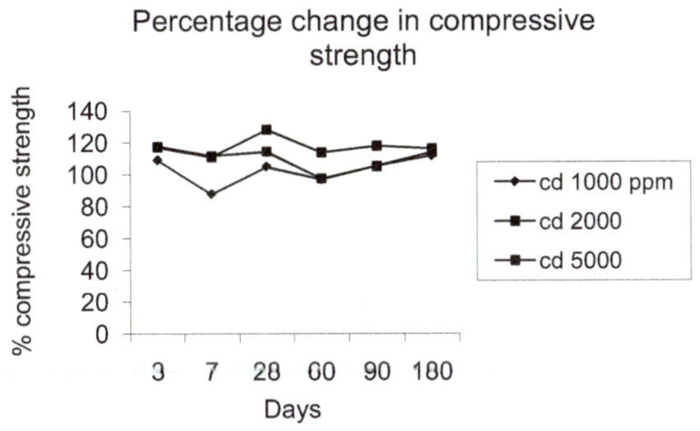

Figure 3: Percentage change in compressive strength of cadmium containing samples in comparison with control.

Microscopic analysis of the cement paste

The added cadmium salts have rather minor effect on cement hydration. When Cadmium Nitrate salt added with ordinary Portland cement cadmium ions begins to precipitated in the form of $Cd(OH)_2$. As clinker begins to be dissolved and hydroxide ion concentration increases. Apart from the rather minor acceleration of C-S-H hydration. The Cd salts have a little effect on the cement matrix. There are minor differences in the cement matrix at 28 days of curing specially an increase in the proportion of the C-S-H gel phase (Figures 4-7) as the C-S-H gel phase is responsible for strength development in the C-S-H gel phase very little free Cd^{2+} is available in solution so cadmium can not be readily leached because it is neither very accessible to water nor in a very soluble chemical form. Cadmium was assumed to be dispersed in the C-S-H matrix as it trapped with in the C-S-H pores. Interactions of cadmium salt with C-S-H were governing with the following Equation.

C-S-H+ metal   metal-C-S-H

C-S-H+ metal   metal –C-S-H+ Ca^{2+}

In C-S-H gel the addition of cadmium salts precipitated during solidification may accelerate or retard the hydration process. However when metal waste species are present as hydroxide or silicate salts they are more compatible with ordinary Portland cement because they from low solubility precipitate whose leachability reduced during solidification process. Cd formed a double compound $CdCa(OH)_4$ and it formed a precipitate present on the surface C-S-H matrix and inside the cement pores.

Leaching studies

The results of the leaching studies of the samples containing cadmium ions are reported in the Table 5. The data shows that the concentration leached out in the leachate are very very small and almost negligible. The concentration leached out in the samples containing 1000 ppm of cadmium ions are 2.5 ppm while the leaching concentration for the samples containing concentration 2000 ppm cadmium ions is 3.0 ppm and for 5000 ppm it is found only 5.0 ppm analysed by the Atomic Absorption Spectrophotometer. So almost all the concentration of added cadmium ions get adsorbed at the surface of cement. And as the concentration increases in the cement samples the concentration leached out becomes lower. So concentration

Added metal concn. (ppm)	No. of days	% change in compressive strength in comparison of control					% change in compressive strength in comparison of control					
	3	7	28	60	90	180	3	7	28	60	90	180
Control	323.4	436.0	486.0	570.0	570.0	620.0	-----	-----	-----	-----	-----	-----
Cd (1000)	352.5	382.5	510.0	552.0	600.0	692.0	108.9	87.7	104.9	96.8	105.3	111.6
Cd (2000)	380.0	486.6	556.0	556.0	600.0	706.0	117.5	111.6	114.4	97.5	105.2	113.8
Cd (5000)	380.0	484.0	620.0	648.0	672.0	720.0	117.5	111.0	128.0	113.7	117.8	116.1

Table 4: Compressive strength (kg/cm²) of mortar samples.

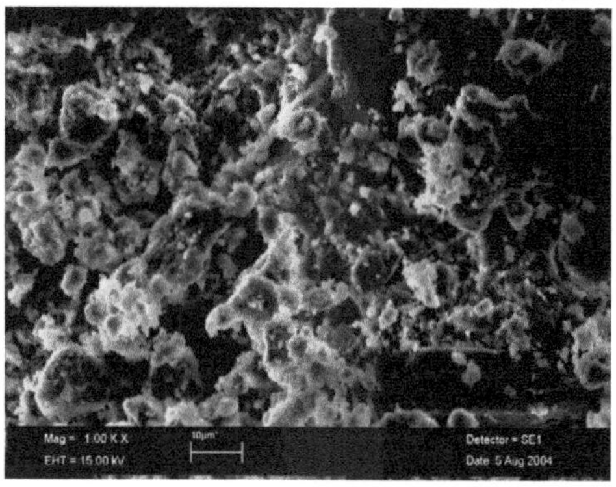

Figure 4: Sem image of control at 28 day of curing.

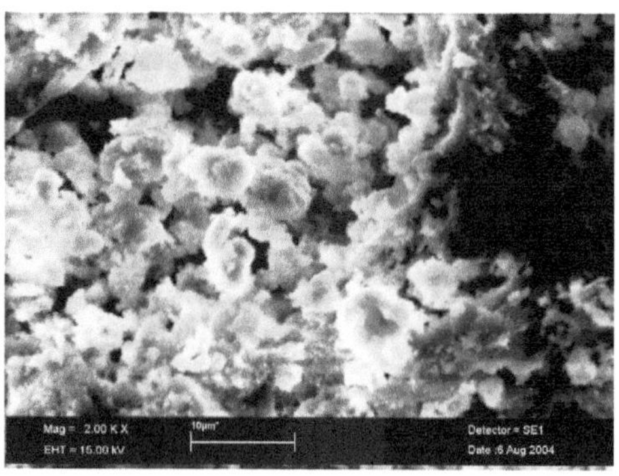

Figure 5: Sem image of Cadmium 1000 ppm at 28 day of curing.

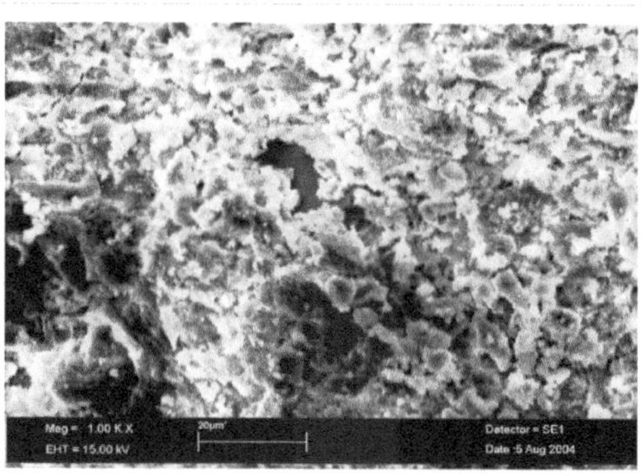

Figure 6: Sem image of cadmium 2000 ppm at 28 day of curing.

Cadmium ions	Results
Conc. Added (ppm)	Conc. Leached out (ppm)
1000.0	2.5
2000.0	3.0
5000.0	5.0

Table 5: Leaching Studies of cadmium ions with TCLP.

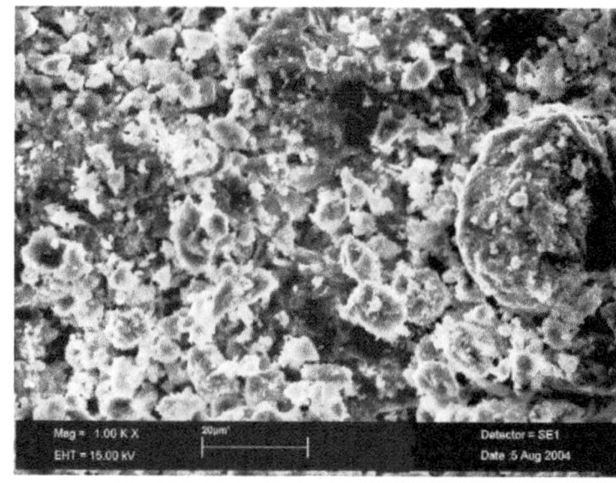

Figure 7: Sem image of cadmium 5000 ppm at 28 day of curing.

leached out of added cadmium ions is inversely proportional to the added cadmium ions.

Conclusions

On the basis of the above experimental study the following conclusions can be drawn:

1. Cadmium increased the compressive strength at different intervals of time to an appreciable extent as the concentration increased.

2. Different concentration of added cadmium ions increased the bulk density values and it is found the directly proportional to the added cadmium ions.

3. Leaching behaviuor of cadmium ions from the solidified mortar samples are negligible which shows that all the cadmium ions gets adsorbed on the surface of cement.

4. Scanning electron microscopic studies states that the C-S-H gel phases are denser of the cadmium containing samples than that of the control samples and C-S-H phases are responsible for the strength development of cement mortar samples.

So according to the following studies it is confirmed that the cadmium improved the above physical properties of the ordinary Portland cement so the industrial waste containing cadmium ions can be used as a material for the improvement of physico-chemical properties of ordinary Portland cement.

Acknowledgement

The authors are thankful to the Director, Central Building Research Institute, Roorkee. We also acknowledge the help of Dr. L.P. Singh for scanning electron microscopy. The help of Mr. Jaswinder Singh, Technical Officer is also acknowledged for the arrangement of required Chemicals & Materials and apparatus in the present research study.

References

1. Cote' PO, Bridle TR, Hamilton DP (1984) Evaluation of Pollutant Release from Solidified Aqueous Waste Using a Dynamic Leaching Test. Hazardous Waste and Environmental Emergencies, Houston TX.

2. Cote' P, Constable T, Stegemann J, Dayal R, Sawell S, et al. (1988) Guide

for the Selection of leaching tests (in preparation), prepared for USEPA, Risk Reduction Engineering Laboratory, Cincinnati, OH, by the Waste Water Technology Center, Burlington, Ontario.

3. Cullinane MJ, Jones LW, Malone PG (1986) Handbook for Stabilization/ Solidification of Hazardous Waste, EPA/540/2-86-001, USEPA, Risk Reduction Engineering Laboratory, Cincinati, OH.

4. Jones LW, Malone PG (1982) Physical Properties and leach Testing of Solidified / Stabilized industrial Waste,"EPA-600/2-82-099 (NTIS PB83-147983), USEPA, Municipal Environmental Research Laboratory, Cincinati, OH.

5. Mehta PK (1987) Concrete Structure, Properties and Materials, Prentice Hall, Inc., Englewood Cliffs, NJ.

6. Myers TE (1986) A Simple Procedure for Acceptance Testing of Freshly Prepared Solidified Waste, Hazardous and Industrial Solid Waste Testing. Fourth Symposium, ASTM STP 886, JK Petros Jr, WJ Lacy, Conway RA, American Society for Testing and Materials, Pheladelphia, PA, USA.

7. U.S. Environmental Protection Agency (1986) Test Method for Evaluating Solid Wastes: Physical/Chemical methods. (3rd edn), Washington DC, USA.

8. Woods H (1968) Durability of Concrete. ACI Monograph 4, American Concrete Institute, Detroit, MI, USA.

9. Jones LW (1989) Interference Mechanism in Waste Solidification/Stabilization Processes. EPA-600/S-89/067 (NTIS No. PB90-156209/AS), USEPA, Risk Reduction Engineering Laboratory, Cincinati, OH, USA.

10. Deja J (2002) Immobilization of Cr6+, Cd2+, Zn2+ and Pb2+ in alkali activated slag binders. Cem Concr Res 32: 1971-1979.

11. Park C (2000) Hydration and solidification of hazardous wastes containing heavy metals using modified cementitious materials. Cem Concr Res 30: 429-435.

12. Loebenstein JR (1992) In: Proceedings, Workshop on Removal, Recovery, Treatment and Disposal of Arsenic and Mercury, USEPA Report EPA/600/R-92/105: 8-9.

13. Landrett RE (1986) Guide to the disposal of chemically stabilized and solidified waste. US Environmental Protection Agency.

14. TEPA (2001) Study to Minimize the Materials for Solidification of Hazardous Fly Ash. Taiwan Environmental Protection Administration, Taipei, Taiwan.

15. Artiola JF, Zabcik D, Johnson SH (1990) In situ treatment of arsenic contaminated soil from a hazardous industrial site: Laboratory studies. Waste Manage 10: 73-78.

16. Dutre V, Vandecasteele C (1995) Solidification/stabilization of arsenic-containing waste: leach tests and behaviour of arsenic in the leachate. Waste Manage 151: 55-62.

17. Rosseti VA, Medici F (1995) Inertization of toxic metals in cement matrices: effects on hydration and hardening. Cem Concr Res 25: 1147-1152.

18. Wang S, Vipulanandan C (2000) Solidification/Stabilization of Cr(VI) with cement: Leachability and XRD analysis. Cem Concr Res 30: 385-389.

19. Corner JR (1990) Chemical Fixation and Solidification of Hazardous wastes. Van Nostrand- Reinhold, New york.

20. Nocun W, Melolepsxy J (1997) Studies on Immobilization of heavy metals in Cement Paste-C-S-H leaching behavior. Procedings of the 10th international congress on the chemistry of cement, Goteborg, Sweden.

21. Macias A, Kindness A, Glasser FP (1997) Impact of carbon Dioxide on the immobilization potential of cemented wastes: Chromium. Cem Concr Res 27: 215-225.

22. Sprung S (1988) Trace elements-concentration build-up and measure for reduction, Zem-Kalk-Gips (Wisebaden) 41.

23. Taylor HFW (1964) The Chemistry of Cement, Academic press, New York, USA.

24. Halim CE, Amal R, Beydaun D, Scott JA, Low G (2004) Implication of the structure of cementitious wastes containing Pb(II), Cd(II) As(V) and Cr(VI) on the leaching of metals. Cem Concr Res 34: 1093-1102.

25. Minocha AK, Kumar P, Singh J, Goyal MK, Verma CL (2004) Immobilization of Toxic Metals in cement based system- case studies. New Build Mater Const World 10: 44-51.

26. Minocha AK, Kumar P, Singh J, Goyal MK, Aggarwal LK (2005) Influence of copper (II), lead(II) and cadmium(II) metal ions on the setting time of ordinary Portland cement. Indian J Env Prot 25: 365-368.

27. Minocha AK, Kumar P, Singh J, Goyal MK, Aggarwal LK, et al. (2004) Effect of molybdate (II), chromium (III) and (VI) metal ions on the setting time of ordinary Portland cement. Indian J Env Prot 24: 771-774.

28. Gervais C, Ouki SK (2002) Performance study of cementitious systems containing zeolite and silica fume: Effects of 4 metal nitrates on the setting time, strength and leaching characteristics. J Hazard Mater 93: 187-200.

29. Sadeghi F, Khani F, Azandaryani AH, Mansouri Y, Mehrabadi ZS, et al. (2013) Synthesis, characterization, and experimental investigation of surface activity of SERS substrates using neodymium oxide (Nd2O3). Journal of Nanostructure in Chemistry 3: 40.

30. Ghosh B, Gupta S, Kumari S, Salimath A (2013) Novel design of combinational and sequential logical structures in quantum dot cellular automata. Journal of Nanostructure in Chemistry 3: 15.

The Bioeconomy of Microalgal Carotenoid-Rich Oleoresins Produced in Agroindustrial Biorefineries

Gabriela Rigo Roso[1], Maria Isabel Queiroz[2], Nívia Maria Streit[2], Cristiano Ragagnin de Menezes[1], Leila Queiroz Zepka[1] and Eduardo Jacob-Lopes[1*]

[1]Department of Food Science and Technology, Federal University of Santa Maria, Roraima Avenue, 1000, 97105-900, Santa Maria, RS, Brazil
[2]School of Chemistry and Food, Federal University of Rio Grande, Itália Avenue, 96201-900, Rio Grande, RS, Brazil

Abstract

The techno-economic evaluation of the obtaining process of a natural mixed carotenoid-rich oleoresin from microalgae dried biomass is presented in this paper. The process is based on solvent extraction on industrial scale, and the oleoresin obtained is suspended in soybean oil at a concentration of 20%. The oleoresin is compound of a mix of *trans* and *cis* isomers of carotenoids, having as the major carotenoid the all-*trans*-β-carotene, in amounts close to 37%. The experimental data were used to estimate the costs of an industrial plant that has the potential to generate 569,016 tons of microalgal biomass per year, in which 107,902.5 kilograms per year are represented by total carotenoids. Based on the determination of the cost analysis, it was demonstrated that the natural mixed carotenoid-rich oleoresin in soybean oil has a production cost estimated as USD 146.9 per kilogram.

Keywords: Cost analysis; Microalgae; Oleoresin; Wastewater

Introduction

Microalgae-based systems for the production of chemicals are an emergent area, representing, therefore, great promise for industrial application. Several processes have primarily demonstrated capabilities forthe food and feed industries, pigments and additives production and the cosmetics industries [1]. These microorganisms have a metabolic versatility that enables the biomass production based on organic sources without commercial value, such as industrial wastes [2].

Such possibilities become attractive for bioprospecting and exploitation as commercial sources in a wide range natural pigments primarily when their feedstock comes from biorefinery systems [3-9]. The biorefinery approach consists of a sustainable processing of biomass into a wide range of valuable bioproducts in an integrated process. The use of these strategies may provide an inexpensive alternative to the conventional technological routes of natural pigments production, e.g.carotenoids [10].

Commercially, carotenoids are used as food colorants and nutritional supplements, with an estimated global market of USD 935 million in 2005 [11]. The growing worldwide market value of carotenoids is projected to reach over USD 1 billion by the end of the decade [12]. This market was predicted to achieve USD 1.2 billion by 2009, and is expected to approach USD 919 million by 2015. Increased competition is the reason for a lower market value than previously predicted [13].

The process of synthesis and purification of carotenoids requires the use of techniques that would make production on a commercial scale very difficult and extremely expensive [14]. This process is highly complicated and involves different organic solvents and multiple steps for the purification [15]. Consequently, the purified form of carotenoids is not easily scaled up to an efficient commercial scale, wherein disposal considerations of various solvents play an important role in the overall feasibility of the process [16]. It is necessary, therefore, to find a more affordable way for commercializing this product, for instance fractions of different carotenoids in oily form in the same extract.

The oily extracts of pigments from different sources have a rather varied carotenoid composition, and according to Rios et al. [17] they are able to provide different tonalities from yellow to red, which are sufficiently concentrated to enable their large-scale commercial use (low doses are sufficient to achieve the desired color in a large amount of foodstuff). When carotenoids are extracted from natural sources and the solvent is evaporated, the residue is called oleoresin. According to these authors, the oleoresins are commercially available as food grade, and the natural carotenoids may also be manufactured as oil suspensions. Palm oil carotenes and carotenoids from *Dunaliella salina* and *Blakeslea trispora* are traded as 20-30% suspensions in vegetable oil. In the oil suspensions, the esters maybe hydrolyzed, and the free carotenoids may be suspended in vegetable oil to give a less viscous product than the isolated oleoresin [18].

A key issue on the viable production of the natural carotenoids is the general absence of low-cost processing technology. The biotechnological production of carotenoids originated of microalgae can circumvent the majority of these limitations, since the biomass carotenoid-rich production can be supported in agroindustrial wastes. These bioresources have low chemical risks, are potentially available on a large scale, and can generate feedstocks at a competitive cost [6,19]. Moreover, the utilization of these substrates might reduce the environmental and energetic problems related to their disposal [20].

In this regard, the aim of this study was to perform a techno-economic evaluation of a natural mixed carotenoid-rich oleoresin extracted from microalgae biomass produced in an agroindustrial biorefinery. The study focused on the carotenoid-rich biomass production, in the determination of the cost analysis and in the evaluation of the applicability of the process.

***Corresponding author:** Eduardo Jacob-Lopes, Department of Food Science and Technology, Federal University of Santa Maria, Roraima Avenue, 1000, 97105-900, Santa Maria, RS, Brazil

Material and Methods

Standards

The standards of all-*trans*-violaxanthin, all-*trans*-lutein, all-*trans*-zeaxanthin, all-*trans*-zeinoxanthin, all-*trans*-lutein, all-*trans*-α-carotene, all-*trans*-β-carotene were donated by DMS Nutritional Products (BASEL, Switzerland) with purities ranging from 95.0% to 99.9%, as determined by HPLC-PDA. Methanol (MeOH), methyl tert-butyl ether (MTBE), hexane and potassium hydroxide (KOH) were obtained from Sigma Aldrich (St. Louis, MO, USA).

Microorganisms and culture media

Axenic cultures of *Phormidium autumnale* were originally isolated from the Cuatro Cienegas desert (26°59′N, 102°03′W-Mexico). Stock cultures were propagated and maintained in solidified agar-agar (20 g/L), containing synthetic BG11 medium [21]. The incubation conditions used were 25°C, a photon flux density of 15 μmol/m^2/s and a photoperiod of 12/12 hour light/dark.

Microalgal biomass production in a biorefinery

The biomass production was made in heterotrophic conditions, using the slaughterhouse wastewater as the culture medium. The cultivations were performed in a bubble column bioreactor (height/diameter (h/D) ratio equal to 1.3 and a nominal working volume of 5 L). The dispersion system of the reactor consisted of a 1.5 cm diameter air diffuser located inside the bioreactor.

The average composition of the wastewater, during one year of sampling, has the following composition (mg/L): pH of 5.9 ± 0.05, chemical oxygen demand of 4100 ± 874, total nitrogen of 128.5 ± 12.1, total phosphorus of 2.84 ± 0.2, total solids of 3.8 ± 2.7, suspended solids of 1.9 ± 0.8, volatile solids of 2.9 ± 1.4 and fixed solids of 0.9 ± 0.3. The operational conditions of the continuous process were previously optimized in order to define a pH adjusted to 7.6, temperature of 20°C, volumetric airflow rate per volume unit of 1 VVM (volume of air per volume of wastewater per minute), absence of light, and a dilution rate of 0.6/d.The wet biomass was separated from the wastewater by centrifugation and then dried on a tray dryer at 60°C. The cultivations were performed twice, and in duplicate.

Carotenoid extraction and carotenoid-rich oleoresin production

The carotenoids were extracted from the dried biomass based on an extraction phase composed by hexane/potassium hydroxide/methanol, which simultaneously affects an alkaline treatment to saponify susceptible lipids and extract the intended carotenoids [22-23]. The carotenoids extract was solubilized in soybean oil, at a concentration of 20%, and stabilized with antioxidant tert-butylhydroquinone (TBHQ) at a concentration of 0.02% (v/v). The final product obtained was a natural mixed carotenoid-rich oleoresin in soybean oil.

Sampling and analytical methods

Samples were collected aseptically in a laminar flow hood; the cell biomasswas monitored every 24 hours during the growth phase of microorganism; the analyses were performed in triplicate and the data refer to the average of six repetitions.

The cell biomass was gravimetrically evaluated by filtering anestablished volume of culture medium through a 0.45μm membrane filter (Millex FG®, Billerica-MA, USA), drying at 60°C for 24h.

The dissolved oxygen concentration in the wastewater was determined by a polarographic oxygen sensor (Mettler-Toledo, Zurich, Switzerland).

The carotenoid extract was analyzed by a high performance liquid chromatography HPLC-PDA-MS/MS (Shimadzu, Kyoto, Japan) equipped with quaternary pumps (model LC-20AD), online degasser, and injection valve with a 20 μL loop (Rheodyne, Rohnert Park, CA, USA). The equipment was connected in series to a PDA detector (model SPD-M20A) and a mass spectrometer with an ion-trap analyzer and atmospheric pressure chemical ionization (APCI) source (model Esquire 4000, Bruker Daltonics, Bremen, Germany). The carotenoid separation was performed on a C_{30}YMC column (5 μm, 250 × 4.6 mm) (Waters, Wilmington, DE, USA). HPLC-PDA-MS/MS parameters were set as previously described by De Rosso and Mercadante [24]. The mobile phase consisted in a mixture of methanol and MTBE. A linear gradient was applied from 95:5 to 70:30 in 30 min, to 50:50 in 20 min. The flow rate was 0.9 mL.min^{-1}. The identification was performed according to the following combined information: elution order on C_{30} HPLC column, co-chromatography with authentic standards, UV-visible spectrum (λ max, spectral fine structure, peak *cis*intensity), and mass spectra characteristics (protonated molecule ($[M+H]^+$) and MS/MS fragments), compared with data available in the literature [1,24-27].The carotenoids were also quantified by HPLC-PDA, using five-point analytical curves of all-*trans*-violaxanthin, all-*trans*-zeaxanthin, all-*trans*-zeinoxanthin, all-*trans*-lutein, all-*trans*-β-carotene and all-*trans*-α-carotene. All other xanthophyll and carotene contents were estimated using the curve of all-*trans*-lutein and all-*trans*-β-carotene, respectively. The *cis*-isomers were estimated by using the curve of the corresponding all-*trans*-carotenoid. The total carotenoid content was calculated as the sum of the contents of each individual carotenoid separated on the C_{30} column.

Scale-up of the microalgal biorefinery

The theoretical scale-up of the process was performed using the criteria of a constant oxygen transfer rate through the constant volumetric mass transfer coefficient (KL_a) method [28]. The estimation of large-scale process was based on an industrial plant operating at a wastewater flow rate of 10,000 m^3/day, working 24 h/day and 336 days/year, and performing one extraction of carotenoids per day.

Cost analysis methodology

The production cost of the mixed carotenoids-rich oleoresin was initially based on the description of the flowchart of the process (Figure 1). A list of the equipment used, its size, and the consumables of the process has also been given.The methodology applied to determine the total capital investment was established based on an estimation of the total capital investment (TCI), which is the somatory of the fixed capital investment (FCI) and the working capital (WC) [29].

The fixed costs were estimated through the factors of the major equipment costs (MEC) and all the required additional costs necessary to build the plant (e.g. installation of the equipment). These additional costs are related to the MEC through certain factors taken from the literature (Lang factors).

The different items were estimated as a percentage of the MEC, multiplying the corresponding Lang factors according to the nature of the item. The estimated cost for each piece of equipment was obtained from a website (www.matche.com) that estimates engineering prices in 2014 FOB in USD [30].

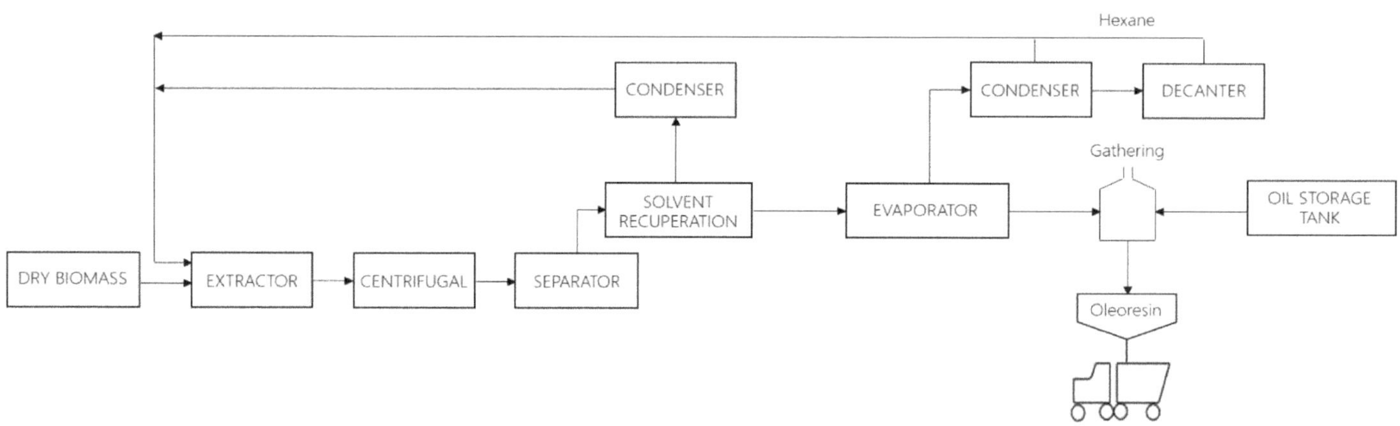

Figure 1: Flowchart diagram of the mixed carotenoid-rich oleoresin in soybean oil extraction process

The total of the operating capital represents the costs that are directly dependent on the production rate. It consists of the cost of raw materials (CRM) as well the cost of the solvent lost during the process, known asthe cost of utilities (CUT), which represents the demand for water that is required for the evaporator and condenser, electricity, waste treatment, and the cost of operational labor (COL).Within operating capital, the direct production costs included raw materials, utilities, labor costs and others (supervision, payroll charges, maintenance, operating supplies, general plant overheads, tax, and contingency). A percentage method was employed to calculate the different items [31]. The amount of the required raw materials was calculated from mass balances, whereas the consumption of utilities was estimated from the power consumption of the process that considered a value of 2% of the plant's capital for an overall utility cost [32,33]. The costs of raw materials were obtained through the selling prices of the market.

The direct labor costs were calculated estimating five workers, during three shifts a day, working 8 h/day and earning USD 8.50 per hour. This value was multiplied by two to include labor charges and then the costs were totaled.

Results and Discussion

Carotenoid-rich biomass production

The microalgae biomass is the primary bioproduct of a microalgal biorefinery and, considering that, the carotenoids are intracellular products, and the biomass productivity is a key performance indicator of the pigments production by micro-algae (Table 2). Thus, the microalgal biomass productivity in cultivation on wastewater was 0.63 $kg/m^3/d$, which enables the prediction of an annual production of 569,016 tons on an industrial scale (F=10.000 m^3/d). This biomass has a total carotenoids concentration of 183.03 mg/kg with the possibility of an annual production of 107,902.5 kg/year.

Qualitatively, Figure 2 shows the carotenoids profile of the microalgal biomass. Twenty carotenoids were found, the majority of which werethe isomers of β-carotene (42.7%), followed by the isomers of echinenone (19.4%), isomers of zeaxanthin (14.6%) and the isomers of lutein (13.1%). Others minority carotenoids comprised by 10.1% of the total.

Moreover, *Phormidium autumnale* biosynthesized some unique types of ketocarotenoids and glycosylated carotenoids (Figure 3), wherein all-*trans*-canthaxanthin and all-*trans*-myxoxanthophyll are exclusively present in the microalgal, besides the already reported isomers of echinenone. The potential bioactivity of these carotenoids should be considered in addition to the coloring capacity.

Determination of the cost analysis

The cost estimate of a natural mixed carotenoids-rich oleoresin production facility was determined using the description of the process proposed. The equipment that was utilized, including its size and type,is described in Table 1. The most costly of the MEC was the evaporator, followed by the extractor. The total amount of the MEC totals USD 25,796,500.00.

The installation expenses are shown, including the deployment, instrumentation, piping, and other elements necessary, resulting in a total fixed capital investment of USD 70,424,445.00. Considering a lifetime of 10 years, the annual fixed capital, required to keep the facility in operation, was estimated as USD 7,977,116.23 per year (Table 3).

Within the total operating capital, the direct production costs such as raw materials, utilities and labor were the main entries. Table 3 shows that the total of the raw materials was summarized as USD 19,218,814.90, wherein the consumption of hexane utilized for extraction was the major cost. The utilities expenses, based on water demand, power consumption and wastes treatment, were estimated as USD 31,683,528.60. In addition, other costs (labor, supervision, payroll charges, maintenance, operating supplies, overheads, taxes and contingencies) reached USD 14,826,276.16. In this sense, the total operating capital was estimated as USD 73,705,735.90 per year.

Regarding the analysis of the major costs of the process, the MEC showed that the evaporator represents an amount close to 54% of the total facility, followed by the extractor with 35%. The fixed capital investment, over 10 years of depreciation, contributed to approximately 10.8% to the cost of the process. The remaining 89.2% of the production cost originated the total operating capital. Depreciation charges contributed an approx. 9.2% to the annual production cost, and raw materials and utilities 26% and 43%, respectively, to the production cost.

Based on the determination of cost analysis, the calculation basis of the industry in analysis (1,693,500 kg per day of biomass), and considering the biomass microalgae formation, it is possible to predict a cost of USD 0.03 cents per kilogram of the dried biomass [34]. The natural mixed carotenoid-rich oleoresin in soybean oil cost production was estimated as USD 146.9 per kilogram.

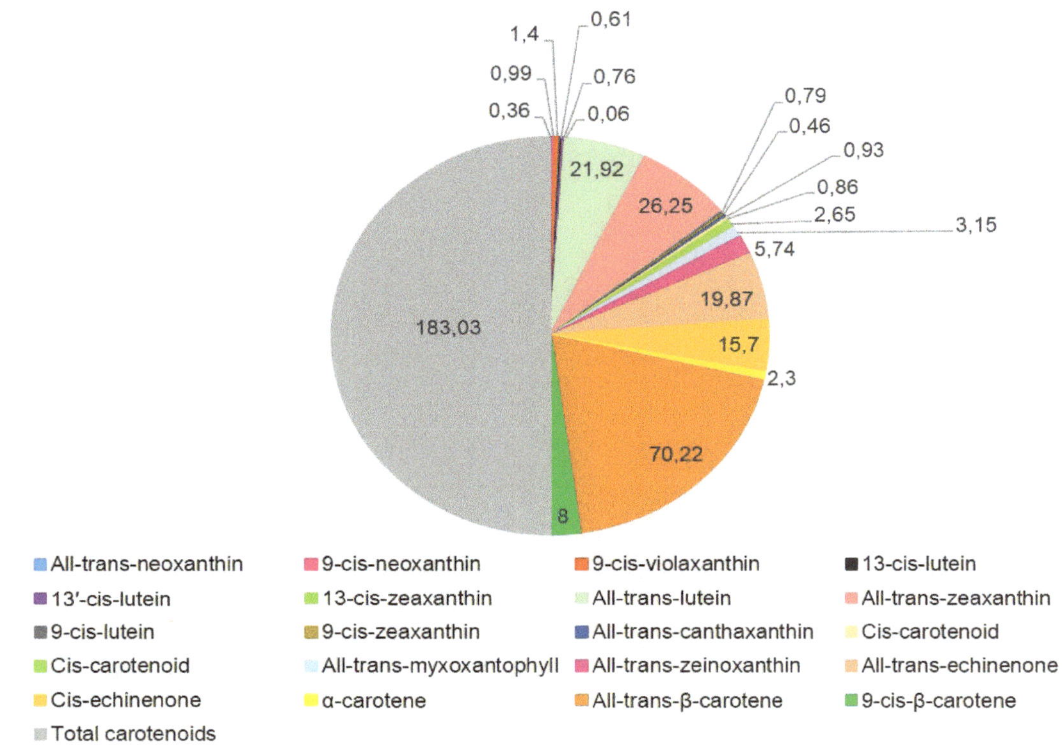

Figure 2: Identification of the carotenoids obtained by HPLC–PDA–MS/MS and their contents in μg/g

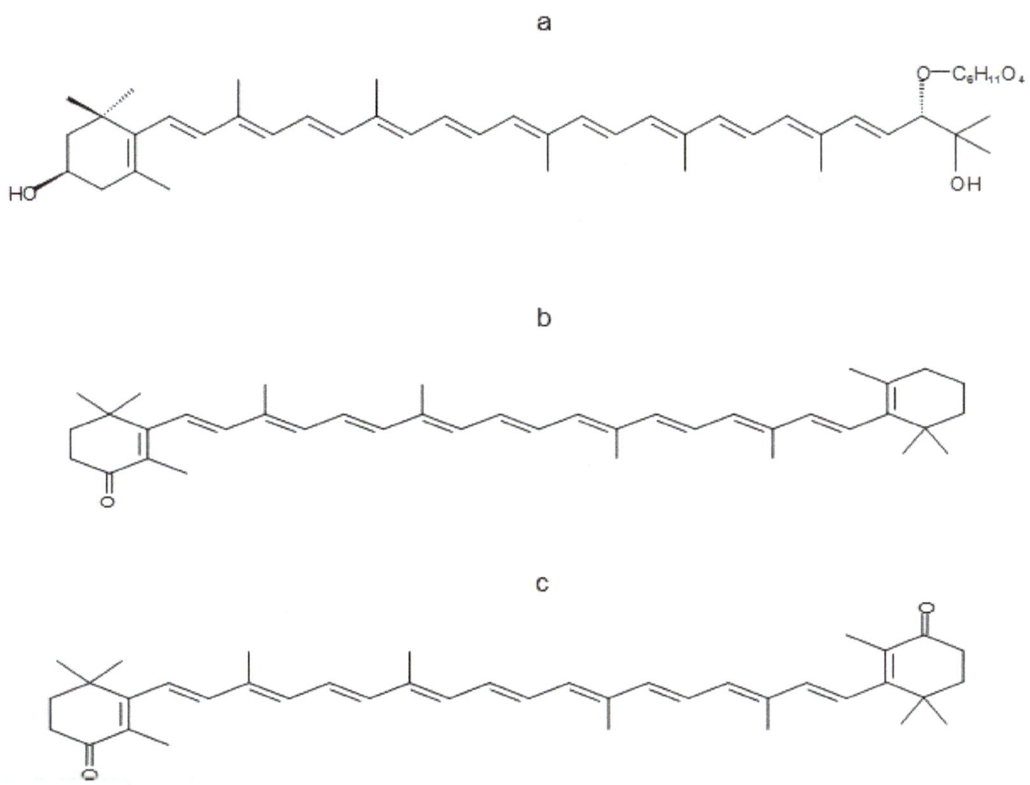

Figure 3: Ketocarotenoids and glycosylated carotenoids present in *Phormidium autumnale*. (a) all-*trans*-myxoxantophyll (b) all-*trans*-echinenone (c) all-*trans*-canthaxanthin.

Items	Units	Cost USD
Extractor (7,790.1 m³)	1	9,108,600.00
Centrifuge (13.5 m)	1	1,368,500.00
Separator (13.5 m)	1	190,000.00
Evaporator (6,397.6 m²)	1	13,946,800.00
Condenser (887.1 m²)	1	850,500.00
Storage tank hexane (6,774 m³)	1	85,000.00
Storage tank methanol (25.4 m³)	1	43,200.00
Storage tank oil (1.6 m³)	1	33,700.00
Storage tank water (11,290 m³)	1	90,400.00
Centrifugal pump (416.6 m³/h)	2	79,800.00
Total MEC (USD)		**25,796,500.00**

Table 1: Major equipment costs used in the extraction process of the oleresin

Parameter	Value
Biomass volumetric productivity (g/m³/d)	630
Biomass production (ton/year)	569,016
Total carotenoids concentration ($\mu g_{carotenoids}/g_{biomass}$)	183.03
Total carotenoids production (kg/year)	107,902.5

Table 2: Kinetic parameters and mass balance for microalgal biomass and carotenoids production in a microalgal biorefinery

Fixed capital investment		
Items	Factor	Cost (USD)
1. MEC	1.0	25,796,500.00
2. Instalattions	0.2	5,159,300.00
3. Instrumentation and control	0.4	10,318,600.00
4. Piping	0.4	10,318,600.00
5. Eletrical	0.09	2,321,685.00
6. Buildings	0.11	2,837,615.00
7. Services	0.14	3,611,510.00
8. Land	0.06	1,547,790.00
9. Engineering and supervision	0.13	3,353,545.00
10. Contractor's fee (0.05 Σ items 1-8)	0.05	3,095,580.00
11. Contingency	0.08	2,063,720.00
Total fixed capital, *A* (USD)		**70,424,445.00**
Depreciation (Σ items 1-7, 9-11)/10 years		6,825,103.50
Property tax (0.01 depreciation)	0.01	68,251.03
Purchase tax (0.16 items 1-10/10)	0.16	1,083,761.70
Total fixed capital per year, *B* (USD)		**7,977,116.23**
Total operating capital		
Raw materials	Total quantity	Cost (USD)
12. Hexane (USD 0.41/kg)[a]	29,466.9 m³	12,081,429.00
13. Methanol (USD 0.42/kg)[a]	25.4 m³	3,584,448.00
14. KOH (USD 0.40/kg)[b]	2,133,801.6 kg	853,520.60
15. Antioxidant (USD 28.66/kg)[c]	113,803.2 kg	3,261,599.70
16. Soybean oil (USD 0.54/kg)[d]	539.5 m³	291,338.20
Total, *C* (USD)		**19,218,814.90**
Utilities		
17. Power consumption (0.02 FCI)[e]	kWh	1,408,488.90
18. Water (USD 0.0003/kg)[f]	3.8×10^9 m³	1,138,032.00
19. Wastewater treatment (USD 2.99/m³)[g]	11,315.4 m³	11,367,903.50
20. Solid-waste disposal (USD 31.81/ton)[h]	558,601.2 tons	17,769,104.20
Total, *D* (USD)		**31,683,528.60**
Others		
21. Labor (USD 8.50/h, 3 shifts)	5 workers	685,440.00
22. Supervision (0.2 labor)		137,088.00
23. Payroll charges (0.25 labor + supervision)		205,632.00
24. Maintenance (0.04 MEC)		1,031,860.00

25. Operating supplies (0.004 C)			768,752.60
26. General plant overheads			1,019,913.40
(0.55 labor + supervision + maintenance)			
27. Tax (0.16 items 12-20, 24 and 25)			8,432,472.98
28. Contingency (0.05 items 12-20)			2,545,117.18
Total, *E* (USD)			**14,826,276.16**
Total production cost, *F* (*B* + *C*+ *D* + *E*) (USD)			**73,705,735.90**

[a](Koutinas et al., 2014); [b](Tabernero et al., 2012); [c](Almeida-Doria and Regitano-D'arce, 2000); [d](Glisic and Orlović, 2014); [e](Anderson, 2009); [f](Qureshi et al., 2013); [g](Buyukkamaci and Koken, 2010); [h](Meyers, 2004).

Table 3: Economic parameters used in the process

Comparatively, the products commercially sold today, characterized as natural mixed carotenoid, are Betatene®, Betanat®, Caromin® and Tocomin®. All of these products are a mixed suspension of natural carotenoids (preferentially *trans* and *cis* isomers of carotenes and xanthophylls) in vegetable oil, marketed in different concentrations. These products are isolated from different matrices as the alga *Dunaliella salina* (Betatene®), fungal *Blakeslea trispora* (Betanat®), palm fruit *Elaeis guineenses* (Caromin®), and crude palm oil (Tocomin®). Two of these products described here in above are more specifically similar with the product developed and presented in this study (Betatene® and Caromin®). The selling prices of these products are USD 12,774 and USD 12,642 per kilogram [35], respectively.Therefore, the new technological route presented in this paper could represent substantial savings per kilogram of natural mixed carotenoids-rich oleoresin produced, potentiating the economic viability of the process.

Applicability of the process

The major criteria for judging the feasibility of the process are the preliminary design and economic potential estimation, which are to be attained, and knowing the price of the final product is necessary for covering the costs involved [36]. The feasibility of the process was determined based on the techno-economic analysis a in a global scenario of the mixed carotenoid-rich oleoresin production, conducted based on a relationship of a benefit-cost ratio. In the present study, the main feasibility indicators were related, such as the economic equilibrium, profitability, rentability, and period of return on investment (Table 4).

Taking into account that the commercial products sold in the market as natural mixed carotenoids reach USD 12,800 per kilogram, the production cost of mixed carotenoid-rich oleoresin demonstrated in this study (USD 146.9 per kilogram) is shown to be extremely low. This occurs because the sources that are commercially available today are extracted of matrices that are highly expensive and difficult to obtain, handle, and extract. However, the oleoresin extracted in our process, is a product supported in a feedstock of negligible costs.

In addition, if our natural mixed carotenoid-rich oleoresin was sold at a value of USD 12,000/kg, the annual revenue would be more than USD 6 billion/year and with a profit margin of 98%. On the other hand, taking into consideration that the feedstock utilized has a negligible cost (USD 0.03 cents/kg of the dried biomass), the oleoresin may be quietly sold at a price of USD 500/kg, and yet, have a net profit estimated as USD 177,164,564.00 with a profit margin of 70.6%. The profitability of the process reports that, each year, the industry recovers approximately 251.5% of the amount invested, and when the revenue reaches the value of USD 89,144,867.10 the payment of the total costs is made. The time of return on investment was estimated as 0.39 years, which means when this period of operation is achieved the industry recovers the invested capital. These values are highly attractive, since the most companies use a value of 12% as minimum acceptable rate of return [37]. This rate is usually determined by evaluating existing opportunities in operations expansion, rate of return for investments, and other factors deemed relevant by management. However, companies operating in industries with more volatile markets might use a slightly higher rate in order to offset risk and attract investors [38]. In this sense, the feasibility of the process demonstrated that the natural mixed carotenoid-rich oleoresin obtained from the microalgae biomass of low production cost has a wide economic margin to explore industrial and commercially.

Parameter	Value
Economic equilibrium (USD)	89,144,867.10
Profitability (% per year)	70.62
Rentability (% per year)	251.50
Period of return on investment (year)	0.39

Table 4: Economical feasibility indicators of process.

Conclusion

The techno-economic modeling of the process demonstrated that the production cost of natural mixed carotenoid-rich oleoresin in soybean oil was USD 146.9/kg.

The feasibility analysis for the industrial applicability of the technology proposed showed that if the commercial value of mixed carotenoid-rich oleoresin is estimated as USD 500/kg, it is possible to obtain a 70.6% profit margin.

Accordingly, the oleoresin production in biorefinery systems can contribute to the technology consolidation of waste-pigment-utilization.

References

1. Rodrigues DB, Flores EMM, Barin JS, Mercadante AZ, Jacob-Lopes E, et al. (2014) Production of carotenoids from microalgae cultivated using agroindustrial wastes. Food Research International 65: 144-148.
2. Perez-Garcia O, Escalante FM, de-Bashan LE, Bashan Y (2011) Heterotrophic cultures of microalgae: metabolism and potential products. Water Res 45: 11-36.
3. Borowitzka MA (2013) High-value products from microalgae-their development and commercialization. Journal of Applied Phycology 25: 743-756.
4. Brennan L, Owende P (2010) Biofuels from microalgae-a review of technologies for production, processing, and extractions of biofuels and co-products. Renewable and sustainable energy reviews 14: 557-577.
5. Dufosse L, Galaup P, Yaron A, Arad SM, Blanc P, et al. (2005) Microorganisms and microalgae as sources of pigments for food use: a scientific oddity or an industrial reality? Trends in Food Science & Technology 16: 389-406.
6. Jacob-Lopes E, Franco TT (2013) From oil refinery to microalgal biorefinery. Jornal of CO2 Utilization 2: 1-7.
7. Octave S, Thomas D (2009) Biorefinery: Toward an industrial metabolism. Biochimie 91: 659-664.
8. Ohara H (2003) Biorefinery. Appl Microbiol Biotechnol 62: 474-477.

9. Queiroz MI, Hornes MO, Manetti AGS, Zepka LQ, Jacob-Lopes E (2013) Fish processing wastewater as a platform of the microalgal biorefineries. Biosystems Engineering 115: 195-202.

10. IEA Bioenergy (2009) Biorefineries: co-production of food, feed, chemicals, materials, fuels, power and heat from biomass. Cape Town: IEA Bioenergy.

11. Marova I, Carnecka M, Halienova A, Certik M, Dvorakova T, et al. (2012) Use of several waste substrates for carotenoid-rich yeast biomass production. J Environ Manage 95 Suppl: S338-342.

12. Del Campo JA, García-González M, Guerrero MG (2007) Outdoor cultivation of microalgae for carotenoid production: current state and perspectives. Appl Microbiol Biotechnol 74: 1163-1174.

13. Marz U (2000) The global market for carotenoids. Business Communications.

14. U. S. Patent 7,691,406. Zeaxanthin formulations for human ingestion.

15. W.O Patent 2,011,145. Process for isolation and purification of carotenoids.

16. Indian Pat 2,216,88. A method of the extraction and purification of natural mixed carotenoid from microalgae.

17. Rios JJ, Fernandez-Garcia E, Minguez-Mosquera MI, Perez-Galvez A (2008) Description of volatile compounds generated by the degradation of carotenoids in paprika, tomato and marigold oleoresins. Food chemistry 106: 1145-1153.

18. Britton G, Khachik F (2009) Carotenoids in food. Birkhauser Basel. Carotenoids 5: 45-66

19. Jacob-Lopes E, Franco TT (2010) Microalgae-based systems for carbon dioxide sequestration and industrial biorefineries. In M. N. B. Momba (Ed.) Biomass 135-146.

20. Simova ED, Frengova GI, Beshkova DM (2004) Synthesis of carotenoids by Rhodotorula rubra GED8 co-cultured with yogurt starter cultures in whey ultrafiltrate. J Ind Microbiol Biotechnol 31: 115-121.

21. Rippka R, Deruelles J, Waterbury JB, Herdman M, Stanier RY (1979) Generic assignments, strain histories and properties of pure cultures of cyanobacteria. Journal of General microbiology 111: 1-61.

22. Fernandez-Sevilla JM, Acien-Fernandez FG, Perez-Parra J, Magan Canadas JJ, Granado-Lorencio F, et al. (2008) Large-scale production of high-content lutein extracts from S. almeriensis. Proceedings of 11th International Conference on Applied Phycology.

23. Fernández-Sevilla JM, Acién Fernández FG, Molina Grima E (2010) Biotechnological production of lutein and its applications. Appl Microbiol Biotechnol 86: 27-40.

24. de Rosso VV, Mercadante AZ (2007) Identification and quantification of carotenoids, by HPLC-PDA-MS/MS, from Amazonian fruits. J Agric Food Chem 55: 5062-5072.

25. Britton G1 (1995) Structure and properties of carotenoids in relation to function. FASEB J 9: 1551-1558.

26. van Breemen RB, Dong L, Pajkovic ND (2012) Atmospheric Pressure Chemical Ionization Tandem Mass Spectrometry of Carotenoids. Int J Mass Spectrom 312: 163-172.

27. Zepka LQ, Mercadante AZ (2009) Degradation compounds of carotenoids formed during heating of a simulated cashew apple juice. Food Chemistry 117: 28-34.

28. Garcia-Ochoa F, Gomez E (2009) Bioreactor scale-up and oxygen transfer rate in microbial processes: an overview. Biotechnol Adv 27: 153-176.

29. Peters MS, Timmerhaus KD (2003) Plant design and economics for chemical engineers. New York: McGraw-Hill.

30. http://www.matche.com (12nd January 2014).

31. Tabernero A, Martin del Valle EM, Galan MA (2012) Evaluating the industrial potential of biodiesel from a microalgae heterotrophic culture: Scale-up and economics. Biochemical Engineering Journal 63: 104-115.

32. Acién FG, Fernández JM, Magán JJ, Molina E (2012) Production cost of a real microalgae production plant and strategies to reduce it. Biotechnol Adv 30: 1344-1353.

33. Anderson J (2009) Determining manufacturing costs. Chemical Engineering Progress 105: 27-31.

34. Roso GR, Queiroz MI, Zepka LQ, Jacob-Lopes E (2015) Techno-economic modeling of microalgal heterotrophic bioreactors applied to agroindustrial wastewater treatment. Environmental Technology .

35. http://www.zauba.com/import-natural+mixed+carotenoids/hs-code-3203-hscode.html (9nd June 2014).

36. Salehi E, Madaeni SS, Shamsabadi AA, Laki S (2014) Applicability of ceramic membrane filters in pretreatment of coke-contaminated petrochemical wastewater: Economic feasibility study. Ceramics International 40: 4805-4810.

37. Park CS (2007) Contemporary engineering economics (4th Edition). London: Prentice Hall.

38. Lang HJ, Merino DN (1993). The selection process for capital projects. New York: J. Wiley & Sons.

Modeling of *Moringa Oleifera* Oil Solubility in Supercritical Carbon Dioxide

Hoang N. Nguyen[1*], Pag-asa D. Gaspillo[1], Julius B. Maridable[1], Roberto M. Malaluan[2], Hirofumi Hinode[3], Chris Salim[3] and Ha K.P. Huynh[4]

[1]*Department of Chemical Engineering, College of Engineering, De La Salle University, 2401 Taft Avenue, 1004 Manila, Philippines*
[2]*Department of Chemical Engineering Technology, School of Engineering Technology, Iligan Institute of Technology, Mindanao State University, Andres Bonifacio Avenue, Tibanga, 9200 Iligan City, Philippines*
[3]*Department of International Development Engineering, Graduate School of Science and Engineering, Tokyo Institute of Technology, 2-12-1-I4-1, Ookayama, Meguro-ku, Tokyo 152-8550, Japan*
[4]*Department of Inorganic Chemistry, Faculty of Chemical Engineering, Ho Chi Minh City University of Technology, 268 Ly Thuong Kiet Street, District 10, Ho Chi Minh City, Vietnam*

Abstract

Moringa oleifera (MO) oil solubility behavior in supercritical carbon dioxide (SC-CO_2) was investigated at temperature rang e from 35 to 60°C and pressure from 15 to 30MPa. It was observed that pressure played key role on solubility while temperature has minor effect on solubility relative to that of pressure. Solubility was covariant with pressure. Solubility effect with temperature shifted to opposite behavior when pressure has reached the crossover pressure point at 26.38MPa. At $p < 26.38$MPa, an increased temperature has reduced the solubility. While at $p \geq 26.38$MPa, the increased temperature led to an enhanced solubility. Fractionation of the oil during extraction period was observed too. Short chain fatty acids (C14:0, C16:0, C16:1) reduced by 21.87, 7.94 and 9.49% while long chain fatty acids (C20:0, C20:1, C22:0, C24:0) increased by 22.43, 11.42, 21.76 and 35.71% respectively. All C18 fatty acids (stearic, oleic, linoleic, linolenic) were not fractionated significantly during extraction period. However, the effect of the fractionation on the solubility was insignificant. Furthermore, modeling of MO oil solubility in SC-CO_2 was conducted. Del Valle-Aguilera model with root mean square percentage deviation of 6.18% is recommended for high oleic oil solubility in SC-CO_2.

Keywords: Modeling; *Moringa oleifera* oil; Solubility; Supercritical carbon dioxide

Introduction

Moringa oleifera (MO) is well known in Asian countries, Africa, South America, the Caribbean and Oceania [1,2]. Its popularity stems from the fact that the plant is a source of high nutritional food [3,4], medicine [5-8], cosmetic [9] and environmental application [10].

In recent years, supercritical fluids extraction using supercritical carbon dioxide (SC-CO_2) has been wide attended by researchers and their successes cited in the fields of pharmaceuticals, food processing, environment, polymer and cosmetic. The green technology was applied widely because not only carbon dioxide is a non-toxic, cheap and non-flammable gas with low critical temperature of 35°C and relatively critical pressure of 7.4MPa, but its solvation power can also be varied widely with the change in pressure and temperature. An increase in pressure may results to increase in density, consequently, enhances the solvation power of SC-CO_2. Thereby, solute solubility increases. The elevation of temperature leads to reduce solvation power due to decrease in density. Also, an contrast behavior is observed that the increased temperature may lead to enhance vapor pressure, consequently, improves solubility [11-15]. Solubility is commonly measured either dynamic or static method, in which, solvent is directly in contact with either liquid oil or ground material.

In this work, the dynamic method was employed where SC-CO_2 was in direct contact with the ground MO kernels. The solubility behavior of MO oil in SC-CO_2 was studied. The effects of pressure in the range from 15 to 30MPa and temperature from 35 to 60°C on the solubility behavior were investigated. Furthermore, modeling of solubility of the oil in SC-CO_2 was conducted.

Materials and Methods

Materials

Air dried *Moringa oleifera* seeds (humidity 8.7%w/w) were used and sourced from Ilocos Norte and Ilocos Sur, Southern part of the Philippines. The seeds were de-coated and the good quality kernels, those are not moldy or breached were selected. MO kernels were ground few hours before extraction to minimize the effect of oxidation of the oil. Average particle size of the kernels at 0.64mm was used in the study.

Commercial CO_2 was supplied by SUGECO (a local company). For fatty acid analysis, standard fatty acid methyl ester of myristic acid (C14:0), palmitic acid (C16:0), palmitoleic acid (C16:1), stearic acid (C18:0), oleic acid (C18:1), linoleic acid (C18:2), linolenic acid (C18:3), arachidic acid (C20:0), cis-11-eicosenoic acid (C20:1), behenic acid (C22:0), tetracosanoic acid (C24:0) and other chemicals such as n-hexane (GC analytical grade), methyl acetate (99.5%), acetic acid (99.7%), dehydrated diethyl ether (99.5%), sodium methoxide (0.5M) in dry methanol, solid sodium were all supplied by Sigma-Aldrich, Inc. and Wako Pure Chemical Industries, Ltd, Japan.

***Corresponding author:** Hoang N. Nguyen, Department of Chemical Engineering College of Engineering, De La Salle University, 2401 Taft Avenue, 1004 Manila, Philippines

Experiments

The extraction of MO oil using SC-CO_2 was carried out in the supercritical fluid extraction pilot plant (Akico, Japan) at the Department of Chemical Engineering Technology at MSU-IIT in Northern Mindanao, Philippines.

Figure 1A illustrates the multiple stages packed bed loaded with 70 g of the ground kernels in a 500 ml extractor arranged in four layers at 25 mm in thickness and 10 mm gap in between layers. The schematic of the pilot plant is shown in Figure 1B. The CO_2 from CO_2 tank was initial liquified prior to entering the high pressure pump (with maximum capacity of 35MPa). The CO_2 liquid was then heated by a heat transfer apparatus (maximum capacity of 100°C) to become SC-CO_2. In this work, the pressure of the SC-CO_2 was varied from 15 to 30 MPa and temperature was from 35 to 60°C. The SC-CO_2 flowed in the extractor and was directly in contact with the ground kernels. The CO_2 - oil solution from the extractor passed through the expansion valve where the oil was separated from CO_2. The oil was collected via a collector vessel and the CO_2 was released to the atmosphere at the flow rate of 0.45kg/h adequate to maintain the solvent saturated. For solubility determination, each experiment was performed in duplication and for about 20min in the first stage (the linear stage) of the supercritical fluids extraction. The solubility was based on the amount of oil and CO_2 by weight measurement.

GC analysis

To analyze the fatty acid components of MO oil, the oil was converted to fatty acid methyl esters using the method of Christie [16]. Up to 2mg of oil was dissolved by mixing 0.5mL sodium-dried diethyl ether, 20μL methyl acetate and 40μL sodium methoxide. After five minutes, the reaction was stopped by adding 2μL acetic acid. Nitrogen was used to evaporate the solvent and 1mL hexane was then added. The solid precipitate was separated from the solution by using centrifuge Hitachi CRG Series R20A2 at 4000 r/min for five minutes. The solution was injected into the capillary column SP-2560 (100m x 0.25mm ID, 0.25μm film) of the Gas Chromatograph GC-17A Shimadzu with helium as carrier gas. Operating temperature program for column was set at 140°C for 5 minutes then increased to 240°C at 4°C/min and withheld for 15 minutes.

Results and Discussion

Effect of pressure on solubility

Figure 2 indicates the effect of pressure on MO oil solubility in SC-CO_2. Pressure was varied from 15MPa to 30MPa while temperature was fixed. Three isothermal curves were generated at temperatures of 35°C, 47.5°C and 60°C.

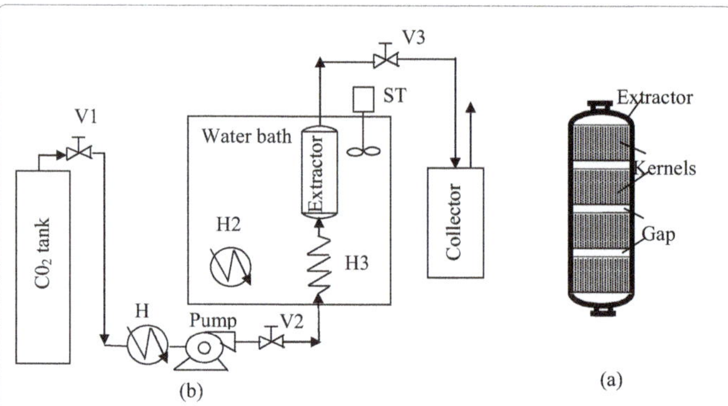

Figure 1: (a) The packing arrangement in the extractor unit. (b) Flow scheme of the CO_2 SFE pilot plant (V1, V2, V3: gas valve; H1: freezer; H2: heater; H3: heat exchanger; ST: stirrer).

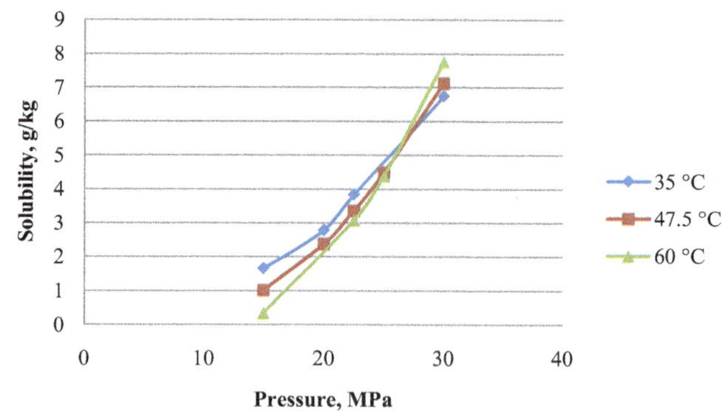

Figure 2: Effect of pressure on *Moringa oleifera* oil solubility in SC-CO_2.

It can be observed, that solubility increases at elevated pressure up to 30 MPa which can be attributed to enhance density, and therefore, improves the solvation power [12].

There was a convergence of the isothermal curves as exhibited in Figure 2. The interaction point of the isotherms termed as crossover pressure point is approximately at 26.38MPa. The result agreed with the previous studies. Stahl and Quirin [17] reported that the crossover pressure point of solubility of soy-bean, rich in C18 fatty acid content, in SC-CO_2 was below 30MPa. The crossover pressure point of high oleic sunflower oil in SC-CO_2 was higher than 25MPa, reported by Kiriamiti [18]. In addition, Ozkal et al. [19] reported that the crossover pressure point of Hazelnuts oil (79% oleic acid) in SC-CO_2 was in between 15 and 30MPa.

Effect of temperature on solubility

The effect of temperature on MO oil solubility in SC-CO_2 is shown in Figure 3. The temperature was varied from 35°C to 60°C at constant pressure. Three isobaric curves were plotted at pressure of 15MPa, 22.5MPa and 30MPa.

At pressure lower than the crossover pressure point, 26.38MPa, the solubility was contra-variant with temperature change as illustrated in the isobaric curves at 15 and 22.5MPa. However, at the pressure of 26.38MPa and higher, the solubility increased at elevated temperature as indicated in the isobaric curve at 30MPa. This could be attributed to density decrease as temperature increases. Thereby, solvent solvation power decreases and consequently, reduces solubility. In other hand, increasing temperature improves the vapor pressure therefore solubility increases [20].

Moreover, the temperature contributed less impact on solubility compared with pressure. At low pressure, 15MPa, and increasing temperatures, the highest solubility achieved was lower than 2g/kg. While, in the care of high pressure, 30MPa, the lowest solubility was higher than 6g/kg.

Fractionation

Fractionation of oil in SC-CO_2 during the extraction period was reported by Sovova et al. [21] and Norulaini et al. [22]. Fatty acid composition of palm kernel oil varied with fractionation performed at pressure of 48.3MPa and temperature of 80°C [22]. The percentage of short chain fatty acids (C8:0, C10:0, C12:0) in the oil was reduced during the extraction period. However, the amount of myristic acid (C14:0) did not change significantly. In the care of longer chain fatty acids (C16:0, C18:0, C18:1, C18:2), its yield was enhanced. Sovova et al. [21] reported that at temperature of 40°C and pressure up to 29MPa, the percentage of free fatty acids, monoglycerides and diglycerides in grape seed oil decreased from first to last fractionation but opposite phenomenon on triglycerides was observed, it was enhanced with time.

Ozkal et al. [19] reported that at pressure of 30 MPa and temperature of 40, 50 and 60°C, the hazelnut oil was not fractionated with time. The main fatty acid composition of hazenut oil such as stearic (C18:0, 2%), oleic (C18:1, 79%), linoleic (C18:2, 11%), palmitic (C16:0, 6%) and palmitoleic (C16:1, 0.15%) remained constant during the extraction period.

In this work, three fractions were collected in the first 30 min, between 30 and 160min, and from the first 160 min until the end of extraction after 7 hours. The experiments were conducted in duplication at pressure of 30 MPa, temperature of 60°C and CO_2 flow of 0.45kg/h. The fatty acid composition of MO seed oil in SC-CO_2 of three fractions is shown in Table 1. Oleic (C18:1) was found as a main component of the oil with percentage of approximately 70%. The variation in the fatty acid composition of MO seed oil upon fractionation is highly noticeable. The percentage of deviation in fatty acids during extraction period was counted as the variation of those between the first fraction and last fraction. Short chain fatty acids such as myristic (C14:0), palmitic (C16:0) and palmitoleic (C16:1) decreased by 19.04, 8.40 and 12.87% during extraction time, respectively. All C18 fatty acids (stearic, oleic, linoleic and linolenic) did not change significantly. Longer chain fatty acids such as arachidic (C20:0), cis 11 eicosenoic (C20:1), behenic (C22:0) and tetracosanoic (C24:0) increased by 22.43, 16.15, 21.76 and 30.95%, respectively. In general, the results of this work agree with Norulaini et al. [22] where the shorter chain fatty acids were found to be more soluble than the longer one. The difference in the variation with fractionation of fatty acid in particular case between this work and others resulted in the difference in proportion of fatty acids, free fatty acid, monoglycerides, diglycerides and triglycerides in the oils. In addition, the difference in experimental conditions (pressure, temperature) may possibly led to the difference in solvent density and solute vapor pressure and consequently, generated variation in solubility of components.

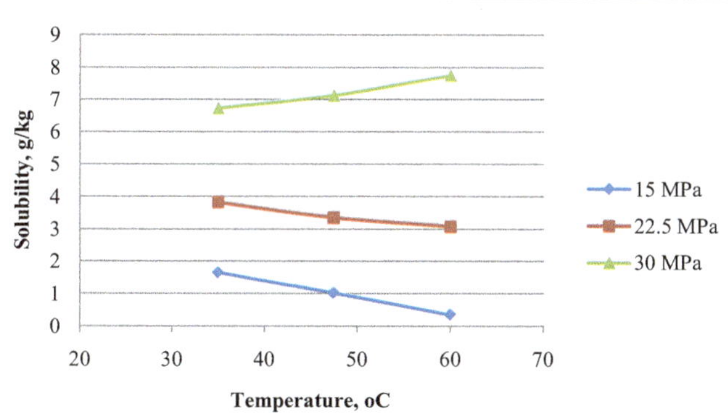

Figure 3: Effect of temperature on *Moringa oleifera* oil solubility in SC-CO_2.

Fatty acid	Fraction I (%)	Fraction II (%)	Fraction III (%)	Variation [a] (%)
Myristic (C14:0)	0.21	0.19	0.17	-19.04
Palmitic (C16:0)	6.78	6.63	6.21	-8.40
Palmitoleic (C16:1)	2.02	1.95	1.76	-12.87
Stearic (C18:0)	5.98	5.91	5.72	-4.34
Oleic (C18:1)	70.75	69.70	69.01	-2.45
Linoleic (C18:2)	0.82	0.83	0.86	4.87
Linolenic (C18:3)	0.28	0.28	0.29	3.57
Arachidic (C20:0)	3.70	4.19	4.53	22.43
Cis 11 eicosenoic (C20:1)	2.60	2.71	3.02	16.15
Behenic (C22:0)	6.02	6.64	7.33	21.76
Tetracosanoic (C24:0)	0.84	0.97	1.10	30.95

[a]Variation value is percentage of chance in fatty acid between fraction I and fraction III.

Table 1: Fatty acid composition of MO oil in SC-CO_2 of three fractions at 30MPa and 60°C

Though the variation of some fatty acids composition with fractionation is significant, the impact on solubility is however minimal because the unchanged C18 fatty acids were predominant in the oil [21].

Model of solubility

Chrastil [23] presented a model of solubility based on the proposition linear relationship between the logarithmic solubility and that of solvent density. Chrastil model also expressed the dependence of solubility on temperature. The expression of Chrastil model is:

$$c = d^{a_1} \exp\exp\left(a_2 + \frac{a_3}{T}\right) \quad (1)$$

where c is oil solubility (kg/m³), d is solvent density in kg/m³ and T is absolute temperature (K).

In 1983, Adachi-Lu [24] modified Chrastil model by improving the effect of density on solubility as follows:

$$c = d^{a_1 d^2 + a_2 d + a_3} \exp\exp\left(a_4 + \frac{a_5}{T}\right) \quad (2)$$

Adachi-Lu model's parameters units are same as in the Chrastil.

In 1988, Del Valle and Aguilera [25] developed Chrastil model to be more flexible with variation of temperature as shown below:

$$c = d^{a_1} \exp\exp\left(\frac{a_2}{T^2} + \frac{a_3}{T} + a_4\right) \quad (3)$$

Del Valle-Aguilera model expresses solvent density (d) in kg/dm³ while c is in kg/m³ and T (K) is absolute temperature.

In this work, density CO_2 was computed based from Peng-Robinson equation of state as follows:

$$\mathrm{p} = \frac{RT}{V-b} - \frac{a\alpha}{V^2 + 2bV - b^2} \quad (4)$$

$$\mathrm{a} = \frac{0.45724 R^2 T_c^2}{\mathrm{p_c}} \quad (5)$$

$$\mathrm{b} = \frac{0.0778 R\, T_c}{\mathrm{p_c}} \quad (6)$$

$$\alpha = (1 + \begin{pmatrix} 0.37464 + 1.54226\omega \\ -0.26992\omega^2 \end{pmatrix}(1 - T_r^{0.5}))^2 \quad (7)$$

$$\mathrm{T_r} = \frac{T}{T_c} \quad (8)$$

where, p is pressure (MPa), T is absolute temperature in K, V is molar volume (L/mol), T_c, p_c are critical temperature (K) and pressure (MPa), respectively, R is ideal gas constant and ω is acentric factor of CO_2 [26].

The exponent of density in Eqs. (1,2 and 3) is the slope of the log-log plot of MO oil concentration in SC-CO_2 vis-à-vis that of CO_2 density. The slope varied from 8.162 to 8.922 as indicated in Figure 4. The combination of the slope value and a set of experimental data in the models resulted the coefficients of the models as shown in Table 2.

The three models above were used to describe the solubility behavior of MO oil in SC-CO_2. The agreement of the models versus experimental data was evaluated by means of the root mean square percentage deviation (RMSD). The RMSD is given as

$$\mathrm{RMSD} = 100\left[\frac{1}{n}\sum_{i=0}^{n}\left(\frac{(c_{cal}-c_i)}{c_i}\right)^2\right]^{0.5} \qquad (9)$$

where c_{cal} stand for the calculated solubility in kg/m³, c_i is experimental solubility in kg/m³ and n is number of experiment points.

The calculated results of del Valle-Aguilera, Adachi-Lu and Chrastil models in comparation with experimental data of MO oil solubility are presented in Figure 5, 6 and 7, respectively. Del Valle-Aguilera model showed the best fit with experimental data with RMSD of 6.18%. While those of Chrastil and Adachi-Lu model are 7.17% and 6.31%, respectively. Moreover, in Figure 5 the crossover pressure point was indicated clearly between 25MPa and 30MPa as the convergent point of the isotherms of Del Valle-Aguilera model. While, the isotherms of Chrastil and Adachi-Lu models did not converge in all cases. It suggests that Del Valle-Aguilera model should be applied for high oleic oil solubility in SC-CO_2.

Conclusion

Effect of process parameters on *Moringa oleifera* oil solubility in supercritical carbon dioxide were studied within the range of temperature from 35 to 60°C, range of pressure from 15 to 30MPa. Pressure played key role on solubility. In relative to pressure, temperature has minor effect on solubility. Increasing the pressure resulted to increase in solubility. For temperature, two opposing results on solubility were observed either an increase or decrease in solubility occurred depending on the relative between the pressure applied and

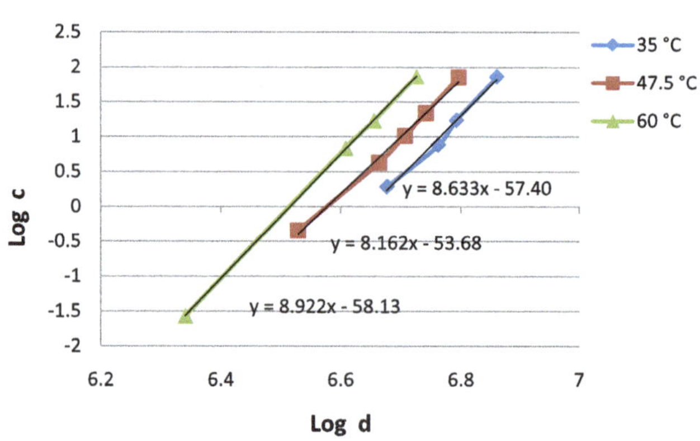

Figure 4: *Moringa oleifera* oil solubility in SC-CO_2 (c, kg/m³) in dependence on the CO_2 density (d, kg/m³).

Model	a_1	a_2	a_3	a_4	a_5
Chrastil	8.922	-40.326	-4.952*10³	-	-
Del Valle-Aguilera	8.922	59.264	-3.116*10⁴	4.059*10⁶	-
Adachi-Lu	8.772	-2.963*10⁻¹⁰	-3.207*10⁻¹²	-40.229	-4.943*10³

Table 2: Model constants.

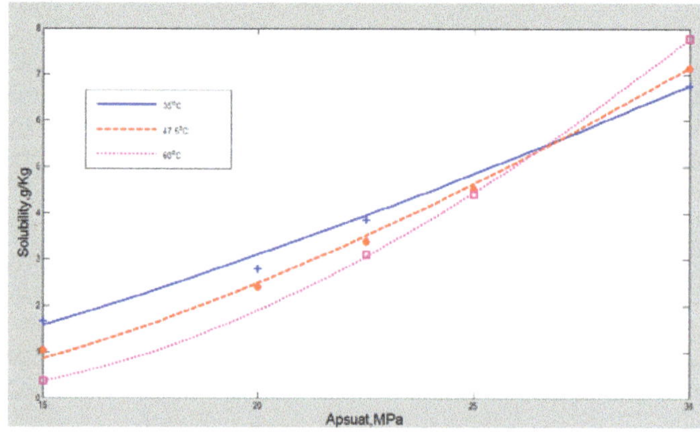

Figure 5: Del Valle-Aguilera model of *Moringa oleifera* oil solubility and experimental data, (+) 35°C, (*) 47.5°C, (□) 60°C.

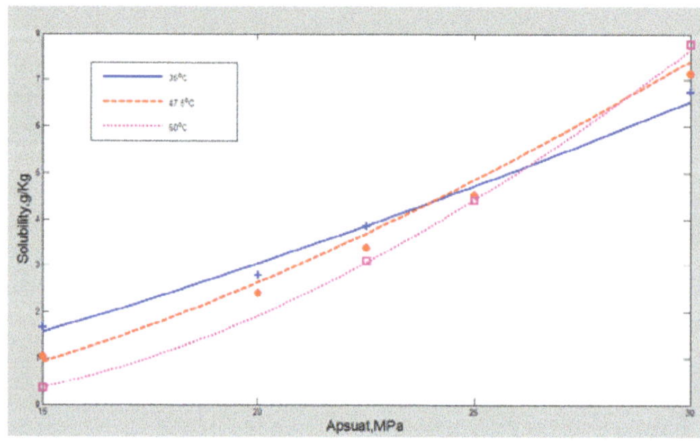

Figure 6: Adachi-Lu model of *Moringa oleifera* oil solubility in SC-CO_2 and experimental data, (+) 35°C, (*) 47.5°C, (□) 60°C.

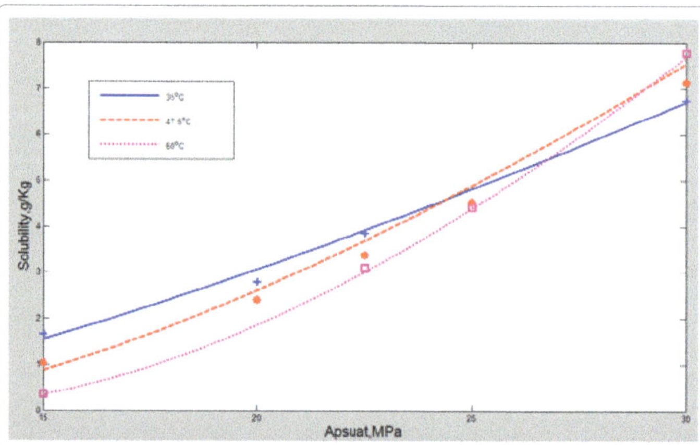

Figure 7: Chrastil model of *Moringa oleifera* oil solubility in SC-CO_2 and experimental data, (+) 35°C, (*) 47.5°C, (□) 60°C.

the crossover pressure point (26.38MPa). Fractionation of the oil during extraction period was observed. Short chain fatty acids such as myristic, palmitic and palmitoleic decreased by 19.04, 8.40 and 12.87%, respectively. All C18 fatty acids (stearic, oleic, linoleic and linolenic) did not fractionate significantly. While, long chain fatty acids such as arachidic, cis 11 eicosenoic, behenic and tetracosanoic increased by 22.43, 16.15, 21.76 and 30.95%, respectively. However, the effect of the fractionation on the solubility was insignificant. Three models were used to describe MO solubility behavior in SC-CO_2. Del Valle-Aguilera model showed the best fit with experimental data with RMSD of 6.18%. While those of Chrastil and Adachi-Lu model are 6.31 and 7.17% respectively. Del Valle-Aguilera model is recommended for high oleic oil solubility in SC-CO_2.

Acknowledgements

The authors gratefully acknowledged the financial support received from the ASEAN UNIVERSITY NETWORK FOR SOUTHEAST ASIA Engineering Education (AUN/SEED-Net) program thru the Japan International Co-operation Agency (JICA). Acknowledgement is also extended to De la Salle University-Manila, Mindanao State University-Illigan Institute of Technology (MSU-IIT) and Tokyo Institute of Technology for the laboratory facilities.

References

1. Olson E, Carlquist S (2001) Stem and root anatomical correlations with life form diversity, ecology, and systematics in Moringa (Moringaceae). Botanical Journal of the Linnean Society 135: 315-348.

2. Iqbal S, Banger I (2006) Original Article - Effect of season and production location on antioxidant activity of Moringa oleifera leaves grown in Pakistan. J Food Composition and Analysis 19: 544-55.

3. Richter N, Siddhuraju P, Becker K (2003) Evaluation of nutritional quality of moringa (Moringa oleifera Lam.) leaves as an alternative protein source for Nile tilapia (Oreochromis niloticus L.). Aquaculture 217: 599-611.

4. Amaglo K, Bennett N, Curto L, Rosa S, Turco L, et al. (2010) Profiling selected phytochemicals and nutrients in different tissues of the multipurpose tree Moringa oleifera L., grown in Ghana. Food Chemistry 122: 1047-1054.

5. Guevera P, Vargas C, Sakurai H, Fujiwara Y, Hashimoto K, et al. (1999) An antitumor promoter from Moringa oleifera Lam, Mutation Research/Genetic Toxicology and Environmental Mutagenesis. 440: 181-188.

6. Costa-Lotufo V, Khan H, Ather A, Wilke V, Jimenes C, et al. (2005) Studies of the anticancer potential of plants used in Bangladeshi folk medicine. J Ethnopharmacolog 99: 21-30.

7. Fakurazi S, Hairuszan I, Nanthini U (2008) Moringa oleifera Lam prevents acetaminophen induced liver injury through restoration of glutathione level. Food and Chemical Toxicology 46: 2611-2615.

8. Sashidhara V, Rosaiah N, Tyagi E, Shukla R, Raghubir R et al. (2009) Short communication - Rare dipeptide and urea derivatives from roots of Moringa oleifera as potential anti-inflammatory and antinociceptive agents. European Journal of Medicinal Chemistry 44: 432-436.

9. Kleiman R, Ashley A, Brown H (2008) Comparison of two seed oils used in cosmetics, moringa and marula. Industrial Crops and Products 28: 361-364.

10. Bhuptawat H, Folkard K, Chaudhari S (2007) Innovative physico-chemical treatment of wastewater incorporating Moringa oleifera seed coagulant. J Hazardous Materials 142: 477-482.

11. Mchugh A, Krukonis J (1994) Supercritical Fluid Extraction - Principles and Practice, (2ndedn). London: Butterworth-Heinemann.

12. Brunner G (2005) Supercritical fluids: technology and application to food processing. J Food Engineering 67: 21-33.

13. Reverchon E, Marco D (2006) Review - Supercritical fluid extraction and fractionation of natural matter. J. Supercritical Fluids 38: 146-166.

14. Herrero M, Cifuifuentes A, Ibanez E (2006) Sub- and supercritical fluid extraction of functional ingredients from different natural sources: Plants, food-by-products, algae and microalgae - A review. Food Chemistry 98: 136-148.

15. Pourmortazavi M, Hajimirsadeghiv S (2007) Review - Supercritical fluid extraction in plant essential and volatile oil analysis. J Chromatography A 1163: 2-24.

16. Christie W (1993) In Advances in Lipid Methodology-Two. (1stedn) The Oily Press, Dundee.

17. Stahl E, Quirin W (1983) Dense gas extraction on a laboratory scale: A survey of some recent results. Fluid phase equilibria 10: 269-278.

18. Kriamiti K, Rascol E, Marty A, Condoret S (2001) Extraction rates of oil from high oleic sunflower seeds with supercritical carbon dioxide. Chemical Engineering and Processing 41: 711-718.

19. Ozkal G, Salgin U, Yener E (2005) Supercritical carbon dioxide extraction of hazelnut oil. J Food Engineering 69: 217-223.

20. Kopcak U, Mohamed S (2005) Caffeine solubility in supercritical carbon dioxide/co-solvent mixtures. J Supercritical Fluids 34: 209-214.

21. Sovova H, Zarevucka M, Vacek M, Stransky K (2001) Solubility of two vegetable oils in supercritical CO_2. J Supercritical Fluids 20: 15-28.

22. Norulaini N, Zaidul M, Anuar O, Omar M (2004) Supercritical enhancement for separation of lauric acid and oleic acid in palm kernel oil (PKO). Separation and Purification Technology 35: 55-60.

23. Chrastil J (1982) Solubility of solids and liquids in supercritical gases. J Physical Chemistry 86: 3016-3012.

24. Adachi Y, Lu Y (1983) Supercritical fluid extraction with carbon dioxide and ethylene. Fluid Phase Equilibria 14: 147-156.

25. Del-Valle M, Aguilera M (1988) An improved equation for predicting the solubility of vegetable oils in supercritical CO_2. Ind Eng Chem Res 27: 1551-1553.

26. Brunner G (1994) Gas extraction – An introduction to fundamentals of supercritical fluids and the application to separation processes. (1stedn), Springer, New York.

Agriculture and Food Applications of Rhamnolipids and its Production by *Pseudomonas Aeruginosa*

Jean Paul Sinumvayo*, Nestor Ishimwe

School of Biotechnology, Jiangnan University, Wuxi, P. R. China

Abstract

Rhamnolipids (RLs) are known as very efficient surface-active which are the most investigated biosurfactants usually produced by *Pseudomonas aeruginosa* strains. Other various bacterial species promising producers of rhamnolipids are shown herein. Researchers on rhamnolipids production concentrate their efforts on conventional mutagenesis and metabolic engineering of different strains especially *P. aeruginosa*. Several modifications of metabolic pathways in combination with optimization of fermentation strategies were suggested to increase the rhamnolipid yields. Unique properties of rhamnolipids, including sequestering, detergency, demulsifying, foaming, emulsifying, thickening, metal, solubilizing, wetting, vesicle forming and phase dispersion make them suitable to be used in a wide range of industrial applications such as cosmetics, food, pharmaceutical formulations and bioremediation of pollutants. The present review shows different rhamnolipid yields obtained and covers the potential applications of rhamnolipid, mainly focusing on food processing, storing, numerous applications in agriculture and farming.

Keywords: Agriculture; Biosurfactants; Food industry; *Pseudomonas aeruginosa;* Rhamnolipids

Introduction

Microorganisms showed a high capacity of synthesizing a wide range of surface-active compounds, generally called biosurfactants. These compounds are mainly classified according to their molecular weight, physico-chemical properties and mode of action [1,2].

Many surfactants including rhamnolipids alter the surface properties of liquids, even when present in small quantities [3]. Like other surfactants, rhamnolipids present a lyophilic group and a lyophobic group which facilitate the orientation of the surfactant to reduce the liquid surface free energy and increase surface viscosity [4,5].

Some chemical surfactants decontaminated and maintain membrane performance as well as membrane integrity under mild operation conditions [6]. Undefined cleaning activity together with its low biodegradability and high cost most possibly determine the much less frequent use of chemical surfactants in the industrial applications [7]. Rhamnolipids are potent natural glycolipid biosurfactants often biosynthesized by *Pseudomonas aeruginosa* strains through the fermentation process, with cheaper agro-based substrates and waste materials. They can be good substituents for chemical surfactants, in different industrial fields [8]. Rhamnolipids are classified as mono-rhamnolipids and di-rhamnolipids, depending on one or two rhamnose sugars linked to a dimer of β-hydroxyacids (primarily β-hydroxydecanoate) which can rise up to three hydroxyl fatty acids containing 8–14 carbons [9].

Non-toxic nature, excellent biodegradability, high surface/interfacial activity, low toxicity, high thermal/chemical stability, production from renewable resources and the ability to form microemulsions [10,11], make rhamnolipids to be potentially used as emulsifiers and stabilizers in a great number of food processing and from bioremediation to food additives [12]. The usefulness of rhamnolipids in cleaning ultrafiltration (UF) membrane fouled by protein is due to their high capability of binding the protein with high affinity as well as increasing the wettability of solid surface [6].

In the present review, we intend to summarize rhamnolipids production strategies by *Pseudomonas Aeruginosa*, its agriculture and food industrial applications are also discussed. Apart from their uses based on their special properties, rhamnolipids yields produced by *Pseudomonas Aeruginosa* in 15 years ago are reviewed and shown in Table 1.

Production of rhamnolipids

Rhamnolipids are well-studied glycolipids with high potential industrial applications [13]. *Pseudomonas* strains are the best producers of glycolipid containing rhamnose and 3-hydroxy fatty acids. However, the most promising bacteria strains, *P. aeruginosa,* was investigated and recommended as the best microorganism to produce two classes of rhamnolipids: monorhamnolipids and dirhamnolipids with excellent surface activity [14,15].

Therefore, their properties and potential application, have encouraged many researchers to improve their production and have become the most investigated biosurfactants [3].

The properties showed by rhamnolipids depend on their homologues composition and distribution that are determined by the bacterial strain, culture conditions and medium composition [14]. High prices of raw-material and low productivities have been found as potential inhibitors of rhamnolipids for industrial scale production. However, strain-engineering is now a promising tool for achieving accurate production so that large-scale production of rhamnolipids becomes economically feasible [16].

Rhamnolipids production by different *P. Aeruginosa* strains

Rhamnolipid are produced by fermentation process similar to

***Corresponding author:** Jean Paul Sinumvayo, School of Biotechnology, Jiangnan University, Wuxi, P. R. China

Low cost or waste raw material	Producer microbial strain	Maximum yields (g/l)	Reference
Rapeseed oil	*Pseudomonas* species DSM 2874	45	[36]
Sunflower, soybean oil and glycerol	*Pseudomonas aeruginosa* DS10-129	4.31, 2.98 and 1.77	[37]
Waste frying oils (sunflower and olive oil)	*Pseudomonas aeruginosa* 47T2 NCIB 40044	2.7	[38]
Soybean soapstock waste	*Pseudomonas aeruginosa* LBI	11.72	[39]
Sunflower oil soapstock waste	*Pseudomonas aeruginosa* LBI	15.9	[40]
Soybean oil	*Pseudomonas aeruginosa* O-2-2	70.56	[33]
Corn steep liquor (10% (v/v)) and molasses (10% (w/v))	*Pseudomonas aeruginosa* strain #112	3.2	[22]
Waste frying oils(olive/ sunflower (50:50; v/v),)	*Pseudomonas. aeruginosa* 47 T2	8.1	[41]
Glycerol and ammonium nitrate	*Pseudomonas aeruginosa* DAUPE 614	3.9	[24]
Glucose and glycerol	*Pseudomonas aeruginosa* EM1	7.5 and 4.9	[26]
1% glucose+0.25% stearic acid (C18)	*Pseudomonas aeruginosa* ATCC 9027	2.1	[31]
Sunflower oil	*Pseudomonas aeruginosa* san-ai	3	[36]
Brazilian Nut (Bertholletia excelsa) and Passion Fruit oils	*Pseudomonas aeruginosa* LBI	9.9 and 9.2	[14]
Clarified blackstrap molasses	*Pseudomonas aeruginosa* mutant strain	*1.50*	[42]
Glucose and glycerol	*Pseudomonas aeruginosa* TMN	0.3 and 0.25	[43]
Soybean oil, safflower oil, and glycerol	**Pseudomonas aeruginosa** DS10-129	4.31, 2.98, and 1.7	[44]
Soybean oil	Pseudomonas aeruginosa MA01	12	[45]

Table 1: Rhamnolipids production in different *Pseudomonas aeruginosa* strains and their corresponding carbon sources.

those used to produce beer or other fermented products. Bacteria are added to a fermentation tank and provided nutrients source, and under properly controlled conditions the result will be rhamnolipids [17]. During biosurfactant production, the downstream processes are critical steps which involve recovery, concentration, and purification. The appropriate approach for downstream processing relies on the type and nature of the substrates, fermentation protocol (batch/fedbatch and growing/resting cells), and type and physicochemical properties of the resulted biosurfactants. Their high ratio of hydrophobic-to-hydrophilic characteristics, a number of traditional techniques such as precipitation, crystallization, centrifugation, and solvent extraction have been used to recover the most biosurfactants from the culture medium [18].

Most of the microorganisms investigated for the production of rhamnolipids are *P. aeruginosa strains*, however, by now, only a handful of strains are potentially relevant for industrial production processes, mainly due to vast differences in biosurfactant yields and achievable maximum concentrations [19].

The specific reason which makes *P. aeruginosa* a favorable biosurfactant-producing or oil-degrading strain may be found in its wide spread occurrence in contaminated environments,

rapid growth, easy for isolation and screening as well as the high biosurfactant production and crude oil degradation capacity [20].

In *P. aeruginosa*, both mono- and di-rhamnolipids biosurfactants are the main products of the convergence of two metabolic pathways: the biosynthesis of dTDP-L-rhamnose and the diversion of the βhydroxydecanoyl-ACP intermediate from the FASII cycle by RhlA to synthesize the fatty acid dimer moiety of rhamnolipids and free 3-[3-hydroxyalkanoyloxy) alkanoic acid (HAA) (Figure1). The rhamnosyltransferases RhlB and RhlC catalyse the transfer of dTDP-L-rhamnose to either HAA, or a previously generated mono-rhamnolipid, respectively [21].

Recently, biosurfactant production by a *P. aeruginosa* with very low-cost substrates showed a highest biosurfactant production of (3.2 g/l) when a culture medium containing corn steep liquor (10% (v/v)) and molasses (10% (w/v)) is used [22]. By inserting rhlAB genes responsible for production of mono-rhamnolipids, into *P. aeruginosa* PA14 and *E. coli* TG2 has not showed any improvement for effective production of rhamnolipids [23]. *P. aeruginosa* DAUPE 614 produced rhamnolipids (3.9 g/l) in 216 h when cultivated on a medium containing glycerol as the sole carbon source and ammonium nitrate with a C:N ratio of 55:1 [24].

Sarachat et al. [18] confirmed the production of rhamnolipid biosurfactants by *P. aeruginosa* SP4 and its concentration using the foam fractionation technique. The *P. aeruginosa* strain BN10 [25], isolated indigenous strain *P. aeruginosa* EM1 originating from an oil-contaminated site [26] and *P. aeruginosa* IFO3924 [27] have been reported to produce rhamnolipid-type biosurfactants.

Two rhamnolipid biosurfactants, L-rhamnopyranosyl-L-rhamnopyranosyl-b-hydroxydecanoyl-b-hydroxydecanoate or Rha-Rha C10-C10 and L-rhamnopyranosyl-L-rhamnopyranosyl-b-hydroxydecanoyl-b-hydroxydodecanoate or Rha–Rha C10-C12 have been reported to be produced by *P. aeruginosa B189* strain [28].

The capacity of producing rhamnolipid biosurfactants from a single specific strain with variation of only one carbon source was emphasized by the usage of *P. aeruginosa* strain NY3. This strain appeared to be much more productive from other rhamnolipid-producing *Pseudomonas* strains due to its capability of producing many more minor components of rhamnolipid when grown with glycerol as the sole carbon source instead on glucose [29]. In 15 years ago, *P. aeruginosa* 57RP was isolated and its capability for rhamnolipid production was assessed on 2% (w/v) when mannitol or naphthalene as carbon sources. Results showed that in the mannitol culture, 2.311 g of purified rhamnolipid extract were obtained after 14 days of incubation, whereas about 180 mg of rhamnolipid extract were recovered from the naphthalene culture after 7 days of incubation [30]. *P. aeruginosa* ATCC 9027 have been demonstrated as other rhamnolipid-producing strain. It was reported that 1% glucose+0.25% stearic acid (C18) produced the greatest yield (2.1 g/l) compared to other glucose-fatty acid combinations (0.8–1.8 g/l) [31]. Evaluation of waste cooking oil (CWO) as low-cost substrate to produce rhamnolipids surfactants for industrials application have been performed by using another potential *P. aeruginosa* 12-1[32]. Other strain *Pseudomonas aeruginosa*

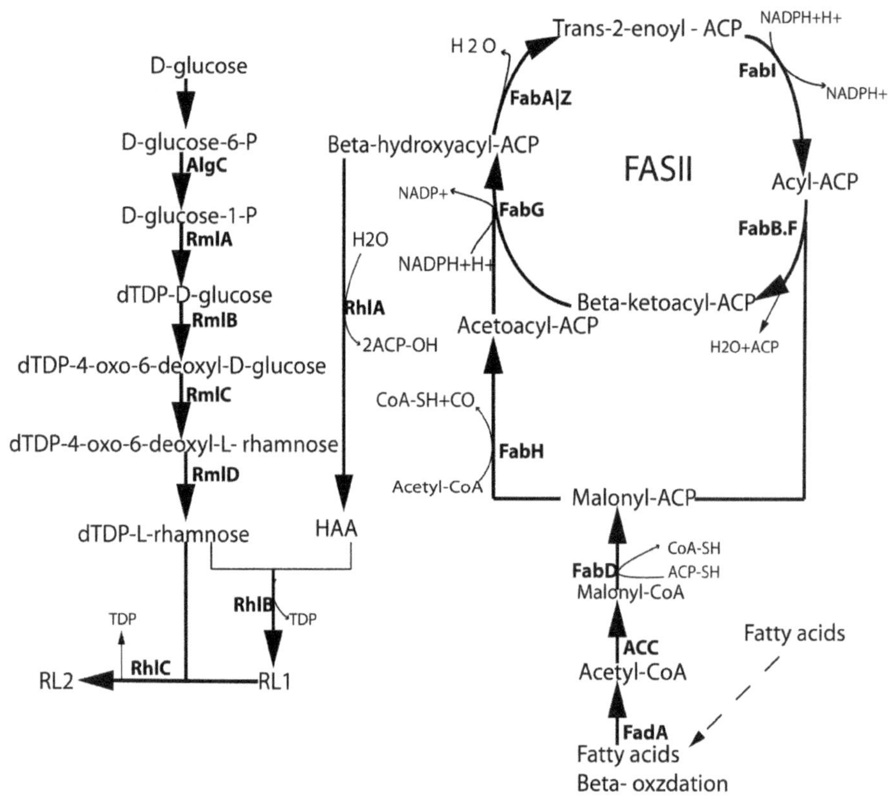

AlgC: phosphomannomutase, (RmlA, EC 2.7.7.24): glucose-1-phosphate thymidylyltransferase catalyzes the conversion of thymidylmonophosphate nucleotide to glucose-1-phosphate, (RmlB, EC 4.2.1.46): dTDP-D-glucose 4,6-dehydratase which catalyzes an oxidation of the C4 hydroxyl group of the D-glucose residue, then dehydration, ending to the formation of dTDP-4-oxo-6-deoxyl-D-glucose, (RmlC, EC 5.1.3.13) :dTDP-4-dehy-drorhamnose 3,5-epimerase which catalyzes a double epimerization reaction at the C3 and C5 positions of the 4-oxo-6-deoxyl-D-glucose ring, (RmlD, EC 1.1.1.133): dTDP-4-dehydrorhamnose reductase which reduces the C4 keto group of the 4-oxo-6-deoxyl-L-mannose moiety and leads to the formation of dTDP-L-rhamnose, FabD: malonyl-CoA:ACP transacylase, FabB, FabH, and FabF: β-ketoacyl-ACP synthetases, FabG: NADPH-dependent β-ketoacyl-ACP reductase, FabA, FabZ: β-hydroxyacyl-ACP dehydratases, FabI: The NADH-dependent enoyl-ACP reductase, HAA: 3-(3-hydroxyalkanoyloxy)alkanoic acid,: RhlA: 3-(3-hydroxyalkanoyloxy) alkanoate synthetases, RhlB: rhamnosyltransferase 1, and RhlC: rhamnosyltransferase 2. All required enzymatic steps are font bolded.

Figure 1: Biosynthesis pathway of RL1, monorhamnolipid; RL2, dirhamnolipid from fatty acids and glucose in *Pseudomonas Aeruginosa*.

O-2-2 has been used in order to improve the yield of rhamnolipid surfactants for reaching industrial requirements. The pH optimization led increased rhamnolipid production to 28.8 g/l, an improvement of 19.7%, and more substrate was converted to rhamnolipids rather than to biomass [32].

In *P. aeruginosa*, both direct and indirect factors like quorum sensing, nutritional status or stress response influence rhamnolipid production [33]. The nature of the carbon source and the nitrogen source as well as the C:N ratio, chemical and physical parameters such as temperature, nutritional limitations, dissolved oxygen, divalent cations and pH have been demonstrated to influence the amount of biosurfactants produced and the type of polymer formed [34]. Production of rhamnolipids can be inhibited by the presence of NH_4^+, glutamine, asparagine, and arginine as nitrogen source and contrary promoted by NO_3^-, glutamate and aspartate [35].

Biosynthesis of rhamnolipids in *P. aeruginosa*

P. aeruginosa produces a glycolipidic biosurfactants consisting of one or two hydrophilic L-rhamnose molecules (monorhamnolipids and dirhamnolipids,) and that of a hydrophobic fatty acid moiety [46]. Furthermore, *P. aeruginosa* produces extracellular rhamnolipid surfactants including the synthesis of rhamnolipids which is controlled by quorum sensing system [47].

The biosynthesis of biosurfactants, including rhamnolipids, can be either proceeded, (1) when both moieties are synthesized independently of the growth substrate (*de novo*), or (2) with a hydrophobic carbon source such as fatty acids and triglycerides, where the lipid moieties are directly derived from the carbon source, but the sugar is synthesized de novo. Another possibility (3) is when the sugar moiety is directly derived from the carbon source, but the lipid component is synthesized de novo [48].

It has been found that two genes rhlA and rhlB are organized in a single operon whereas another gene necessary for synthesis of di-rhamnolipids, *rhlC*, has also been identified and localized in another region of *P. aeruginosa* genome and which forms an operon with a gene of unknown function [13]. Those three kinds of genes *rhlA*, *rhlB* and *rhlC*, have been also investigated in various *Burkholderia* species where they are grouped within one putative operon, and have been shown to be required for rhamnolipid production [49]. The RhlC (rhamnosyltransferase II) catalyses the addition of the second rhamnose moiety to mono-rhamnolipids forming di-rhamnolipids [51]. Like *rhlA* and *rhlB*, *rhlC* is thought to be an ancestral gene controlled by the same quorum sensing system as *rhlA* and *rhlB*. The rhamnose moiety for mono- and di-rhamnolipids is derived from AlgC activity and the RmlABCD pathway, encoded on the *rml BCAD* operon [50]. In rhamnose synthesis, AlgC produces glucose-1-

phosphate which is converted to dTDP-D-glucose by RmlA followed by its conversion to dTDP-4-oxo-6-deoxyl-D-glucose by RmlB, the latter is converted to dTDP-6-deoxyl-L-deoxyl-4-rhamnose by RmlC. dTDP-6-deoxyl-L-deoxyl-4-rhamnose is converted to dTDP-L-rhamnose by RmlC [51]. Finally, dTDP-L-rhamnose, the key substrate of two rhamnosyltransferases RhlB and RhlC, is converted in one (monorhamnolipid) or two (di-rhamnolipid) molecules [52].

Genetic engineering for rhamnolipid production

Different direct and indirect factors such as nutritional status, quorum sensing and stress response can influence rhamnolipid production in *P. aeruginosa* [12]

Efforts have been concentrated to interfere with these regulatory mechanisms by introduction of genes responsible for rhamnolipid production into some microorganism's chromosome.

It has been proved that AlgR which controls various factors in *P. aeruginosa,* involves directly in rhlAB operon expression by binding to the promoter [53].

The deletion of AlgR PSL317 strain shows an increased rhamnolipid production compared with wild-type PAO1-strain [54]

The model for transcription activation of rhlAB (encoding rhamnosyltransferase 1) and rhlC (encoding rhamnosyltransferase 2) during biofilm-grown PAO1 and PSL317 increased the concentration of butanoyl-homoserine lactone (C4-HSL) to the transcription activator RhlR. The latter can work either as an activator or inhibitor for the operon transcription depending on concentration level of C4-HSL which is an autoinducer that binds to RhlR [46].

Therefore, the absence of C4-HSL influences RhlR to work as an inhibitor. Engineered strains by deleting lasR, essential gene for C4-HSL production did not produce any type of rhamnolipids [55].

Other regulators such as PtxR, RsaL, RsmA, DksA and various genes encoding sigma factors like RpoN and RpoS also influence the expression levels of the rhlAB operon [56,57], and several potential researches lie on manipulations of this complex regulatory system for rhamnolipids production.

Another strategy for enhancing rhamnolipid production was done by studying *P. aeruginosa* (NRRL B-771) and its mutant strain, PaJC, harboring the Vitreoscilla hemoglobin gene vgb, and it was reported as an effective strategy to increase rhamnolipid production where an increase up to 8.373 g/lwas obtained with PaJC mutant [58].

By knocking-out RhlC gene that is located in another operon with an upstream unknown gene (PA1131) in *P. aeruginosa* PAO1, and which is not organized with RhlAB, only resulted in production of monorhamnolipids [59].

A recombinant *E. coli* strain expressing *P. aeruginosa rhlAB* operon has been reported. The availability of dTDP-L-rhamnose a substrate of RhlB found in *E. coli* showed that metabolic engineering strategy leads to an increased monorhamnolipids production in heterologous host [60].

An insertion through transposome-mediated chromosome integration of the RhlAB gene into *P. aeruginosa* PAO1-rhlA⁻ and *Escherichia coli* BL21 (DE3) which could not produce rhamnolipid, showed a significant production of rhamnolipid in *P. aeruginosa* PEER02 and *E. coli* TnERAB recombinant [61]. Also, an introduction of rhlAB operon from *P. aeruginosa* PAO1 into *Burkholderia kururiensis* KP23, showed an increase of 6-fold [62].

Using *Burkholderia kururiensis* as expression platform of two *Pseudomonas aeruginosa* biosynthetic enzymes RhlA applied in fatty acid synthesis to generate the HAA and RhlB which catalyzes the transfer of dTDP-L-rhamnose to β-hydroxy fatty acids in the biosynthesis of rhamnolipids, an engineered *Burkholderia kururiensis* showed an increase over 600% comparing to the wild type [63].

An rhlABRI cassette contained genes for rhamnolipid synthesis was cloned and integrated into *P. putida* KT2440 chromosome random transposon vector with absence of antibiotic-resistance marker. This generate a genetically engineered microorganism named *P. putida* KT2440-rhlABRI, which showed a good expression of rhlABRI cassette and produced rhamnolipid at a yield of 1.68 g [64].

A recently engineered *P. stutzeri* Rhl was constructed for heterologous production of rhamnolipid under anaerobic conditions where the rhlABRI genes responsible for rhamnolipid biosynthesis were cloned into a facultative anaerobic strain *P. stutzeri* DQ1 and 1.61 g/l rhamnolipid was obtained [65].

The *rhlAB*-genes from *P. aeruginosa* PAO1 were expressed in *P. putida* KT2440 followed by removing of a competing pathway leads to the of formation polyhydroxyalkanoates (PHAs) and the highest yield of 0.15 $g/g_{glucose}$ was obtained [66].

Other producers of rhamnolipids

In last few decades, researchers concentrated their work on finding alternative strains for the production of rhamnolipids by using non-pathogenic bacteria, [67] demonstrated the potential use of mahua oil cake as substrate for the production of rhamnolipid using *S. rubidaea* SNAU02. Unlike *P. aeruginosa*, both *Acinetobacter* and *Enterobacter* non-pathogenic strains produced rhamnolipids that exhibited excellent emulsification activity with aromatic and aliphatic hydrocarbons. Several host strains such as *P. fluorescens*, *P. putida*, *P. oleovorans* and *E. coli* by introducing the rhlAB operon (rhamnosyltransferase), produced the highest rhamnolipid concentration of 60 mg/l for *P. putida* [68]. Compared to other rhamnolipids producers, P. *putida* can be considered as a potential rhamnolipid producer. *C* NCIM 2112 (Pd 2112) was screened for rhamnolipid production, however, the medium supplemented with sucrose, mannitol, or glycerol did not increase the yield. The medium supplemented with Mannitol and hexadecane produced more rhamnolipid compared to the medium supplemented with only mannitol [69]. The studies conducted on *P. chlororaphis* showed this strain as a potential non-pathogenic rhamnolipid producer due to its energy saving property [70]. The researchers have been carried out on several other bacterial species such as *Burkholderia mallei, B. pseudomallei, Serratia rubidea* and other non-pathogenic *B. thailandensis*, and by using optimized medium [71].

Potential industrial applications

Rhamnolipids, potential and current biosurfactants also exhibits many properties of, i.e. detergency, emulsifying, metal sequestering, demulsifying, solubilizing, wetting, thickening, foaming, environmental compatibility, and vesicle forming and phase dispersion. All of these properties are associated with the amphiphilic character of rhamnolipid molecules and confer upon them the ability to accumulate between fluid phases, therefore, reducing surface and interfacial tensions. The range of applications of rhamnolipids, including environmental biodegradation, food processing, agricultural, cosmetic and pharmaceutical uses, is potentially as extensive as their properties [19,72].

Applications of rhamnolipids in food processing and preservative

Nowadays, rhamnolipids produced by *P. aeruginosa* are the most promising class of biosurfactants, since the U.S. Environment Protection agency has approved their use in food products, and other industrials applications. Due to their high antimicrobial activity and physicochemical properties, some can be exploited by food industry particularly in increasing of food shelf life without concern to consumer health, excluding the need of adding synthetic comfits, which, in most of case, are harmful. In combination with niacin they extended shelf life and inhibited hemophilic spores in UHT soymilk. Niacin with rhamnolipids in salad has extended its shelf life and inhibited mold growth. Natamicin, nisin, and rhamnolipids combined together in cottage cheese have extended shelf life by inhibiting mold and bacterial growth, particularly gram-positive and spore-forming bacteria. Therefore, they can be used to avoid food contamination directly, as food additive, or indirectly, as a detergent formulation to clean surfaces that come in contact with the food [73].

Rhamnolipids can be used in bread, had rolls, soft rolls, hamburger buns, flat bread, baguettes, pizza, chinese steam breads, croissants, argentine breads, schnittbotchen, cake and sponge cake to improve their dough or batter stability ,volume and shape, structure, dough texture, width of the cut and microbiological conservation. Furthermore, they are used to improve the properties of butter cream, decoration cream and non-dairy cream filling for croissants, Danish pastries, and other fresh or frozen fine confectionery products [74]. They also serve as a source of rhamnose for the synthesis of food flavors [16].

Apart from their obvious role as agents that decrease surface and interfacial tension, rhamnolipids can have other several functions in food where they improve texture and shelf-life of starch-containing products, control the agglomeration of fat globules, stabilize aerated systems, modify rheological properties of wheat dough and improve stability, consistency and texture of oils and fat-based products and inhibits separation. They help in the general mixing of ingredients and can also slow the growth of molds and some bacteria in food [75]. In ice cream and bakery formulations, rhamnolipids can be used to control consistency, retard staling, solubilize flavor oils, stabilize fats, and reduce spattering [15]. It has been demonstrated that rhamnolipids can be explored to control the attachment and to disrupt biofilms of individual and mixed cultures of the food-borne pathogens [76].

Applications of rhamnolipids in agriculture and farming

Deep insight into the physiochemical effects of rhamnolipids and their biological importance would reveal new dimensions in the fields of research like agriculture. Rhamnolipids have the potential to be a part of alternative strategies in order to reduce or replace pesticides in agriculture. So, nowadays, they play a great importance for the efficiency of new biopesticides.

They are involved in non-specific immunity in plants, induce resistance in plants and are also active in other plant species. Rhamnolipids are capable of stimulating defense genes in tobacco, also potent protectors in monocotyledonous plants against biotrophic fungi [77]. Rhamnolipids are found to be useful in removal of polyaromatic hydrocarbons and pentachlorophenol from soil. They can facilitate the absorption of nutrients and fertilizers through the roots [78].

Stability constants were established by an ion-exchange resin technique. Due to the anionic nature of rhamnolipids, they are able to remove toxic metals from agriculture land and ions such as, lanthanum, cadmium, copper, lead and zinc due to their complexation ability. Cations of lowest to highest affinity for rhamnolipid are described as $K^{+}<Mg^{2+}<Mn^{2+}<Ni^{2+}<Co^{2+}<Ca^{2+}<Hg^{2+}<Fe^{3+}<Zn^{2+}<Cd^{2+}<Pb^{2+}<Cu^{2+}<Al^{3+}$. The affinities were approximately the same or higher than those of organic acids (acetic, citric, fulvic and oxalic acids) towards metals, thus indicating the potential of the rhamnolipid for metal remediation [79].

The success of rhamnolipids in increasing recovery of heavy metal contaminants from soils will also depend on the amount of rhamnolipid present in the aqueous phase [80]. Particularly, Rhamnolipid type I and type II, with surface tensions of 29 mN/m are suitable for heavy metals removal from contaminated agriculture areas [81].

Rhamnolipids with a fertilizer (Inipol EAp- 22) have a high capacity of enhancing the biodegradation of aromatic and aliphatic compounds in aqueous phase and soil reactors.

Those kinds of biosurfactants, have demonstrated inhibition of zoospore forming plant pathogens that have acquired resistance to commercial chemical pesticides and another investigation has shown that rhamnolipid can stimulate plant immunity which is considered as an alternative strategy to reduce the infection by plant pathogens.

Recent investigation has also established rhamnolipid as an insecticidal compound [82]. Agriculture land containing weathered petroleum hydrocarbons which are important land contaminants can be remedied by an introduction of rhamnolipids into soil due to their high solubilization and increased bioavailability properties on some inaccessible compounds [83].

As reviewed by [84], biosurfactants, such as rhamnolipids, are being widely used in soil remediation and also confirmed that the removal of phenanthrene from contaminated soil using rhamnolipid and short-chain organic acids (SCOA) gradually increased as the SCOA concentration increased up to a concentration of 300 mmol/l. They are also potential surfactants for removing dibenzothiophenes compound from soil [85]. Rhamnolipids, have been found to be able to remove chromium and arsenic oxyanions from spiked soils or mine tailings, conducted a series of dynamic column elution tests, suggested that rhamnolipids at a high concentration (5.0 g/l) could remove ~70% of the pyrene in soil. On the other side, polycyclic aromatic hydrocarbons (PAH) and chemical spills are known as most limiting factors to soil fertility and crop productivity; Therefore, their desorption from contaminated soils is assisted by rhamnolipids which improve the bioavailability of different pollutants [4]. Other application of rhamnolipid biosurfactants-enhanced soil washing coupled with activated carbon adsorption have been promised as alternative to remediate soils contaminated with hydrocarbon organic contaminants (HOC) [86].It has been reported that continuous application of 150 mg/l of rhamnolipid biosurfactant regulate 100% of disease caused by *P. capsici* (e.g. Phytophtora blight) in pepper plants (*Capsicum annuum*) [87].

Recently, researches indicated that integration of *R. glutinis* in combination with rhamnolipids at low concentrations might be an efficient and safe strategy in field application to control *A. alternate,* which is a serious pathogenic to certain cultivars of tomatoes, causing postharvest black rot at high frequencies [88]. Application of rhamnolipids have been suggested to be a good way for wastewater treatment and prevention of overwatering during irrigation due to their wettability properties which facilitates the breaking down of impenetrable barriers to water, allowing water to easily reach the soil and spread more evenly throughout the soil profile [89].

In addition, many techniques using rhamnolipids have been developed for cleaning up petroleum contaminated soil. Addition of rhamnolipids enhances the solubility and elimination of these contaminants by improving oil biodegradations rates [90].

Different comparisons of biosurfactants (aescin, saponin, lecithin, rhamnolipid and tannin) for washing crude oil contaminated soil was carried out where rhamnolipids showed a high removal capacity up to 80% oil [91]. Oil washing experiments done by a combination of 10 g/l NaCl, 5.0 g/l n-butyl alcohol and 2.0 g/l rhamnolipid provided a highest oil extraction rate of 74.55% [91]. Even if rhamnolipids are the best enhancers for petroleum hydrocarbon soil pollutants removing and their potential to facilitate the bioremediation of soil contaminated by hydrocarbons, Zheng et al. and Kaczorek et al. [91] suggested that their application must be evaluated carefully to reduce their exhibition on antimicrobial activity.

Outlook

This review outlined the successful development ways of the production and application of a better known rhamnolipid biosurfactant.

Nowadays, research on rhamnolipids biosurfactants is becoming easier due to the availability of several different strains screening methods and the ability to purify diverse amphiphilic molecules. Rhamnolipids have several applications in agriculture, agrochemical industries, food processing and food preservation relating to their potential properties as they increase solubility, foaming capacity and lower surface tensions. Regarding too many researches reviewed herein, we admire the *P. aeruginosa* strains to be the most potential microorganisms which can be involved in large-scale production of rhamnolipids. Most of the studies conducted on rhamnolipids production up to now, they confirm and suggest the exploitation of modern approach such as functional metagenomics which is the utmost essential and may even lead to discovery of novel green surfactants including rhamnolipids biosurfactants. Therefore, intense work on green surfactants is a priority to prevent the adverse effects of synthetic surfactants largely employed in many commercial sectors including agro- chemical industries. Considering the social and technological backgrounds, utilization of biosurfactants, which are environmentally friendly and highly functional, rhamnolipids have become more and more important. Therefore, their application in washing, bioremediation and microbial enhanced oil recovery basing on their physical chemical properties and environmental conditions must be considered. Like other surfactants, attention should be paid on rhamnolipids application on plants as they could be affected in many different ways. Thus, the improved study on their negative impact on crop yield or other important agronomical traits should be a crucial role to avoid any impact on plant growth or metabolism, while improving plant immunity.

References

1. Radhika C, Jun S, Minmin C, Kanaji M, Jain AK, et al. (2014) Properties and characterization of biosurfactant in crude oil. Fuel 122: 140-148.
2. Banat IM, Satpute SK, Cameotra SS, Patil R, Nyayanit NV (2014) Cost effective technologies and renewable substrates for biosurfactants' production. Front Microbiol 5: 697.
3. Zgola-Grzeskowiak A, Kaczorek E (2011) Isolation, preconcentration and determination of rhamnolipids in aqueous samples by dispersive liquid-liquid microextraction and liquid chromatography with tandem mass spectrometry. Talanta 83: 744-750.
4. Mao X, Jiang R, Xiao W, Yu J (2015) Use of surfactants for the remediation of contaminated soils: a review. J Hazard Mater 285: 419-435.
5. Gujar RS, Hamde VS (2012) Separation and characterization of biosurfactant from P.aeruginosa sp1 isolated from oil mill area MIDC, Parbhani (M.S.). Int Multidiscip Res J 2:13-15.
6. Long X, Meng Q, Zhang G (2014) Application of biosurfactant rhamnolipid for cleaning of UF membranes. J Membr Sci 457:113-119.
7. Muller MM, Kugler JH, Henkel M, Gerlitzki M, Hormann B, et al. (2012) Rhamnolipids--next generation surfactants. J Biotechnol 162: 366-380.
8. Lovaglio RB, dos Santos FJ, Jafelicci M Jr, Contiero J (2011) Rhamnolipid emulsifying activity and emulsion stability: pH rules. Colloids Surf B Biointerfaces 85: 301-305.
9. Heyd M, Kohnert A, Tan TH, Nusser M, Kirschhofer F, et al. (2008) Development and trends of biosurfactant analysis and purification using rhamnolipids as an example. Anal Bioanal Chem 391: 1579-1590.
10. Long X, Zhang G, Shen C, Sun G, Wang R, et al. (2013) Application of rhamnolipid as a novel biodemulsifier for destabilizing waste crude oil. Bioresour Technol 131: 1-5.
11. Nguyen TT, Sabatini DA (2011) Characterization and emulsification properties of rhamnolipid and sophorolipid biosurfactants and their applications. Int J Mol Sci 12: 1232-1244.
12. Reis RS, Pereira AG, Neves BC, Freire DM (2011) Gene regulation of rhamnolipid production in Pseudomonas aeruginosa-a review. Bioresour Technol 102: 6377-6384.
13. Wittgens A, Tiso T, Arndt TT, Wenk P, Hemmerich J, et al. (2011) Growth independent rhamnolipid production from glucose using the non-pathogenic Pseudomonas putida KT2440. Microb Cell Fact 10: 80.
14. Siddhartha GVAOC, Nitschke M, Haddad R, Eberlin MN, Contiero J (2006) Production of Pseudomonas aeruginosa LBI rhamnolipids following growth on Brazilian native oils. Process Biochem 41: 483-488.
15. Irfan-Maqsood M, Seddiq-Shams M (2014) Rhamnolipids- Well-Characterized Glycolipids with Potential Broad Applicability as Biosurfactants. Ind Biotechnol 10: 285-291.
16. Muller MM, Kugler JH, Henkel M, Gerlitzki M, Hormann B, et al. (2012) Rhamnolipids-next generation surfactants. J Biotechnol 162: 366-380.
17. Zhu L, Yang X, Xue C, Chen Y, Qu L, et al. (2012) Enhanced rhamnolipids production by Pseudomonas aeruginosa based on a pH stage-controlled fed-batch fermentation process. Bioresour Technol 117: 208-213.
18. Sarachat T, Pornsunthorntawee O, Chavadej S, Rujiravanit R (2010) Purification and concentration of a rhamnolipid biosurfactant produced by Pseudomonas aeruginosa SP4 using foam fractionation. Bioresource Technol 101: 324-330.
19. Nitschke M, Costa SGVAO, Contiero J (2011) Rhamnolipids and PHAs: Recent reports on Pseudomonas-derived molecules of increasing industrial interest. Process Biochem 46: 621-630.
20. Zhang X, Xu D, Zhu C, Lundaa T, Scherr KE (2012) Isolation and identification of biosurfactant producing and crude oil degrading Pseudomonas aeruginosa strains. Chem Eng J 209: 138-146.
21. Reis RS, Pacheco GJ, Pereira AG, Freire DMG (2013) Biosurfactants: Production and Applications 3: 124.
22. Gudiña EJ, Rodrigues AI, Alves E, Domingues MR, Teixeira JA, et al. (2015) Bioconversion of agro-industrial by-products in rhamnolipids toward applications in enhanced oil recovery and bioremediation. Bioresour Technol 177: 87-93.
23. Kryachko Y, Nathoo S, Lai P, Voordouw J, Prenner EJ, et al. (2013) Prospects for using native and recombinant rhamnolipid producers for microbially enhanced oil recovery. Int Biodeter Biodegr 81: 133-140.
24. Monteiro SA, Sassaki GL, de Souza LM, Meira JA, de Araujo JM, et al. (2007) Molecular and structural characterization of the biosurfactant produced by Pseudomonas aeruginosa DAUPE 014. Chem Phys Lipids 147: 1-13.
25. Terziyski I, Alexandrova L, Stoineva I, Christova N, Todorov R, et al. (2014) Foam and wetting films from rhamnolipids produced by Pseudomonas aeruginosa BN10. Colloids Surf A Physicochem Eng Asp 460: 299-305.

26. Wu JY, Yeh KL, Lu WB, Lin CL, Chang JS (2008) Rhamnolipid production with indigenous Pseudomonas aeruginosa EM1 isolated from oil-contaminated site. Bioresour Technol 99: 1157-1164.

27. Hori K, Ichinohe R, Unno H, Marsudi S (2011) Simultaneous syntheses of polyhydroxyalkanoates and rhamnolipids by Pseudomonas aeruginosa IFO3924 at various temperatures and from various fatty acids. Biochem Eng J 53:196-202.

28. Thanomsub B, Pumeechockchai W, Limtrakul A, Arunrattiyakorn P, Petchleelaha W, et al. (2006) Chemical structures and biological activities of rhamnolipids produced by Pseudomonas aeruginosa B189 isolated from milk factory waste. Bioresource Technol 97: 2457-2461.

29. Nie M, Yin X, Ren C, Wang Y, Xu F, et al. (2010) Novel rhamnolipid biosurfactants produced by a polycyclic aromatic hydrocarbon-degrading bacterium Pseudomonas aeruginosa strain NY3. Biotechnol Adv 28: 635-643.

30. Deziel E, Lepine F, Dennie D, Boismenu D, Mamer OA, et al. (1999) Liquid chromatography/mass spectrometry analysis of mixtures of rhamnolipids produced by Pseudomonas aeruginosa strain 57RP grown on mannitol or naphthalene. Biochim Biophys Acta 1440: 244-252.

31. Zhang L, Pemberton JE, Maier RM (2014) Effect of fatty acid substrate chain length on Pseudomonas aeruginosa ATCC 9027 monorhamnolipid yield and congener distribution. Process Biochem 49: 989-995.

32. Costa SGVAO, Nitschke M, Lepine F, Deziel E, Contiero J (2010) Structure, properties and applications of rhamnolipids produced by Pseudomonas aeruginosa L2-1 from cassava wastewater. Process Biochem 45: 1511-1516.

33. Geys R, Soetaert W, Van Bogaert I (2014) Biotechnological opportunities in biosurfactant production. Curr Opin Biotechnol 30: 66-72.

34. Fakruddin M (2012) Biosurfactant: Production and Application. J Pet Environ Biotechnol 3: 124.

35. Milena R, Gojgic-Cvijovic G, Vrvic M, Karadzic I (2012) Production and characterization of rhamnolipids from Pseudomonas aeruginosa san-ai. J Serb Chem Soc 77: 27-42.

36. Trummler K (2003) An integrated microbial/enzymatic process for production of rhamnolipids and l-(+)-rhamnose from rapeseed oil with Pseudomonas sp DSM 2874. Eur J Lipid Sci Tech 105: 563-571.

37. Rahman KS, Rahman TJ, McClean S, Marchant R, Banat IM (2002) Rhamnolipid biosurfactant production by strains of Pseudomonas aeruginosa using low-cost raw materials. Biotechnol Prog 18: 1277-1281.

38. Haba E, Espuny MJ, Busquets M, Manresa A (2000) Screening and production of rhamnolipids by Pseudomonas aeruginosa 47T2 NCIB 40044 from waste frying oils. J Appl Microbiol 88: 379-387.

39. Nitschke M, Costa SG, Haddad R, Gonçalves LA, Eberlin MN, et al. (2005) Oil wastes as unconventional substrates for rhamnolipid biosurfactant production by Pseudomonas aeruginosa LBI. Biotechnol Prog 21: 1562-1566.

40. Benincasa M, Contiero J, Manresa MA, Moraes IO (2002) Rhamnolipid production by Pseudomonas aeruginosa LBI growing on soapstock as the sole carbon source. J Food Eng 54: 283-288.

41. Saharan BS, Sahu RK, Sharma D (2011) A review on biosurfactants: Fermentation, current developments and perspectives. Genet Eng Biotechnol J 29: 1-14.

42. Raza ZA, Ahmad N, Kamal S (2014) Multi-response optimization of rhamnolipid production using grey rational analysis in Taguchi method. Biotechnol Rep 3:86-94.

43. Moussa TAA, Mohamed MS, Samak N (2014) Production and characterization of di-rhamnolipid produced by Pseudomonas aeruginosa TMN. Braz J chem eng 31: 867-880.

44. Abbasi H, Hamedi MM, Lotfabad TB, Zahiri HS, Sharaf H, et al. (2012) Biosurfactant-producing bacterium, Pseudomonas aeruginosa MA01 isolated from spoiled apples: physicochemical and structural characteristics of isolated biosurfactant. J Biosci Boeng 113: 211-219.

45. Bazire A, Dufour A (2014) The Pseudomonas aeruginosa rhlG and rhlAB genes are inversely regulated and RhlG is not required for rhamnolipid synthesis. BMC Microbiol 14: 160.

46. Lequette Y, Greenberg EP (2005) Timing and localization of rhamnolipid synthesis gene expression in Pseudomonas aeruginosa biofilms. J Bacteriol 187: 37-44.

47. Ahmad Mohammad AM, Rudolf H, Francois L, Markus M, Eric D (2011) Rhamnolipids: Detection, Analysis, Biosynthesis,Genetic Regulation,and Bioengineering of Production. In: Soberón-Chávez ASE (ed.) Biosurfactants: From Genes to Applications. Microbiology Monographs, Springer, Münster, Germany 20: 13-55.

48. Abdel-Mawgoud AM, Lépine F, Déziel E (2010) Rhamnolipids: diversity of structures, microbial origins and roles. Appl Microbiol Biotechnol 86: 1323-1336.

49. Rahim ROUA, Olvera C, Graninger M, Messner P, Lam JS, et al. (2001) Cloning and functional characterization of the Pseudomonas aeruginosa rhlC gene that encodes rhamnosyltransferase, an enzyme responsible for di-rhamnolipid biosynthesis. Mol Microbiol 40: 708-718.

50. Olvera C, Goldberg JB, Sánchez R, Soberón-Chávez G (1999) The Pseudomonas aeruginosa algC gene product participates in rhamnolipid biosynthesis. FEMS Microbiol Lett 179: 85-90.

51. Miller DJ, Zhang YM, Rock CO, White SW (2006) Structure of RhlG, an essential beta-ketoacyl reductase in the rhamnolipid biosynthetic pathway of Pseudomonas aeruginosa. J Biol Chem 281: 18025-18032.

52. Aguirre-Ramírez M, Medina G, González-Valdez A, Grosso-Becerra V, Soberón-Chávez G (2012) The Pseudomonas aeruginosa rmlBDAC operon, encoding dTDP-L-rhamnose biosynthetic enzymes, is regulated by the quorum-sensing transcriptional regulator RhlR and the alternative sigma factor ÏƒS. Microbiology 158: 908-916.

53. Okkotsu Y, Tieku P, Fitzsimmons LF, Churchill ME, Schurr MJ (2013) Pseudomonas aeruginosa AlgR phosphorylation modulates rhamnolipid production and motility. J Bacteriol 195: 5499-5515.

54. Morici LA, Carterson AJ, Wagner VE, Frisk A, Schurr JR, et al. (2007) Pseudomonas aeruginosa AlgR represses the Rhl quorum-sensing system in a biofilm-specific manner. J bacteriol 189: 7752-7764.

55. Medina G, Juárez K, Díaz R, Soberón-Chávez G (2003) Transcriptional regulation of Pseudomonas aeruginosa rhlR, encoding a quorum-sensing regulatory protein. Microbiology 149: 3073-3081.

56. Carty NL, Layland N, Colmer-Hamood JA, Calfee MW, Pesci EC, et al. (2006) PtxR modulates the expression of QS-controlled virulence factors in the Pseudomonas aeruginosa strain PAO1. Mol Microbiol 61: 782-794.

57. Rampioni G, Schuster M, Greenberg EP, Zennaro E, Leoni L (2009) Contribution of the RsaL global regulator to Pseudomonas aeruginosa virulence and biofilm formation. FEMS Microbiol Lett 301: 210-217.

58. Kahraman H, Erenler SO (2012) Rhamnolipid production by Pseudomonas aeruginosa engineered with the Vitreoscilla hemoglobin gene. Prikl Biokhim Mikrobiol 48: 212-217.

59. Cabrera-Valladares N, Richardson AP, Olvera C, Treviño LG, Déziel E, et al. (2006) Monorhamnolipids and 3-(3-hydroxyalkanoyloxy)alkanoic acids (HAAs) production using Escherichia coli as a heterologous host. Appl Microbiol Biotechnol 73: 187-194.

60. Wang Q, Fang X, Bai B, Liang X, Shuler PJ, et al. (2007) Engineering bacteria for production of rhamnolipid as an agent for enhanced oil recovery. Biotechnol Bioeng 98: 842-853.

61. Tavares LF, Silva PM, Junqueira M, Mariano DC, Nogueira FC, et al. (2013) Characterization of rhamnolipids produced by wild-type and engineered Burkholderia kururiensis. Appl Microbiol Biotechnol 97: 1909-1921.

62. Cao L, Wang Q, Zhang J, Li C, Yan X, et al. (2012) Construction of a stable genetically engineered rhamnolipid-producing microorganism for remediation of pyrene-contaminated soil. World J Microbiol Biotechnol 28: 2783-2790.

63. Zhao F, Shi R, Zhao J, Li G, Bai X, et al. (2015) Heterologous production of Pseudomonas aeruginosa rhamnolipid under anaerobic conditions for microbial enhanced oil recovery. J Appl Microbiol 118: 379-389.

64. Wittgens A, Tiso T, Tarndt T, Wenk P, Hemmerich J, et al. (2011) Growth independent rhamnolipid production from glucose using the non-pathogenic Pseudomonas putida KT2440. Microb Cell Fact 10: 80.

65. Nalini S, Parthasarathi R (2014) Production and characterization of rhamnolipids produced by Serratia rubidaea SNAU02 under solid-state fermentation and its application as biocontrol agent. Bioresour Technol 173: 231-238.

66. Ochsner UA, Reiser J, Fiechter A, Witholt B (1995) Production of Pseudomonas aeruginosa Rhamnolipid Biosurfactants in Heterologous Hosts. Appl Environ Microbiol 61: 3503-3506.

67. Jadhav M, Kalme S, Tamboli D, Govindwar S (2011) Rhamnolipid from Pseudomonas desmolyticum NCIM-2112 and its role in the degradation of Brown 3REL. J Basic Microbiol 51: 385-396.

68. Gunther NW 4th, Nuñez A, Fett W, Solaiman DK (2005) Production of rhamnolipids by Pseudomonas chlororaphis, a nonpathogenic bacterium. Appl Environ Microbiol 71: 2288-2293.

69. Kalyani A, Naga Sireesha G, Aditya A, Girija Sankar G, Prabhakar T (2014) Production Optimization of Rhamnolipid Biosurfactant by Streptomyces coelicoflavus (NBRC 15399T) using Plackett-Burman design. Eur J Biotechnol Biosci 1: 07-13.

70. Mukherjee S, Das P, Sen R (2006) Towards commercial production of microbial surfactants. Trends Biotechnol 24: 509-515.

71. Ozdemir G, Malayoglu U (2004) Wetting characteristics of aqueous rhamnolipids solutions. Colloids Surf B Biointerfaces 39: 1-7.

72. Freire DMG, Araujo Vd, Kronemberger FdA, Nitschke M (2009) Biosurfactants as Emerging Additives in Food Processing. In: Ribeiro CP (ed) Innovation in Food E New Techniques and Products engineering 685-705.

73. Van HI, Vanzeveren E (2004) Rhamnolipids in bakery products. US 20060233935 A1.

74. Nitschke M, Costa SGVAO (2007) Biosurfactants in food industry. Trends Food Sci Tech 18: 252-259.

75. Zezzi do Valle Gomes M, Nitschke M (2012) Evaluation of rhamnolipid and surfactin to reduce the adhesion and remove biofilms of individual and mixed cultures of food pathogenic bacteria. Food Control 25: 441-447.

76. Vatsa P, Sanchez L, Clement C, Baillieul F, Dorey S (2010) Rhamnolipid biosurfactants as new players in animal and plant defense against microbes. Int J Mol Sci 11: 5095-5108.

77. Mulligan CN (2005) Environmental applications for biosurfactants. Environ Pollut 133: 183-198.

78. Sachdev DP, Cameotra SS (2013) Biosurfactants in agriculture. Appl Microbiol Biotechnol 97: 1005-1016.

79. Asci Y, Nurbas M, Acikel YS (2008) A comparative study for the sorption of Cd(II) by soils with different clay contents and mineralogy and the recovery of Cd(II) using rhamnolipid biosurfactant. J Hazard Mater 154: 663-673.

80. Hashim MA, Mukhopadhyay S, Sahu JN, Sengupta B (2011) Remediation technologies for heavy metal contaminated groundwater. J Environ Manage 92: 2355-2388.

81. Chrzanowski L, Dziadas M, Lawniczak L, Cyplik P, Bialas W, et al. (2012) Biodegradation of rhamnolipids in liquid cultures: effect of biosurfactant dissipation on diesel fuel/B20 blend biodegradation efficiency and bacterial community composition. Bioresource Technol 111: 328-335.

82. An CJ, Huang GH, Wei J, Yu H (2011) Effect of short-chain organic acids on the enhanced desorption of phenanthrene by rhamnolipid biosurfactant in soil-water environment. Water Res 45: 5501-5510.

83. Urum K, Grigson S, Pekdemir T, McMenamy S (2006) A comparison of the efficiency of different surfactants for removal of crude oil from contaminated soils. Chemosphere 62: 1403-1410.

84. Wan J, Chai L, Lu X, Lin Y, Zhang S (2011) Remediation of hexachlorobenzene contaminated soils by rhamnolipid enhanced soil washing coupled with activated carbon selective adsorption. J Hazard Mater 189: 458-464.

85. Raudales RE, Parke JL, Guy CL, Fisher PR (2014) Control of waterborne microbes in irrigation: A review. Agr Water Manage 143: 9-28.

86. Yan F, Xu S, Chen Y, Zheng X (2014) Effect of rhamnolipids on Rhodotorula glutinis biocontrol of Alternaria alternata infection in cherry tomato fruit. Postharvest Biol Tec 97: 32-35.

87. Renfro TD, Xie W, Yang G, Chen G (2014) Rhamnolipid surface thermodynamic properties and transport in agricultural soil. Colloids Surf B Biointerfaces 115: 317-322.

88. Millioli V, Servulo EL, Sobral L, De carvalho D (2009) Bioremediation Of Crude Oil-Bearing Soil: Evaluating The Effect Of Rhamnolipid Addition To Soil Toxicity And To Crude Oil Biodegradation Efficiency. Global Nest J 11:181-188.

89. Urum K, Pekdemir T (2004) Evaluation of biosurfactants for crude oil contaminated soil washing. Chemosphere 57: 1139-1150.

90. Zheng C, Wang M, Wang Y, Huang Z (2012) Optimization of biosurfactant-mediated oil extraction from oil sludge. Bioresour Technol 110: 338-342.

91. Kaczorek E (2012) Effect of External Addition of Rhamnolipids Biosurfactant on the Modification of Gram Positive and Gram Negative Bacteria Cell Surfaces during Biodegradation of Hydrocarbon Fuel Contamination. Pol J Environ Stud 21: 901-909.

Response Surface Methodology for Adsorption of Fluoride Ion Using Nanoparticle of Zero Valent Iron from Aqueous Solution

Ali Fakhri* **and Saeideh Adami**

Department of Chemistry, Shahre-Qods Branch, Islamic Azad University, Tehran, Iran

Abstract

This study is removal of fluoride ion from aqueous solution by nano zero valent iron (nZVI). Effects of the factor variables (temperature, nZVI dose and pH) and their interactions on adsorption of fluoride ion were investigated by response surface methodology (RSM) based on Box-Behnken design (BBD). Optimized values of temperature, nZVI dose and pH for fluoride sorption were found as 313K, 0.5 g, and 4, respectively. The effect of initial fluoride concentration on the adsorption amount was investigated by a batch experiment. For study the fluoride removal mechanism, various adsorption isotherms such as Langmuir, Freundlich and Temkin were fitted. The results showed that the Freundlich isotherm gave high fit for fluoride adsorption. The time of adsorption reaction was rapid and subordinated pseudo-second-order kinetics. The values of thermodynamic parameters indicated that adsorption was spontaneous and endothermic in nature. The nano zero valent iron (nZVI) could be used as a potential adsorbent for fluoride ion containing aqueous solution.

Keywords: Adsorption; Fluoride; Box-Behnken design; Nanoparticle zero valent iron; Response surface methodology

Introduction

Fluoride is a health affecting substance. The physiologic effects of fluoride ingestion on public health have been studied extensively [1]. The acceptable fluoride concentration in drinking water is generally in the range of 0.5 to 1.5 mgl^{-1} [2]. High concentration has the effects on the metabolism of elements such as Ca, P in human body and lead to dental and skeletal fluorosis. The fluoride content of soils varies from under 20 to several thousand ppm, the higher records being mostly from areas with bedded phosphate on fluoride deposits [3]. The natural presence of fluoride generally occurs through soil and rock formation in the form of fluorapatite, fluorspar and amphiboles, geochemical deposits, natural water systems and earth crust [4,5]. In addition to this fluoride can also be found in various Industrial work, chiefly semiconductor, electroplating, glass, steel, ceramic and fertilizers industries [6]. Because of these reasons the pollution of ground water by fluoride contamination has been a major concern. The problems in connection with fluoride ion pollution could be reduced or minimized by ultra-filtration, precipitation, reverse osmosis, electrode-deposition, etc., but these processes have flaws such as high cost, generation of secondary pollutants and low removal efficiency. Adsorption experiment has been found to be an effective and economic method with high potential for the removal, recovery and recycle of fluoride ions from aqueous solution [7], although desorption is an issue.

Zero valent iron (ZVI) was proposed as a reactive material in permeable reactive barriers (PRGs) due to its great ability in reducing and the stabilization pollutants different [8]. Nowadays, there has been an increasing interest in synthesizing this material on nanoscale for to enhance its ability restore by probity of the increase in the surface area and surface reactivity of the particles [9]. Nano zero valent iron (nZVI), a recently discovered technology, is being used to successfully treatment various organic compounds (e.g. [10-12]) and various heavy metal ions in aqueous solutions (e.g. [13-18]).

Design of experiments (DOE) and response surface methodology (RSM) is largely used for modeling mechanism parameters, especially in adsorption or removal process [19-32]. Because RSM contains a lower number of experiments, it is flaws over conventional methods available. It is suitable for multi-factor experiments and searches the common connection between various factors for the determined of most favorable or unfavorable conditions of the processes. Response surface methodology has different model types such as central composite design (CCD), Doehlert matrix (DM) and Box-Behnken design (BBD). The objective of this study was the application of the RSM combined with Box-Behnken design as a statistic method in optimizing adsorption mechanism of fluoride ion using nano zero valent iron. The Langmuir, Freundlich and Temkin isotherm models were used to define the equilibrium data. The adsorption mechanisms of fluoride ion from aqueous solutions onto nZVI were also evaluated in terms of kinetics and thermodynamic parameters.

Materials and Methods

Raw materials

Sodium fluoride salt (NaF) (molecular weight, 41.98871 g/mol) was supplied by Merck Co. (Germany) (maximum purity available > 99%). Doubly distilled deionized water (HPLC grade 99.99% purity) was obtained from Sigma Aldrich Co. (Germany).

Synthesis of nanoparticle zero valent iron (nZVI)

The nZVI material was synthesized by drop wise addition of 1.6 M NaBH4 (sodium borohydride) aqueous solution to a Ne gas-purged 1 M $FeCl_3.6H_2O$ (Ferric chlorid hexahydrate) aqueous solution at 23°C with magnetic stirring as described by Wang and Zhang [33]. Ferric iron (Fe^{3+}) was reduced according to the reaction [34]: $4Fe^{+3} + 3BH_4 + 9H_2O \rightarrow 4Fe^0 + 3H_2BO_3- + 12H^+ + 6H_2$.

***Corresponding author:** Ali Fakhri, Department of Chemistry, Shahre-Qods Branch, Islamic Azad University, Tehran, Iran

The solution was stirred for 20 min and centrifuged at 6000 rpm for 2 min, and the supernatant solution was replaced by acetone. Acetone-washing prevented the immediate rusting of nZVI during purification leading to a fine black powder product after freeze-drying. X-ray diffractometer (XRD, Philips X'Pert) was used to investigate the material structure of zero valent iron nanoparticles. In addition, the surface morphology of nZVI was determined using transmission electron microscopy (TEM, JEM-2100F HR, 200 kV).

Adsorption experiment

The adsorption of fluoride onto nZVI was investigated using batch experiments. In these studies 1000 mg/L stock solution was prepared by dissolving 1 g of NaF in 1000 mL distilled water. Different concentrations (20-200 mg/L) of fluoride solutions were prepared by this stock solution. Solutions were evacuated to flasks of 100 ml. Then adsorbent in the range of dosage 0.05-0.5 g was added and placed in the water bath shaker after pH adjustments made in the range of 2-10. The suspensions were shaken at 2000 rpm for 12 min at room temperature. Samples from shaker were filtered with filter paper, and then remaining fluoride levels were measured using a fluoride electrode (Orion, 9606BNWP). The final adsorption amount was calculated from the equation

$$q_e = \frac{(C_o - C_e)V}{W} \qquad (1)$$

where, qe (mg/g) is the equilibrium adsorption amount, C_e is the fluoride concentration at equilibrium (mg/l), V is the volume of solution (l) and w is the weight of adsorbent (g).

Response surface methodology

The three-level, three-factorial Box-Behnken experimental design with categorical factor of 0 was engaged to optimize the treatment process based on the adsorption amount of the nZVI for fluoride ion. The design was composed of three levels (low, medium and high, being coded as (-1, 0 and +1) and a total of 17 runs were carried out in repetitious to optimize the level of chosen variables, such as temperature, nZVI dosage and pH. For the aim of statistical calculations, the three factor variables were denoted as x_1, x_2, and x_3, respectively. According to the preparatory experiments, the range and levels used in the experiments are selected and listed in Table 1. The main effects and in connection between factors were determined. Figure 1 illustrates the mean of the experimental results for the respective low, medium and high levels of temperature, pH and nZVI dosage of scattering. The experimental design matrix by the Box-Behnken design is tabulated in table 2 and corresponding experiments were performed. The results were analyzed by applying the response plots and analysis of variance (ANOVA). For RSM, the most commonly used second-order polynomial equation developed to fit the experimental data can be written as:

$$Y = \beta_0 + \sum_{i=1}^{k} \beta_i x_i + \sum_{i=1}^{k} \beta_{ii} x_i^2 + \sum_{1 \le i \le j}^{k} \beta i_j x_i x_j + \varepsilon \qquad (2)$$

where Y represents the predicted response, i.e. the adsorption amount

Factor	Low level (-1)	Medium level (0)	High level (+1)
Temperature (X_1)	283 K	297 K	313 K
nZVI dosage (X_2)	0.25 g	0.50 g	0.75 g
pH (X_3)	4	8	12

Table 1: Factors and levels used in the factorial design.

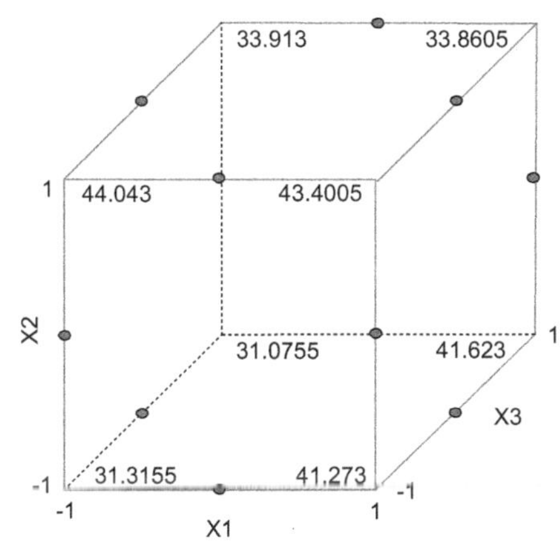

Figure 1: Cube plots for Y.

Runs	X_1	X_2	X_3	Y_{Exp} (mg/g)
1	0	0	0	22.48
2	0	0	0	22.55
3	+1	+1	0	30.84
4	+1	0	-1	45.20
5	0	0	0	22.53
6	0	0	0	22.38
7	-1	0	1	31.39
8	-1	+1	0	32.37
9	+1	0	+1	37.82
10	-1	-1	0	25.24
11	0	-1	-1	25.48
12	0	+1	+1	26.51
13	0	0	0	22.43
14	0	+1	-1	33.56
15	-1	0	-1	39.36
16	0	-1	+1	28.32
17	+1	-1	0	34.31

Table 2: BBD and results for the study of three experimental variables in coded units.

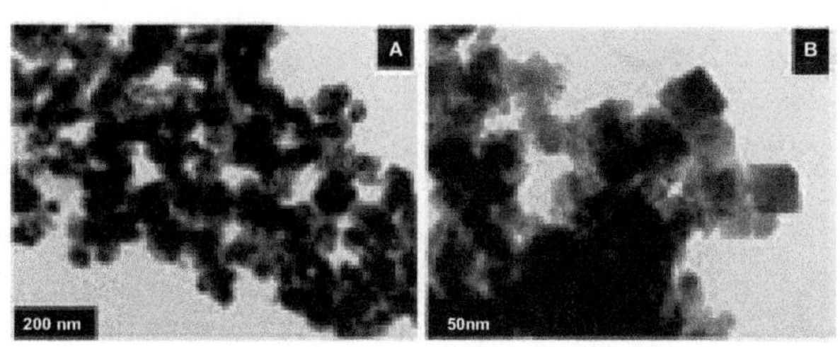

Figure 2: TEM images of (A) nZVI and (B) nZVI after adsorption of fluoride ion.

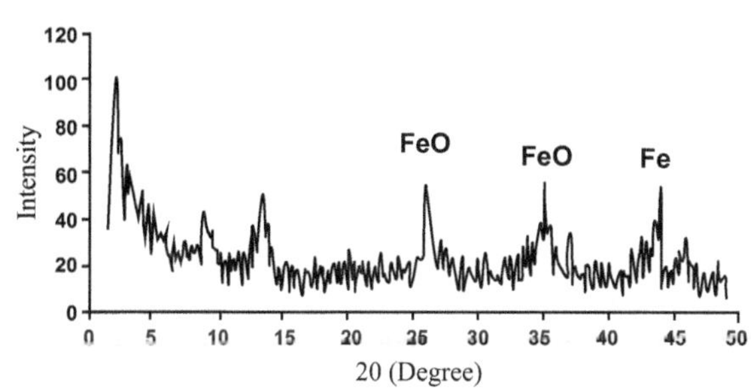

Figure 3: X-Ray diffraction pattern of Nanoparticle zero valent iron (nZVI) surface.

for fluoride ion by the nZVI (mg/g), β_0, the constant coefficient, β_i, the ith linear coefficient of the input factor x_i, β_{ii}, the ith quadratic coefficient of the input factors x_i, β_{ij}, the different interaction coefficients between input factors x_i and x_j ($i = 1–3$, $j = 1–3$ and $i / = j$), and ε, the error of the model [35]. The equation expresses the in connection between the predicted response and factor variables in coded values according to Tables 1 and 2.

Results and Discussion

Characterization of nZVI

Figure 2 shows the transmission electron microscope image of freshly synthesized iron nanoparticles. Surface morphology shows that there exist two layers in the nZVI particle. The layer of intrant core Indicative the Fe^0, and the external layer surrounding on the Fe^0 was iron-oxide(s). Figure 2B shows that the fluoride molecules into the nZVI surface are covered. The X-Ray diffraction of nZVI surface composition under ambient conditions is shown in Figure 3. The wide peak reveals the being of an amorphous phase of iron. The characteristic broad peak at 2θ of 45° indicates that the zero valent iron is predominantly present in the sample.

Statistical analysis

The optimum terms for adsorption of fluoride onto nZVI surface were determined by means of the BBD under RSM. The results were displayed in Tables 3. In this manner, the fluoride uptake by nZVI could be described using the following equation: $Y = 22.474 + 2.476 X_1 + 1.241 X_2 - 2.445 X_3 - 2.650 X_1X_2 + 0.147 X_1X_3 - 2.472 X_2X_3 + 9.095 X_{12} - 0.879 X_{22} + 6.873 X_{23}$ (3)

The state of the fitted model was declared by the coefficient of determination. The R^2 coefficient gives the relation of the total variation in the response predicted by the model and a high R^2 value (close to 1) is desirable. Eq. (3) Indication that the model is well fitted, considering the determination coefficient (R^2= 97.71%). The estimated effects and coefficients for model are listed in Table 3. Model terms were evaluated by the P-value (probability) with 95% confidence level. The P-values were used to estimate whether F was large enough to indicate statistical significance and used to check the significance of each coefficient. The P-values lower than 0.05 indicated that the model and model terms were statistically significant. All the factors and their square interactions ($P < 0.05$) except for interaction of temperature–temperature (X_1^2) and pH-pH (X_3^2) were significant at the 95% confidence level. nZVI dose was the most significant factor that affect the removal of fluoride. Also, the quadratic effect of nZVI dose - nZVI dose (X_2^2) was found larger than effect of nZVI dose, and the removal of fluoride significantly decreased. Figure 4 shows that the data were well distributed near to a straight line (R^2=0.9726), which proposed a good relationship between the experimental and predicted values of the response, and the underlying assumptions of the above analysis were appropriate.

Source	Sum of Squares	df	Mean Square	F Value	p-value Prob > F
Model	738.36	9	82.04	29.94	<0.0001
Temperature(X_1)	49.05	1	49.05	17.90	0.0039
nZVI dosage (X_2)	12.33	1	12.33	4.50	0.0716
pH (X_3)	47.82	1	47.82	17.45	0.0041
Temperature-nZVI dosage (X_1X_2)	28.09	1	28.09	10.25	0.0150
Temperature-pH (X_1X_3)	0.087	1	0.087	0.032	0.8636
nZVI dosage-pH (X_2X_3)	24.45	1	24.45	8.92	0.0203
Temperature-Temperature (X_1^2)	348.33	1	345.33	127.13	<0.0001
nZVI dosage- nZVI dosage (X_2^2)	3.26	1	3.26	1.19	0.3117
pH-pH (X_3^2)	198.90	1	198.90	72.59	<0.0001
Residual	19.18	7	2.74		
Lack of Fit	19.16	3	6.39		
Pure Error	0.020	4	4.930E-003		
Cor Total	757.55	16			

Table 3: Analysis of variance for the response of the adsorption capacity for fluoride ion.

3D response surface plot

The 3D response surface plot are useful in investigation both the main and interaction effects of the factors [36,37]. The response surface plots are presented in Figure 5. This figure also shows the estimated Y parameter as a function of the normalized factor variables, the height of the surface represents the value of Y. After executing a screening of factors using a BBD, the surface plots of the response (Y) indicated the same results as observed in the interaction plot (Figure 5).

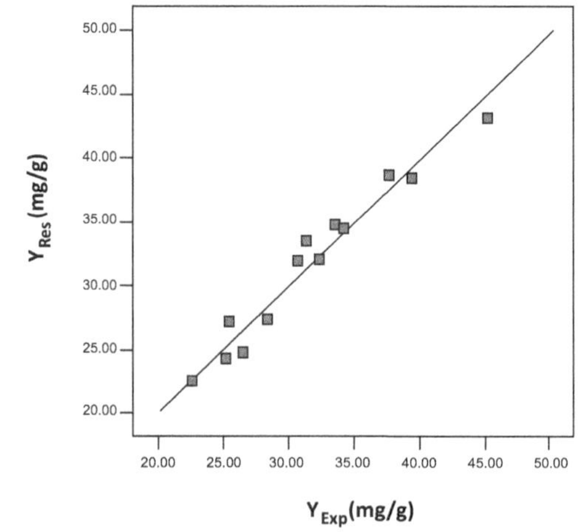

Figure 4: Plot of the experimental and predicted responses.

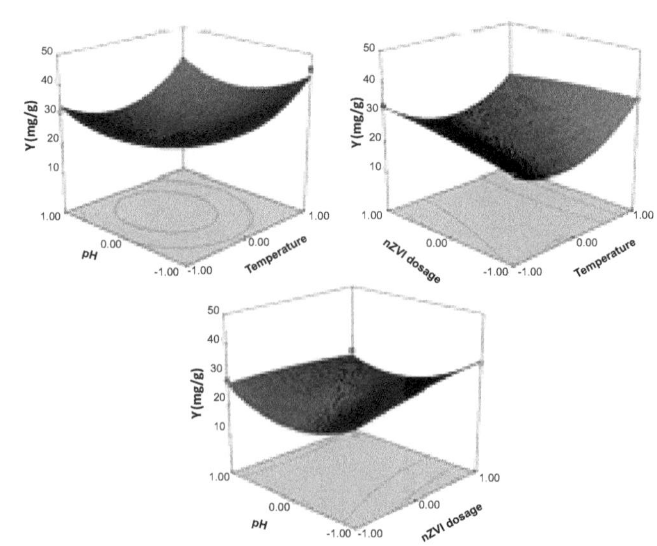

Figure 5: Response surface plots of adsorption capacity versus the effect of three variables.

Adsorption isotherms

The mechanism of fluoride adsorption from aqueous solutions onto nZVI is evaluated using adsorption isotherm. In this study, isotherm data were applied to four adsorption models and the results of their linear regressions were used to find the model with the best fit. Values of resulting parameters and regression coefficients (R^2) are listed in Tables 4 and 5.

The R^2 value for the Freundlich isotherm was 0.9990, which is higher than the values obtained from the Langmuir and Temkin isotherm models. The experimental data fit very well to this isotherm model, and indicates that fluoride adsorption occurs on heterogeneous surfaces, which is similar to the conclusion reached for nZVI [38-40].

Adsorption kinetics

Various models have been used to investigation of the adsorption mechanisms and potential rate. Effects of contact time on adsorption are investigated, as shown in Figure 6. The adsorption process was quite rapid and reached equilibrium in 30 min (Figure 6). In this study, the different kinetics models such as pseudo-first-order, pseudo-second-order, and intra-particle diffusion models were used.

Isotherm model	Equation
Langmuir	$\frac{C_e}{q_e} = \frac{1}{K_L q_m} + \frac{C_e}{q_m}$
Freundlich	$ln\, q_e = ln\, K_F + \frac{1}{n} ln\, C_e$
Temkin	$q_e = B_1 \ln K_T + B_1 \ln C_e$

Table 4: Summary of equilibrium isotherms (K_L, K_F, B_1, K_T: Langmuir, Freundlich and Temkin constants; n: heterogeneity coefficient; q_m: maximum adsorption capacity; q_e: uptake at equilibrium; C_e: equilibrium concentration; b: activity coefficient related to mean sorption energy).

Langmuir model			
K_L(L/g)	q_m(mg/g)		R^2
0.1053	129.87		0.9902
Freundlich			
K_F(mg/g)/(mg/L)$^{1/n}$	n		R^2
14.2235	1.4212		0.9990
Temkin			
K_T(L/mg)	B_1	b=RT/B_1	R^2
1.1161	27.943	88.367	0.9876

Table 5: Langmuir, Freundlich and Temkin isotherm constants for adsorption of Fluoride ion.

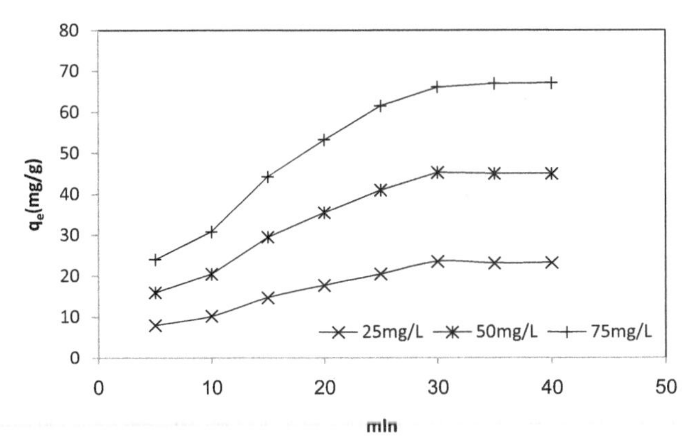

Figure 6: Optimisation of contact time for fluoride removal (initial pH: 4.0; adsorbent dose: 0.5 g/L).

The pseudo-first-order rate equation is given as [41]: $\log (q_e - q_t) = \log (q_e) - \left(\frac{k_1}{2.303}\right) t$ (4)

where q_e and q_t are the amounts of fluoride adsorbed (mg/g) at equilibrium and at time t (min), respectively, and k_1(L/min) is the adsorption rate constant of first-order adsorption. A straight line of $\log(q_e-q_t)$ versus t (Figure not shown) suggests the applicability of this kinetic model. q_e and k_1 were determined from the intercept and slope of the plot which were shown in Table 6. From the data, q_e (calculated) and q_e (experimental) values are not in agreement with each other. Therefore, that indicates the adsorption of fluoride on nZVI was not a first-order reaction.

In addition, the experimental data was also applied to the pseudo-second-order kinetic model Equation [42]:

$$\frac{t}{qt} = \frac{1}{k_2 q_e^2} + \frac{t}{q_e} \quad (5)$$

where k_2 is the rate constant of pseudo-second-order chemisorptions (g/(mg min)). The plot t/q_t versus t giving a straight line which is shown in Figure 7 and the constant calculated from the slop and intercept of the plots are given in Table 6. Figure 7 shows that R^2 values are higher than those obtained from the first-order kinetics. In addition, theoretical and experimental q_e values are in agreement. Therefore, it is

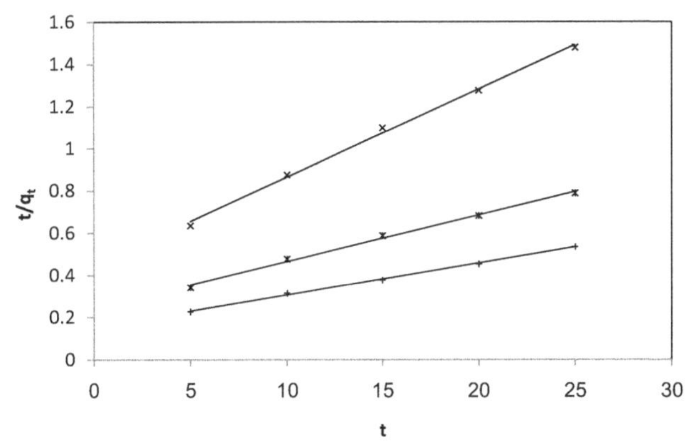

Figure 7: Second-order kinetic plot for fluoride adsorption on nZVI at different initial concentration.

Fluoride ion concentration (mg/L)	25	50	75
$q_{e,exp}$ (mg/g)	23.50	45.31	66.12
Pseudo-first-order model			
q_e (mg/g)	20.73	25.85	36.71
k_1 (min^{-1})	0.0291	0.0301	0.0302
R^2	0.9826	0.9801	0.9828
Pseudo-second-order model			
q_e (mg/g)	23.92	45.45	66.22
k_2 (g.mg^{-1}.min^{-1})	0.0039	0.0019	0.0014
R^2	0.9967	0.9957	0.9981
Intra-particle diffusion model			
k_i (mg.g^{-1}.min^{-1})	4.8227	9.4463	13.871
C	3.7547	0.0723	9.3529
R^2	0.9843	0.9850	0.9841

Table 6: Kinetic parameters and experimental adsorption capacities for fluoride onto nZVI.

possible to prove that the adsorption process using nZVI followed the second-order kinetic model.

The intra-particle diffusion equation can be described as:

$$q = k_i t^{1/2} + C \quad (9)$$

where ki is the intra-particle diffusion rate constant (mg/g min). The data (Figure not shown) shown have an initial curved portion, followed by a linear portion. The curved portion of the plot is due to the diffusion of fluoride through the solution to the external surface of nZVI, or boundary layer diffusion. However, the extrapolated linear regions at different initial concentrations did not pass through the origin and that suggests that the intra-particle diffusion was not the rate-controlling step [43].

Thermodynamic of adsorption studies

Thermodynamic parameters connected to the adsorption reaction, i.e., Gibbs free energy change ($\Delta G°$, kJ mol^{-1}), enthalpy change ($\Delta H°$, kJ mol^{-1}), and entropy change ($\Delta S°$, J mol^{-1} K^{-1}) are defined by the following equations: $\Delta G° = -RT\ln Kc$ (8)

$$\Delta G° = \Delta H° = \Delta H° - T\Delta S° \quad (9)$$

where K_c is the equilibrium constant, which can be obtained from Langmuir isotherm, R is the universal gas constant, 8.314 $Jmol^{-1}$ K^{-1}, and T is absolute temperature (K). $\Delta H°$ and $\Delta S°$ were obtained from the slope and intercept of the plot of Gibbs free energy change, $\Delta G°$ vs. temperature, T (Figure 8).

The negative values of $\Delta G°$ (-5.746, -6.793 and -8.270 KJ mol^{-1} for 283, 297 and 313K, respectively) confirm the feasibility of the process and the spontaneous nature of adsorption with a high preference for fluoride onto nZVI. The standard enthalpy and entropy changes determined from plot in Figure 8. The value of $\Delta H°$ is positive (18.165 KJ mol^{-1}), indicating that the adsorption reaction is endothermic. The positive value of $\Delta S°$ (84.300 J $mol^{-1}K^{-1}$) reflects an increase in the randomness at the solid/solution interface during the adsorption process [44].

Conclusions

The statistical design of the experiments was applied in optimizing the conditions of maximum adsorption of the fluoride onto nZVI. The result data from ANOVA demonstrates that the model was highly significant. nZVI dose was the most significant factor affecting fluoride removal. Therefore, it is apparent that the response surface methodology not only gives valuable information on interactions between the factors but also helps to the recognition of possible optimum values of the studied factors. Adsorption experiments show that the adsorption equilibrium can be achieved within 30 min. The kinetics studies of fluoride on nZVI indicated that the adsorption kinetics of fluoride on nZVI followed the pseudo-second order at different initial concentration. The results of Isotherm data showed that the removal of fluoride followed Freundlich isotherm. Thermodynamic adsorption studies indicated that the adsorption fluoride using nZVI aqueous solution was a spontaneous, endothermic.

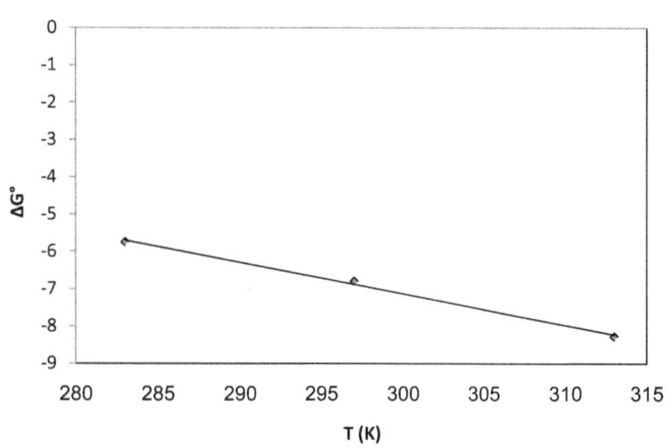

Figure 8: Plot of Gibbs free energy change vs. temperature for fluoride ion adsorption onto nZVI.

Acknowledgment

The authors gratefully acknowledge supporting of this research by the Islamic Azad University Shahre-Gods Branch.

References

1. Zhu CS, Bai GL, Liu XL, Li Y (2006) Screening high-fluoride and high-arsenic drinking waters and surveying endemic fluorosis and arsenic in Shaanxi province in western China Water Research. 40: 3015-3022.
2. WHO (2008) Guidelines for Drinking Water Quality, World Health Organization, Geneva.
3. Bishnoi M, Arora S (2007) Potable groundwater quality in some villages of Haryana, India: Focus on fluoride J Environ Biol 28: 291-294.
4. Shailaja K, Johnson MEC (2007) Fluorides in groundwater and its impact on health. J Environ Biol 28: 331-332.
5. Tripathy SS, Bersillon JL, Gopal K (2006) Removal of fluoride from drinking water by adsorption onto alum-impregnated activated alumina. Sep Purif Technol 50: 310-317.
6. Toyoda A, Taira A (2000) IEEE Trans, Semiconductor Manufacture. 13: 305-309.
7. Bailey SE, Olin TJ, Bricka RM, Adrian DD (1999) A review of potentially low-cost sorbents for heavy metals. Water Res 33: 2469-2479.
8. Blowes DW, Ptacek CJ, Benner SG, McRae Che WT, Bennett TA, et al. (2000) Treatment of inorganic contaminants using permeable reactive barriers. J Contam Hydrol 45: 123-137.
9. Sun Y, Li X, Cao J, Zhang W, Wang HP (2006) Characterization of zero-valent iron nanoparticles. Adv Colloid Interf Sci 120: 47-56.
10. Liu Y, Majetich SA, Tilton RD, Sholl DS, Lowry GV (2005) TCE dechlorination rates, pathways, and efficiency of nanoscale iron particles with different properties. Environ Sci Technol 39: 1338-1345.
11. Joo SH, Feitz AJ, Sedlak DL, Waite TD (2005) Quantification of the oxidizing capacity of nanoparticulate zero-valent iron. Environ Sci Technol 39: 1263-68.
12. Li XQ, Brown DG (2007) Zhang WX, Stabilization of biosolids with nanoscale zero-valent iron (nZVI). J Nanoparticle Res 9: 233.
13. Li XQ, Zhang WX (2007) Sequestration of metal cations with zerovalent iron nanoparticles-a study with high resolution X-ray photoelectron spectroscopy (HR-XPS). J Phys Chem 111: 6939-6946.
14. Uzum C, Shahwan T, Eroglu AE, Hallam KR, Scott TB, et al. (2009) Synthesis and characterization of kaolinite-supported zero-valent iron nanoparticles and their application for the removal of aqueous Cu2+ and Co2+ ions. Appl Clay Sci 43: 172-181.
15. Elebi OC, Uzum C, Shahwan T, Erten HNA (2007) A radiotracer study of the adsorption behavior of aqueous Ba2+ ions on nanoparticles of zero-valent iron. J Hazard Mater 148: 761-767.
16. Kanel SR, Manning B, Charlet L, Choi H (2005) Removal of arsenic (III) from groundwater by nanoscale zero-valent iron. Environ Sci Technol 39: 1291-1298.
17. Jegadeesan G, Mondal K, Lalvani SB (2005) Arsenate remediation using nano sized modified zero valent iron particles. Environ Prog 24: 289-296.
18. Li XQ, Zhang WX (2006) Iron nanoparticles: the core–shell structure and unique properties for Ni(II) sequestration. Langmuir 22: 4638-4642.

19. Kaushik P, Malik A (2011) Process optimization for efficient dye removal by Aspergillus lentulus FJ172995. J Hazard Mater 185: 837-843.

20. Ghorbani F, Younesi H, Ghasempouri SM, Zinatizadeh AA, Amini M, et al. (2008) Application of response surface methodology for optimization of cadmium biosorption in an aqueous solution by Saccharomyces cerevisiae. Chem Eng J 145: 267-275.

21. Abdel-Ghani NT, Hegazy AK, El-Chaghaby GA, Lima EC (2009) Factorial experimental design for biosorption of iron and zinc using Typha domingensis phytomass. Desalination 249: 343-347.

22. Cestari AR, Vieira EFS, Oliveira IA, Bruns RE (2007) The removal of Cu(II) and Co(II) from aqueous solutions using cross-linked chitosan-Evaluation by the factorial design methodology. J Hazard Mater 143: 8-16.

23. Hasan SH, Srivastava P, Talat M (2009) Biosorption of Pb(II) from water using biomass of Aeromonas hydrophila: Central composite design for optimization of process variables. J Hazard Mater 168: 1155-1162.

24. Annadurai G, Juang RS, Lee DJ (2002) Factorial design analysis for adsorption of dye on activated carbon beads incorporated with calcium alginate. Advances in Environmental Research 6: 191-168.

25. Ravikumar K, Krishnan S, Ramalingam S, Balu K (2007) Optimization of process variables by the application of response surface methodology for dye removal using a novel adsorbent. Dyes Pigments 72: 66-74.

26. Cestari AR, Vieira EFS, Mota JA (2008) The removal of an anionic red dye from aqueous solutions using chitosan beads-The role of experimental factors on adsorption using a full factorial design. J Hazard Mater 160: 337-343.

27. Bingol D, Tekin N, Alkan M (2010) Brilliant Yellow dye adsorption onto sepiolite using a full factorial design. Applied Clay Science 50: 315-321.

28. Jabasingh SA, Pavithra G (2010) Response Surface Approach for the Biosorption of Cr6+ Ions by Mucor racemosus. CLEAN-Soil, Air, Water 38: 492-499.

29. Zolgharnein J, Shahmoradi A, Sangi MR (2008) Optimization of Pb(II) biosorption by Robinia tree leaves using statistical design of experiments. Talanta 76: 528-532.

30. Zolgharnein J, Adhami Z, Shahmoradi A, Mousavi SN, MR Sangi (2010) Multivariate optimization of Cd(II) biosorption onto Ulmus tree leaves from aqueous waste. Toxicological and Environmental Chemistry 8: 1461-1470.

31. Bingöl D (2011) Removal of Cadmium (II) from Aqueous Solutions using a Central Composite Design. Fresenius Environmental Bulletin 10: 2704-2709.

32. Das D, Das N (2011) Response Surface Approach for the Bisorption of Ag(I) by Macro fungus Pleurotus platypus. CLEAN – Soil, Air, Water 2: 157-161.

33. Wang CB, Zhang W (1997) Synthesizing nanoscale iron particles for rapid and complete dechlorination of TCE and PCBs. Environ. Sci. Technol. 31: 2154-2156.

34. Glavee GN, Klabunde KJ, Sorensen CM, Hadjipanayis GC (1995) Chemistry of borohydride reduction of iron(II) and iron(III)ions in aqueous and nonaqueous media. formation of nanoscale Fe, FeB, and Fe2B powders. Inorg Chem 34: 28-35.

35. Benyounis KY, Olabi AG, Hashmi MSJ (2005) Effect of laser welding parameters on the heat input and weld-bead profile. J Mater Process Technol 164: 978-980.

36. Cojocaru C, Trznadel GZ (2007) Response surface modeling and optimization of copper removal from aqua solutions using polymer assisted ultrafiltration. J Membr Sci 298: 56-70.

37. Arbizu IP, Luis Pérez CJ (2003) Surface roughness prediction by factorial design of experiments in turning processes. J Mater Process Technol 143: 390-396.

38. Ponnusami V, Krithika V, Madhuram R, Srivastava SN (2007) Biosorption of reactive dye using acid-treated rice husk: factorial design analysis. J Hazard Mater 142: 397-403.

39. Mathialagan T, Viraraghavan T (2005) Biosorption of pentachlorophenol by fungal biomass from aqueous solutions: a factorial design analysis. Environ Technol 6: 571-579.

40. Sujana MG, Thakur RS, Rao SB (1998) Removal of fluoride from aqueous solution by using alum sludge. J Colloid Inter Sci 206: 94-101.

41. Lagergren S (1898) About the theory of so called adsorption of soluble substances. Vetenskapsad Handl 24: 1-39.

42. Ho YS, Mckay G (1999) Pseudo-second order model for sorption process Process Biochem 34: 451-465.

43. Gupta VK, Ali I, Saini VK (2007) Defluoridation of wastewaters using waste carbon slurry. Water Res 41: 3307-3316.

44. Ngah WSW, Fatinathan S (2008) Adsorption of Cu(II) ions in aqueous solution using chitosan beads, chitosan-GLA beads and chitosan-alginate beads. Chem Eng J 143: 62-72.

Honeycomb Structures of TiO_2-modified Hydroxyapatite Composite for Microbial Filtration Application

Naruporn Monmaturapoj*, Witoon Thepsuwan, Suda Wanakitti, Charusporn Mongkolkachit, Katanchalee Mai-ngam, Sawinee Ngernpimai, Wattana Klinsukhon and Chureerat Prahsarn

National Metal and Materials Technology Center, 114 Thailand Science Park, Pathumthani 12120, Thailand

Abstract

Honeycomb structures of TiO_2-modified hydroxyapatite composite (HA/TiO_2) were fabricated using extrusion technique. The mixture formulations of HA/TiO_2 extrusion pastes (S1-S4) were investigated to optimize the process. The filtering effectiveness of the HA/TiO_2 honeycombs was evaluated by photocatalytic activity measurement and a bactericidal test. The effects of sintering temperature and honeycomb thickness on photocatalytic and antibacterial activities were additionally studied. The S3 and S4 honeycomb samples, sintered at 650°C, exhibited a similar trend in reducing methylene blue concentration. At 800°C, the S3 honeycomb sample showed a slightly faster reduction in methylene blue concentration compared to the S4 honeycomb. Honeycomb samples of 2 and 5 cm in thickness showed significantly greater photocatalytic activity than that of 1 cm such that methylene blue concentrations rapidly decreased after UV exposure for 24 hours. The S3 and S4 honeycomb samples also exhibited decomposition of both gram-negative *E. coli* and gram-positive *S. aureus*. Approximately 50% of gram-positive *S. aureus* and gram-negative *E. coli* were decomposed by the honeycombs in the sample-thickness dependent manner after 2 hours of UV exposure. Low survival ratios of bacteria (5% - 10%) were observed after 6 hours of UV exposure and the 2 mm and 5 mm thick honeycomb samples showed the greatest and most similar antibacterial activities.

Keywords: Honeycomb; Hydroxyapatite; Titanium dioxide; Antibacterial; Photocatalytic activity

Introduction

Hydroxyapatite (HA, $Ca_{10}(PO_4)_6(OH)_2$), a calcium phosphate based compound which closely resembles natural bones and teeth, has been extensively used for biomedical applications, particularly for replacing human hard tissue [1-6]. It possesses excellent osteoconductivity, bioactivity, ionic exchange property and high stability under reducing and oxidizing situations leading to the adsorption ability for microbial bacteria and viruses [7-10]. HA is capable of adsorbing molecules on its surface but not decomposing molecules nor making microbes harmless, therefore, saturation will be reached over time [11].

Nanoparticle titanium dioxide (TiO_2) is a basic material extensively utilized in daily life, for example, as a pigment for paints, cosmetics and foodstuffs because of its effective performance as a photocatalyst. It is of great interest due to its potential in the oxidative decomposition of organic compounds, strong oxidizing power under UV, chemical stability, non-toxicity and long-term stability against photo and chemical corrosion [7-12]. However, TiO_2 can decompose only substances that come into direct contact, and the decomposition requires the presence of light, especially in the ultraviolet range [13-15]. By modifying the HA with TiO_2, their advantages can be combined-the HA adsorbs molecules that will then later be decomposed by TiO_2. The HA/TiO_2 composite has, therefore, the potential to be used for solving environmental problems such as water and air pollution due to its photocatalytic, antibacterial and antivirus properties. Nishikawa et al. reported that HA/TiO_2 composite exhibited higher photocatalytic activity than commercial TiO_2 and showed the rapid and complete oxidative decomposition of acetaldehyde [16]. The use of HA/TiO_2 composite as a bone scaffold coupled with collagen type I in the form of nanofiber mats fabricated by the electro spinning technique has also been reported [17]. Coating of hydroxyapatite on titania has been reported for bone substitute application [18]. Moreover, the composite of HA/TiO_2 can be applied in the field of sonocatalysis to damage bovine serum albumin (BSA) under ultrasonic irradiation [19].

At present, for the filtration application, most research has concentrated on coating the thin film of HA/TiO_2 composite onto substrate materials. However, the photocatalytic activities of the HA/TiO_2 particles coated on the substrate decreased compared to those in powder form, due to a decrease in the surface area of the particles. In addition, coating the HA/TiO_2 composite onto fibrous substrates for microbial filter application was also reported to have limitations in use. In the coating, particles can either deposit onto the fibers' surfaces or penetrate into the inter-fiber space which may not be fully exposed to UV light. The latter could lead to a decrease in photocatalytic activity [20]. By forming HA/TiO_2 composites into porous structures, which provide higher surface area and increase exposure area to UV light, it is possible to increase photocatalytic efficiency.

Honeycomb is a complex structure composed of several small unit cells, which have a thin wall between each cell. Honeycomb ceramics are widely used as a catalytic converter in automobile gas exhausts and a catalyst carrier [21]. Ceramic honeycomb structures provide the high surface areas with least volume. The geometric surface area of the carrier due to the characteristics of the porous thin wall greatly increased and improved thermal shock resistance [22,23]. Industrially, extrusion is one of the most popular techniques with which to fabricate honeycomb ceramics. Many factors influencing the properties of the final honeycomb products, particularly the compositions of raw materials, need to be controlled to achieve the desirable products.

***Corresponding author:** Naruporn Monmaturapoj, National Metal and Materials Technology Center, 114 Thailand Science Park, Pathumthani 12120

In this study, extrusion technique was employed to fabricate HA/TiO_2 composite at a ratio between HA to TiO_2 of 50:50 by weight. Four different mixture formulations of extrusion pastes were investigated in order to fabricate HA/TiO_2 honeycombs. The effects of sintering profiles and honeycomb thickness on photocatalytic activity and bactericidal performance were investigated and discussed in terms of their potential uses as microbial filters.

Experimental Procedure

Preparation and characterization of pure hydroxyapatite (HA)

Pure hydroxyapatite was prepared using a solid-state reaction method. Mixture powders of calcium carbonate ($CaCO_3$) and calcium phosphate dibasic ($CaHPO_4$) reacted as shown in equation (1).

$$4CaCO_3+6CaHPO_4 \rightarrow Ca_{10}(PO_4)_6(OH)_2+2H_2O+4CO_2 \quad (1)$$

All raw materials were used as received. $CaCO_3$ (Carlo Erba Reagenti, Italy) and $CaHPO_4$ (Sigma-Aldrich, Germany) were ball-milled with zirconia balls in the presence of deionized water as a medium for 48 hours, before drying. The mixture was then calcined at 900°C for 2 hours to produce pure hydroxyapatite powder. The obtained powders were ground and sieved to attain a particle size of less than 10 µm. Characterization of HA powders after calcination was done using x-ray diffraction: XRD (JEOL JDX 3530) with CuK_α source (K_α = 1.5406 nm) operating at 30 mA and 50 kV. XRD measurement was conducted at 20-60° 2θ, scan speed of 2°/min with a step size of 0.02°. The XRD spectra were analyzed using JADE software and ICDD cards. Particle size analysis was measured by laser diffraction technique (Mastersizer 2000, Malvern instruments, UK).

Preparation and characterization of HA/TiO2 composite

Anatase phase TiO_2 (Sigma-Aldrich, Germany) was used as received. HA/TiO_2 composite was prepared by dry ball milling the mixture between HA and TiO_2 powders at the composition ratio of 50:50 (HA: TiO_2 50:50), using ZrO_2 balls for 3 hours. The obtained HA/TiO_2 composite was then calcined at 650°C for 2 hours. The composite was then wet milled for another 24 hours using deionized water as the medium under ZrO_2 balls to achieve a particle size of less than 5 micron, followed by drying at 100°C. The particle size of HA/TiO_2 composite was analyzed by laser diffraction technique (Mastersizer 2000, Malvern instruments, UK). Analysis of their phases was carried out using XRD: JEOL JDX 3530.

Preparation and characterization of honeycomb structure

Four different mixture formulations of the extrusion pastes (S1-S4) as summarized in Table 1 were prepared for fabricating the HA/TiO_2 honeycombs. The paste compositions consisted of HA/TiO_2 composite powder as the main component, processing additives which were water as medium, methylcellulose (MC) and starch as binders, polyethylene glycol (PEG) and flour as plasticizers, sodium carbonate as a dispersant as well as some lubricants such as oil. Each formulation was uniformly mixed using a high speed mixer before being further transferred to a 3-roll mill for homogenous mixing. These paste batches were then de-aired and extruded using a screw extruder (FM-30, Miyazaki Iron Work, Japan) into cylindrical honeycomb structures of 50 mm in diameter with a cell density of 72 cell/inch^2 and having three different thicknesses (1, 2 and 5 cm). The extruded HA/TiO_2 honeycomb samples were covered with semi-dry cloth to let them slowly dry at room temperature. After drying, the honeycomb samples of selected formulations, S3 and S4, were sintered at 400°C for 2 hours and then ramped up to three different temperatures (650, 800 and 1000C) held for 4 hours, followed by furnace cooling to room temperature at a rate of 50°C/h. Phases and microstructures of the obtained HA/TiO_2 honeycombs were analyzed using XRD: JEOL JDX 3530 and scanning electron microscopy (SEM: JEOL JSM-5410).

Physical and mechanical properties

Five specimens each of S3 and S4 honeycomb structures, sintered at 650°C and 800°C, were used to measure bulk density and total porosity based on Archimedes' principle. The honeycombs were dried in an oven at 100°C and cooled to room temperature before testing. In accordance with Archimedes' principle, a sample was completely dried and then weighed . The weight was recorded as W_d. After weighing, the sample was placed in a beaker with the deionized (DI) water to allow the specimen to be soaked in the water for 24 hours. Within 24 hours, the specimen was assumedly saturated and then a vacuum was applied to force the water into the open pores of the honeycomb structures until no air bubbles emerged from the structure. After that, the beaker was placed on a stand over a balance. A density kit was hung below the water level but not touching any sides of the beaker in a position ready to determine the density of the sample. The sample was lifted onto the wire and the suspended weight was recorded as W_s. Finally, the sample was removed from the water, rapidly blotted, and weighed to attain the saturated weight, W_w. The bulk density of the sample, D_b, was calculated as

$$D_b = \frac{W_d \times \rho_w}{W_w - W_s} \quad \text{where } \rho_w = 1 \text{ g/cm}^3 \text{ at } 25°C$$

The total porosity of sample, P_t was evaluated as

$$P_t = \frac{(D_t - D_b)}{D_t}$$

where D_t is the theoretical density of HA/TiO_2 (50:50) composite (3.468 g/cm^3)

For mechanical properties testing, the specimen was prepared by extruding HA/TiO_2 composites into bars (6 mm wide x 5 mm long) and sintered at 650°C and 800°C. The 3-point bending strength of the sintered specimen was tested using a universal testing machine (Instron 55R4502, USA) with 10kN load cells. The crosshead speed was set at 0.5 mm/min and the span length to 20 mm. Five specimens were tested to obtain average values and standard deviations.

Investigation of photocatalytic activity

The photocatalytic activities of HA/TiO_2 honeycomb structures were investigated by measuring the degradation of methylene blue (MB) under UV irradiation. For comparison, MB solution without honeycomb sample was used as a blank. The evaluation was conducted

Sample	HA/TiO_2	DI water	MC	PEG	Na_2CO_3	Oil	Starch	Flour
S1	63.45	22.46	5.08	2.54	0.76	5.71	-	-
S2	59.95	21.58	9.59	0.36	0.24	5.28	2.99	-
S3	57.47	22.99	10.34	0.92	0.23	5.17	2.87	-
S4	55.07	25.88	8.81	0.66	0.22	4.96	-	4.40

Table 1: Formulations of extrusion paste studied in this work (%wt).

in a closed chamber equipped with UV-A lamps (315-380 nm). 250 mL of 5 ppm MB solution was added into a beaker, and the sample was put into the prepared MB solution. The beaker was placed into chamber and dark closed (no UV light) for 1 hour for initialization. Then, the MB solution was sampled out and measured for UV absorption (600-700 nm) on a UV spectrophotometer (JASCO-V530) and the absorption at peak (665 nm) was used to determine MB concentration at initial time (t_0). UV lamps in the chamber were then turned on to allow photocatalytic activity. After 1 hour, the lamps were turned off and MB concentration at the testing time of 1 hour (t_1) was determined. The same procedure was continued for other exposure times (1, 3, 6, and 24 hours) to determine the corresponding MB concentrations at t_1, t_3, t_6, and t_{24}.

Antibacterial Activity Test

Two types of bacteria, *E. coli* ATCC 8739, a Gram-negative bacterium, and *S. aureus* ATCC 6538, a Gram-positive bacterium, were chosen as model microorganisms in this study. Both bacteria, stored in glycerol at -80°C, were plated on nutrient agar, a mixture of nutrient broth (NB) and Bacto agar (Difco), cultured overnight at 37°C for 24 hours and stored at 4°C. Prior to each experiment, the strains were inoculated in sterile test tubes in NB and cultured at 37°C for 12 hours in order to reach the stationary growth phase. The cultures were harvested by centrifugation and resuspended in phosphate buffered saline (PBS, pH 7.4) to OD600 of 0.4, corresponding to an approximate concentration of 10^8 colony-forming units (cfu)/mL.

Honeycomb specimens were autoclaved and kept separate until use in a sterile beaker. The bacterial suspension (40, 60 and 100 mL) was added to the honeycomb samples sized 1, 2 and 5 cm, respectively and kept under 6-watts UV lamp illumination for 2, 4 and 6 hours with an approximate distance of 65 cm between the specimens and the lamp. *E. coli* suspension was removed at 2, 4 and 6 hours. Viable concentration of *E. coli* was enumerated using the spreading plate method on nutrient agar (NA) after a series of dilutions of the sample in normal saline. A 100 µl bacterial suspension was removed at 4-hour intervals and spread onto the nutrient agar medium and incubated for 24 hours to determine the number of viable cells. The efficacy of the antimicrobial activity is expressed as: survival ratio (%) = N/N_0 x 100, where N is number of survival bacteria and N_0 is the number of original bacteria. Experiments with bacterial cultures without honeycomb sample were conducted as controls. The mean and standard deviation from triplicate samples were indicated. In addition, a survival ratio curve for a 48-hour culture period was completed by combining two data sets in order to avoid working at night.

Results and Discussion

Effect of paste formulations on Hydroxyapatite/Titania honeycomb structures

In extrusion of HA/TiO_2 honeycomb structures, processing additives such as binders, plasticizers and lubricants were added to improve the plasticity and extrudability of the non-plastic HA/TiO_2 powders. By employing different paste formulations (Table 1), it was found that different pressure was required for extruding the samples due to the different textures in each formulation. The S3 and S4 formulations, which yielded complete shapes of honeycomb structure with sufficient green strength and good handling characteristics, were chosen for further investigation in the next step. Example images of final honeycomb structures of S3 samples after sintering at 800°C for 4 hours with different thickness are shown in Figure 1.

Characterization and mechanical properties

Figure 2 shows the XRD patterns of as-synthesized HA, TiO_2, and HA/TiO_2 honeycombs (as-prepared, and after sintering at 650, 800 and 1000°C). XRD patterns confirmed presence of all expected major hydroxyapatite (JCPDS No. 09-0432) and anatase TiO_2 (JCPDS No. 21-1272) crystallizations in the synthesized HA and TiO_2 used as raw materials, respectively.

The mixture of hydroxyapatite and anatase TiO_2 phases can be indexed in the as-prepared honeycombs and after sintering at 650°C. No additional phase associated with either decomposition of HA to β-tricalcium phosphate (β-TCP) or transformation of anatase to rutile TiO_2 was observed in the honeycomb sintered at this temperature. As expected, for HA/TiO_2 honeycomb sintered at 800°C, few small peaks of β-TCP due to the decomposition of HA into β-TCP as well as the peaks of the anatase phase of TiO_2 were observed. Generally, HA can decompose into β-TCP at a sintering temperature range of 800-1350°C [2,24,25]. With increasing sintering temperature, HA tended to completely decompose into β-TCP coupled with more transformation of anatase to rutile of TiO_2. The peaks related to $CaTiO_3$ were also observed due to the reaction between TiO_2 and Ca^{2+} in the hydroxyapatite site, particularly at a sintering temperature of 1000°C. As a photocatalyst, anatase TiO_2 exhibited higher photocatalytic activity than other phases of TiO_2 (rutile and brookite) [7-12]. The high intensity of peaks relating to the anatase TiO_2 observed in the prepared HA/TiO_2 composite powder implied superior crystalline size and high crystallinity. The optimal sintering temperature range of 600°C to 950°C for TiO_2 had been reported to avoid the transition of anatase to

Figure 1: Honeycomb structures of S3 of different thicknesses after sintering at 800°C for 4 hours.

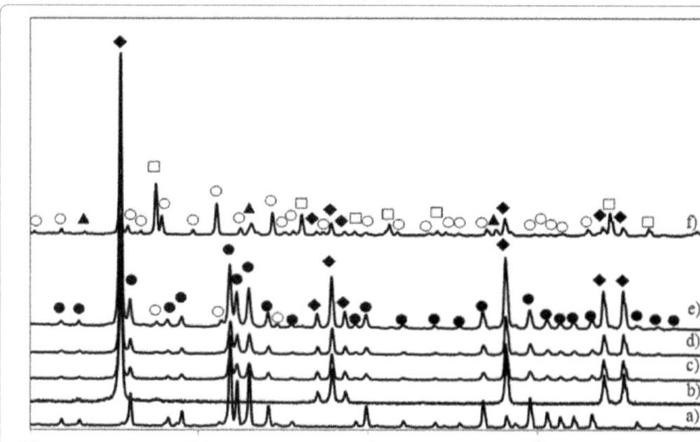

Figure 2: XRD patterns of a) as-synthesized HA, b) anatase, TiO_2, c) as-prepared HA/Ti composite, honeycomb sintering at d) 650°C, e) 800°C and f) 1000°C. • Hydroxyapatite, ♦ anatase TiO_2, □ rutile, TiO_2, o β-TCP and ▲ $CaTiO_3$.

rutile [26-28]. Therefore, the sintering temperatures 650°C and 800°C were selected for further investigation in this study.

The influence of sintering temperature on the morphology of HA/TiO_2 honeycombs was investigated. In this step, S3 and S4 formulations, which yielded complete shapes of honeycomb structure, were employed. Figure 3 illustrates the morphology of the honeycomb structure. Overall, irregular shapes of HA and TiO_2 crystals were observed in which small crystals of anatase TiO_2 were distributed and embedded in larger HA crystals with micropore sizes in the range of 0.23-0.28 μm. Similar morphologies were observed in honeycombs sintered at 650 and 800°C. Meanwhile, honeycomb samples sintered at 1000°C exhibited larger pore sizes (~0.40 μm) compared to those sintered at 650°C and 800°C. HA/TiO_2 grain size increased with sintering temperature, thus resulting in the decrease in porosity.

From the results shown earlier (Figure 2), honeycombs sintered at 1000°C exhibited transformation from anatase to rutile TiO_2 whereas those sintered at 650°C and 800°C did not. Therefore, the honeycombs sintered at 650°C and 800°C were chosen for further investigation on the properties of HA/TiO_2 honeycombs such as strength, photocatalytic activity, and anti-bacterial performance.

Table 2 presents the bulk density and total porosity of the S3 and S4 honeycomb structures of different sintering temperature and thickness. A small discrepancy in density (~30%) and porosity (~60%) was observed among S3 and S4 samples sintered at 650°C and 800°C. It seems that samples sintered at 800°C exhibited a slightly higher density, particularly in the thick samples (5 cm). It might be difficult to clearly determine the density and porosity of the honeycomb structure using Archimedes' principle, compared to SEM images, due to the limitations of the technique and the complexity of the honeycomb structure itself. Therefore, neither a trend of increasing nor decreasing for both density and porosity could be clearly identified with sintering temperature.

The bending strength of honeycomb structures sintered at 650°C and 800°C are reported in Table 3. The honeycombs sintered at 800°C possessed a greater bending strength than those sintered at 650°C due to higher densification and less porosity, in micro scale, as evidenced by the SEM images (Figure 3). This implies that the strength of the honeycomb could be improved by increasing the sintering temperature. At a higher sintering temperature (1000°C), however, the phase transformation from anatase to rutile occurred and could reduce the desired photocatalytic activity.

Photocatalytic activity

The photocatalytic activities of S3 and S4 HA/TiO_2 honeycombs of thickness 1, 2, and 5 cm sintered at 650°C and 800°C were evaluated using 5 ppm MB solution. According to the results, the

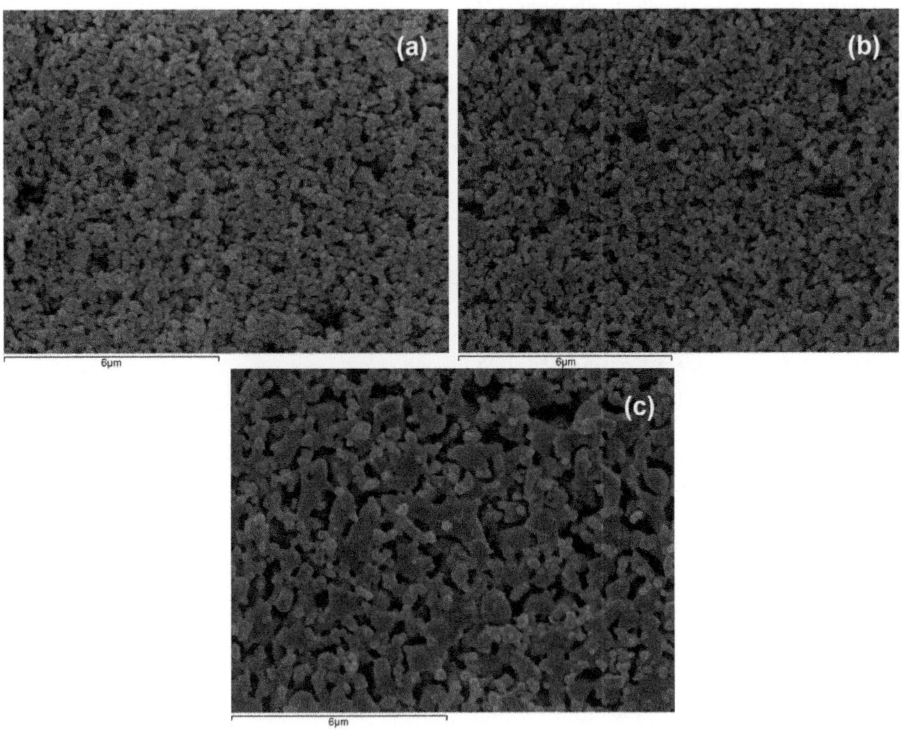

Figure 3: Morphology of honeycomb samples after sintering at a) 650°C, b) 800°C and c) 1000°C (heating rate of 50°C/h).

Formulation	Temperature (°C)	Density (%) ± SD. Thickness (cm)			Porosity (%) ± SD. Thickness (cm)		
		1	2	5	1	2	5
S3	800	33.97 ± 1.26	34.23 ± 1.34	36.21 ± 0.82	62.72 ± 1.38	62.43 ± 1.47	60.26 ± 0.90
S4	650	32.62 ± 0.78	35.30 ± 1.17	34.78 ± 1.34	64.20 ± 0.86	61.26 ± 2.24	61.83 ± 1.47
	800	33.44 ± 1.41	31.81 ± 0.90	34.89 ± 0.64	63.30 ± 1.55	65.09 ± 0.99	61.71 ± 0.71

Table 2: Density and porosity of honeycomb samples.

Formulation	Maximum Flexural Strength (MPa)	
	800°C / rate 50 °C/h	650°C / rate 50°C/h
S1	7.17 ± 1.79	-
S2	9.51 ± 1.00	-
S3	9.29 ± 1.01	4.84 ± 0.33
S4	6.91 ± 1.55	4.38 ± 0.28

Table 3: Bending strength of all samples.

MB concentration decreased with UV exposure time in the presence of HA/TiO_2 honeycombs and the rate of MB decomposition was rapid during the first 3 hours, whereas the blank (without HA/TiO_2 sample) showed no significant change in concentration (Figures 4a and 4b). Overall, HA/TiO_2 honeycombs sintered at 800°C tended to show slightly higher rates of MB decomposition than those sintered at 650°C. This may be related to the higher intensities of anatase TiO_2 peaks observed in the HA/TiO_2 honeycomb sintered at 800°C (Figure 2e). The difference, however, was not significant. The S3 and S4 honeycombs exhibited comparable rates of MB decomposition. At the same sintering temperature, HA/TiO_2 honeycombs of thickness 2 and 5 cm exhibited comparable rates of MB decomposition, but slightly greater than that of the 1-cm thick honeycomb ($p<0.07$). The lower initial MB concentration (at 0 hr) of 5-cm thick honeycomb, compared to those of the 1 and 2 cm thick samples, was thought to be due to its greater area for MB adsorption during initialization. After 24 hours of UV exposure, the MB solutions in the presence of the HA/TiO_2 honeycombs were effectively decomposed as the solution became clear.

Antibacterial activity

A photocatalytic decomposition reaction of our HA/TiO_2 has been previously reported to be applicable to deactivate microorganisms under weak UV light [20]. In this study, the *in vitro* antibacterial activities of the S3 and S4 HA/TiO_2 honeycombs of different thicknesses under room ambience with UV light were investigated. The S3 and S4 HA/TiO_2 honeycombs were prepared from two similar compositions of extrusion paste that differed in the additive selection (Table 1). Figures 5a and 5b, respectively, show the curves of survival ratio (%) versus culture time for the HA/TiO_2 honeycombs against gram-positive *S. aureus* and gram-negative *E. coli*. The smaller the survival ratio, the higher the antibacterial activity of the sample. The survival ratios of both bacterial strains decreased with incubation time for all tested HA/TiO_2 honeycombs. The S3 and S4 honeycombs exhibited comparable antimicrobial activity as their survival curves with similar sample lengths could almost be superimposed for both *S. aureus* and *E. coli*. This may be explained by both the starch and flour respectively selected for the S3 and S4 honeycombs as processing additives being generally neither bactericidal nor bacteriostatic agents. Moreover, their other compositions were almost quantitatively and qualitatively similar.

At the UV exposure time of 2 hours, the survival ratios of both bacterial strains found for the S3 and S4 honeycombs significantly decreased as the sample length increased ($p < 0.003$). The explanation for this is that the longer the sample, the more the HA/TiO_2 photocatalyst is incorporated into the sample and the faster its bactericidal activity is initiated. At longer exposure times (4 and 6 hours), only 5 - 15 % of bacteria remained in a proliferative state and the 1-cm thick honeycomb samples, containing the least amount of HA/TiO_2, still showed the highest survival ratios, as compared to the corresponding 2 and 5 cm thick samples ($p<0.003$ for *S. aureus* and $p<0.001$ for *E. coli*). However, the bactericidal activities of the 2 and 5 cm thick samples after 4 and 6 hours exposure became statistically comparable ($p<0.088$ for *S. aureus* and $p<0.700$ for *E. coli*).

Conclusion

The honeycomb structures of TiO_2-modified hydroxyapatite (HA/TiO_2) composite were successfully fabricated through the extrusion method. The complete honeycombs of 50 mm in diameter and cell density of 72 cell/inch^2 were extruded in three different thicknesses (1, 2 and 5 cm). Fabrication of honeycomb structure requires a slow and consistent drying process before sintering. The appropriate sintering temperatures were 650 or 800°C. At these temperatures, the crystallizations of hydroxyapatite and anatase TiO_2, which were required for better photocatalytic and antibacterial activities as well as stronger samples, were achieved.

HA/TiO_2 honeycombs sintered at 800°C exhibited slightly higher photocatalytic activity, compared to those sintered at 650°C, due to the greater amount of anatase TiO_2. The S3 and S4 honeycombs exhibited comparable rates in reducing methylene blue concentration. For the same sintering temperature, the 2- and 5-cm thick honeycombs showed greater photocatalytic activity than that of the 1-cm thick honeycomb.

Antibacterial activity testing results reported that 50% of bacteria were decomposed after 2 hours of testing. With increasing culture time, less numbers of bacteria could survive -around 5-10% of bacteria survived after 6 hours of culture time. The S3 honeycomb sample did not exhibit significantly higher decomposition of both gram-negative *E. coli* and gram-positive *S. aureus* compared to S4. The thickness of the honeycomb samples also influenced the antibacterial performance of the samples for both *E. coli* and *S. aureus*. It has been found that 2cm and 5cm honeycomb samples showed greater antibacterial activity than the 1cm-honeycomb through the culture period of time.

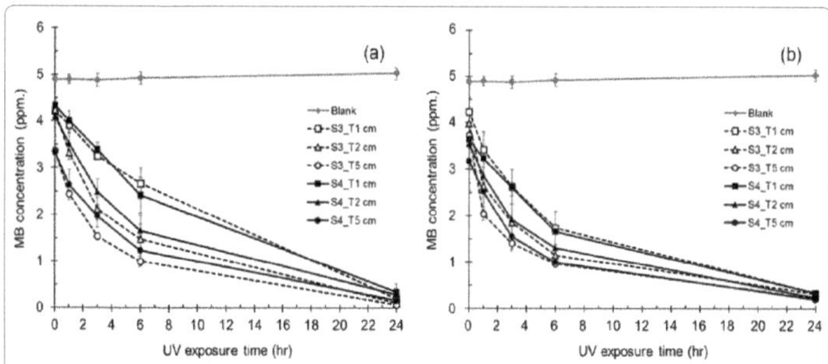

Figure 4: Methylene blue concentration after different UV exposure times in the presence of S3 and S4 HA/TiO_2 honeycombs with different thicknesses sintered at a) 650°C and b) 800°C

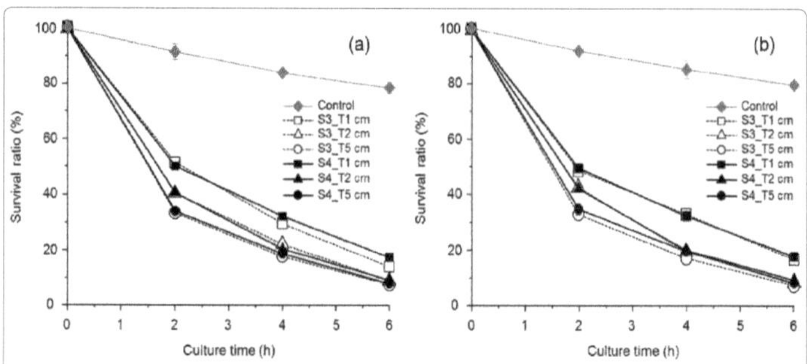

Figure 5: Antibacterial activities of the S3 and S4 HA/TiO_2 honeycombs with different thicknesses a) *E. coli* and b) *S. aureus*.

Acknowledgements

This work was performed with the support of the National Metal and Materials Technology Center, the Ministry of Science and Technology of Thailand, Project No. P-1200966.

Conflict of Interest Statement

There is no financial or other relationship that might be perceived as leading to a conflict of interest.

The authors declare that there are no conflicts of interest.

References

1. Hench LL (1991) Bioceramics: from concept to clinic. J Am Ceram Soc 74: 1487-1510.
2. Legeros RZ, Legeros JP (1993) Dense hydroxyapatite: An Introduction to Bioceramics. (1stedn), World Scientific, London.
3. Huang J, Best SM, Bonfield W, Buckland T (2010) Development and characterization of titanium-containing hydroxyapatite for medical applications. Acta Biomater 6: 241-249.
4. Wang CK, Ju CP, Lin JH (1998) Effect of doped bioactive glass on structure and properties of sintered hydroxyapatite. Mater Chem Phy 53:138-149.
5. Ramay HR, Zhang M (2003) Preparation of porous hydroxyapatite scaffolds by combination of the gel-casting and polymer sponge methods. Biomaterials 24: 3293-3302.
6. Saiz E, Gremillard L, Menendez G, Miranda P, Gryn K, et al. (2007) Preparation of porous hydroxyapatite scaffolds. Mat Sci and Eng C 27: 546-550.
7. Wakamura M (2005) Photocatalysis by calcium hydroxyapatite modified with Ti (IV). FUJITSU Sci Tech J 41: 181-190.
8. Fujishima A, Honda K (1971) Electrochemical evidence for the mechanism of the primary stage of photosynthesis. Bull Chem Soc Jpn 44: 1148-1150.
9. Fujishima A, Honda K (1972) Electrochemical photolysis of water at a semiconductor electrode. Nature 238: 37-38.
10. Ji S, Murakami S, Kamitakahara M, Ioku K (2009) Fabrication of titania/hydroxyapatite composite granules for photo-catalyst. Mater Res Bull 44: 768-774.
11. Anmin H, Ming L, Chengkang C, Dali M (2007) Preparation and characterization of a titanium-substituted hydroxyapatite photocatalyst. J Mol Catal A Chem 267: 79-85.
12. Ye FX, Ohmori A,Tsumura TT, Nakata K, Li CJ (2007) Microstructural analysis and photocatalytic activity of plasma-sprayed titania-hydroxyapatite coatings. J Therm Spray Tech 16: 776-782.
13. Lee MS, Hong SS, Mohseni M (2005) Synthesis of photocatalytic nanosized TiO_2-Ag particles with sol-gel method using reduction agent. J Mol Catal A Chem 242: 135-140.
14. Liu Y, Liu CY, Wei JH, Xiong R, Pan CX, et al. (2010) Enhanced adsorption and visible-light-included photocatalytic activity of hydroxyapatite modified Ag-TiO_2 powders. App Sur Sci 256: 6390-6394.
15. Nosaka Y, Matsushita M, Nishino J, Nosaka AY (2005) Nitrogen-doped titanium dioxide photocatalysts for visible response prepared by using organic compounds. Sci Tech Adv Mater 6: 143-148.
16. Nishikawa H, Kato S, Ando T (2005) Rapid and complete oxidation of acetaldehyde on TiO_2 photocatalytic filter supported by photo-induced activated hydroxyapatite. J Mol Catal A Chem 236: 145-148.
17. Kim HM, Chae WP, Chang KW, Chun S, Kim S, et al. (2010) Composite nanofiber mats consisting of hydroxyapatite and titania for biomedical applications. J Biomed Mater Res B Appl Biomater 94: 380-387.
18. Nayak S, Satpati B, Shukla KR, Dhawan A, Bhattacharjeea S, et al. (2010) A Facile synthesis of nanostructured hydroxyapatite–titania bio-implant scaffolds with different morphologies: their bioactivity and corrosion behavior. J Mater Chem 20: 4949-4954.
19. Jin X, Guo Y, Wang J, Wang Z, Gao J, et al (2011) The preparation of TiO_2/hydroxylapatite (TiO2/HA) composite and sonocatalytic damage to bovine serum albumin (BSA) under ultrasonic irradiation. J Molecular Catalysis A: Chemical 341: 89-96.
20. Monmaturapoj N, Thepsuwan W, Mai-Ngam K, Ngernpimai S, Klinsukhon W, et al (2014) Preparation and properties of hydroxyapatite/titania composite for microbial filtration application. Adv Appl Ceram 113: 267-274.
21. Suarez S, Yates M, Avila P, Blanco J (2005) New TiO_2 monolithic supports based on the improvement of the porosity. Catal Today 105: 499-506.
22. Ceramic Honeycomb Substrate
23. Ceramic honeycomb
24. Ou SF, Chioub SY, Ou KL (2013) Phase transformation on hydroxyapatite decomposition. Ceram Inter 39: 3809-3816.
25. Wang PE, Chaki TK (1993) Sintering behavior and mechanical properties of hydroxyapatite and dicalciumphosphate. J Mater Sci: Mater In Med 4: 150-158.
26. Suresh C, Biju V, Mukundan P, Warrier KGK (1998) Anatase to rutile transformation in sol-gel titania by modification of precursor. Polyhedron 17: 3131-3135.
27. Zhang Q, Gao L, Guo J (2000) Effects of calcination on the photocatalytic properties of nanosized TiO_2 powders prepared by $TiCl_4$ hydrolysis. Appl Catalysis B: Environmental 26: 207-215.
28. Hofer M, Penner D (2011) Thermally stable and photocatalytically active titania for ceramic surfaces. J Eur Ceram Soc 31: 2887-2896.

Art and Science: Thermodynamic Functions via Art

Abraham Tamir*

Department of Chemical Engineering, Ben-Gurion University of the Negev, Beer-Sheba, Israel

Abstract

The basis of thermodynamics is four laws that form an axiomatic basis. The laws are quantitatively defined on the basis of "Thermodynamic Functions" such as temperature, heat, work, entropy etc. that are derived from the thermodynamic definition of the specific function and demonstrated in this article by artworks.

Keywords: Thermodynamic functions; Heat; Work; Entropy; Temperature; Pressure; Internal energy; Free energy; Enthalpy; Entropy; Potential energy; Kinetic energy; Dead state

Thermodynamics, in general, is concerned with those physical and chemical phenomena that involve heat, work and other functions elaborated in the following. From the practical definition, thermodynamics is the theory of converting heat to work and understanding the role of energy and other properties of matter in this conversion process. The four laws are: The zero law that is the definition of thermodynamic equilibrium; the 1st law is the law of energy conservation; the 2nd law with its popular statement that there is no free lunch and the 3rd law stating that all processes cease as temperature approaches zero. Among the "thermodynamic functions" are "state functions" that are properties whose values do not depend on the path taken to reach that specific value but depend on the state of the system. For example, internal energy, enthalpy and entropy are *state quantities* because they describe quantitatively an equilibrium state of a thermodynamic system, irrespective of how the system arrived in that state. In contrast, mechanical work and heat are process quantities because their values depend on the specific transition (or path) between two equilibrium states. Figures 1-7 demonstrate the "state functions". Figure 1 demonstrates the "absolute temperature" T where Leonardo Da Vinci painted the original artwork entitled "The Vitruvian Man". Figure 2 demonstrates the "pressure" P. On the left is the artwork of Hanoch Piven, an Israel Illustrator and Caricature Artist that demonstrates a relatively high pressure. On the right is a demonstration of an atmospheric pressure. Figure 3 demonstrates by an artwork of Rene Magritte, Belgium Surrealist, the function "volume" V. Figure 4 demonstrates the "internal energy" U. The left artwork was painted by the English illustrator Walter Hodges demonstrating man's brain. It is well known that the internal energy consumption of the brain is very high although its mass constitutes only 2% of the entire body weight. On the right is an artwork by an unknown artist demonstrating the inside of human's body that is the storage of our internal energy. Figure 5 demonstrates the "Helmholz free energy" G in a system. Figure 6 is a combination of artworks that demonstrate the function "enthalpy" H that is the sum of the internal energy of the system plus the product of its volume multiplied by the pressure exerted on it by its surroundings. Figure 7 demonstrates the "entropy" state function S defined as the quantitative measure of disorder in a system. Left bottom is the artwork of the French artist Ives Klein that demonstrates minimal entropy. Top left and right are paintings of the Dutch artist Piet Mondrian where on right bottom are a painting of the Italian artist Gino Severini that demonstrates a highest disorder, namely maximal entropy. And now we discuss Figures 8-12 that present "non-state functions" because their values depend on the specific path between two equilibrium states. Figure 8 demonstrates the "gravitational potential energy" P.E. On the left is an artwork of the German artist Quint Buchholz where on the right is an artwork of Magritte. Figure 9 demonstrates the "kinetic energy" K.E. On the left K.E=0 where on the right, by Magritte, K.E>0. Figure 10 demonstrates the "heat" Q. The two paintings on the left for Q>0 and Q=0 were painted by De Es Schwertberger an Austrian artist. On the right is a surrealistic painting by Magritte. Figure 11 by Magritte demonstrates the "work" function W. On the right is the original painting by Magritte for which W>0 where on the left W=0. And finally the function "dead state" that is the state when a system has zero availability is demonstrated in Figure 12 painted by the Italian painter Enzo Cucchi. To conclude the article the authors believe that the artistic demonstrations of the thermodynamic functions make them clearer and more understandable. And finally it should be emphasized that all information in the article as well as artworks is based on Google as source reference.

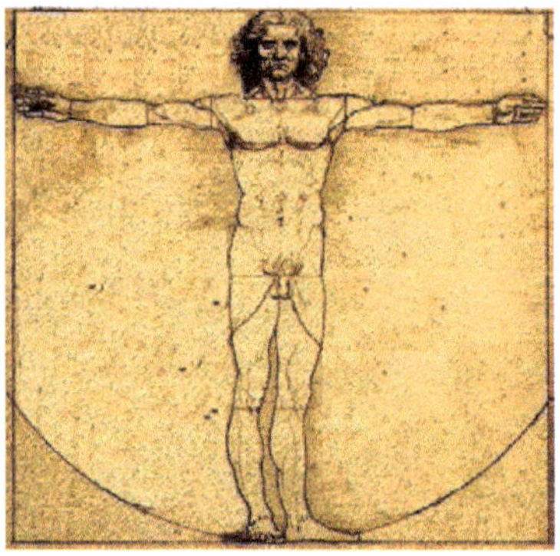

Figure 1: Absolute temperature.

***Corresponding author:** Abraham Tamir, Department of Chemical Engineering, Ben-Gurion University of the Negev, Beer-Sheba, Israel

Figure 2: Pressure.

Figure 3: Artwork of Rene Magritte.

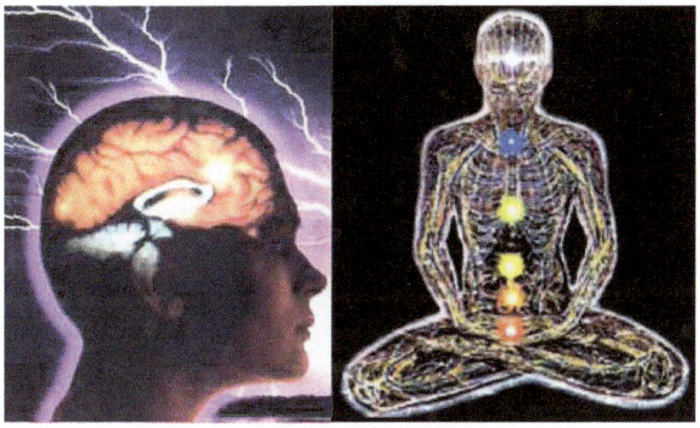

Figure 4: Internal energy.

Figure 5: Helmholz free energy.

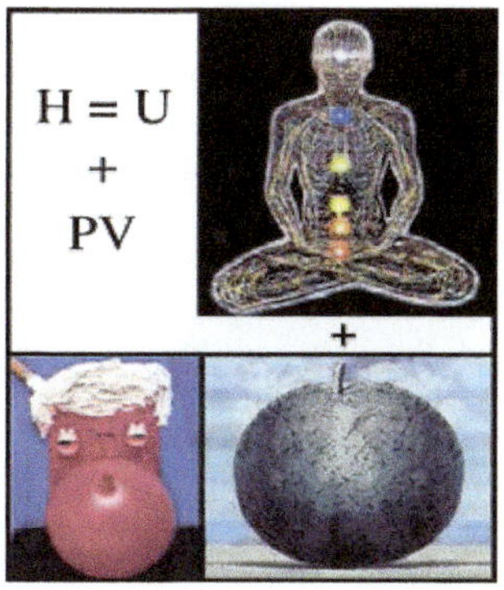

Figure 6: Enthalpy.

Figure 7: Entropy.

Figure 8: Gravitational potential energy.

Figure 9: Kinetic energy.

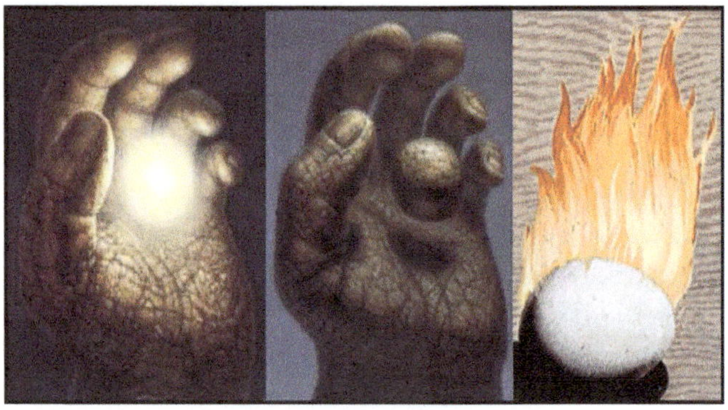

Figure 10: Heat.

Figure 12: Zero availability.

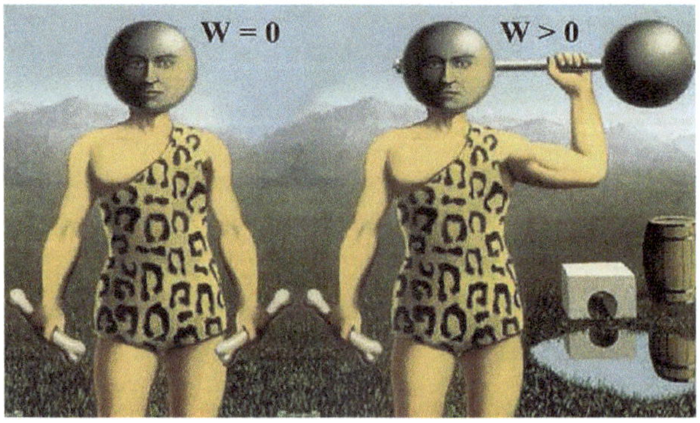

Figure 11: Work function.

Optimization of ABS Latex Coagulation using Taguchi Method and Mathematical Modeling for Percent Particles Size Distribution

Karimi L, Nouroozi S* and Rasoulifard MH

Department of Chemistry, Faculty of Science, University of Zanjan, Zanjan, Iran

Abstract

In this paper, the optimization of particles size distribution (PSD) in the coagulation processes of Acrylonitrile-Butadiene- Styrene (ABS) latex was studied using H2SO4 as coagulant agent. The effects of parameters such as; temperature, agitation speed, coagulant agent concentration, the ratio of latex to coagulant agent and residence time on coagulation process were investigated and optimized using Taguchi experimental design. Based on Taguchi method, sixteen experiments were designed and performed to get optimum conditions of effective factors. The optimum conditions for the selected factors were obtained as: temperature (95°C); coagulant concentration (2% wt); residence time (10 min); agitation speed (180 rpm) and latex to coagulant volume ratio (1:1). According to the results, latex to coagulant volume ratio was the main factor having significant effects on the particles size distribution of the ABS. In the optimal conditions, coagulated particles with narrow in particles size distribution (220-280 μm) were prepared. An equation with 5 main parameters for predicting particles size distribution in the range of 220-280 μm with good compatibility and correlations between its results and experimental data was derived.

Keywords: Acrylonitrile–Butadiene–Styrene(ABS); Coagulation; Taguchi; Mathematical modeling.

Introduction

Coagulation is one of the important processes in the polymer production. Polymer latex made from emulsion polymerization often needs to be converted into dry powder by coagulation. Graft copolymer comprising an elastomeric base polymer such as poly butadiene and grafted chains of a thermoplastic, non-elastomeric polymer such as styrene-acrylonitrile are coagulated for isolation from aqueous emulsion. The ABS polymer which is based on three monomers; acrylonitrile (A), Butadiene (B) and Styrene (S); represent the industrially most important thermoplastic two-phase systems with an amorphous structure [1,2].

ABS latex coagulation process is a post processing stage in the manufacturing of ABS resins where the particles are recovered from colloidal latex prepared by emulsion polymerization. Since the surfaces of the colloidal particles in the latex are negatively charged by adding emulsifiers in the polymerization step, the latex is a stable colloidal solution by electrostatic repulsion between particles. The particles begin to coagulate into bigger flocs in the coagulator where the repulsive forces are weakened by addition of a coagulant such as sulfuric acid [3].

Basically, coagulation and flocculation can be done by double layer compression, charge neutralization, bridging and colloid entrapment [4]. Coagulation will occur if the kinetic energy of the particles is sufficiently high to overcome the potential energy barrier. Energy barrier is the sum of the Vander Walls attraction energy and the electrostatic repulsion energy [5]. A general rule in the coagulation process is that the Zeta potential of the particles which is a good approximation of surface potential is less than 14 mV [6]. Coagulation is the process of encounters between two or more particles by transport mechanisms, such as Brownian motion, field shear, etc., and coalescence of these particles due to short range of interfacial forces [7]. The coagulation process thus modifies the size distribution of the suspended particles in dispersive systems and controls the behavior of suspensions. The particle-particle interaction leading to any such coagulation is one of the major factors determining the design parameters in latex processing operation.

Several techniques such as chemical, evaporative, shear and freeze coagulation have been developed to destabilize polymeric emulsions and cause the desired coagulation [8,9]. Rapid and controlled coagulation of latex, induced by chemical coagulant, mechanical shear and heat, has been found to be the most effective process for the recovery of polymer from the latex. Chemical coagulation is the most common technique for coagulation, and is done by adding dilute salt or acid solution such as NaCl , $MgSO_4$, $CaCl_2$, $Al_2(SO_4)_3$ and H_2SO_4 to the colloidal solutions of polymers [10].

A major concern in the ABS latex processing is the excessive content of fine particles in the slurry coming out of the coagulator tank. The fine particles may be lost in the downstream recovering process or may lead to troubles in resin powders recovering. Thus, major efforts for process improvement have been directed to the reduction of fine particles in the coagulated slurry. The mean particles size in latex is in the range of 1-500 nm and after the coagulation process, the solid state have a mean particles size in the range of about 100-500 μm. Powder structure, particle size and particle size distribution (PSD) can be affected some parameters such as coagulant concentration, the ratio of latex to coagulant, agitation speed, residence time and temperature [2-4,10,11].

In this paper, optimization of coagulation process in the emulsion preparation of ABS latex using sulfuric acid as coagulant agent was investigated and experimental conditions to obtain a narrow particle diameter distribution with good morphology are studied based on

***Corresponding author:** Nouroozi S, Department of Chemistry, Faculty of Science, University of Zanjan, P.O. Box 45371-38791, Zanjan, Iran

Taguchi experimental design [12]. Because of industrial problems such as filtering, color stability and finishing process, the mean particles size in the range of 220-280 µm are desired and was selected in this work.

Experimental

Apparatus and reagents

The equipment's used in coagulation process are shown in Figure 1. The experiments were conducted in a 500 mL beaker that equipped with a mixer by a blade shaft and Thermostat (Lauda RE 104) for temperature controlling. ABS latex used in this experiment was prepared from Tabriz Petrochemical Company. The coagulant was aqueous solution of H_2SO_4 (98%, Merck, Germany). NaOH (98%, Fluka, Switzerland) was used to adjust pH of the solutions. Particles size distribution (PSD) and morphology of the ABS resin were studied by images resulted from Scanning Electron Microscopy (SEM, VEGA TESCAN, HV 20.0 KV). Ventilation oven (Memmert, UFB, 230V) was used for drying the products.

Recommended procedure

According to the set up in Figure 2, experiments were initiated by mixing the H_2SO_4 solution and latex in 75°C for 10 seconds at 600 rpm to form uniform dispersion of the coagulant and latex. Coagulation then continued with mild stirring by adjustment the pH of solution in the range of 3-4 by addition of NaOH (10% wt) [3]. The final slurry is neutralized and treated at 80-95°C to obtain the solidified particles. At the end, stirring was stopped and coagulated particles were collected by filtration through a 200 µm mesh size filter and dried in a ventilation oven at 30°C for about 24 hours. Images of the prepared ABS particles were obtained by SEM and measurement-Nano software was used for determination of particles size distribution.

Results and Discussion

Design of the experiment by Taguchi method

According to Taguchi experimental design, we can assume independent behavior of all effecting input factors on the process. On the other hand, the effect of an independent variable on the performance parameter does not depend on the different levels of other independent variables [13].

According to the previous works on ABS latex coagulation process [2-4,7] and also many preliminary experiments, the main operational factors and their levels were selected which are shown in Table 1. Residence time, temperature, agitation speed, concentration of coagulant and latex to coagulant volume ratio were selected as factors which have more effect in the coagulation process of ABS latex. Normally, in the case of five factors with four levels it is necessary to take 1024 (4^5) experiments in the optimization process. By Taguchi method, the standard orthogonal array, namely L_{16} reduces the number of experiments to 16. The designed L_{16} is an array of 16 experiments with the specified combination of the levels of factors. Factors are in columns and each row determines the levels of factors which must be used to run every individual experiment. The structure of Taguchi's orthogonal robust design and the results of measurements are shown in Table 2. The last column in this table shows calculated signal to noise ratio (S/N) from the measured results. The S/N ratios are different according to the type of characteristic and are classified into three groups [13].

Type N (nominal is the best):

$$\eta = S/N_N = 10\log\frac{\left(\frac{1}{n}\right)(((\sum yi)^{\wedge}2/n)-(\sum yi^2-((\sum yi)^{\wedge}2/n)/n-1))}{(\sum yi^2-((\sum yi)^{\wedge}2/n)N-1} \quad (1)$$

Type S (the smallest is the best):

$$\eta = S/N_s = -10\log 1/n\,(\sum yi\wedge 2) \quad (2)$$

Type B (the biggest is the best):

$$\eta = S/N_B = -10\log\left(\frac{1}{n}\sum\frac{1}{yi^{\wedge}2}\right) \quad (3)$$

Where y_i is the characteristic property (PSD% in this study) and n is the number of measurements in each experiment. All S/N ratios in Table 2 were calculated according to the equation 3, the biggest is the best, in order to maximize the PSD in the range of 220-280 µm.

The average S/N ratios for each level of the parameters are summarized and are shown in Table 3. The average of S/N was calculated by comparing the average responses (η) obtained at different levels (j) for each factor. For example, average response of B at level 1 is obtained from the results of experiments 1, 5, 9 and 13 (Table 2).

$$M_{B1} = \frac{\eta_1+\eta_5+\eta_9+\eta_{13}}{4} = 26.74 \quad (4)$$

Similar calculations led to: M_{B2} = 27.57, M_{B3} = 28.34 and M_{B4} = 30.53. Figure 3 shows the effects of each factor on PSD based on the average S/N. From these results, the optimum levels for each factors can be obtained as: A_3, B_4, C_1, D_2 and E_4 (i.e. A at level 3, B at level 4, C at level 1, D at level 2 and E at level 4).

The results for term L_{max}-L_{min} are shown in the last column of Table 3 which resembles the highest value for factor E. According to this result the volume ratio of latex to coagulant is the most important factor in the coagulation process.

Analysis of variance

The purpose of Analysis of variance (ANOVA) is to investigate which factor significatly affect the particles size distribution. This is accomplished by separating the total variability of the S/N ratio, which is measured by the sum of the squared deviations (SS) from the total mean of the S/N ratio (S_m), into contributions by each factor and the error. Equations for ANOVA calculations are as follows [13,14].

Total degree of freedom

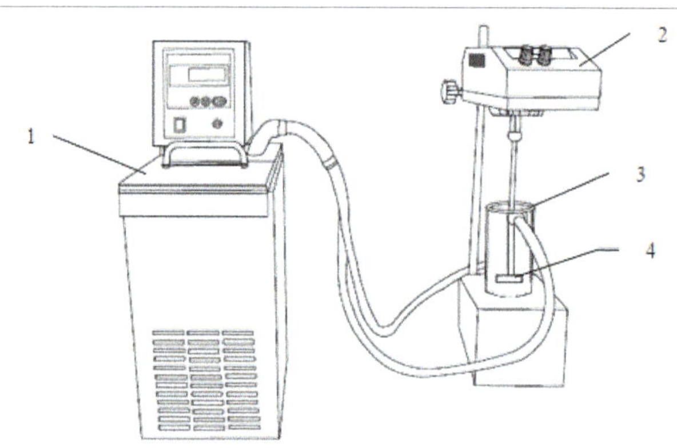

Figure 1: Experimental set up. 1) Thermostat 2) Variable speed motor 3) Agitated tank 4) impeller.

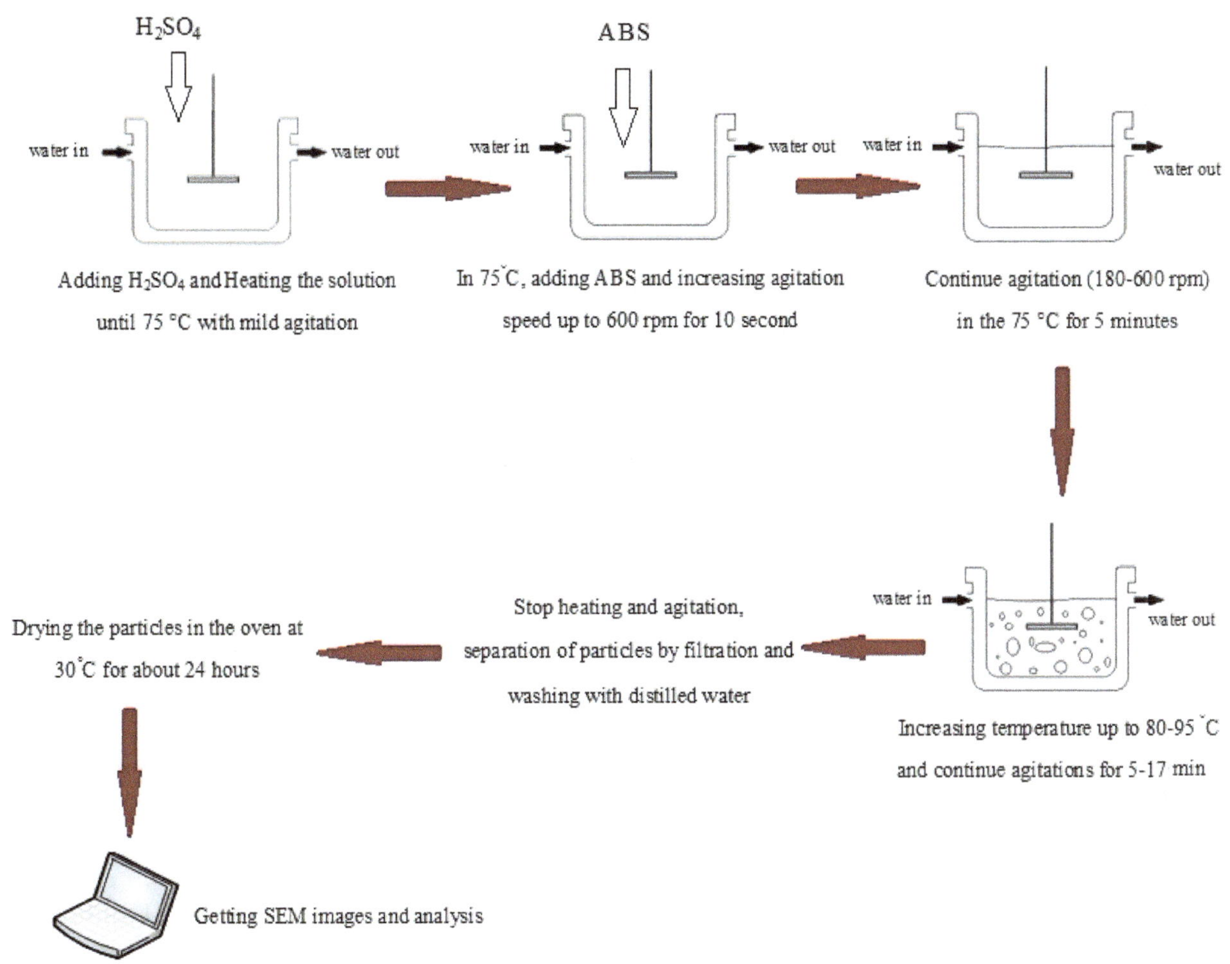

Figure 2: Sequential steps of coagulation process.

Parameter	Level1	Level 2	Level3	level4
A. Time (min)	5	7.5	10	17.5
B. Temperature (°C)	80	85	90	95
C. Agitation speed (rpm)	180	350	450	600
D. Coagulant concentration agent (wt %)	1	2	3	4
E. latex to coagulant Volume ratio	3:1	2:1	3:2	1:1

Table 1: Factors and their values corresponding to the levels in the experiments.

$$f_T = N - 1 = 16 - 1 = 15 \quad (5)$$

Where, N is the total number of result.

Degree of freedom for a factor

$$f_A = K_A - 1 = 4 - 1 = 3 \quad (6)$$

Where, K_A is the number of level for factor A and f_A is the degree of freedom of factor A.

Degree of freedom for error:

$$f_e = f_T - f_A - f_B - f_C - f_D - f_E \quad (7)$$

$$S_m = \text{Correction factor} = \frac{(\Sigma_{i=1}^{16} S/Ni)^\wedge 2}{16} = (452.72)^\wedge 2 \quad (8)$$

$$SS_i = \text{Sum of squares for factor I} = \frac{(4\times mi1)^2 + (4\times mi2)^2 + (4\times mi3)^\wedge 2 + (4\times mi4)^\wedge 2}{4} - S_m \quad (9)$$

Where, m_{i1}, m_{i2}, m_{i3} and m_{i4} are the average S/N ratio for level 1, 2, 3 and 4 of factor I, respectively.

$$S_T = \text{Total sum of squares} = \Sigma_{j=1}^{16}(\frac{S}{Ni})^\wedge 2 - S_m = 166.53 \ (DF = 15) \quad (10)$$

$$MS_j = \text{Mean sum of squares for factor i} = \frac{ssi}{DF \text{ of } i} \quad (11)$$

$$S_e = \text{Error sum of squares} = S_T - \Sigma SSi \quad (12)$$

$$F_i = \text{Variance ratio for factor i} = \frac{MSi}{MS \text{ of error}} \quad (13)$$

F_A, F_B, F_C, F_D, F_E and F_e is indeterminate since V_e=0

$$P\ (\%) = \text{the percentage contribution for factor I} = \frac{SSi - (DF \text{ of error})}{ST} \times 100 \quad (14)$$

The results for ANOVA calculations are shown in Table 4. According to the results, latex to coagulant volume ratio has the most effect on the coagulation processes. Temperature and coagulant concentration has the next largest effect on the coagulation process, respectively.

However, "error" could not be calculated directly (degree of freedom for error was obtained zero), therefore the "Pooling technique"

Parameter Exp. No.	A	B	C	D	E	Res ults , PSD% (220-280 µm)			S/N
						Run1	Run 2	Avrage of results	
1	5	80	180	1	3:1	21.81%	18.46%	20.13%	25.99
2	5	85	350	2	2:1	22.53%	26.08%	24.30%	27.64
3	5	90	450	3	3:2	24.52%	30.76%	27.64%	28.66
4	5	95	600	4	1:1	40.47%	33.33%	36.90%	31.22
5	7.5	80	350	3	1:1	31.66%	25.80%	28.73%	29.03
6	7.5	85	180	4	3:2	24.24%	25.81%	25.02%	27.95
7	7.5	90	600	1	2:1	23.28%	23.33%	23.30%	27.35
8	7.5	95	450	2	3:1	29.72%	24.24%	26.98%	28.49
9	10	80	450	4	2:1	16.77%	16.29%	16.53%	24.36
10	10	85	600	3	3:1	17.00%	13.13%	15.06%	23.34
11	10	90	180	2	1:1	60.31%	51.51%	55.91%	34.87
12	10	95	350	1	3:2	42.31%	48.48%	45.39%	33.08
13	17.5	80	600	2	3:2	22.22%	26.00%	24.11%	27.56
14	17.5	85	450	1	1:1	36.84%	36.95%	36.89%	31.34
15	17.5	90	350	4	3:1	11.62%	16.21%	13.91%	22.51
16	17.5	95	180	3	2:1	31.82%	27.27%	29.54%	29.33

Table 2: Experimental layout us ing the L16 orthogonal array and experimental res ults for percentage particles s ize dis tribution and S/N ratio.

Factor	Level 1	Level2	Level 3	level 4	Lmax-Lmin
A. Time (min)	28.38	28.20	**28.91**	27.69	1.22
B. Temperature (°C)	26.74	27.57	28.34	**30.53**	3.79
C. Agitation s peed (rpm)	**29.54**	28.07	28.21	27.36	2.17
D.Coagulant concentration (wt%)	29.44	**29.64**	27.59	26.51	3.13
E. Latex to coagulant volume ratio	25.08	27.17	29.31	**31.61**	6.53

Table 3: Average of the S/N ratios for each level of the factors.

is used to estimate the approximate error. By pooling, residence time is excluded as a factor that has minimum effect on the PSD. The results of pooled calculations are shown in Table 5.

Prediction and confirmation of the optimum conditions

The final step in the Taguchi method is predicting and verifying the improvement of the quality characteristic using the optimal levels of the factors. The predicted S/N ratios using the optimal levels of the factors can be calculated as follows [12,13]; predicted $S/N = S/N_m + \{(S/N_1 - S/N_m) + (S/N_2 - S/N_m) + (S/N_3 - S/N_m) + (S/N_4 - S/N_m) + (S/N_5 - S/N_m)\}$ Where S/N_m is the total mean S/N ratio and S/N_1, S/N_2, S/N_3, S/N_4, S/N_5 is the mean S/N ratio at the optimal level. For PSD in the desired range, the calculated value of S/N_m from Table 2 is 28.30. Also, S/N ratio for A3, B4, C1, D2 and E4 can be obtained from Table 3 and the values are 28.91, 30.53, 29.54, 29.64 and 31.61, respectively. Using these values, the above equation can be written as predicted S/N=28.30+ {(28.91-28.30) + (30.53-28.30) + (29.54-28.30) + (29.64-28.30) + (31.61-28.30)}. Therefore, the predicted S/N ratio (37.095) for desired PSD can then be obtained and the corresponding estimated PSD can also be calculated by the following equation; $37.05 = -10 \log \frac{1}{yi^{\wedge}2}$. Table 6 shows the comparison of the predicted PSD and the experimental results using the optimum conditions. The results show good agreement between the predicted and experimental PSD. The SEM images of the obtained particles under the optimum conditions are shown in Figure 4 in which the sizes of the particles are in the desired ranges (220-280 µm).

Effect of pH

In order to investigate the effect of pH, many experiments were done in the different levels of pH. Best results were obtained in the pH levels between 3 and 4 in the coagulation process. In the high pH value, the latex could not coagulate completely because Zeta potential and electrostatic repulsion between particles were high [6]. In low pH value, rapid coagulation will take place and the coagulated particles take an irregular shape.

Effect of coagulant concentration

The concentration of coagulant agent has more effect on the coagulation process. In the high concentration of coagulant agent, pH is less than 2, thus particles exceed the energy barrier. In this case rapid coagulation occurs and consequently particles have poor morphology with wide diameter distribution and so higher amount of fine particles were produced. On the other hand, when the coagulant concentration is low, coagulated particles are small in size because electrostatic repulsion between particles is high. In the coagulant concentration slightly less than the critical coagulant concentration (CCC), the rate of coagulation is decreased and mixing rate can be used to control aggregated size and shape [15], but in these experiments it is found that high concentration of coagulant agent cause quick coagulation process and so there is not special effect from the mixing rate. According to the result which was shown in Figure 3A and Table 3, the optimum coagulant concentration was obtained as 2 wt%.

Effect of temperature

Temperature of the solution is the important parameter in the coagulation process. Coagulated particles with the required diameter could be obtained at a suitable temperature and it is usually 10°C below the glass transition temperature, Tg [3]. Glass transition temperature of ABS is about 102°C [8]. It was found that the average diameter of particles was increased with increasing the operating temperature. More

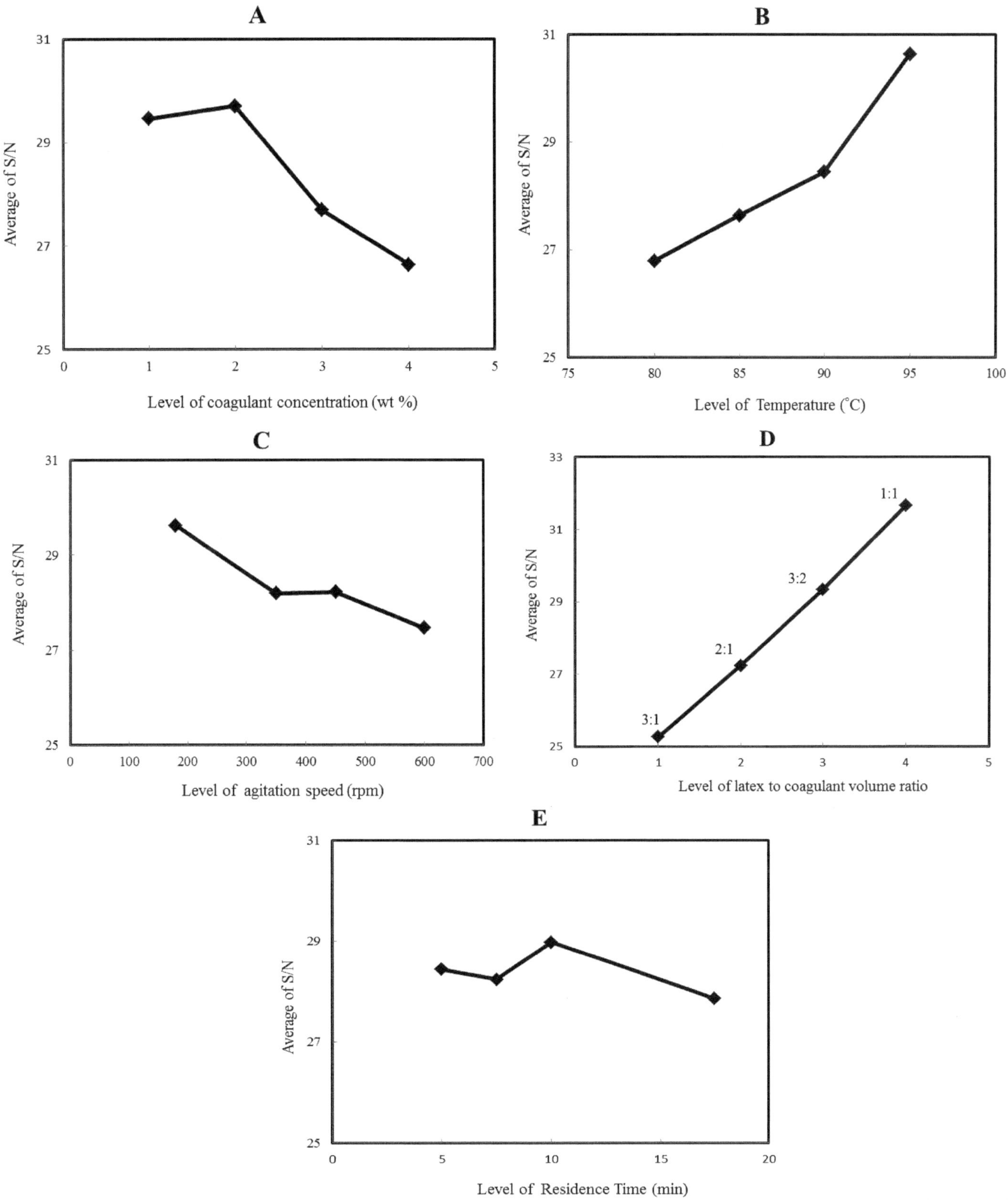

Figure 3: Average S/N ratio for different levels of the factors: (A) coagulant concentration, (B) temperature, (C) agitation speed, (D) latex to coagulant volume ratio and (E) residence time.

Factor	DF	SS	MS	F -Ratio	Pure s um S"	P%
A. Time(min)	3	3.07	1.02	----	3.70	1.84
B. Temperature(°C)	3	31.79	10.60	----	31.79	19.10
C. Agitation s peed (rpm)	3	9.83	3.28	----	9.83	5.91
D. Coagulant concentration (%wt)	3	27.16	9.05	----	27.16	16.32
E. Latex to coagulant volume ratio	3	94.55	31.52	----	94.55	56.82
Total	15	166.40				100%

Table 4: ANOVA res ults for S/N ratio of the PSD.

Factor	DF	SS	MS	F -Ratio	Pure s um S"	P%
A. Time(min)	(3)	(3.07)	1.02	Pooled	-----	-----
B. Temperature(°C)	3	31.79	10.60	10.36	28.72	17.26
C. Agitation s peed (rpm)	3	9.83	3.28	3.20	6.76	4.06
D. Coagulant concentration (%wt)	3	27.16	9.05	8.85	24.10	14.48
E. Volume ratio of latex to coagulant	3	94.55	31.52	30.80	91.48	54.97
Error	3	3.07	1.02			9.23
Total	15	166.40			100%	

Table 5: Pooled ANOVA res ults for S/N ratio of the PSD.

	Levels	PSD (%)	S/N ratio
Prediction	A3,B4,C1,D2,E4	71.25	37.05
Confirmation	A3,B4,C1,D2,E4	70.88	37.01

Table 6: Prediction and confirmation res ults for PSD.

Figure 4: SEM images in the optimum conditions (temperature, 95 °C; coagulant concentration, 2% wt; residence time, 10 min; agitation speed, 180 rpm and latex to coagulant volume ratio, 1:1).

flocs appeared with increasing the temperature and thus the resulted particles were larger. When interfacial stress increases and Brownian motion speeds up with increasing temperature, latex particles are then easy to form with an irregular shape and loose structure [7]. From the result which was shown in Figure 3B and Table 3, the optimum temperature for coagulation of ABS was obtained as 95°C.

Effect of agitation speed

High agitation speed provides more favorable conditions for breaking up the particles than for coagulation. The size of particles is governed by the equilibrium of the attractive forces among the latex particles and the shear stress provided by the blades. The average diameter of the particles becomes smaller with increasing the agitation speed. According to results from Figure 3C and Table 3 the optimum agitation speed was obtained as 180 rpm

Effect of latex to coagulant volume ratio

The volume ratio of latex to coagulant agent has the most important role in the coagulation process. Reduction of this ratio leads to high PSD in the desired size. Increase of the slurry content in the latex leads to the formation of more particles. In this condition, polymer content increases with low energy dissipation but increase the content of fine and smaller coagulated particles with a wide distribution in size. From the results which are shown in Figure 3D and Table 3, the 1:1 ratio was obtained as the optimum latex to coagulant volume ratio.

Effect of residence time

Distribution of the particles diameter is affected by residence time because the attractive forces between the coagulated particles are weak and the particles are easily broken. So, the diameter of particles becomes larger with increasing the residence time. According to results from Figure 3E and also Table 3 the optimum residence time was obtained as 10 min.

Mathematical Model

Driving an equation according to the results of the optimized parameters could be useful in prediction of PSD%. From the results of 16 experiments and Data fit software equation 15 was derived as a result of mathematical modeling.

$$Y = \exp\left[a * X1 + b * X2 + c * X3 + d * X4 + e * X5 + f\right] \quad (15)$$

The results related to modeling calculations are shown in Table 7 and Figure 5. Variables and their values for derived equation are also shown in Table 8. This model could be used to predict percent particles size distribution in the desired size range (220-280 μm) as a function

	PSD%	Agitation s peed (rpm) (X5)	Volume ratio (latex/coagulant) (X4)	Coagulant Conc. (%wt) (X3)	Temperature (°C) (X2)	Time (min) (X1)
Run 1	20.13	180	3:1	1	80	5
Run2	24.30	350	2:1	2	85	5
Run3	27.64	450	3:2	3	90	5
Run4	36.90	600	1:1	4	95	5
Run5	28.73	350	1:1	3	80	7.5
Run6	25.02	180	3:2	4	85	7.5
Run7	23.30	600	2:1	1	90	7.5
Run8	26.98	450	3:1	2	95	7.5
Run9	16.53	450	2:1	4	80	10
Run10	15.06	600	3:1	3	85	10
Run11	55.91	180	1:1	2	90	10
Run12	45.39	350	3:2	1	95	10
Run13	24.11	600	3:2	2	80	17.5
Run14	36.89	450	1:1	1	85	17.5
Run15	13.91	350	3:1	4	90	17.5
Run16	29.54	180	2:1	3	95	17.5

Table 7: Data fit res ults for percentage particles s ize dis tribution in different levels of the factors.

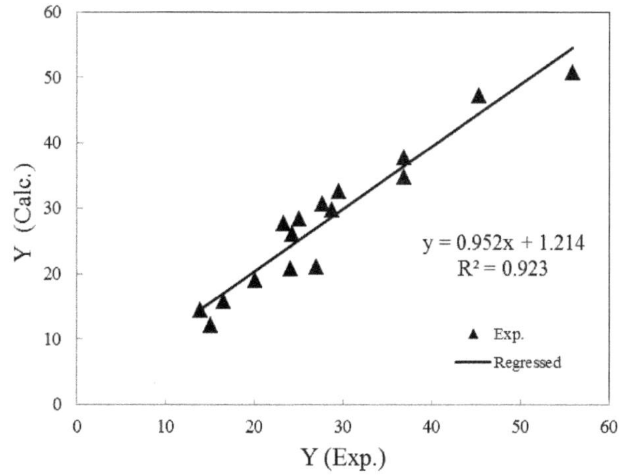

Figure 5: Correlation between the experimental and calculated (according to the proposed mathematical modeling) results.

Variable	Value
a	-5.8338577032605E-03
b	3.02411611742547E-02
c	-0.114074699106494E-04
d	-0.407078317729991
e	-7.51809709863315E-04
f	2.02365206866783

Table 8: Values for the variables in the propos ed mathematical equation.

of the factors according to Table 1. Prediction results obtained by this equation have 92% matches by the experimental results.

Conclusion

Coagulation process of ABS was performed using H_2SO_4 as chemical coagulant agent and Taguchi method wad used for optimizing the effective factors to obtain desired characteristics for particles sizes. The effects of various factors on the particles size distribution were studied and optimized conditions were obtained. According to the result, latex to coagulant volume ratio was the main factor having significant effects on the particles size distribution of the ABS. In the optimal conditions, coagulated particles with narrow in particle size distribution (220-280 µm) were prepared. An equation with 5 main parameters for predicting particles size distribution in the range of 220-280 µm with good compatibility and correlations between its results and experimental data was derived.

Acknowledgements

The authors would like to express their gratitude to the University of Zanjan and Tabriz Petrochemical Company, for financial support and assistance.

Abbreviations

Particles Size Distribution (PSD)

Acrylonitrile-Butadiene-Styrene (ABS)

Scanning Electron Microscopy (SEM)

Critical Coagulant Concentration (CCC)

Glass Transition temperature (Tg)

References

1. Alder RJ, Gardner N (1997) Ind Eng Chem 36: 2156-2162.
2. Kobe HY, Himeji YM, Kobe WO (1998) U.S. patent 4:767-803.
3. Ullmann's, "Encyclopedia of industrial chemistry", 6th edition, Chap. 5, Willy-VCH, 2002.
4. Chang CB, Park SH, Han IS (1998) AIChE J 44: 1256-1265.
5. Kemmere MF, Meulijk J, Drinkenburg AAH, German AL (1998) Aspects of coagulation during emulsion polymerization of styrene and vinyl acetate. J Appl Polym Sci 69: 2409-2421.
6. Ravina L, Zeta-Meter (1993) Inc1-44.
7. Li Y, Xu G, Chen M, Wang K (1997) IChemE 75: 81-86.

8. Ghosh AK, Linpt JT (1990) Processing of polymer latex emulsion: Batch coagulation. J Appl Polym Sci 39: 1553-1568.

9. Sadhan CJ, Iren D, Gary SB (2000) U.S. patent 6156,875.

10. Henton DE, Midland M (1989) U.S. patent 4831,116.

11. Zhang D, Luam M, Lin Q , Gu Q, cui Z, et al. (2007) B Polym Int 56: 195-199.

12. Kamyabi GA, Sheikh AM (2010) J Iron Steel Res Int 17: 45-52.

13. Kamyabi-gol A, Zebarjad SM, Sajjadi SA (2009) Colloid Surface A 36: 69-74.

14. Tang SH, Tan YJ, Sapuan SM, Sulaiman S, Ismail N,et al. (2007) J Mat Proces Tech 182: 418-426.

15. Edward Kostansek-Rohm and Hass Company (2004) JCT research 1:41- 43.

Treatment of the Olive Mill Industry Wastewater with Ultrasound and Some Nano-Sized Metal Oxides

Delia Teresa Sponza* and Rukiye Oztekin

Department of Environmental Engineering, Engineering Faculty, Dokuz Eylül University, Tınaztepe Campus, 35160, Buca, İzmir, Turkey

Abstract

The effects of nano-sized metal oxides namely titanium dioxide (TiO_2), nickel oxide (NiO) and zinc oxide (ZnO) on the ultrasound of olive mill effluent wastewater (OMW) in Turkey were investigated. 150 minutes ultrasound alone provided 61%, 50%, 61% and 66% dissolved chemical oxygen demand (COD_{dis}), color, total phenol and total aromatic amines (TAAs) removals, respectively, at 25°C. The maximum TAAs (90%), total phenol (97%), color (94%) and COD_{dis} (97%) removals were obtained with 5 mg/L nano-sized ZnO, 4 mg/L nano-sized TiO_2, 4 mg/L nano-sized TiO_2 and 4 mg/L nano-sized TiO_2, respectively, through out ultrasound.

Keywords: Olive mill effluent; Ultrasound; Nano-sized titanium dioxide metal oxides; Nano-sized nickel oxide metal oxides; Nano-sized zinc oxide metal oxides

Introduction

Agro-industrial wastewaters such as olive-oil mill effluent wastewaters (OMW) are among the strongest industrial effluents and increasing concern has been expressed about their treatment and safe disposal, since they cause considerable environmental problems (coloring of natural waters, a serious threat to aquatic life, pollution in surface and ground waters, alterations in soil quality, phytotoxicity and odor nuisance) particularly in the Mediterranean Sea region due to its high organic chemical oxygen demand (COD), polyphenol, aromatic amines concentration, and organic content [1]. The concentration of phenolic compounds in the OMW may vary from as low as 0.05–0.2 g/L to as high as 10 g/L depending on the type and origin of the effluent [2]. The TAAs in the OMW are known to be carcinogenic and toxic. Some aromatic amines containing the azo bonds (–N═N–) have complex structure and are resistant to biodegradation under aerobic conditions [2,3].

Significant numbers of studies were focused on the efficient treatment of the OMW including various chemical, physical, physicochemical and biological treatments or combinations of them [1,4]. Usually, the OMW is inappropriate for direct biological treatment and the alternative treatment technologies mentioned above did not give sufficient removals for pollution parameters (COD_{dis}, total phenol, color and aromatic amines). Recently, significant interest has been shown in the application of ultrasound for the degradation of the OMW [5]. Hydrophobic compounds with high volatility are easily and directly destroyed inside the cavitation bubbles [6]. Hydrophilic organic compounds are indirectly decomposed mainly through the reaction with hydroxyl radicals (OH•) that is produced during cavitation process. The highly reactive OH• could diffuse from the cavitation bubbles to the interfacial region and bulk solution when large temperature gradient exist [6]. There are three potential reaction zones in sonochemistry; i.e. inside of the cavitation bubble, interfacial liquid region between cavitation bubbles and bulk liquid, and in the bulk solution [6]. The collapse of cavitation bubbles near the micro-particle surface will generate high-speed microjets of liquid in the order of 100 m/s [6]. This will subsequently produce ultrasonic asymmetric shock wave upon implosion of cavitation bubbles which may cause direct erosion (damage) on the particle's surface and de-aggregation of particles to hinder agglomeration. Consequently, it will experience a decrease in particle size and an increase in reactive surface area available for the subsequent reaction. The nano-particles with the size less than that of cavitation bubbles have higher cavitation erosion resistant and are easier to approach the interfacial region (bubbles surface) during the expansion cycles of ultrasound [6]. It was observed synergetic effects with the addition of various metal oxides with ultrasound to enhance the rate of degradation efficiency of organic pollutants via increasing the OH• radicals. Thus increasing the rate of degradation of the organic compouns in wastewaters. The sonication of organic pollutants in the present of some metal oxides (this reaction could be named as heterogenous sonication) can easily ocur in the interfacial region where very high concentration of OH• is achieved after the bubbles collapse [6]. Ultrasound (US) will induce the splitting of water molecules with the presence of dissolved oxygen [6]. In these reactions, ')))' denotes the ultrasonic irradiation. Thermal dissociation of water and dissolved oxygen molecules in the cavities will convert them into reactive species such as OH•, hydrogen atoms (H•), O atoms and hydroperoxyl radicals (OOH•) in a cavitation bubble reactions [6].

Previously, the sonocatalytic degradation of some organic pollutants in aqueous solution adopting nano-sized metal oxides powder (TiO_2, ZnO and NiO) has been reported [6]. The nano-sized TiO_2 and nano-sized ZnO as photocatalyst have been widely used with ultraviolet (UV) radiation in the degradation of organic dyes due to their stability of the chemical structure, non-toxicity, optical and electrical properties. Although, these nano-sized metal oxides are non-toxic, stable, and cheap, the utilization of UV increase the cost of the treatment. The nano-sized NiO nanoparticles are new class of advanced materials with very interesting features and capacity such as superior magnetic, electrical and selective adsorption properties which also has been utilized in water treatment [6]. Chowdhury et al. [7] and Zhang et al. [8]

***Corresponding author:** Dr. Delia Teresa Sponza, Department of Environmental Engineering, Engineering Faculty, Dokuz Eylül University, Tınaztepe Campus, 35160, Buca, İzmir, Turkey

reported that nanoporous materials such as nano-sized NiO and nano-sized TiO_2 have led to higher adsorption and catalytic efficiency with sizes varying between 17-21 nm range and relatively high surface areas up to 83 m^2/g. Chowdhury et al. [7], found 70% methylene blue and procion red yields using nano-sized NiO surface as adsorbent. Ahmed [9] found 85% removal in methylene blue dye by nano-sized NiO/TiO_2 nanocomposites with photodegradation. Abu Tarboush and Husein [10] found that the asphaltenes from a heavy oil could be adsorbed with nano-sized NiO nanoparticles with a yield of 76%. Zhang et al. [8] investigated the adsorption of congo red on nano-sized NiO hollow microspheres with an adsorption capacity of 526.3 mg/g resulting a congo red removal of 78%. However, to date, the use of nano-sized metal oxides namely, TiO_2, ZnO and NiO in the ultrasound of the OMW for the removal of pollutants from this wastewater was not studied yet.

Parameters	Values		
	Minimum	Medium	Maximum
pH_0	4.0 ± 0.1	4.5 ± 0.1	4.9 ± 0.1
DO_0 (mg/L)	0.01 ± 0.0004	0.05 ± 0.002	0.09 ± 0.003
ORP (mV)	120 ± 4.2	126.0 ± 4.4	132 ± 4.6
TSS (mg/L)	53.6 ± 1.87	53.7 ± 1.8	53.8 ± 1.8
TVSS (mg/L)	34.8 ± 1.2	35.6 ± 1.2	36.4 ± 1.2
COD_t (mg/L)	98780 ± 3457	116632 ± 4432	121560 ± 4920
COD_{dis} (mg/L)	85400 ± 2989	109444 ± 3831	113500 ± 4323
TOC (mg/L)	58510 ± 2048	66488 ± 2327	80450 ± 2816
BOD_5 (mg/L)	63800 ± 2233	81254 ± 2844	99130 ± 3470
BOD_5 / COD_{dis}	0.5 ± 0.02	0.7 ± 0.026	0.9 ± 0.032
Total N (mg/L)	194 ± 6.7	248 ± 8.6	300 ± 10.5
NH_4-N (mg/L)	23.4 ± 0.8	30 ± 1.05	36.6 ± 1.2
NO_3-N (mg/L)	39 ± 1.3	50 ± 1.7	61 ± 2.1
NO_2-N (mg/L)	17.6 ± 0.6	22.6 ± 0.7	27.5 ± 0.9
Total P (mg/L)	492 ± 17	630 ± 22.05	768.6 ± 26.9
PO_4-P (mg/L)	350 ± 12.2	448 ± 16	546.2 ± 19.1
Total phenol (mg/L)	1332 ± 117	2990 ± 143	3250 ± 180.2
Phenol metabolites			
Catechol (mg/L)	2	23	26
Tyrosol (mg/L)	7	42	45
Quercetin (mg/L)	9	21	28
Caffeic acid (mg/L)	19	30	42
4- methyl catechol (mg/L)	9	15	29
2-PHE (mg/L)	2	4	6
3-PHE (mg/L)	2	9	14
TFAs (mg/L)	3050 ± 141	3900 ± 182	4344 ± 222
TAAs (mg/L)	1190 ± 82	1990 ± 105	2200 ± 128
Individual TAAs (mg/L)			
2,4,6 trimetylalanin (mg/L)	47	125	187
Aniline (mg/L)	43	83	163
o-toluidine (mg/L)	29	110	156
o-ansidine (mg/L)	45	110	129
Dimethylalanine (mg/L)	13	68	87
Ethylbenzene (mg/L)	21	110	116
Durene [1,2,4,5-tetramethyl-benzene] (mg/L)	34	100	123
Color (1/m)	99.7 ± 3	99.8 ± 3.4	99.9 ± 3.4
Oil (mg/L)	564 ± 19	640 ± 22.4	775 ± 27.1

Table 1: Characterization of the OMW at pH=5.4 (n=3, mean values ± standard deviation).

In İzmir, Turkey the OMW treatment plants consisting of a combined chemical and biological process could not provide efective treatment removals. Therefore, in the peresent study, the effects of increasing ultrasound times (60, 120 and 150 min) and ultrasound temperature (30°C and 60°C) on the removals of COD_{dis}, total phenol, color and TAAs were investigated. The main ultrasound mechanisms for the pollution parameters given above were investigated. The poly-phenols were identified and their metabolites were detected. The effects of some nano-sized metal oxides (TiO_2, NiO and ZnO) on the removals of COD_{dis}, color, total phenol and TAAs in the OMW were investigated in a sonicator with a power of 640 W and a frequency of 35 kHz.

Material and Methods

Raw wastewater

The characterization of raw OMW taken from the influent of an olive oil production industry in İzmir, Turkey is given in table 1. This plant was operated with a three phase olive oil extraction process.

Configuration of sonicator

A BANDELIN Electronic RK510 H sonicator was used for ultrasound of the OMW samples. Glass serum bottles in a glass reactor were filled to volumes of 5-500 mL with OMW and were placed in a water bath. They were then closed with teflon coated stoppers throughout the measurement of the OMW. Ultrasonic waves for 35 kHz frequency were emitted from the bottom of the reactor through a piezoelectric disc (4-cm diameter) fixed on a pyrex plate (5-cm diameter). The evaporation losses of volatile compounds were estimated to be 0.01% in the reactor and, therefore, assumed to be negligible. The serum bottles were filled with 0.1 mL methanol (CH_3OH) in order to prevent adsorption on the walls of the bottles and minimize evaporation. The temperature in the sonicator was monitored continuously and it was adjusted electronically in the sonicator with an automatic heater. The stainless steel sonicator reactor was sealed with a water jacket to maintain the desired temperature and to prevent the temperature losses. The schematic configuration of the sonicator used in this study was given elsewhere [11].

Operational conditions

Fresh solutions of nano-sized metal oxides TiO_2 (2, 4 and 8 mg/L), NiO (4, 7 and 10 mg/L) and ZnO (1, 3 and 5 mg/L) were added to the OMW by an peristaltic pump (Watson-Marlow Bredel pumps, USA) with a flow rate of 0.1 mL/min through 5 min before ultrasound was begun at pH=5.4. Sonicated samples were taken at 60th, 120th and 150th min of ultrasound time and were kept in refrigerator with a temperature of + 4 °C for experimental analysis. Deionized pure water (R ¼ 18 MΩ/cm) was obtained through a SESA Ultrapure water system.

All experiments were in batch mode by using an ultrasonic transducer (horn-type), which has five adjustable active acoustical vibration areas of 12.43, 13.84, 17.34, 26.4 and 40.69 cm^2, with diameters 3.98, 4.41, 4.7, 5.8 and 7.2 cm, respectively, and a maximum input power of 640 W. Five ultrasound intensities (15.7, 24.2, 36.9, 46.2 and 51.4 W cm^{-2}) were chosen to identify the optimum intensity for maximum removal of pollutant parameters (COD_{dis}, color, total phenol and TAAs) in the OMW while the sonicator power was 640 W. Samples were taken after 60, 120 and 150 min of ultrasound time and they were analyzed immediately as mentioned in the recent studies [11].

Nano-sized metal oxides TiO_2, NiO and ZnO were purchased from Merck (Germany), while reagent grade perfluorohexane (C_6F_{14})

was taken from Fluka (Germany). Aniline (99%), 2-PHE (99%), 3-PHE (99%), 2, 4, 6 trimetylaniline (99%), dimethylaniline (99%) and o-toluidine (99%) was purchased from Aldrich (USA).

Analytical methods

BOD_5 and COD were monitored following standard methods 5210 B and 5220 D, respectively [12]. Total-N, NH_4-N, NO_3-N, NO_2-N, total-P and PO_4-P were measured with cell test spectroquant kits (Merck) in a spectroquant NOVA 60 (Merck) spectrophotometer (2003). Oil, Na^{+1} and Cl^{-1}, total suspended solid concentration (TSS), total volatile suspended solid concentration (TVSS), dissolved oxygen (DO), pH, temperature T (°C) and oxydation reduction potential (ORP, mV) were determined following Standard Methods 5520 B, 3550, 2540 C, 2540 E, 2550, 2580, respectively [12]. The measurement of color was carried out following the methods described by Olthof and Eckenfelder [13]. In order to identify the TAAs, OMW (25 mL) was acidified at pH=2.0 with a few drops of 6 N HCl and extracted three times with 25 mL of ethyl acetate. The pooled organic phases were dehydrated on sodium sulphate, filtered and dried under vacuum. The residue was sylilated with bis(trimethylsylil)trifluoroacetamide (BSTFA) in dimethylformamide and analyzed by GC-MS. Mass spectra were recorded using a VGTS 250 spectrometer equipped with a capillary SE 52 column (0.25 mm ID, 25 m) at 220°C with an isothermal program for 10 min. TAAs were measured using retention times and mass spectra analysis. The total phenol was monitored as follows: 40 mL of OMW was acidified to pH=2.0 by the addition of concentrated HCl. Phenols were then extracted with ethyl acetate. The organic phase was concentrated at 40°C to about 1 mL and silylized by the addition of N,O-bis(trimethylsilyl) acetamide (BSA). The resulting trimethylsilyl derivatives were analysed by GC-MS (Hewlett-Packard 6980/HP5973MSD). 2-PHE and 3-PHE concentrations were determined with Shimadzu CLASS-VPV6.14SP2 with Phenomenex Hyperclone 125 mm×4.6 mm×5 µm HPLC column using an ultraviolet (UV) method [14]. Aniline, 2,4,6 trimetylaniline and dimethylaniline measurements were carried out using a high pressure liquid chromatograph (HPLC) (Agilent-1100) with a C-18 reverse phase HPLC column (25 cm×4.6 mm, 5 µm (Ace 5C-18)) following the method developed by EPA [15]. Detection was performed at 280, 214 and 216 nm wavelengths using a UV detector for aniline, 2,4,6 trimetylaniline and dimethylaniline, respectively. *o*-toluidine was determined using a HPLC (Agilent-1100) with a Spectra system model SN4000 pump and Asahipak ODP-506D column (150 mm×6 mm×5 µm). o-anisidine was measured in a HPLC (Agilent-1100) with UV detector at a mobile phase of 35% acetonitrile/65% H_2O at a flow rate of 1.2 mL min^{-1}. The column was 50 cm×2 mm ID stainless steel packed with µ-Bondapak C-18. Ethylbenzene and durene (1,2,4,5-tetramethylbenzene) measurements were performed with a GC-MS (GC-MS Agilent Technologies, 7890A (G3440A) GC System, 5975C Inert MSD7683B Series Injector) containing inlet split/splitless ratio of 50:1 with a HP-1 polymethylsiloxane 30 m×0.32 mm×ID 0.25 µm film thickness column (Hewlett-Packard) at 240°C. The carrier was helium (He) with a flowrate of 1.4 mL min^{-1} and column 60 m×0.25 m×ID 0.25 µm. The steady-state OH ion concentration [OH•] throughout sonication was calculated following the method proposed by [16] while the second-order reaction kinetics of aromatic amines and phenols were found by [14] and [15].

Results And Discussion

Role of ultrasound frequency, ultrasound power, ultrasound intensity and power density on the removals of COD_{dis}, color, total phenol and TAAs removals in OMW

Five different ultrasound frequencies (25, 35, 132, 170 and 350 kHz) was researched under ambient conditions (25°C), at constant ultrasound power (640 W) to determine the optimum ultrasound frequency for maximum COD_{dis} removals in the OMW (Table 2). Among the frequencies used in the ultrasound process (25, 35, 132, 170 and 350 kHz) it was found that an ultrasound frequency of 35 kHz is the optimum frequency for maximum COD_{dis} removal (61%) in the OMW after 150 min ultrasound at 25°C (Table 2). Increasing the ultrasound frequencies >35 kHz did not increase the number of OH• radicals [17]. A more energetic implosion of cavitation bubbles is expected to occur at low frequency rather than at high frequency because of a larger bubbles radius observed at low frequency [17]. The size of bubbles formed in H_2O, is inversely proportional to the frequency of the wave. 120, 350, 640 and 3000 W ultrasound powers was researched under ambient conditions (at 25°C), at constant ultrasound frequency (35 kHz) to determine the optimum ultrasound power for maximum COD_{dis} removals in the OMW (Table 2). Among the powers used it was found that a ultrasound power of 640 W is the optimum power for maximum COD_{dis} removals (61%) in the OMW after 150 min ultrasound at 25°C (Table 2). In general, when ultrasonic irradiation is used, the degradation ratio gradually becomes higher when the output power of the ultrasound is increased from 120 to 640 W. The sono-degradation of pollutants increased with increasing applied power. As the power increased, the number of collapsing cavities also increased, thus leading to enhanced degradation rates, thus resulting in high OH• production and COD_{dis} removal. Because the formation of cavitation bubbles and the extent of bubble collapse depends on the sound intensity, various ultrasound intensities were compared for their impacts on the COD_{dis} removals. In this study, the optimum cavitation threshold was 51.4 W/cm² for the maximum removals of COD_{dis} (61%) at 25°C after 150 min ultrasound time (Table 2). For the maximum COD_{dis} removal, the specific energy was 11.5 kWh/kg COD_{dis} in the influent (Table 2).

Effect of increasing ultrasound time and temperature on the COD_{dis}, total phenol, color and totatal aromatic amine (TAA) removal efficiencies in the OMW

The OMW samples were sonicated at 30°C and at 60°C during 60, 120 and 150 min ultrasound times (Table 3). Increasing the ultrasound time from 60 to 120 min enhanced the COD_{dis} removals under ambient conditions (at 25°C). The maximum COD_{dis} removal was 61% after 150 min ultrasound time, at 25°C and at pH=5.4 (Table 3). 16%, 42% and 63% COD_{dis} removals were measured after 60, 120 and 150 min ultrasound time, respectively, at pH=5.4 and at 30°C (Table 3). The maximum COD_{dis} removal was 67% after 150 min ultrasound time, at pH=5.4 and at 60°C (Table 3). An initial increase in solution temperature increase the rate of reaction leading to a greater fraction of OH ion production and volatile compounds partitioning into the cavity. A further increase in solution temperature leads to a decrease in the rate of reaction [11]. The treatment by ultrasound converts COD_{dis} in the OMW to much smaller sonodegraded compounds. Short ultrasound times (60 min) did not provide high degradation removals for refractory COD

Ultrasound parameters	Values				
Ultrasound frequency (kHz)	25	35	132	170	350
Ultrasound power (W)	120	640	350	3000	5000
Power density (W/mL)	0.1	2.14	0.9	1.65	1.9
Ultrasound intensity (W/cm²)	15.7	51.4	24.2	36.9	46.2
Specific energy (kWh/kg $COD_{influent}$)	2.4	11.5	3.1	4.1	5.1
COD removal efficiency (%)	45	61	47	53	58

Table 2: Sonicational parameters and corresponding values of sonication process in this study at pH=5.4 after 150 min ultrasound time for maximum COD_{dis} yields under ambient conditions (at 25°C).

SET	Parameters	Removal Efficiencies (%)								
		25°C (Control)			30°C			60°C		
		60. min	120. min	150. min	60. min	120. min	150. min	60. min	120. min	150. min
1	COD_{dis}	13	42	61	16	42	63	18	47	67
2	Total phenol	30	57	61	45	62	80	48	70	88
3	Color	36	48	50	39	53	75	45	70	84
4	TAAs	41	52	66	47	56	69	62	71	79

Table 3: Effect of increasing ultrasound time (60, 120 and 150 min) and increasing temperature (at 25°C, at 30°C and at 60°C) on the COD_{dis}, total phenol, color and TAAs removal efficiencies in OMW, at an initial COD_{dis} concentration=109444 mg/L, at an initial total phenol concentration=2990 mg/L, at an initial color concentration=99.8 1/m, at an initial TAAs concentration=1990 mg benzidine/L, at ultrasound power=640 W, at ultrasound frequency=35 kHz and at pH=5.4 (n=3, mean values).

throughout ultrasonic irradiation. COD_{dis} was not completely removed under the ultrasonic action even with a long ultrasound time (150 min) in OMW. This was probably caused by insufficient levels of OH• formation. Hence, COD_{dis} do not have the ability to attack to the low concentrations of OH•, thus leading to decreased sonodegradation rates of COD_{dis} as reported by [17].

The COD_{dis} removals obtained in the OMW were higher than those of the recent studies obtained at 25°C with ultrasound: Canizares-Macias et al. [18] 56% COD_{dis} removal found at 20 kHz frequency, and at a power of 400 W, after 60 min ultrasound time by an ultrasound system assisted with Rancimat Method. In a study performed by Atanassova et al. [19] a COD_{dis} removal of 45% was found after 240 min ultrasound time at a power of 150 W and a frequency of 80 kHz at 25°C. 40% COD_{dis} removal was provided in the OMW containing 10 mg/L alpechin at 40 kHz frequency, at 700 W power, after 15 min ultrasound time, at 50°C [20]. In this study, the COD_{dis} removal (61%) is higher than the data obtained by [20] and by Canizares-Macias et al. [18] mentioned above. In this study, it was found that the total phenol removal efficiencies in the OMW increased from 30% up to 57% as the ultrasound time was increased from 60 to 120 min at 25°C at pH=5.4. The maximum total phenol removal efficiency was 61% after 150 min ultrasound time, at pH=5.4 and at 25°C (Table 3, SET 2). 48%, 70% and 88% total phenol removals were found after 60, 120 and 150 min ultrasound time, respectively, at pH=5.4 and at 60°C (Table 3, SET 2). The maximum total phenol removal was 88% after 150 min ultrasound time, at pH=5.4 and at 60°C (Table 3, SET 2). The Kruskal-Wallis test statistics showed that the effects of increasing ultrasound time (from 60 up to 120 and 150 min) and temperature (from 25°C to 30 and 60°C) on the total phenol removals are not differed and these differences are not significant (Mann–Whitney *U*-test statistic=2.98, $p < 0.10$) after 60 and 120 min at all increasing temperatures. Although, the increase in ultrasound times increase the total phenol removals, the temperature increse from 25 up to 60°C did not contribute to total phenol removals after 60 and 120 min ultrasound time. However, the increase in temperature contributed to the total phenol removals after 150 min ultrasound time. The total phenol removals is significantly differed (Mann–Whitney *U*-test statistic=11.02, $p < 0.1$).

The degradation of phenol occurs in the bulk liquid medium due to hydroxylation reaction induced by OH• generated from cavitation bubble. This is a consequence of low vapor pressure of phenol (due to which it does not evaporate into the cavitation bubble) and the hydrophilic nature of the phenol molecule. Higher ultrasound times are needed for complete mineralization of phenol. Short ultrasound times (60 min) did not provide high degradation removals for refractory part of phenolic organics throughout ultrasonic irradiation since the OH• radicals can not have enough time to binding to the phenol. Phenol was not completely removed under the ultrasonic action even with a long ultrasound time (150 min) in the OMW.

Around 36, 48 and 50% color removals were obtained at 60 and 150 min ultrasound time, respectively, at an initial color level of 99.8 1/min in the OMW at ambient conditions (at 25°C) (Table 3). 39%, 53% and 75% color removals were found after 60, 120 and 150 min ultrasound time, respectivley, at pH=5.4 and at 30°C. 45%, 70% and 84% color removals were observed after 60, 120 and 150 min ultrasound time, respectively, at pH=5.4 and at 60°C (Table 3). The maximum color removal was 84% after 150 min ultrasound time, at pH=5.4 and at 60°C.

Although, the color yields increased as the ultrasound time were increased from 60 to 120 min, the contribution of low temperature on the color yields was found to be not significant. Kruskal-Wallis test statistics showed that the color removals are not differed significantly after 60 and 120 min ultrasound times, at 30°C (Mann–Whitney *U*-test statistic=2.38, p=0.05) Low color removals in the OMW effluents showed that low ultrasound temperature (25°C) is not able to achieve the complete removal of the colored organics (e.g. lignin, organic acids, etc.) and the phenolic compounds (e.g. tannins, antocyanins, catechins, etc.). This might be due to the generation of high concentration of intermediate compounds that cannot be further oxidized by OH• and consequently are accumulated in the system. It was found that high temperature such as 60°C increased significantly the color yields after 150 min ultrasound time. The Kruskal-Wallis test statistics showed that the color removals are differed after 150 min ultrasound time at a temperature of 60°C and these differences are significant (Mann-Whitney *U*-test statistic=12.63, p=0.05). The increasing of temperature is likely to facilitate the buble formation due to an increase of vapour pressure of a medium and so lead to easier cavitation with less violent collapse.

41%, 52% and 66% TAAs removal efficiencies were observed at an influent TAAs concentration of 1990 mg/L after 60, 120 and 150 min ultrasound time, respectively, at pH=5.4 and at 25°C (Table 3). As the temperature increased to 30°C the TAAs removals increased to 47%, 56% and 69% after 60, 120 and 150 min ultrasound time, respectively, at pH=5.4 (Table 3). 62%, 71% and 79% TAAs removal efficiencies were found after 60, 120 and 150 min ultrasound time, respectively, at pH=5.4 and at 60°C (Table 3). The maximum TAAs removal efficiency was 79% after 150 min ultrasound time, respectively, at pH=5.4 at 60°C (Table 3).

Although, the increasing of ultrasound time contribute to the TAAs removals the Kruskal–Wallis test statistics showed that the TAAs removals are not differed significantly after 60 min ultrasound time, at 30 and at 60°C (Mann–Whitney **U**-test statistic=1.09, **p**=0.05). The increasing of temperature and ultrasound time from 25 to 30 and to 60°C and from 120 to 150 min ultrasound time, respectively, increased the TAAs removals. The Kruskal-Wallis test statistics showed that the TAAs removals are differed after 150 min ultrasound time versus increasing temperature and these differences are significant (Mann–Whitney *U*-test statistic=13.88, p=0.05).

Conditions	COD_{dis} removal		Color removal		Aromatic amines removal		Phenol removal	
	H_2O_2[a]	$OH^{\cdot}$[a]	H_2O_2[a]	$OH^{\cdot}$[a]	H_2O_2[a]	$OH^{\cdot}$[a]	H_2O_2[a]	$OH^{\cdot}$[a]
H_2O_2 concentration (mg/L) in deionized water (60°C) at pH=7.0 at the beginning	177	0	177	0	0	0	209	0
H_2O_2 concentration (mg/L) in deionized water (60°C) at pH=7.0 after 150 min ultrasound time	176	0	176	0	0	0	198	0
H_2O_2 concentration (mg/L) in OMW (60°C) at pH=5.4 at the beginning	34	$1*10^{-92}$	49	$1*10^{-82}$	175	$1*10^{-272}$	98	$25*10^{-82}$
H_2O_2 concentration (mg/L) in OMW (60°C) after 30 min ultrasound time at pH=5.4	156	$10*10^{-62}$	160	$1*10^{-52}$	174	$1*10^{-262}$	185	$10*10^{-62}$
H_2O_2 concentration (mg/L) in OMW (60°C) after 120 min ultrasound time at pH=5.4	92	$13*10^{-15}$	73	$9*10^{-12}$	174	$1*10^{-252}$	145	$2*10^{-52}$
H_2O_2 concentration (mg/L) in OMW (60°C) after 150 min ultrasound time at pH=5.4	11	$16*10^{-7}$	5	$9*10^{-5}$	173	$1*10^{-242}$	14	$43*10^{-7}$
[a] Concentration (mg/L)								

Table 4: H_2O_2 and $OH^{\cdot}$ ion concentrations in the OMW at 60°C, after 30, 120 and 150 min ultrasound time, at an initial COD_{dis} concentration=109444 mg/L, at an initial color concentration=99.81/m and at an initial TAAs concentration=1990 mg/L, respectively, at pH=5.4 at ultrasound power=640 W and ultrasound frequency=35 kHz (n=3, mean values).

Removal mechanisms of COD_{dis}, total phenol, color and total aromatic amines (TAAs) in the OMW throughout ultrasound

The H_2O_2 measurements were performed in the absence and in the presence of OMW throughout ultrasound of TAAs, color, total phenol and COD_{dis} at 30°C after 150 min ultrasound time, in order to detect the ultrasound mechanisms of the parameters in suspension. In the absence of OMW (in deionized water) the H_2O_2 was accumulated (176 mg/L). Similarly the H_2O_2 concentration remains as formed after 150 min ultrasound time (Table 4). The initial rate of H_2O_2 formation associated with the COD_{dis} treatment by ultrasound in the OMW decreased significantly with increasing ultrasound time at 60°C (Table 4). The H_2O_2 concentration was 34 mg/L at the beginning of the ultrasound in the OMW then increased to 156 mg/L with the production of $OH^{\cdot}$ ion since H_2O_2 is formed by the recombination of $OH^{\cdot}$ throughout sonolysis. Afterwards, the H_2O_2 concentrations decreased to 92 and 11 mg/L with the $OH^{\cdot}$ ion increase. The $OH^{\cdot}$ ion concentrations increased from $10*10^{-62}$ to $16*10^{-7}$ mg/L after 150 min ultrasound time in the OMW. This showed that hydroxylation is the main mechanism for the removal of COD_{dis}.

In the absence of the OMW (in deionized water) the H_2O_2 concentration was measured as 209 mg/L at pH=7.0 while the H_2O_2 concentration decreased sligtly and remained around 198 mg/L after 150 min ultrasound time, at pH=5.4 throughout ultrasound of phenols (Table 4). The H_2O_2 concentration was 98 mg/L at the beginning of the ultrasound in the OMW and then increased to 185 mg/L after 30 min ultrasound time with the production of $OH^{\cdot}$ ion since H_2O_2 is formed by the recombination of $OH^{\cdot}$ throughout sonolysis. Afterwards, the H_2O_2 concentrations decreased to 145 and 14 mg/L with the $OH^{\cdot}$ ion increase in the OMW after 120 and 150 min ultrasound time, respectively. The H_2O_2 values in the OMW are much lower compared to those in deionized water (at a neutral pH=7.0) due to the fact that in the former case many of the $OH^{\cdot}$ produced by ultrasound reacted with phenol before they could combine to form $OH^{\cdot}$. The $OH^{\cdot}$ ion concentrations reached $25*10^{-82}$ mg/L from $43*10^{-7}$ mg/L after 150 min ultrasound time in the OMW. This showed that hydroxylation is the main mechanism for the removal of phenol. In this study, in the OMW the most sonogenerated $OH^{\cdot}$ reacted with 93% maximum total phenol removal and radical recombination to produce H_2O_2 after 150 min ultrasound time. Different phenol removal mechanisms and metabolites were reported throughout ultrasound: Pétrier et al. [21] investigated the sonolysis of phenol in aqueous solution and found hydroquinone, catechol, and benzoquinone as reaction products. These reaction products are attributed to the attack of the $OH^{\cdot}$ on phenol. Similarly, Anju et al. [22] showed that the $OH^{\cdot}$ was involved in the degradation pathways of phenol. They reported that $OH^{\cdot}$ generated from the dissosiation of H_2O during ultrasonic irradiation process. very high concentration of $OH^{\cdot}$ is achieved after the bubbles collapse. The concentration that could be reached is estimated to be as high as 4 mM. However, Currell et al. [23] found that the ultrasound of phenol was sono-degraded partially by pyrolysis in the bubble phase. This could be due to the reactor configuration and geometry, to the wastewater characteristics, to the ultrasound intensity, frequency and the presence of matrix components in the wastewater researched.

Throughout ultrasound of the colored organics and phenols in the OMW, the H_2O_2 production decreased while the $OH^{\cdot}$ ion concentration increased. Since the sonooxidation of the COD_{dis} and the colored organic substances comprised 63% and 75% of the total sonodegradation process at 30°C after 150 min ultrasound time, $OH^{\cdot}$ is the major process for complete degradation of these parameters. Throughout the sonodegradation of the TAAs the rate of $OH^{\cdot}$ production was zero. This accomplished with H_2O_2 concentration as low as $1*10^{-272}$ mg/L (Table 4). Therefore, it can be seen that $OH^{\cdot}$ is not the major process for complete degradation of the aromatic amines. In other words, in this study, the contribution of $OH^{\cdot}$ was minor for the ultimate sonodegradation of aromatic amines. The formation of by-products (aniline, o-toluidine, anisidine, dimethylaniline and dimethylaniline) for possible $OH^{\cdot}$ oxidation was not observed in HPLC. Similar results were obtained in the studies performed by Wu and Ondruschka [24]. The contribution of the pyrolysis to the destruction of TAAs is significant. The occupation of the heart and/or the surrounding shell of the bubble by TAA molecules, limiting the sonolysis of H_2O and the formation of $OH^{\cdot}$. This indicates that the main process for the destruction of TAAs is pyrolysis. The addition of 600 mg/L 1-butanol to the OMW very slightly affected (4%) the TAAs removals by the $OH^{\cdot}$ scavenger (data not shown) while about 67-59% inhibition was observed for COD_{dis} and color removal (data not shown). Throughout ultrasound of TAAs some gaseous by-products were observed in the headspace of the ultrasound reactor. 31.2-41.11% CO_2 and 16-18.32% NH_3 were measured after 10 min ultrasound time in the headspace of the sonicator reactor (data not shown). It can be assumed that these gases are primarily formed as soon as TAAs penetrate the cavitation bubbles in order to be pyrolyzed. The oxidation of TAAs with $OH^{\cdot}$ appears to be of relatively minor importance, because of the low inhibition of the sono-degradation by means of 1-butanol and since the expected steady-state $OH^{\cdot}$ concentration in the interfacial region of the cavitation bubble, where TAAs accumulate, is lower and not sufficient for a complete degradation of TAAs. The TAAs are then expected to be mainly localized in the heart and/or in the surrounding shell of the bubble, inhibiting the production of $OH^{\cdot}$ and hence the oxidation pathway.

In this study, the ultrasound mechanism of the AAs in the OMW are attributable to the priority of cleavage of azo links by pyrolysis. Throughout pyrolysis, $OH^{\cdot}$ initially diffuse into their long-lived cavity bubbles to undergo pyrolytic destruction inside the collapsing bubble then the cleavage of azo groups open the N═N bonds. The cleavage of methyl, ethyl and C-H-O bonds can destroyed the aromatic structures in TAAs namely o-toluidine, dimethylaniline, ethylbenzene and durene.

Effects of nano-sized metal oxides (TiO_2, NiO and ZnO) on the sonodegradation of COD_{dis}, total phenol, color and TAAs in the OMW

Keeping the conditions of 35 kHz, 640 W and 21°C, ultrasonic degradation of COD_{dis}, TAAs, total phenol and color in the OMW were investigated in the presence of nano-sized metal oxides TiO_2, NiO and ZnO. Table 5 shows the effect of nano-sized metal oxides on the removal of the pollutant parameters given above in the OMW samples. The contribution of nano-sized TiO_2 on COD_{dis}, TAAs, total phenol and color yields was found to be significant. Although, 2 mg/L nano-sized TiO_2 increase significantly the COD_{dis} (92%), TAAs (86%), total phenol (90%) and color (85%) (Table 5). The maximum COD_{dis} (97%), TAAs (88%), total phenol (97%) and color (94%) yields were obtained at 4 mg/L nano-sized TiO_2 concentration. The increasing of nano-sized TiO_2 to 8 mg/L decreased the removals in the aforementioned pullution parameters. Nano-sized TiO_2 particles provides more chances for the organic compounds (COD_{dis}) phenol and TAAs) to adsorb on the surface of the nano-sized TiO_2 particles to be directly composed by holes. Throughout ultrasound of the OWW with semiconductor nano-sized TiO_2, firstly, the formation of the light at wavelentgh below 375 nm occurred. Under these conditions the ligth can excite the semiconductor nano-sized TiO_2 acting as a photocatalyst and $OH^•$ radicals with oxidative performance activate the surface of the nano-sized TiO_2. Secondly, the temperatrure of hot spot produced by ultrasonic cavitation in the sonicator can achieve high temperatures such as 102 and 104°C. As a result, temperatures as high as 104°C bring many holes producing $OH^•$ radicals on the surface of the nano-sized TiO_2. Throughout ultrasound the steps of treatment of the OMW include the following steps: (i) the mass transfer of the polutants (COD_{dis}, phenol, TAAs and color giving organics) from the bulk solution to the surface of nano-sized TiO_2, (ii) adsorption of pollutants onto the nano-sized TiO_2 surface (iii) chemical reactions ($OH^•$ and other radical productions) on the surface of nano-sized TiO_2. Desorption of the products from the nano-sized TiO_2 surface during oscillation and mass transfer of the product from the nano-sized TiO_2 surface to the bulk solution of the OMW in the sonicator [5]. At low nano-sized TiO_2 concentrations the adsorbed pollutant molecules on nano-sized TiO_2 surfaces decrease in the OMW. Therefore, the degradatıon ratio of the pollutants decreased since the organics did not find enough holes in the surface of the nano-sized TiO_2 nano-particles. Thus, it reduces the formation rate and number of $OH^•$ radicals and inhibits the generation of OH. At high nano-sized TiO_2 concentrations the COD_{dis}, total phenol, TAAs and color removals decreased. This could be attributed to the aggregation of nano-sized TiO_2 particles at high concentrations, causing a decrease in the number of active surface sites, and this decrease in the passage of sonic waves through sonication of the OMW samples. Wang et al. [25] also reported that high nano-sized TiO_2 concentrations (≥ 6-12 mg/L) decrease the COD_{dis} removals. The increase in sono-destruction of polutants at 4 mg/L nano-sized TiO_2 may be explained by the cavitation process which produces higher surface area. The ultrasonic waves not only destroy the phenol compounds through cavitation process but also could increases the adsorption process by increasing the surface area of the nano-sized TiO_2 catalyst as reported by Laxmi et al. [26]. During the ultrasound of the OMW with nano-sized TiO_2, more active free radicals, such as $OH^•$ and $O_2H^•$ were produced from the dissociation of the H_2O and H_2O_2 Laxmi et al. [26]. In this study, the maximum TAAs removal (88%) observed in 4 mg/L nano-sized TiO_2 at 30°C, was higher than the removal obtained by Cardoso et al. [27] for 4,4-oxydianiline aromatic amine removal (81%) with 5 mg/L nano-sized TiO_2 after 150 min ultrasound. In a study performed by Khochawala and Gogate [28] it was observed that 2 g/L nano-sized TiO_2 is the optimum concentration for maximum phenol (78%) removal. The removals obtained in our study are comparable higher than that the literatures mentioned above. Among the nano-sized ZnO doses used in this study the maximum COD_{dis} (92%), TAAs (90%), total phenol (93%) and color (93%) were obtained at a nano-sized ZnO concentration of 5 mg/L (Table 5). At low nano-sized ZnO doses the removal efficiencies of the pollution parameters given above decreased. Nano-sized ZnO particles have been used to bind to the pollution parameters in wastewaters throughout ultrasound due to their surface properties, porosity, size distribution, density and surface charges. Throughout ultrasound of the OMW in the presence of nano-sized ZnO particles ultrasound acted as a deaggregator by microstreaming and causing microbubble collapse, which induced surface cleaning of the nano-sized ZnO particles; (1) US acted as a deaggregator by microstreaming and causing microbubble collapse, which induced surface cleaning of the ZnO particles; (2) the presence of an additional liquid–solid interface in the liquid bulk promotes cavitation, and (3) ultrasound (US)-accelerated mass transfer occurs between the solution phase and the nano-sized ZnO surface [22,29]. The nano-sized ZnO nanoparticles help to break up the microbubbles created by ultrasound into smaller ones, thus increasing the number of regions of high temperature and pressure [29]. This leads to increase in the number of $OH^•$ radicals which will interact with the organic molecules present in H_2O and oxidise them, resulting in eventual mineralization. The increase in degradation removals of COD_{dis}, phenol rate can be attributed to the increase in the number of active cavitation bubbles and consequent generation of more $OH^•$ radicals. At high nano-sized ZnO concentration (>10 mg/L) the aggregation of nano-sized ZnO nanoparticles cause decreasing in surface holes [25]. This leads to decrease in the formation of $OH^•$ radicals, resulting in low sonolytic efficiency. Beyond the optimum nano-sized ZnO concentration (3 mg/L), the degradation slows down and thereafter remains more or less steady or even decreases. The enhanced degradation efficiency is probably due to increased number of $OH^•$ radicals and their interactions. Another reason may be the aggregation of nano-sized ZnO particles causing decrease in the number of available nano-sized ZnO surface sites for binding of $OH^•$ radicals. The particles cannot be fully and effectively suspended beyond a particular loading in the sonicator. Anju et al. [22] found 63%, 62% and 73% phenol yields in the present of 10 mg/L nano-sized ZnO under US, UV and combination of UV and US irradiation conditions after 2 h of exposure time. In another study by Drijvers et al. [30], the decomposition of phenol and COD_{dis} were

Nano-sized metal oxides and their concentrations	Sonodegradation removals (%)			
	COD_{dis}	Phenol	TAAs	Color
Control without nano-sized metal oxides	61	61	66	50
TiO_2 = 2 mg/L	92	90	86	85
TiO_2 = 4 mg/L	97	97	88	94
TiO_2 = 8 mg/L	90	94	83	90
NiO = 4 mg/L	66	84	84	83
NiO = 7 mg/L	89	91	88	88
NiO = 10 mg/L	85	88	85	90
ZnO = 1 mg/L	64	86	74	80
ZnO = 3 mg/L	72	90	89	88
ZnO = 5 mg/L	92	93	90	93

Table 5: Effects of some nano-sized metal oxides on the removal of COD_{dis}, total phenol, TAAs and color at 25°C, after 150 min ultrasound time, at a frequency of 35 kHz, at a power of 640 W and at pH=5.4 (n=3, mean values).

investigated using nano-sized metal oxides Al_2O_3, ZnO and NiO under the combined effect of sonolysis at 520 kHz and chemical oxidation with H_2O_2, 67%, 68% and 70% phenol yields was obtained with the metal oxides mentioned above. These COD_{dis} and phenol yields in these studies were lower than that our data. This could be attibuted to the differences of the wastewater, nano-sized ZnO concentrations and sonicational operational conditions. pH affects the nano-sized ZnO surface properties, phenol, TAAs dissociation and OH• formation [25]. Hydroxyl anions (OH^-) in alkaline solution typically act as scavenger of holes on the surface of nano-sized ZnO particles and become OH• with strong oxidation ability after they lose one electron. Consequently, the probability of generating OH increases with pH [25]. Nano-sized ZnO is known to be an effective oxidation catalyst as it could promote the formation of reactive radicals to oxidize organic compounds to lower carbon organic compounds and even to CO_2 and H_2O [22].

In this study, the maximum COD_{dis} (89%), TAAs (88%), total phenol (91%) and color (88%) yields were obtained at a nano-sized NiO concentration of 7 mg/L (Table 5). At low (4 mg/L) and high (10 mg/L) nano-sized NiO doses the removal efficiencies of the pollution parameters given above decreased (Table 5). Dissociation of water in the cavitation bubbles throughout ultrasound will convert it into reactive species such as OH•, H• and OOH• [11]. H_2O_2 will be formed outside the hot bubbles or at the cooler interface as a consequence of OH• and OOH• recombination. On the other hand, the H• and OH• species may further react with H_2O_2. The presence of 7 mg/L nano-sized NiO increase the number of free radicals generated, thereby increasing the rate of degradation of the COD_{dis}, phenol and TAAs in the OMW. The nano-sized NiO nano-particles with the size less than that of cavitation bubbles have higher cavitation erosion resistant and are easier to approach the interfacial region (bubbles surface) during the expansion cycles of ultrasound [31]. Throughout ultrasound for the removals of pollutans from the OMW contained; mass transfer of the polutants from the bulk solution to the surface of nano-sized NiO, adsorption of pollutants onto the nano-sized NiO surface during oscillation, chemical reactions (OH• and other radical productions) on the surface of nano-sized NiO, Desorption of the products from the nano-sized NiO surface during oscillation and mass transfer of the product from the nano-sized NiO surface to the bulk solution steps. A nano-sized NiO system was used as a catalyst for low-temperature oxidation of phenol [31]. Nano-sized NiO is also being used as a catalyst for the oxidation of CO [31]. The catalytic properties of nano-sized NiO depend on the number of factors such as the particle size and surface area, as well the catalyst structure. The maximum conversion of cyclohexane has been observed when nano-sized NiO was used as a catalyst [31]. The utilization of the nano-sized NiO in the treatment of wastewater are limited only with the utilization of this oxide as adsorbent fo the removal of dyes from the textile industry wastewaters: Chowdhury et al. [7] found 78% dye and color yields in a textile industry wastewater using nano-sized NiO as adsorbent. Zhang et al. [8] found a high adsorption of congo red dye onto nano-sized NiO particles with a color removal of 79%. Similarly, Song et al. [32] used nano-sized NiO for the removal of dyes from the wastewaters. Microwave-induced degradation of crystal violet dye in the existence of OH group with an atomic percentage of O:Ni 2.80 to 1 was investigated. This dye was degraded efficiently (73%) under microwave irradiation (2450 MHz) [33].

Variation of poly-phenols in the OMW in the presence of nano-sized metal oxides

The treatment by ultrasound converts the poly-phenols in the OMW to much smaller sonodegraded compounds such as tyrosol, hydroxytyrosol, caffeic acid, quercetin, ferulic acid, catechol, vanillic acid, o-quinone, *p*-coumaric acid, *p*-hydroxybenzaldehyde, 4-methyl catechol, 2-PHE and 3-PHE Wang et al. [25]. Table 6 shows the measured phenol intermediates and their removals in this study (cathecol, tyrosol, quercetin, caffeic acid, 4-methyl catechol, 2-PHE and 3-PHE) in the OMW defined with HPLC before ultrasound at t=0 and after 150 min ultrasound times, at pH=5.4 at increasing concentrations of nano-sized metal oxides TiO_2 (2, 4 and 8 mg/L), NiO (4, 7 and 10 mg/L) and ZnO (1, 3 and 5 mg/L) at 25°C.

As shown in table 6, 2990 mg/L total phenol in the raw wastewater consisted of 23 mg/L catechol, 42 mg/L tyrosol, 21 mg/L quercetin, 30 mg/L caffeic acid, 15 mg/L 4-methyl catechol, 4 mg/L 2-PHE and 9 mg/L 3-PHE before ultrasound at t=0. The phenol metabolite concentrations mentioned above increased to 600 mg/L, 388 mg/L, 350 mg/L, 345 mg/L, 395 mg/L, 247 mg/L and to 268 mg/L with 2 mg/L nano-sized TiO_2 after 120 min ultrasound time, at pH=5.4 at 30°C (data not shown). Catechol, tyrosol, quercetin, caffeic acid, 4-methyl catechol, 2-PHE and 3-PHE reduced to 280, 220, 190, 185, 195, 123 and 99 mg/L after 150 min ultrasound time, at pH=5.4 and at 30°C (data not shown). These phenol metabolites were removed with yields of 62, 61, 68, 64, 63, 64 and 68%, respectively, at a nano-sized TiO_2 concentration of 2 mg/L after 150 min ultrasound time, at pH=5.4 and at 25°C while 299 mg/L total phenol remained in the sonicator reactor. The catechol, tyrosol, quercetin, caffeic acid, 4-methyl catechol, 2-PHE and 3-PHE removals were 84, 83, 82, 85, 84, 84 and 83%, respectively, at nano-sized TiO_2=4 mg/L after 150 min ultrasound time, at 25°C and pH=5.4 with an effluent total phenol concentration of 90 mg/L. The yields of the aforementioned phenol metabolites decreased to 78, 79, 76, 75, 77, 72 and 69%, respectively, as the nano-sized TiO_2 concentration was decreased from 4 to 8 mg/L after 150 min ultrasound time, at pH=5.4 and at 25°C (Table 6). Anju et al. [22] showed that the primary degradation products of phenol were hydroquinone, catechol, and benzoquinone. The differences in the phenol metabolites can be attributed to the differences in wastewater composition, to the operational conditions and to the nano-sized TiO_2 concentration.

Throughout the simultaneous ultrasound in the presence of nano-sized TiO_2 in the OMW, the bonds in nano-sized TiO_2 could be easily attacked by OH and H^+ cations that originate from H_2O_2 and H_2O molecules, respectively, to subsequently form O Ti^+- O^-- OH• and O Ti^+- OH• ions. These ions would combine with each other on the surface of original nano-sized TiO_2 particles. The generation of OH• radicals from the pyrolysis of H_2O during ultrasonic irradiation of phenol was shown in equation (1):

$$C_6H_5-OH+OH^{\bullet} \rightarrow C_6H_5(OH)_2 + phenol\ metabolites \rightarrow CO_2+H_2O \quad (1)$$

The phenol metabolite concentrations increased after 60 and 120 min ultrasound time with 4 mg/L nano-sized NiO addition. 550 mg/L catechol, 377 mg/L tyrosol, 330 mg/L quercetin, 314 mg/L caffeic acid, 344 mg/L 4-methyl catechol, 226 mg/L 2-PHE and 251 mg/L 3-PHE were produced with the addition of nano-sized NiO=4 mg/L after 60 min ultrasound time, at pH=5.4 at 25°C. These phenol metabolites were removed with yields of 59, 56, 64, 61, 60, 61 and 63%, respectively, at the aforementioned nano-sized NiO concentration after 150 min ultrasound time, at pH=5.4 and at 25°C while 478 mg/L total phenol remained in the sonicator reactor. The catechol, tyrosol, quercetin, caffeic acid, 4-methyl catechol, 2-PHE and 3-PHE removal efficiencies increased to 82, 80, 78, 82, 81, 81 and 78% as the nano-sized NiO concentration was increased from 4 mg/L to 7 mg/L after 150 min ultrasound time, at pH=5.4 and at 25°C. The metabolite yields decreased to 70, 76, 72, 71,

			TiO_2=2 mg/L[a] ; TiO_2=4 mg/L[b] ; TiO_2=8 mg/L[c]						
Time (min)	PHE_0[d]	Control	Initial phenol metabolites concentrations [d]						
			2-PHE[d]	3-PHE[d]	Catechol[d]	4–methyl catechol[d]	Tyrosol[d]	Quercetin[d]	Caffeic acid[d]
0	2990	2990	4	9	23	15	42	21	30
Time (min)	PHE_0[d]	PHER[e]	Phenol degradation metabolites						
			2-PHE[e]	3-PHE[e]	Catechol[e]	4–methyl catechol[e]	Tyrosol[e]	Quercetin[e]	Caffeic acid[e]
150[a]	299	90	64	68	62	63	61	68	64
150[b]	90	97	84	83	84	84	83	82	85
150[c]	179	94	72	69	78	77	79	76	75
			NiO=4 mg/L[f] ; NiO=7 mg/L[g] ; NiO=10 mg/L[h]						
Time (min)	PHE_0[d]	PHER[e]	Phenol degradation metabolites						
			2-PHE[e]	3-PHE[e]	Catechol[e]	4–methyl catechol[e]	Tyrosol[e]	Quercetin[e]	Caffeic acid[e]
150[f]	478	84	61	63	59	60	56	64	61
150[g]	269	91	81	78	82	81	80	78	82
150[h]	359	88	69	64	70	73	76	72	71
			ZnO=1 mg/L[j]; ZnO=3 mg/L[k]; ZnO=5 mg/L[m]						
Time (min)	PHE_0[d]	PHER[e]	Phenol degradation metabolites						
			2-PHE[e]	3-PHE[e]	Catechol[e]	4-methyl catechol[e]	Tyrosol[e]	Quercetin[e]	Caffeic acid[e]
150[j]	419	86	62	65	60	61	58	65	62
150[k]	209	93	82	80	83	82	81	80	84
150[m]	299	90	70	66	72	75	77	74	73

PHE_0: Initial total phenol concentration (mg/L),
PHER: Total phenol removal efficiency (%),
2-PHE (%): 2-phenyl-phenol removal efficiency (%),
3-PHE (%): 3-phenyl-phenol removal efficiency (%),
[a]TiO_2=2 mg/L,
[b]TiO_2=4 mg/L,
[c]TiO_2=8 mg/L,
[d]Concentration (mg/L),
[e]Removal efficiency (%),
[f]NiO=4 mg/L,
[g]NiO=7 mg/L,
[h]NiO=10 mg/L,
[j]ZnO=1 mg/L,
[k]ZnO=3 mg/L,
[m]ZnO=5 mg/L.

Table 6: Measurements of phenol metabolites (catechol, 4-methyl catechol, tyrosol, quercetin, caffeic acid, 2-PHE and 3-PHE) in the OMW with the addition of some nano-sized TiO_2, NiO and ZnO oxides with HPLC after 150 min ultrasound time at pH=5.4, at an initial total phenol concentration=2990 mg/L, at ultrasound power=640 W, at ultrasound frequency=35 kHz and at 25°C (n=3, mean values).

73, 69 and 64%, at nano-sized NiO=10 mg/L after 150 min ultrasound time, at 30°C and pH=5.4 (Table 6). The increase in degradation yields of phenol metabolites at a 4 mg/L nano-sized NiO concentration can be attributed to the increase in the number of active cavitation bubbles and consequent generation of more OH• radicals with simultaneous adsorption of metabolites to the aforementioned nano-sized metal oxides.

Catechol, tyrosol, quercetin, caffeic acid, 4-methyl catechol, 2-PHE and 3-PHE were removed with yields of 60, 58, 65, 62, 61, 62 and 65%, respectively, in nano-sized ZnO=1 mg mg/L after 150 min ultrasound time, at pH=5.4 and at 30°C while 419 mg/L total phenol remained in the sonicator reactor. The removal yields in the phenol metatabolites increased to 83, 81, 80, 84, 82, 82 and 80% removals were respectively, in nano-sized ZnO=3 mg/L after 150 min ultrasound time, at 25°C and at pH=5.4. The removal efficiencies decreased to 72, 77, 74, 73, 75, 70 and 66%, respectively, as the nano-sized ZnO concentration was increased from 3 mg/L to 5 mg/L after 150 min ultrasound time, at pH=5.4 and at 30°C (Table 6). Throughout ultrasound the highly excited holes on on the surface of nano-sized ZnO not only directly decompose the phenol to their metabolites but also phenol molecules adsorbed on the surface of nano-sized ZnO nano-particles, but also oxidize the H_2O to produce OH• radicals with high activity and indirectly degrade the phenols molecules to their metabolites in the OMW.

It can be assumed that the total phenol concentrations of 90, 269 and 209 mg/L remaining from the ultrasound at nano-sized TiO_2=4 mg/L (see Table 6), at nano-sized NiO=7 mg/L and at nano-sized ZnO=3 mg/L, respectively, at 30°C, after 150 min ultrasound time and at pH=5.4 could not be cleaved to other phenol metabolites namely, hydroxytyrosol, o-quinone, *p*-coumaric acid and *p*-hydroxybenzaldehyde throughout ultrasound since their concentrations were detected as zero. Probably the phenol remaining from the ultrasound can be the phenol which was not converted to corresponding metabolites given above and to the other metabolites not measured in this study (oleurapein, quercetin, ferulic acid, and vanillic acid) through ultrasound. In the study by [5] 45% catechol, 52% tyrosol, 42% quercetin, 49% caffeic acid and 30% 2-PHE removal efficiencies were found at a 80 kHz frequency with 140 min ultrasound time. The higher phenol and phenol metabolites removals found in our study could be originated from the composition of the OMW, from the operarational conditions such as the ultrasound power, ultrasound intensity, ultrasound frequency and ultrasound temperature.

Conclusions

The maximum COD_{dis}, color, TAAs and total phenol yields were > 90% with 4 nano sized mg/L TiO_2 and 5 mg/L ZnO and 4 mg/L TiO_2. The combination of ultrasound and nano-sized metal oxides appears to have a positive synergistic effect on the reduction of COD_{dis}, total phenol, color, TAAs and phenol metabolites from the the OMW. COD_{dis}, phenol by-products and color are mainly eliminated by $OH^{\bullet}$ radicals while the dominant TAAs degradation mechanism is pyrolysis throughout ultrasound.

Acknowledgements

This research study was undertaken in the Environmental Microbiology Laboratory at Dokuz Eylül University Engineering Faculty Environmental Engineering Department, İzmir, Turkey. The authors would like to thank this body for providing financial support.

References

1. Lafi WK, Shannak B, Al-Shannag M, Al-Anber Z, Al-Hasan M (2009) Treatment of olive mill wastewater by combined advanced oxidation and biodegradation. Sep Purif Technol 70: 141-146.
2. El Hajjouji H, Merlina G, Pinelli E, Winterton P, Revel JC, et al. (2008) 13C NMR study of the effect of aerobic treatment of olive mill wastewater (OMW) on its lipid-free content. J Hazard Mater 154: 927-932.
3. Francioso O, Ferrari E, Saladini M, Montecchio D, Gioacchini P (2007) TG–DTA, DRIFT and NMR characterisation of humic-like fractions from olive wastes and amended soil. J Hazard Mater 149: 408-417.
4. Uğurlu M, Karaoğlu MH (2011) TiO2 supported on sepiolite: Preparation, structural and thermal characterization and catalytic behaviour in photocatalytic treatment of phenol and lignin from olive mill wastewater. Chem Eng J 166: 859-867.
5. Khoufi S, Aouissaoui H, Penninckx M, Sayadi S (2004) Application of electro-fenton oxidation for the detoxification of olive mill wastewater phenolic compounds. Water Sci Technol 49: 97-102.
6. Pang YL, Abdullah AZ, Bhatia S (2011) Review on sonochemical methods in the presence of catalysts and chemical additives for treatment of organic pollutants in wastewater. Desalination 277: 1-14.
7. Chowdhury AN, Rahim A, Ferdosi YJ, Azam MS, Hossain MM (2010) Cobalt–nickel mixed oxide surface: A promising adsorbent for the removal of PR dye from water. Appl Surf Sci 256: 3718-3724.
8. Zhang P, Ma X, Guo Y, Cheng Q, Yang L (2012) Size-controlled synthesis of hierarchical NiO hollow microspheres and the adsorption for congo red in water. Chem Eng J 189-190: 188-195.
9. Ahmed MA (2012) Synthesis and structural features of mesoporous NiO/TiO2 nanocomposites prepared by sol–gel method for photodegradation of methylene blue dye. J Photochem Photobiol A Chem 238: 63-70.
10. Abu Tarboush BJ, Husein MM (2012) Adsorption of asphaltenes from heavy oil onto in situ prepared NiO nanoparticles. J Colloid Interface Sci 378: 64-69.
11. Sponza DT, Oztekin R (2011) Contribution of oxides, salt, and carbonate to the sonication of some hydrophobic polyaromatic hydrocarbons and toxicity in petrochemical industry wastewater in İzmir, Turkey. J Environ Eng 137: 1012-1025.
12. Eaton AD, Clesceri LS, Rice EW, Greenberg AE, Franson MAH (2005) (Ed.) In: M.A.H. Franson (Eds.) twenty first edn, Standard Methods for the Examination of Water and Wastewater. American Public Health Association (APHA), American Water Works Association (AWWA), Water Environment Federation (WEF), American Public Health Association 800 I Street, NW Washington DC, 20001-3770, USA.
13. Olthof M, Eckenfelder Jr WW (1976) Coagulation of textile wastewater. Text Chem Color 8: 18-22.
14. Silva AMT, Nouli E, Carmo-Apolinário AC, Xekoukoulotakis NP, Mantzavinos D (2007) Sonophotocatalytic/H2O2 degradation of phenolic compounds in agro-industrial effluents. Catal Today 124: 232-239.
15. Bertin L, Majone M, Di Gioia D, Fava F (2001) An aerobic fixed-phase biofilm reactor system for the degradation of the low-molecular weight aromatic compounds occurring in the effluents of anaerobic digestors treating olive mill wastewaters. J Biotechnol 87: 161-177.
16. Villeneuve L, Alberti L, Steghens JP, Lancelin JM, Mestas JL (2009) Assay of hydroxyl radicals generated by focused ultrasound. Ultrason Sonochem 16: 339-344.
17. David B (2009) Sonochemical degradation of PAH in aqueous solution. Part I: Monocomponent PAH solution. Ultrason Sonochem 16: 260-265.
18. Canizares-Macias MP, Garcia-Mesa JA, Lugue de Castro MD (2004) Fast ultrasound-assisted method for the determination of the oxidative stability of virgin olive oil. Analyt Chim Acta 502: 161-166.
19. Atanassova D, Kefalas P, Petrakis C, Mantzavinos D, Kalogerakis N, et al. (2005) Sonochemical reduction of the antioxidant activity of olive mill wastewater. Environ Int 31: 281-287.
20. Benedito J, Mulet A, Clemente G, Garcia-Perez JV (2004) Use of ultrasonics for the composition assessment of olive mill wastewater (alpechin). Food Res Int 37: 595-601.
21. Petrier C, Lamy MF, Francony A, Benahcene A, David B, et al. (1994) Sonochemical degradation of phenol in dilute aqueous solutions: Comparison of the reaction rates at 20 and 487 kHz. J Phys Chem 98: 10514-10520.
22. Anju SG, Yesodharan S, Yesodharan EP (2012) Zinc oxide mediated sonophotocatalytic degradation of phenol in water. Chem Eng J 189-190: 84-93.
23. Currell DL, Wilhelm G, Nagy S (1963) The effect of certain variables on the ultrasonic cleavage of phenol and of pyridine. J Am Chem Soc 85: 127-130.
24. Wu Z, Ondruschka B (2005) Roles hydrophobicity and volatility of organic substrates on sonolytic kinetics in aqueous solutions. J Phys Chem A 109: 6521-6526.
25. Wang J, Jiang Z, Zhang Z, Xie Y, Wang X, et al. (2008) Sonocatalytic degradation of acid red B and rhodamine B catalyzed by nano-sized ZnO powder under ultrasonic irradiation. Ultrason Sonochem 15: 768-774.
26. Nalini Vijaya Laxmi P, Saritha P, Rambabua N, Himabindua V, Anjaneyulu Y(2010) Sonochemical degradation of 2-chloro-5-methyl phenol assisted by TiO2 and H2O2. J Hazard Mater 174: 151-155.
27. Cardoso SM, Mafra I, Reis A, Nunes C, Saraiva JA, et al. (2010) Naturally fermented black olives: effect on cell wall polysaccharides and on enzyme activities of Taggiasca and Conservolea varieties. LWT-Food Sci Technol 43: 153-160.
28. Khokhawala IM, Gogate PR (2010) Degradation of phenol using a combination of ultrasonic and UV irradiations at pilot scale operation. Ultrason Sonochem 17: 833-888.
29. Wu CH (2008) Effects of sonication on decolorization of C.I. reactive red 198 in UV/ZnO system. J Hazard Mater 153: 1254-1261.
30. Drijvers D, Langenhove HV, Beckers M (1999) Decomposition of phenol and trichloroethylene by the ultrasound/H2O2/CuO process. Water Res 33: 1187-1194.
31. Srivastava DN, Perkas N, Seisenbaeva GA, Koltypin Y, Kessler VG, et al. (2003) Preparation of porous cobalt and nickel oxides from corresponding alkoxides using a sonochemical technique and its application as a catalyst in the oxidation of hydrocarbons. Ultrason Sonochem 10: 1-9.
32. Song Z, Chen L, Hu J, Richards R (2009) NiO (111) nanosheets as efficient and recyclable adsorbents for dye pollutant removal from wastewater. Nanotechnology 20.
33. He H, Yang S, Yu K, Ju Y, Sun C, et al. (2010) Microwave induced catalytic degradation of crystal violet in nano-nickel dioxide suspensions. J Hazard Mater 173: 393-400.

Effect of Partial Partitions on Natural Convection in Air Filled Cubical Enclosure with Hot Wavy Surface

Farah Zemani, Amina Sabeur-Bendehina* and Mustapha Boussoufi

Laboratoire d'Energie et Propulsion Navale, Faculté de Génie Mécanique, Université des Sciences et de la Technologie Mohamed, Boudiaf d'Oran, B.P. 1505 Oran El-M'naouar, 3-1000 Oran, Algeria

Abstract

Natural convection in cubical enclosure with hot surface geometry and partial partitions has been analyzed. The geometry is a cube with wavy hot surface (three undulations) and three partitions. The investigation has been performed for different partitions lengths and Rayleigh number while the Prandtl number kept constant. This problem is solved by using the partial differential equations which are the equation of mass, momentum, and energy. The results obtained show that the hot wall geometry with partitions affects the flow and the heat transfer rate in the cavity. It has been found also that the mean Nusselt number decreases compared with the heat transfer in the undulated cubical cavity without partitions.

Keywords: Natural convection; Cubical enclosure; Wavy hot surface; Partial partitions

Nomenclature

g: Gravitational acceleration [m/s^2]

L: Cavity side [m]

Nu: Nusselt number

p: Static pressure [Pa]

P: dimensionless pressure

Pr: Prandtl number, ν/α

Ra: Rayleigh number, $g\beta L^3 \Delta T/a\nu$

T: Temperature [K]

T_0: Average temperature, $(T_c+T_f)/2$ [K]

T_h, T_c: Hot and cold temperature [K]

ΔT: Temperature variation, $T_h - T_c$ [K]

U,V,W: Dimensionless fluid velocities

u,v,w: Velocity components in x,y,z directions [m/s]

X,Y,Z: Dimensionless Cartesian coordinates, x/L, y/L, z/L

x,y,z: Coordinates [m]

Greek symbols

α: Thermal diffusivity of air [m^2s^{-1}]

β: Thermal expansion coefficient [k^{-1}]

θ: Dimensionless temperature

ν: Kinematic viscosity [m^2/s]

ρ: air density [Kg/m^3]

subscripts

a: average

Introduction

Natural convection flow analysis in enclosures has many thermal engineering applications, such as cooling of electronic devices, energy storage systems and fire-safe compartment. In the numerical domain, a big progress has been made on the natural convection in complex geometries [1-4].

For many years a lot of attention has been paid to the problems of natural convection in enclosure. However, the limitations of numerical tools and experimental techniques restrict investigators within the approximation of a two dimensional model even though fluid motion is three dimensional in nature. Bessonov et al. [5] have suggested a benchmark numerical solution for the three dimensional natural convection in cubical enclosure. Three dimensional laminar flow has been also studied by Mallinson et al. [6] and Lee et al. [7] for enclosures of the length aspect ratio A_z varying from 2 to the enclosure with A_z =1 and 2 have been considered by Lankhorst and Hoogendoorn [8] who computed steady flows for numbers of Rayleigh ranging from 10^6 to 10^8. Fusegi et al. [9] made a three dimensional flow analysis on natural convection in a differentially heated cubical enclosure; the detailed structures of the fields were scrutinized by using high-resolution computational results over the range of Rayleigh numbers studied, $10^3 \leq Ra \leq 10^6$, they clarified three dimensional structures of flow, vorticity and temperature in the cavity; they also compared their numerical results with the experimental measurements as studied by Bilsk et al. [10].

The complexity of such problems increases when the hot surface of the cubical enclosure becomes wavy. Adjlout et al. [11] have also investigated the two dimensional natural convection in an inclined cavity with hot wavy wall. They simulated the fluid and heat transfer with ADI scheme; one of their interesting results was a decrease of the averaged heat transfer compared with the mean Nusselt number of the square

***Corresponding author:** Amina Sabeur-Bendehina, Laboratoire d'Energie et Propulsion Navale, Faculté de Génie Mécanique, Université des Sciences et de la Technologie Mohamed, Boudiaf d'Oran, B.P. 1505 Oran El-M'naouar, 3-1000 Oran, Algeria

cavity. They showed that the thermal boundary layer is considerably affected by wavy wall and they recommended investigating the hot wall geometry optimisation. Natural convection in three dimensional rectangular enclosures had been analyzed numerically by Sik Lee et al. [12]; the effect of the Rayleigh number was mainly investigated. The results showed that the temperature disturbance imposed on the end wall reinforced the axial flow and magnified the three dimensional effect. Three dimensional structures in laminar natural convection in a cubic enclosure were investigated experimentally by Hiller et al. [13]. The Rayleigh numbers ranged from 10^4 to 2×10^7 and the Prandtl numbers from 5.8 to 6×10^3. They showed the velocity and vorticity fields and compared the experimental observations with numerical calculations found in the literature. Gunes [14] studied numerically in his thesis three dimensional natural convection cooling in enclosures and vertical channels with spatially periodic flush mounted and protruding heat sources. It was established that convection reduces the maximum operating temperature for all configurations. Yu and Joshi [15] carried out a numerical study of a three dimensional laminar natural convection in a vented enclosure. They presented the local and overall heat transfer from the heat source and the substrate, in terms of Nusselt numbers and the surface temperatures to illustrate the vent effects. Akrour et al. [16] made numerically a three dimensional steady flow on natural convection in a differentially heated cubical enclosure. The variation of Nusselt numbers on the hot and cold walls were also presented to show the overall heat transfer characteristics inside the enclosure. They found that the three dimensional data demonstrate reasonable agreement with the experimental measurement. Natural convection heat transfer associated to fluid dynamics phenomena was studied extensively by Corzo et al. [17]. The results presented revealed a good agreement not only in 2D also in 3D and for a wide range of Ra numbers. Kürekci and Özcan [18] made an experimental and numerical study of laminar natural convection in a differentially heated cubical enclosure. They compared numerical and experimental results and were found to be in general agreement with each other. Ahmadi and Dastgerdi [19] carried out a numerical analysis to study the laminar natural convection in a three dimensional enclosure. Results exposed that the average Nusselt number increases by increasing in Ra, the rate of heat transfer from hot and cold sources and maximum component of velocity increase. Lahoucine et al. [20] carried out the three dimensional numerical study of natural convection in a cubical enclosure with two heated square sections submitted to periodic temperatures, filled with air. The obtained results showed significant changes in terms of heat transfer and flow intensity. Sabeur et al. [21] performed a numerical investigation of the influence of the hot surface geometry on a laminar natural convection in a cubical cavity filled with air differentially heated. The results obtained showed that the hot wall geometry affects the flow and the heat transfer rate in the cavity. The mean Nusselt number decreases compared with the heat transfer in the cubical cavity.

Fusegi et al. [22] studied numerically a transient three dimensional natural convection in a differentially heated cubical enclosure at Rayleigh number of 10^6. They showed that the behavior of the heat transfer rate in the enclosure was considerably influenced by the presence of the internal gravity wave motion. They also [23] carried out a three dimensional numerical simulation of periodic natural convection in a differentially heated cubical enclosure; at the Rayleigh number of 8.5×10^6, they found that the period of the oscillations was consistent with the experimental measurements. In another study, the same authors [24] made numerical simulations of natural convection in a differentially heated cubical enclosure with a partition. They scrutinized the effects of the partition geometry on the three dimensional flow properties. Silva and Gosselin [25] evaluated the effect of the aspect ratio and horizontal length of a high conductivity rectangular fin attached to the hot wall of a three dimensional differentially heated cubic enclosure in laminar natural convection, a scale analysis was used to predict the domain in which the fin geometry played a significant role. Frederick [26] made a numerical study of natural convection of air in a differentially heated cubical enclosure with a thick fin placed vertically in the middle of the hot wall; he investigated the variation of overall Nusselt number with Rayleigh number and thermal conductivity ratio. He [27] studied also a numerically natural convection heat transfer in a cubical enclosure with two active sectors on one vertical wall over a wide range of Rayleigh number. He described the flow patterns and temperature distribution and proposed an expression for overall heat transfer. In another study, Frederick and Moraga [28] investigated numerically three dimensional natural convection of air in a cubical enclosure with a fin on the hot wall. It was concluded that for 10^5 Ra 10^6, a fin of partial width is more effective in promoting heat transfer than a fin of full width.

Bocu and Altac [29] studied numerically laminar natural convection heat transfer in 3D rectangular air filled enclosures, with pins attached to the active wall. The Rayleigh numbers considered in their study ranges from 10^5 to 10^7; they concluded that the mean Nusselt number increases with increasing Rayleigh number for a fixed case. The present investigation is an extension of the work already established by Belkadi et al. [30] which they treated similar cavity in two dimensional study. The most part of our simulation is to show the three dimensional effects and the influence of the hot wavy wall with partial partitions on the laminar natural convection.

Analysis

The geometry and coordinate system are illustrated in Figure 1. The geometry is simply represented by a cubical cavity. This latter has a hot surface geometry wavy with partitions introduced at the crests with the constant temperature, T_h. The opposite wall is straight and has a constant colder temperature Tc. The others are thermally insulated. The study has been carried out for different partitions lengths and for two Rayleigh numbers (10^5 and 10^6); the Prandtl number is equal to 0.71.

Governing equations

The viscous incompressible flow inside a cubical enclosure and a temperature distribution are described by the Navier-Stokes and the energy equations. The governing equations for this case are for laminar and three dimensional flow. In the model development, the following assumptions are adopted:

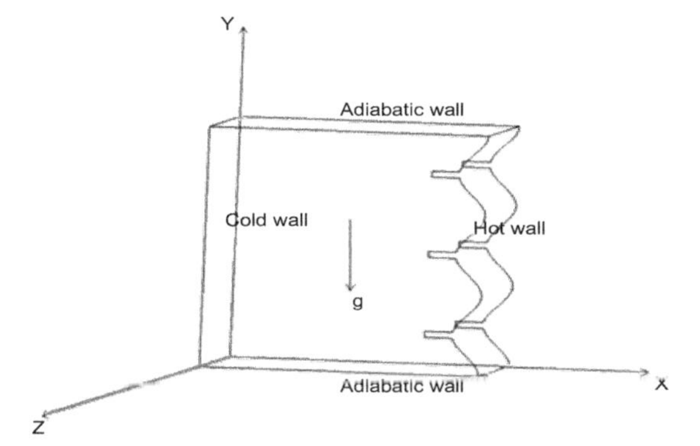

Figure 1: Physical domain.

- The process is in steady state.
- The Boussinesq approximation applies, which implies that except for the density in the gravitational term, all other properties in the governing equations are kept constant.
- There is no source or sink in the system.

Once above assumptions are employed into the conservation equation of mass, momentum and energy and the following dimensionless variable are introduced:

$$X = \frac{x}{L}\ Y = \frac{y}{L}\ Z = \frac{z}{L} \quad (1)$$

$$U = \frac{u}{\nu / L}\ V = \frac{v}{\nu / L}\ W = \frac{w}{\nu / L} \quad (2)$$

$$P = \frac{p + \rho g y}{\rho(\nu / L)^2} \quad (3)$$

$$\theta = \frac{T_0 - T_c}{T_h - T_c}\ T_0 = \frac{T_h + T_c}{2} \quad (4)$$

The dimensionless governing equations in Cartesian coordinates for the present study then take the following forms:

Continuity

$$\frac{\partial U}{\partial X} + \frac{\partial V}{\partial Y} + \frac{\partial W}{\partial Z} = 0 \quad (5)$$

X Momentum

$$U\frac{\partial U}{\partial X} + V\frac{\partial U}{\partial Y} + W\frac{\partial U}{\partial Z} = -\frac{\partial P}{\partial X} + Pr\left(\frac{\partial^2 U}{\partial X^2} + \frac{\partial^2 U}{\partial Y^2} + \frac{\partial^2 U}{\partial Z^2}\right) \quad (6)$$

Y Momentum

$$U\frac{\partial V}{\partial X} + V\frac{\partial V}{\partial Y} + W\frac{\partial V}{\partial Z} = -\frac{\partial P}{\partial Y} + Pr\left(\frac{\partial^2 V}{\partial X^2} + \frac{\partial^2 V}{\partial Y^2} + \frac{\partial^2 V}{\partial Z^2}\right) + Ra\,Pr\,\theta \quad (7)$$

Z Momentum

$$U\frac{\partial U}{\partial X} + V\frac{\partial W}{\partial Y} + W\frac{\partial W}{\partial Z} = -\frac{\partial P}{\partial Z} + Pr\left(\frac{\partial^2 W}{\partial X^2} + \frac{\partial^2 W}{\partial Y^2} + \frac{\partial^2 W}{\partial Z^2}\right) \quad (8)$$

Energy

$$U\frac{\partial \theta}{\partial X} + V\frac{\partial \theta}{\partial Y} + W\frac{\partial \theta}{\partial Z} = \left(\frac{\partial^2 \theta}{\partial X^2} + \frac{\partial^2 \theta}{\partial Y^2} + \frac{\partial^2 \theta}{\partial Z^2}\right) \quad (9)$$

Knowing that $0 \le Y \le 1$ and $0 \le X \le f(Y)$ with $f(Y) = [1 - amp + amp (\cos 2 n Y)]$ and $0 \le Z \le 1$ with n and amp are respectively, the number of undulations and amplitude as proposed by Adjlout et al. [11]. The geometry of the tested cavity is shown in Figure 1.

The study is completed with the definition of the following boundary conditions:

$\theta = 0.5$ U=V=W=0 on the hot wall

$\theta = -0.5$ U=V=W=0 on the cold wall

$\frac{\partial \theta}{\partial Y} = 0$ U=V=W=0 on the adiabatic walls

It can be seen from the above dimensionless equations that the Rayleigh number and Prandtl number are two of the important model parameters for a given flow geometry. The overall heat transfer characteristics are described by the average Nusselt number which is defined over the hot wall as follows:

$$\overline{Nu} = \frac{1}{L}\int_{X=0}^{1}\int_{Z=0}^{1} \frac{\partial \theta}{\partial Y} dX dZ \quad (10)$$

Numerical method and model validation

A 3D uniform and staggered grid is used with a control volume formulation for the discretization. The central difference scheme is employed for the convective diffusive transport variables. Pressure correction and velocity correction are implemented in accordance with the SIMPLE algorithm to achieve a converged solution. The discretized algebraic equations are solved by the tri-diagonal matrix algorithm (TDMA). Relaxation factors of about 0.2-0.7 are used for the velocity components, while relaxation factors of about 0.5-0.8 are adopted for the temperature and pressure corrections. The adiabatic boundary condition is treated by the additional source term method. In the present study, typically 2000-8000 outer iterations are required to achieve convergence. For the convergence criteria, the relative variations of the temperature and velocity between two successive iterations are imposed to be smaller than the specified accuracy levels of 10^{-6}. A grid independency test was carried out, and the results are indicated in Table 1. Three sets of grids 54 × 54 × 54, 60 × 60 × 60 and 67 × 67 × 67 were employed; the case with 67×67×67 grids (Figure 2) was used for taking both the accuracy and convergence rate into account.

Results and Discussion

The streamlines and Isotherms in the computation domain for the case mentioned above and for Ra= 10^5 and 10^6 and different partitions length and in different planes XYZ. It was noticed in Figures 3 and 4 that near the wavy geometry, the thermal boundary layer thickness increases and decreases just before the partitions or just after a trough resulting in a big decrease in the global heat transfer along the hot geometry. When Compared with the cube and the undulated cavity without partitions, the thermal boundary thickness appears to enhance with the presence of partitions. Every partition takes part to the thickening of the thermal

Grid	54 × 54 × 54	60 × 60 × 60	67 × 67 × 67	Bessov [5]
Nu_a	4.458	4.325	4.287	4.339

Table 1: Comparison of average Nusselt Number at several grids for cubical cavity Ra= 10^5.

Figure 2: Grid of the geometry.

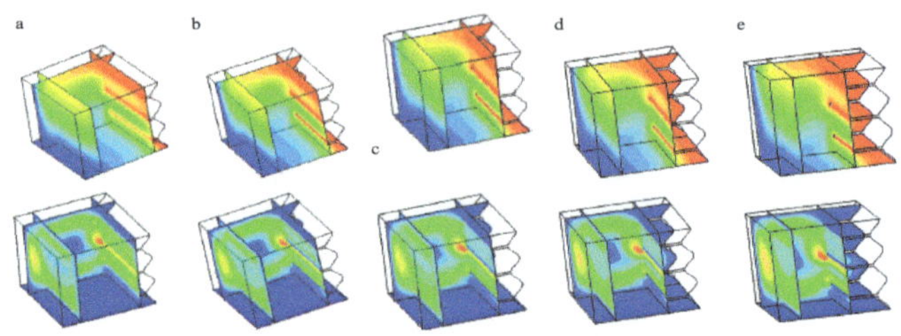

Figure 3: Temperature field and Isosurfaces of the absolute values of the vorticity at Ra=10^5 for all partitions Length and at different planes: X=0.84 Y=0.2 and Y=0.8 a-1 amp, b-2 amp, c-3amp, d-4 amp, e-5 amp.

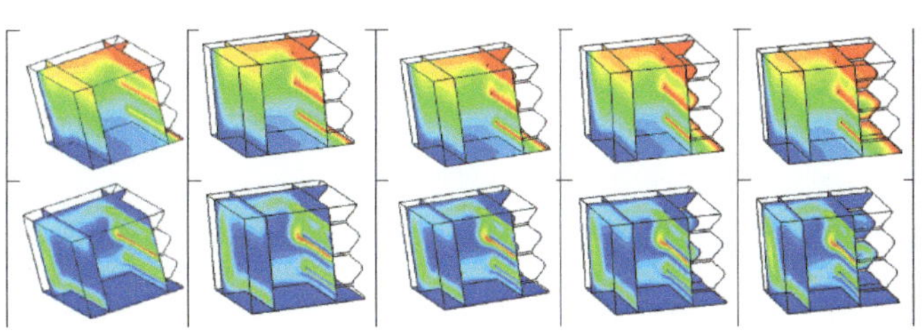

Figure 4: Temperature field and Isosurfaces of the absolute values of the vorticity at Ra=10^6 for all partitions Length and at different planes: X=0.84 Y=0.2 and Y=0.8 a-1 amp, b-2 amp, c-3amp, d-4 amp, e-5 amp.

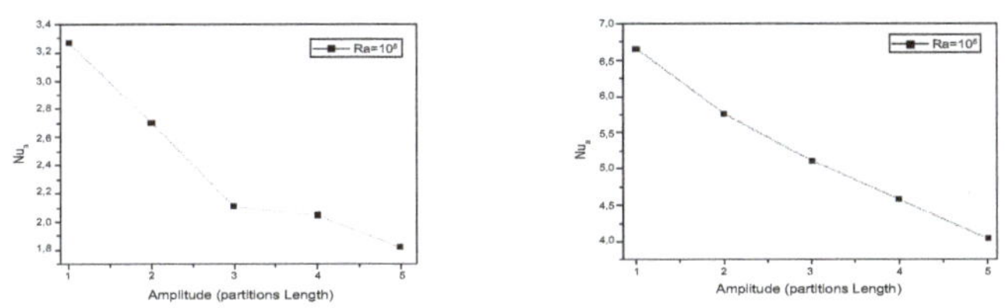

Figure 5: Average Nusselt number variation according to amplitude(partition length) for Ra=10^5 and Ra=10^6.

Figure 6: Local Nusselt Number distributions at the heated wall for a-Cube ,b-undulated cavity without partitions c-undulated cavity with partitions (1amp) d- undulated cavity with partitions (2amp) e-undulated cavity with partitions (3amp) for Ra=10^5.

Geometry (2D)	square	Undulated cavity without partitions	Undulated cavity With partitions
Nu_a	**4.52**	**3.51**	**3.36**
Geometry (3D)	Cube	Undulated cavity without partitions	Undulated cavity With partitions
Nu_a	**4.32**	**3.48**	**3.26**

Table 2: Comparison of the Mean Nusselt Number between the cubical and undulated enclosure (Ra=10^5).

Geometry (2D)	square	Undulated cavity without partitions	Undulated cavity With partitions
Nu_a	**8.82**	**7.55**	**6.97**
Geometry (3D)	Cube	Undulated cavity without	Undulated cavity With

Table 3: Comparison of the Mean Nusselt Number between the cubical and undulated enclosure (Ra=10^6).

Geometry	1 Amp	2 Amp	3 Amp	4 Amp	5 Amp
Nu_a	**3.26**	**2.7**	**2.11**	**2.04**	**1.82**

Table 4: Mean Nusselt Number for tested enclosures for (Ra=10^5).

Geometry	1 Amp	2 Amp	3 Amp	4 Amp	5 Amp
Nu_a	**6.65**	**5.76**	**5.11**	**4.58**	**4.03**

Table 5: Mean Nusselt Number for tested enclosure for (Ra=10^6).

boundary layer. That shows the induction of the partition length on isotherms and streamlines distributions. It was clearly seen that an increase in partitions length seems to increase in the thermal boundary thickness.

Figure 6 presents the local Nusselt Number distributions at the heated wall for undulated cavity with wavy hot geometry with partitions, undulated cavity without partitions and cube (x=1) for Ra=10^5. The wavy tendency of the local Nusselt number is describing the cavities with corrugated wall. It was observed also that for the undulated cavity, the heat transfer decreases notably compared with the undulated cavity without partition and the cube as illustrated in Table 2. The relative decrease of the global heat transfer for the cavity with hot surface geometry and partitions reaches 40% for Ra=10^6 compared with a cube.

The distribution of average Nusselt number for different partition length are represented in Figure 5 for all Rayleigh numbers investigated (Ra=10^5 and 10^6 respectively). It was clearly seen that there is an influence of the partition length on the mean Nusselt number so; the trend of the curve is notably decreasing with an increase in the partition length (Tables 3-5).

Conclusion

The present investigation deals with the effect of the presence of a partition on the hot wall geometry of undulated cubical cavity, air filled, and differentially heated enclosure. These partitions caused decrease of up 40% in the heat transfer with respect to the case at the same Rayleigh number. As Ra grew, boundary layer got progressively thinner, and the effects of the partitions on the temperature field became more and more localized around the partitions. This was confirmed by the results, the percentage heat transfer reduction grows with the partitions length which is a function of amplitude of the waviness hot geometry. As the insulating effect was related to a local distortion of the temperature field, it can be concluded that multiple partitions would be required in such cavities to reach the levels of heat transfer reduction.

References

1. De Vahl Davis G (1983) Natural Convection of air in a square cavity: a benchmark numerical solution, International Journal of Numerical Methods in Fluids 3: 249-264.
2. Catton I (1978) Natural Convection in enclosures. In: Proceeding of the sixth international Heat Transfer Conferences 6: 13-31.
3. Ostrach S (1972) Natural convection in enclosures. Advances in Heat Transfer 8: 161-227.
4. Yang KT (1988) Transitions and bifurcations in laminar buoyant flows in confined enclosures. J Heat Transfer 110: 1191-1204.
5. Bessonov OA, Brailovskaya VA, Nikitin SA, Polezhaev VI (1997) Three dimensional natural convection in a cubical enclosure: A Bench Mark numerical solution. Proceeding of computational heat transfer symposium.
6. Mallinson JD, de Vahl Davis G (1977) Three-dimensional natural convection in a box: a numerical study. J Fluid Mech 83: 1-31.
7. Lee TS, Son GH, Lee JS (1988) Numerical predictions of three-dimensional natural convection in a box. Proc 1st KSME-JSME Thermal and Fluids Engng Conf. 2: 278-283.
8. Lankhorst AM, Hoogendoorn CJ (1988) Three-dimensional numerical calculations of high Rayleigh number natural convective flows in enclosed cavities. Proc Natn Heat Transfer Conf ASME HTD-96, 3: 463-470.
9. Fusegi T, Hyin JM, Kuwahara K (1991) A numerical study of 3D natural convection in a differently heated cubical enclosure. Fluid Dynamics Research 8: 5-6.
10. Bilski SM, Lloyd JR, Yang KT (1986) An experimental investigation of the laminar natural convection velocity in square and partitioned enclosure. In: Proceeding of the eight international Heat Transfer Conference 4: 1531-1518.
11. Adjlout L, Imino O, Azzi A, Belkadi A (2002) Laminar natural convection in an inclined cavity with a wavy wall. International Journal of Heat and Mass Transfer 45: 2141-2152.
12. Lee TS, Son GH, Lee JS (1989) Numerical study on natural convection in three-dimensional rectangular enclosures. KSME Journal 3: 50-55.

13. Hiller WJ, Koch S, Kowalewski TA (1989) Three-Dimensional Structures in Laminar Natural Convection in a Cubic Enclosure. Experimental Thermal and Fluid Science 2: 34-44.

14. Gunes H (1992) Three dimensional thermal convection in enclosures and vertical channels: applications to cooling of electronic equipment. Theses and Dissertations 148.

15. Yu E, Joshi Y (1997) A numerical study of three-dimensional laminar natural convection in a vented enclosure. Int J Heat and Fluid Flow 18: 600-612.

16. Akrour D, Bennacer R, Kalahe D (2000) Natural convection of air in a square cavity. Laboratoire de Mécanique des Fluides, Faculté de Physique.

17. Corzo SF, Damián SM, Ramajo D, Nigro NM (2011) Numerical simulation of natural convection phenomena, Mecánica Computacional 30: 277-296.

18. Kürekci NA, Özcan O (2012) An experimental and numerical study of laminar natural convection in a differentially-heated cubical enclosure. Isi Billimi ve Teknigi Dergisi 32: 1-8.

19. Ahmadi M, Dastgerdi BB (2012) Fluid flow and heat transfer characteristics of natural convection in square 3-D enclosure due to discrete sources. Proc. of the International Conference on Future Trends in Automation and Robotics - FTAR -114.

20. Belarche L, Abourida B, Mediouni T (2013) Three-dimensional Numerical Study of Natural Convection in a Cubical Enclosure with Two Heated Square Sections Submitted to Periodic Temperatures. IJCSI International Journal of Computer Science Issues 10: 8-16.

21. Sabeur-Bendehina A, Imine O, Adjlout L, Imine B (2007) Effect of the hot surface geometry on laminar Natural Convection in Cubical Air Filled Enclosures. Proceedings of the Third International Conference on Thermal Engineering: Theory and Applications.

22. Fusegi T, Hyun JM, Kuwahara K (1991) Transient three-dimensional natural convection in a differentially heated cubical enclosure. Int J Heat Mass Transfer 34: 1559-1564.

23. Fusegi T, Hyun JM, Kuwahara K (1992) Three-dimensional numerical simulation of periodic natural convection in a differentially heated cubical enclosure. Applied Scientific Research 49: 271-282.

24. Fusegi T (1992) Numerical simulations of natural convection in a differentially heated cubical enclosure with a partition. Int J Heat and Fluid Flow 13: 176-183.

25. da Silva AK, Gosselin L (2005) On the thermal performance of an internally finned three-dimensional cubic enclosure in natural convection. International Journal of Thermal Sciences 44: 540-546.

26. Frederick RL (1997) Natural convection heat transfer in a cubical enclosure with two active sectors on one vertical wall. Int Comm Heat Mass Transfer 24: 507-520.

27. Frederick RL (2007) Heat transfer enhancement in cubical enclosures with vertical fins. Applied Thermal Engineering 27: 1585-1592.

28. Frederick RL, Moraga SG (2007) Three-dimensional natural convection in finned cubical enclosures, International Journal of Heat and Fluid Flow 28: 289-298.

29. Bocu Z, Altac Z (2011) Laminar natural convection heat transfer and air flow in three-dimensional rectangular enclosures with pin arrays attached to hot wall. Applied Thermal Engineering 31: 3189-3195.

30. Belkadi M, Aounallah M, Imine O, Adjlout L, Addad Y (2006) Free convection in an inclined square cavity with partial partitions on a wavy hot wall. Progress in Computational Fluid Dynamics 6: 428-434.

Kinetics Modeling of The Column Adsorption for The Dehydration of Ethanol-Water Mixtures using Biomass Adsorbents

Okewale AO[1*], Igbokwe PK[2] and Babayemi KA[3]

[1]*Department of Chemical Engineering, Federal University of Petroleum Resources, P.M.B., 1221, Effurun, Nigeria*
[2]*Department of Chemical Engineering, Nnamdi Azikiwe University, P.M.B., 5025, Awka, Anambra State, Nigeria*
[3]*Department of Chemical Engineering, Anambra State University, P.M.B., 02, Uli, Anambra State, Nigeria*

Abstract

This study has explored the use of biomass - based adsorbents other than the conventional materials as adsorbent for the dehydration of ethanol - water mixtures. The column study was carried out using enzyme modified corn starch and un- modified corn starch for the adsorptive dehydration of ethanol - water mixtures. The effect of bed heights and flow rates on the breakthrough curves revealed that adsorption efficiency decreased with increased flow rate and increased with bed height. The throughput volume of the anhydrous ethanol - water mixtures increased with increase bed height, due to availability of more sorption sites on the adsorbents. Thomas, Yoon and Nelson, and Adams - Bohart kinetic models were used to describe the packed column study. The column kinetics was best described by Yoon and Nelson and Adams - Bohart models. The presence of the hydroxyl O - H stretch group's spectroscopy confirmed the chemical affinity of these adsorbents for water in the ethanol - water mixture. The amorphous or crystallinity nature of the adsorbents was investigated using the X - RD analysis.

Keywords: Starch; X-RD; FTIR; Kinetics; Packed bed column; Adsorption; Ethanol - water mixtures

Introduction

A lot of research efforts continue to focus on the improvement of the dehydration of ethanol - water mixtures. Protection of the global environment and depletion of the conventional hydrocarbon fuel supplies have motivated researchers to develop alternative fuels including hydro, wind, biofuels, solar and geothermal energy [1,2]. Fuel grade ethanol, biodiesel and biogas are the most promising biofuels being explored in recent times [3]. Among these proposed substitutes, biofuels have drawn more attention. Ethanol is commonly used as a fuel itself or an additive that helps enhanced the octane number and combustibility of gasoline [4]. The use of biofuel is beneficial from the standpoint of the environment, energy security, and economic development [2]. One of the main advantages of using biofuels instead of fossil fuels is reduced carbon dioxide emissions, which are responsible for global warming. One of the energy efficient techniques widely used for ethanol dehydration is process of adsorption. Compared to distillation, which required about 2.8×10^6 J/L of ethanol to break the azeotrope using benzene [5] adsorption is energy efficient with estimates of 5.6×10^5 J/L to arrive at 99.6% ethanol from a 95% solution using starch [6] and 2.0×10^6 J/L to go from 85% to 99.5% using corn meal [7,8].

Low operation cost, high efficiency, and vast variety of sorbents contributed to the use of adsorption in the industrial separation processes. Energy saving potential separation processes like the various pressure swings and thermal swing adsorption process has been used as an alternative separation processes. Adsorption can be carried out either in static or dynamic conditions.

Adsorption by dynamic methods gives the possibility of studying the breakthrough curves, to find the adsorption capacity of the material, the length of the working zone (the one with the constant concentration profile), the breakthrough time (life time of the bed), as well as the velocity at which the external mass transfer resistance can be neglected. Fixed bed operations is influenced by equilibrium (isotherm capacity), kinetics (diffusion and convention coefficients), and hydraulic (liquid hold - up, geometric analysis, and mal - distribution factors [9,10], used complementary X-ray data to show that their 110-120°C starch endotherms were caused by the melting of crystalline V - type amylose - lipid complexes. These structures are formed from collapsed amylose helices in which chemical adjuncts are trapped inside [11]. Waxy maize has an A structure and lipids similar to those in maize but amylose is absent [11].

The starch based adsorbents adsorb water by forming hydrogen bond between the hydroxyl groups on the surface of the adsorbents and the water molecules [12].

Wheat straw and wood chips derived from cellulosic based materials used xylans and cellulose as the major adsorbing mechanism instead of amylopectin used in starch materials. The predisposing characteristics of starch based and other polysaccharide adsorbents lie in their chemical composition, hydrophilic nature and their subsequent regeneration ability [12-15]. The corn starch that is made use of in this work is more cost effective and less energy consuming compared to the other conventional materials that has been employed in ethanol - water separation. The present work was carried out to evaluate the starch based materials potential ability to adsorptively dehydrate ethanol - water mixtures in a fixed bed adsorption column. The adsorption process data were analyzed to study the breakthrough profiles and kinetics. Fourier transform infrared was used to identify the functional group while the X-ray diffractogram was used to determine the amorphous or crystallinity nature of the adsorbents. Enzyme modified corn starch

***Corresponding author:** Okewale AO, Department of Chemical Engineering, Federal University of Petroleum Resources, P.M.B., 1221, Effurun, Nigeria

and un- modified corn starch were used for the adsorptive dehydration of ethanol - water mixtures. Three models (Thomas model, Yoon and Nelson model, and Adams - Bohart model) were used to analyze the column performance.

Materials and Method

Un- modified corn starch was procured from Eke - Awka market, Awka, Anambra Sate, Nigeria. It was sun dried and thermally treated in an oven at 110°C for 16hours and thereafter classified into the desired particle size. Analytical grade of ethanol, de - ionized water, α - amylase, sodium azide, sodium phosphate, sodium chloride, and sodium hydroxide were purchased from accredited chemical dealers in Onitsha, Anambra State, Nigeria using a scale with an accuracy of 0.01 g. The starch adsorbents were packed in an air tight plastic container which was kept in a desiccator prior to use. The method of Beery et al. [16] was used for the enzyme modification of the corn starch. Ethanol - water mixture is prepared at the required mass concentrations of 90 wt% ethanol. The fluid phase concentration was measured with the aid of an Abbe refractometer with automatic calibration in the experimental range of concentration.

Characterization of the corn starch

Starch content determination: Calibration curve with a suitable carbohydrate standard for the unknown sample were prepared. Six test tubes were labeled 1 to 6. The carbohydrate standard was dispensed. Distilled water was added to make it up to 0.5 ml. 0.5 ml of 5% phenol solution was added and thoroughly mixed. 2.5 ml of concentrated sulphuric acid was dispersed to each tube and mix thoroughly this is allowed to stand for 20 minutes and reading taking in a spectrophotometer at 470 nm wave length. Unknown samples were treated equally as standard in duplicate and concentrations extrapolated from the predetermined calibration curve.

Determination of pH: The pH was determined using standard test ASTMD 3828 - 80 ASTM, [17].

Moisture content determination: The moisture content of the starchy adsorbents was determined using standard test ASTMD 2867 - 91 ASTM, [18].

Determination of bulk density: 5 gm each of the adsorbents was poured through a short - stemmed glass funnel into a 100 ml graduated measuring cylinder filled with water to the 50 ml mark.

The rise in volume, the volume of water displaced was obtained. The volume of water displaced is equal to the volume of the 5 g of the adsorbent.

Fourier Transform Infrared (FTIR)

The functional groups presents in the starch adsorbents were investigated using SHIMADZU FTIR - 8400S spectrophotometer with the range 500 - 4000 cm^{-1}. KBr was used as background material in the analysis.

X - Ray Diffractogram (X - RD) Analysis

The amorphous and crystallinity nature of the adsorbents was examined using a diffractometer system (EMPYREAN) using radiation Cuα (α1 = 1.540598A° and α1= 1.544426A°) and a secondary graphite monochromator (No), angle 2θ swept and the scan range (-0.002 - 74.99997°).

Experimental procedure

The column studies were carried out using a glass column of 30 cm length and internal diameter 3 cm. The particle size of 600 μm was used for the adsorbents. The prepared adsorbent was packed in the column with glass wool layer at the bottom of the bed. Bed heights of 40 mm, 80 mm, and 120 mm were used. Peristaltic pump was used to supply the concentration of the ethanol - water mixture (adsorbate) used into the adsorption column at constant flow rate of 6 ml/min, 10 ml/min, and 14 ml/min while the initial adsorbate concentration used was 90 wt%. The effluent samples were collected at various time intervals and the resulting concentration determined using refractometry method and the earlier obtained calibration curve. The studies were terminated when the exhaustion of column is reached.

Results and Discussion

Characteristics of the prepared adsorbents

The physico - chemical properties of the starchy adsorbents are shown in Table 1.

Breakthrough Curves

Effect of bed heights on breakthrough curves: The effect of bed heights 40 mm, 80 mm, and 120 mm on the uptake of water figure 1 at an inlet flow rate of 6 ml/min, and inlet ethanol - water mixture of 90 wt% concentration is shown in figures 2.0 and 3.0. The results indicate that there was a decreased in the sorption equilibrium capacity of the starch material as the bed height is increased due to less availability of number of sorption sites. In the batch mode the probability of contact between adsorbate and adsorbent is high compared to the fixed bed mode of operation, which usually results in small sorption equilibrium capacity experienced in column study. Saturation of bed occurs in less time for smaller bed heights compared to the higher bed heights. At higher bed height the availability of the effective surface area of adsorbent is more which offers more active sites to adsorption and it also broadens the mass transfer zone length [19].

Effect of flow rates on breakthrough curves: The effect of flow rate for the sorption of water on to the starch at flow rates of 6 ml/min,

Properties	Modified corn starch (EMCOS)	Native corn starch (NCOS)
pH	6.0	5.6
Moisture content (%)	3.04	2.91
Colour	White	Yellowish white
Starch content (%)	86.5	87.5
Bulk density (g/ml)	1.57	1.37
Micro pore volume ($\mu m^2/g$)	0.2	0.1
Diameter (μm)	7.99	6.68
Oxygen (%)	86.6	87.3
Carbon (%)	13.4	12.7

Table 1.0 Physico-chemical properties of Corn starch.

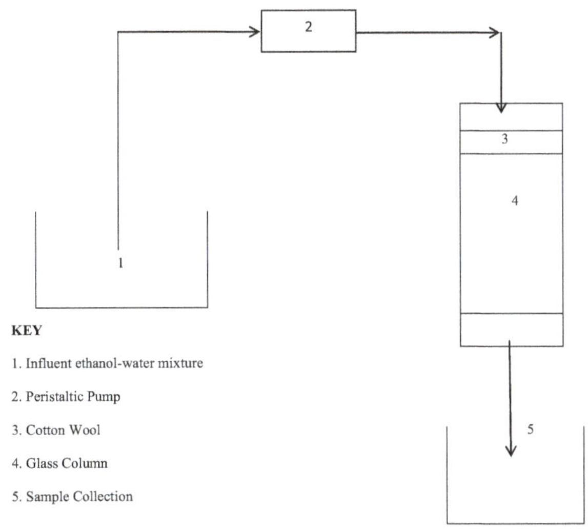

Figure 1: Block Diagram of the Experimental Set – up for Column Adsorption Study.

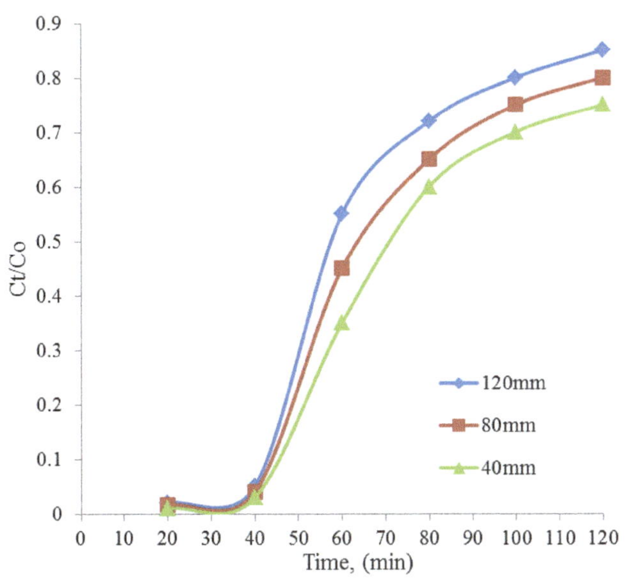

Figure 2: Effect of bed height on breakthrough curve for adsorption of water on EMCOS

10ml/min, and 14 ml/min, at an inlet concentration of 90 wt%, and bed height of 90 mm as shown in figures 4.0 and 5.0. It is seen from the figure that sorption of water is rapid in the initial stages and there was decreased in rate and thereby reaching saturation finally. As the flow rate increase, the breakthrough is reached quickly because at higher flow rate, the time of contact between the ethanol - water mixtures and the waste biomass is reduced.

This corroborates the results obtained in the works of Nwabanne, Sivakumar and Palanissamy [20,21]. It can also be seen from figures 4.0 and 5.0 that breakthrough occurred at a comparatively faster rate at higher adsorbate flow rate and less time was taken to reach the saturation breakthrough. This can be explained from the mass transfer vicinity, the rate of mass transfer gets increased at higher flow rate leading to faster saturation. Increasing the flow rate, the mass transfer rate increases due to the decreased in the boundary layer thickness. At low flow rate, it takes a longer time for the adsorbate (ethanol-water mixtures) to diffuse towards the starch adsorbents because of the thickness of the film and resistance surrounding the starch adsorbents. The effect of flow rate is helpful for the large - scale treatment systems in order to utilize the bed for its maximum capacity with minimal flow rate [21].

Kinetic study

Three models (Yoon and Nelson model, Thomas model and Adams - Bohart model) were employed to analyze the column performance.

Yoon and nelson model: The linearized form of the model is given as equation (1) Kavak and Öztürk,[22]

$$\ln\left(C_e / C_o - C_e\right) = K_{YN}t - \tau K_{YN} \quad \ldots\ldots\ldots\ldots\ldots\ldots (1)$$

Where C_o and C_e are the inlet sorbate (solute) and effluent concentrations respectively. K_{YN} is the Yoon and Nelson kinetic rate constant, τ is the time required for 50% of the adsorbate to breakthrough (minute), t, the breakthrough (sampling) time (min). Based on the Yoon and Nelson model, the amount of water being adsorbed in a fixed bed is half of the total water entering the adsorption bed within 2τ period Aksu and Ttezer, [23].

For a given bed;

$$q_{OYN} = \frac{q(total)}{m} = \frac{1/\,2C_0\left[\left(\frac{r}{1000}\right)\times 2\tau\right]}{m} = \frac{C_0 \times r \times \tau}{1000m} \quad \ldots\ldots (2)$$

Where r is the flow rate, q_{OYN} is the adsorption capacity and m is the mass of the adsorbent.

A plot of $\ln C_e/C_0$ - C_e against t gives a straight line with slope of K_{YN} and intercept of $-\tau K_{YN}$ (Figures 5- 8). The results indicated that the kinetic constant rate increased with flow rate. The adsorption capacity increased with increase in flow rate while it decreased with increase in bed height because it has been established that K_{YN} and τ are inversely proportional. The high values of co-efficient of regression obtained indicate that Yoon and Nelson model fitted well to the experimental data. The time required for 50% breakthrough, τ decreased with increase in flow rate, and bed height respectively. The ethanol -water mixtures has more time to move through the column, which results in reduced adsorption rate experienced at higher bed height.

Thomas model: Thomas model is as given in equation (3) Kavak and Öztürk, [22]

$$\ln\frac{C_0}{C_t} - 1 = \frac{K_{TH}q_0X}{Q} - \frac{K_{TH}}{Q}V \quad \ldots\ldots\ldots\ldots\ldots\ldots\ldots\ldots\ldots\ldots (3)$$

Where C_o and C_e are the inlet solute and effluent concentrations respectively, q_o is the maximum adsorption capacity (g/g), x is the total mass of the adsorbent (g), K_{TH} is the Thomas rate constant (ml/min/g), V is the throughput volume (ml), and Q is the volumetric flow rate (ml/min).

The kinetic co -efficient K_{TH} and adsorption capacity of the column q_o can be determined from a plot of

$\ln\frac{C_0}{C_t} - 1$ Versus V (Figures 9 -12). K_{TH} is dependent on the flow rate. The adsorption capacity, q_o, increased with increase in bed height. The results indicate that the EMCOS adsorbent is capable of holding a maximum of 10.39 g/g of water at 14 ml/min flow rate as shown in Table 2. The values of K_{TH} and R^2 obtained are similar to that of Jayanta et al. [19]. High values of correlation co-efficient determined indicates that the kinetic data also conformed to Thomas model.

Adams - Bohart Model: For the Adams - Bohart model, the mass transfer rates obey the following equations

$$\frac{\partial q}{\partial t} = -K_{AB}C_b \quad \ldots\ldots\ldots\ldots\ldots\ldots\ldots\ldots\ldots\ldots\ldots\ldots\ldots\ldots\ldots \quad (4)$$

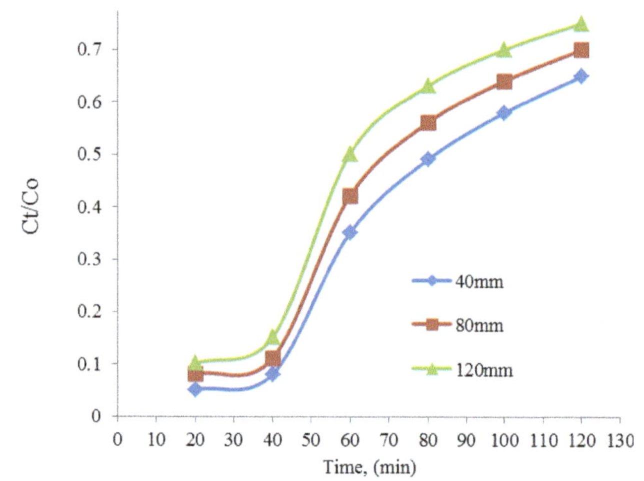

Figure 3: Effect of bed height on breakthrough curve for adsorption of water on NCOS.

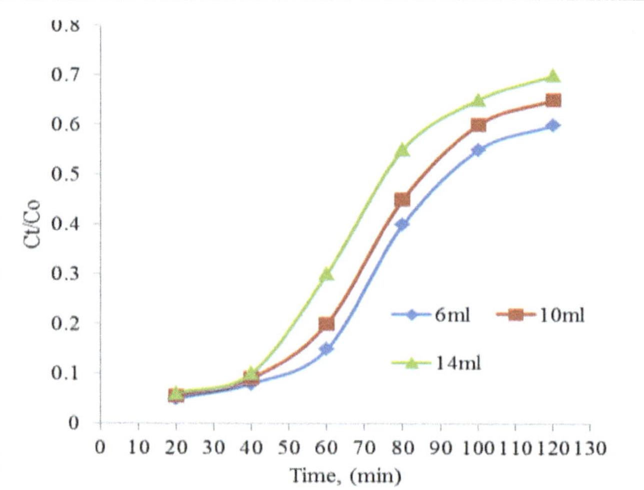

Figure 4: Effect of Flow rate on breakthrough curve for adsorption of water on EMCOS.

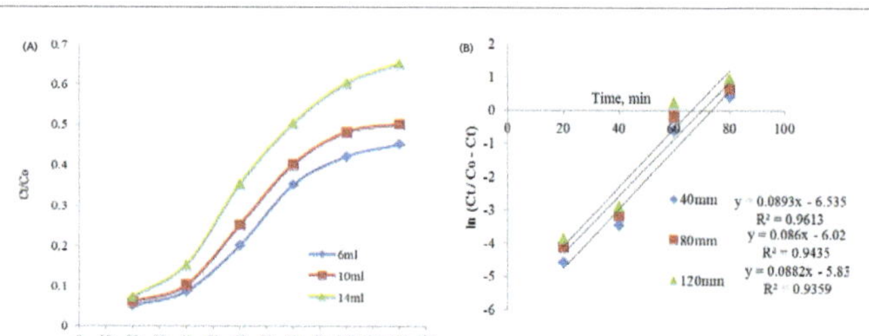

Figure 5: (a) Effect of Flow rate on breakthrough curve for adsorption of water on NCOS (b) Yoon and Nelson plot for the adsorption of water on EMCOS: Effect of Bed height

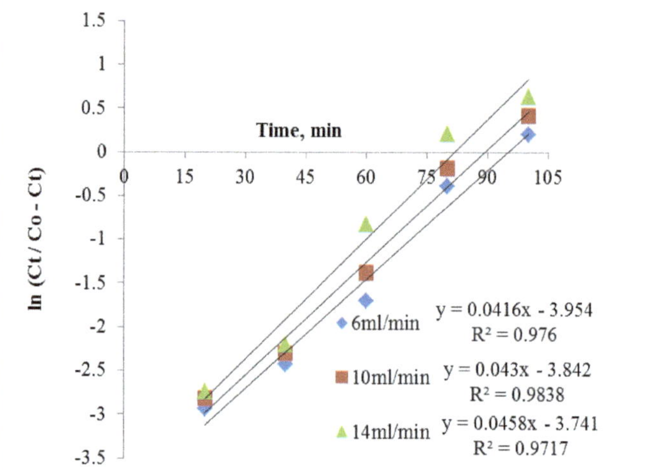

Figure 6: Yoon and Nelson plot for the adsorption of water on EMCOS: Effect of Flow rate.

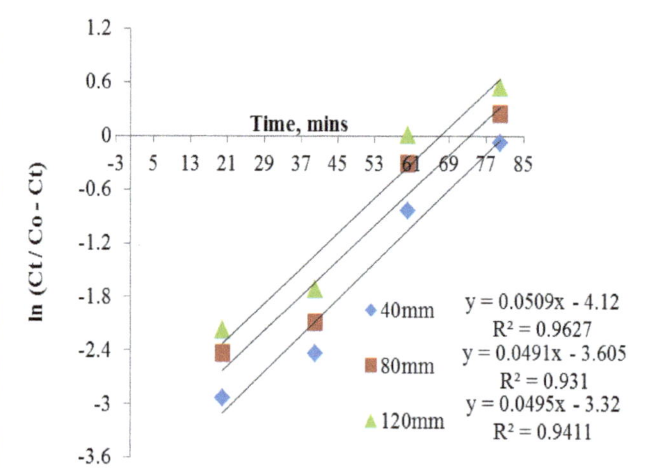

Figure 7: Yoon and Nelson plot for the adsorption of water on NCOS: Effect of Bed height.

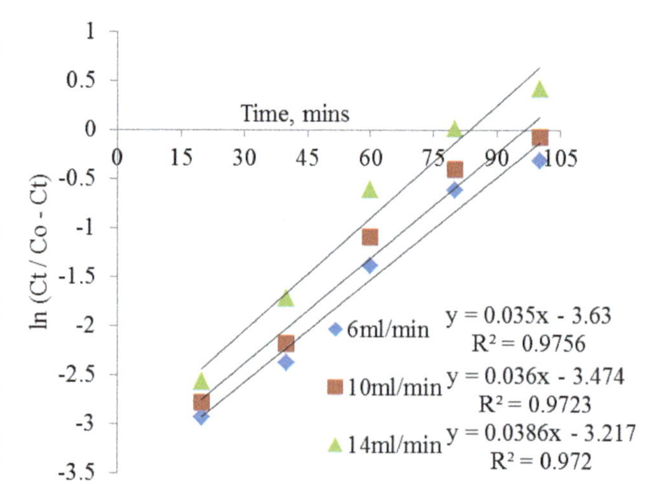

Figure 8: Yoon and Nelson plot for the adsorption of water on NCOS: Effect of Flow rate.

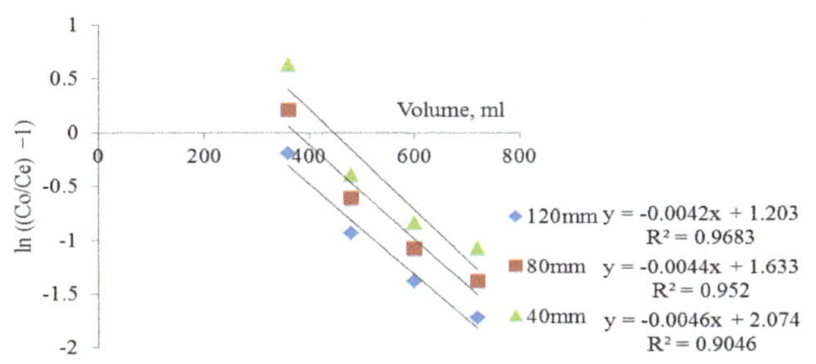

Figure 9: Thomas Kinetic plot for the adsorption of water on EMCOS: Effect of Bed height.

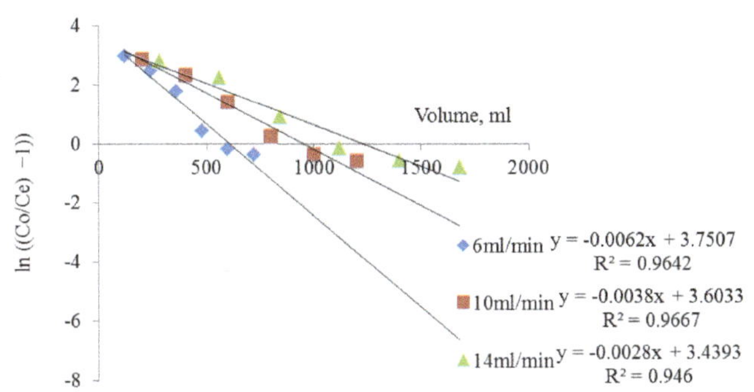

Figure 10: Thomas Kinetic plot for the adsorption of water on EMCOS: Effect of Flow rate.

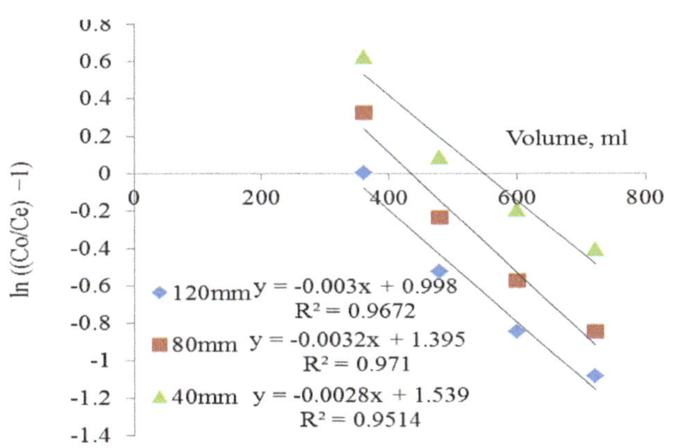

Figure 11: Thomas Kinetic plot for the adsorption of water on NCOS: Effect of Bed height.

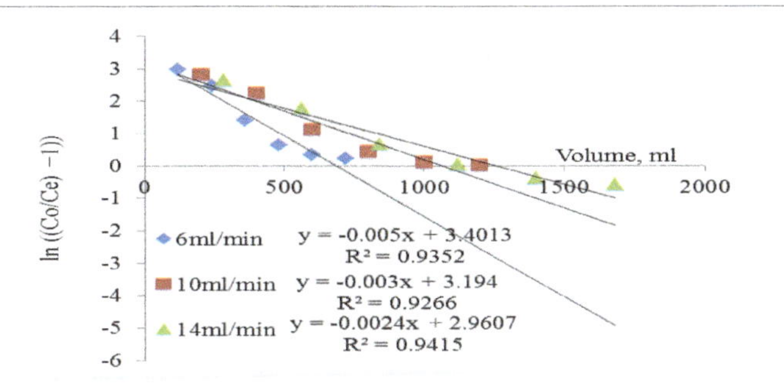

Figure 12: Thomas Kinetic plot for the adsorption of water on NCOS: Effect of Flow rate.

$$\frac{\partial C_b}{\partial Z} = \left(\frac{K_{AB}}{U_0}\right) q\, C_b \quad \text{.........} (5)$$

Two assumptions are made for the solution of these differential equation systems: $t \rightarrow \infty$ and $q \rightarrow N_o$ Quintelas et al., [24]. Solving the differential equations (4) and (5).

The following equation is obtained

$$\ln\left(Ce / C_o\right) = K_{AB} C_o t - K_{AB} N_o \left(Z / U_o\right) \quad \text{.........} (6)$$

The values of K_{AB}, and N_o can be evaluated from a plot of $\ln(Ce/C_o)$ against t (time) at a given bed height (z) and flow rate (U_o). Where K_{AB} is the Adams - Bohart rate constant, and N_o is the adsorption capacity (g/g). It can be seen from Figures 13 -16 that the N_o decreased as the bed height increases and increased as the flow rate increase. This can be explained because at high flow rate the number of adsorbate molecules passing through the adsorbent is more which increases the rate. The K_{AB} rate constant obtained is similar to the rate constant obtained when Thomas and Yoon and Nelson models were used to analyze the column performance. The R^2 obtained showed the data best fitted to Adams - Bohart model.

Determination of the Functional Group in the Starch

Figures 17 and 18 give the FTIR spectroscopy of both the Enzyme modified corn starch (EMCOS) and unmodified corn starch (NCOS). Figure 17 Bands of (2924.18 cm^{-1}, 3206.76 cm^{-1}, 3345.64 cm^{-1}, and 3498.02 cm^{-1}) indicated the presence of hydroxyl group, H bonded OH stretch Okewale et al, [14]. The highest peak value of 1646.3 cm^{-1} in the enzyme modified corn starch suggest the amorphous crystalline nature of the starch which is responsible for the highest water adsorption experienced in EMCOS. Figure 18 broad bands (2929.97 cm^{-1}, 3219.3 cm^{-1}, 3317.67 cm^{-1}, 3401.58 cm^{-1}, and 3489.34 cm^{-1}) exhibited the OH stretch group. 1372.4 cm^{-1} showed OH bend. COH bending was shown in 1025.2 cm^{-1} band. The highest peak of 1649.19 cm^{-1} suggests that water is adsorbed at the amorphous parts of the corn starch.

X - Rays Diffractogram analysis (X-RD)

Figure 19 which is the X - RD patterns for the adsorbents. The strong X - ray diffractogram patterns of the corn starch noticed are; 14.664°, 17.212°, and 22.776° this corroborate the works of Quintero and Cardona, Bertuzzi et al., [25,26] which indicated a typical A type diffraction pattern. The amorphous zone present in the diffractograms is mainly due to amylopectin as reported in the works of Ahmad et al. [26].

Amylopectin α - 1, 6 branched structures has an overlapping hydroxyl groups which are proposed to correspond to more hydroxyl groups per unit area of the starch surface [27,28], reported that corn has an A - type structure, 27% amylose and polar lipids that can complex with the amylose fraction to form V crystals this also corroborate the result of X - RD reported in this work where it was shown that amylopectin is the main constituent of corn starch and it is type A structure. Thus, it was revealed that the greater adsorption capacity noticed in starch based adsorbents was as a result of the amorphous nature of the adsorbents which resulted from the amylopectin structure as revealed in the X - RD analysis carried out on the adsorbents. It was shown from Table 3 that the adsorbents fall into small category which means that the adsorbents can also be said to be mesopores in nature which makes these adsorbents to be able to dehydrate ethanol - water mixtures similar result of this classification was obtained by Lindeboom et al. [29].

Kinetic model	Flow rate 6 ml/min	10 ml/min	14 ml/min	Bed Height 40 mm	80 mm	120 mm
				Yoon and Nelson Kinetics		
K_{YN} (min^{-1})	0.0416	0.043	0.0458	0.089	0.086	0.088
τ (min)	95.05	89.34	81.68	73.43	70.0	66.25
q_0 (g/g)	4.81	7.53	9.64	3.71	3.54	3.35
R^2	0.976	0.9838	0.9717	0.9613	0.9435	0.9359
				Thomas kinetic		
K_{Th} (ml/min/g)	0.372	0.38	0.372	0.276	0.264	0.252
q_0 (g/g)	5.10	7.99	10.39	3.80	3.13	2.42
R^2	0.9642	0.9667	0.946	0.9046	0.952	0.9683
				Adams – Bohart kinetic		
K_{AB} (min^{-1})	0.343	0.354	0.388	0.738	0.678	0.657
N_o (g/g)	2.23	2.49	2.53	3.76	1.85	1.21
R^2	0.9726	0.986	0.9802	0.9464	0.9274	0.92

Table 2: Summary of the Column Kinetics Model for EMCOS.

Kinetic model	Flow rate 6 ml/min	10 ml/min	14 ml/min	Bed Height 40 mm	80 mm	120 mm
				Yoon and Nelson Kinetics		
K_{YN} (min^{-1})	0.035	0.036	0.0386	0.0509	0.049	0.088
τ (min)	103.71	96.5	83.34	80.94	73.42	66.25
q_0 (g/g)	5.24	8.13	9.83	4.09	3.71	3.39
R^2	0.9756	0.9723	0.972	0.9627	0.931	0.9411
				Thomas kinetic		
K_{Th} (ml/min/g)	0.3	0.3	0.336	0.168	0.192	0.18
q_0 (g/g)	5.73	8.97	10.39	4.63	3.67	2.80
R^2	0.9352	0.9266	0.9415	0.9514	0.971	0.9672
				Adams – Bohart kinetic		
K_{AB} (min^{-1})	0.334	0.38	0.338	0.414	0.36	0.336
N_o (g/g)	2.24	2.32	2.58	4.28	2.11	1.36
R^2	0.9907	0.9916	0.9755	0.9336	0.9234	0.9327

Table 3: Summary of the Column Kinetics Model for NCOS.

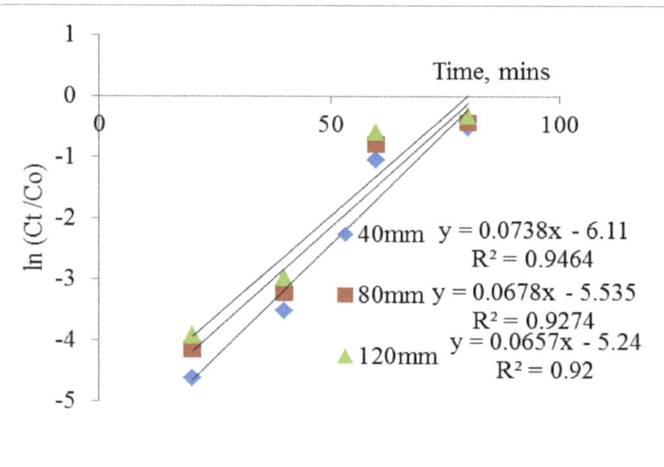

Figure 13: Adams- Bohart plot for the adsorption of water on EMCOS: Effect of Bed height.

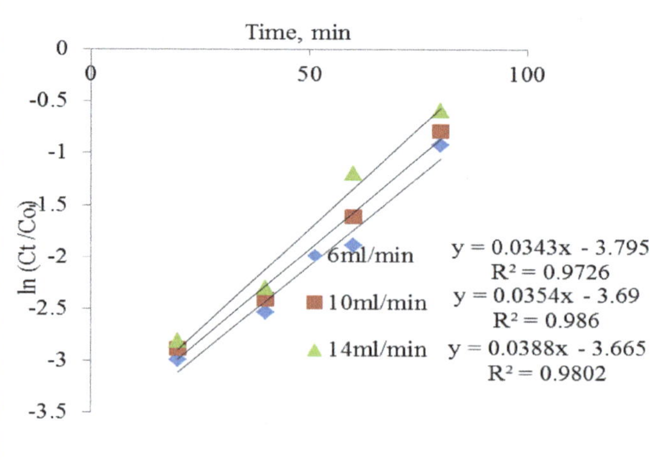

Figure 14: Adams- Bohart plot for the adsorption of water on EMCOS: Effect of Flow rate.

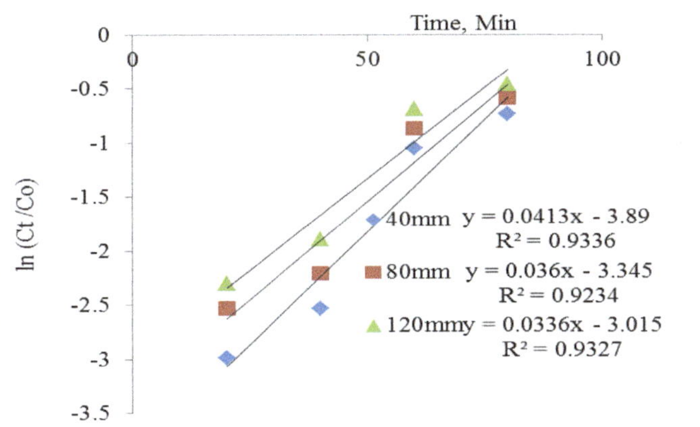

Figure 15: Adams- Bohart plot for the adsorption of water on NCOS: Effect of Bed height.

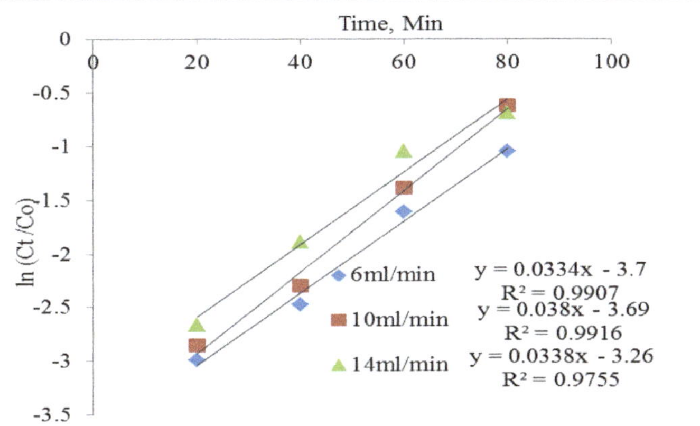

Figure 16: Adams- Bohart plot for the adsorption of water on NCOS: Effect of Flow rate.

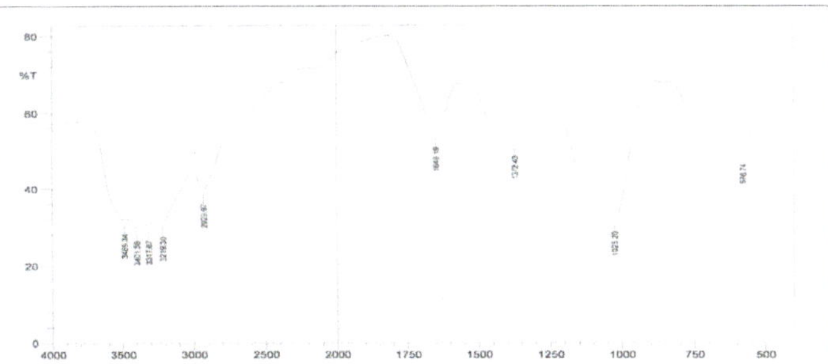

Figure 17: FTIR spectroscopy of the enzyme modified corn starch (EMCOS)

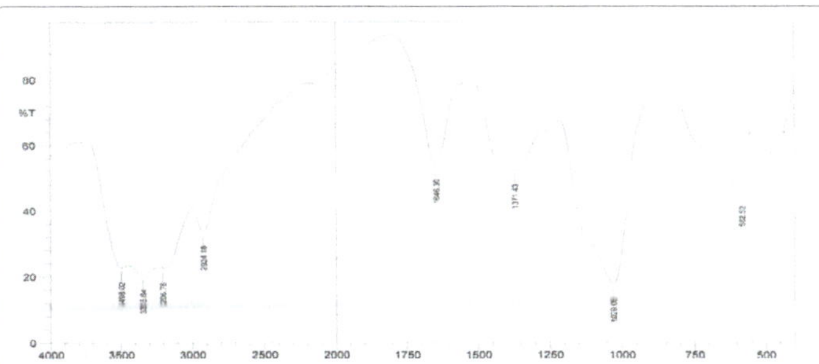

Figure 18: FTIR spectroscopy of the native corn starch (NCOS).

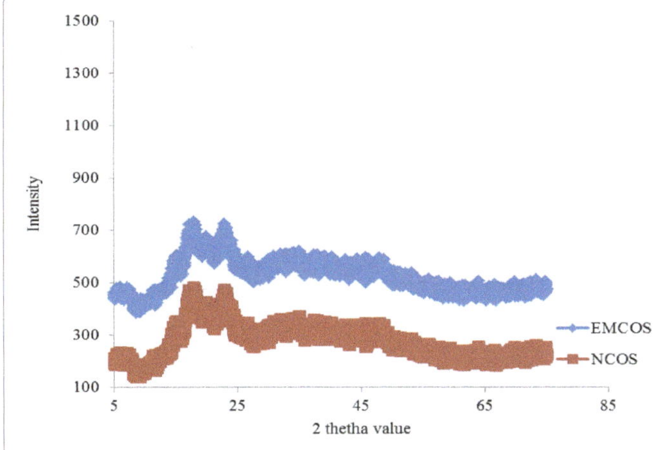

Figure 19: X-RD analysis of the adsorbents.

Conclusion

The flow rate effect on the uptake of water on the corn starch adsorbents at flow rates of 6, 10, and 14 ml/min, an inlet concentration of 90 wt% and bed height of 90 mm has been studied. The effects of flow rate, and bed height on the breakthrough curves obtained showed that the adsorptive dehydration efficiency decreased with increasing flow rate, and increased with increase in bed height. The dehydration efficiency is strongly dependent on bed height and flow rate. The adsorbents were able to hold 10.39 g/g maximum of water. The column kinetics was best described by Yoon and Nelson. The capacity of the adsorption decreased with increase in bed height and flow rate. The FTIR result of these adsorbents also confirmed the chemical affinity of starch to water. The highest peak value of 1646.3 cm^{-1} in the enzyme modified corn starch suggest the amorphous crystalline nature of the starch which is responsible for the highest water adsorption experienced in EMCOS. The A - type and amorphous nature of the adsorbents which resulted from the amylopectin structure was revealed in the X - RD analysis carried out on the adsorbents. The dehydration of water was made possible by the presence of amylopectin in the biomass studied.

References

1. Frolkova AK, Raeva VM (2010) Bio-ethanol dehydration: State of the art. Theor Foundation Chemical Engineering 44: 545- 556.
2. Kumar S, Singh N, Prasad R (2010) Anhydrous ethanol: A renewable source of energy. Renewable Sustainable Energy Rev 14: 1830 - 1844.
3. Ranjbar Z (2012) Ethanol dehydration with protein extracted canola meal in a pressure swing adsorption process. A Thesis submitted to the Department of Chemical and Biological Engineering university of Saskatchewan Saskatoon Canada.
4. Diaz JC, Chavez ID, Liliana G, Juan Carlos MP (2010) Separation of ethanol - water mixture using type -A Zeolite Molecular Sieves. E -Journal of Chemistry 7: 483-495.
5. Katzen R, Ackley WR, Moon GD, Messick JR, Bruch BF, et al. (1980) Low energy distillation systems. ACS Meeting Las Vegas USA.
6. Ladisch MR, Voloch M, Hong J, Bienkowski P, Tsao GT (1984) Corn meal adsorber for dehydrating ethanol vapours. Industrial Engineering Chemical Process Design Development 23: 437-443.
7. Hassaballah AA, Hills JH (1990) Drying of ethanol vapors by adsorption on corn meal. Biotechnol Bioeng 35: 598-608.

8. Lee JY, Westgate PJ, Ladisch MR (1991) Water and ethanol sorption phenomena on starch. AIChE Journal 37 : 1187-1195.

9. Inglezakis VJ, Poulopoulos SG (2006) Adsorption, ion exchange and Catalysis: Design of operation and Environmental Applications, 1st ed. Elsevier Publishers Amsterdam. 16 - 17.

10. Kugimiya BM, Donovan JW, Wong RY (1980) Phase transitions of amylose - lipid complexes in starches: A calorimetric study. Staerke 32: 265-270.

11. Zobel HF, Young SN, Rocca LA (1988) Starch gelatinization: An X - ray diffraction study. Cereal Chemists 65: 443 - 446.

12. Beery KE, Ladisch MR (2001) Adsorption of water from liquid - phase ethanol -water mixtures at room temperature using starch based adsorbents. Industrial Engineering Chemical Resources 40: 2112 - 2115.

13. Okewale AO, Etuk BR, Igbokwe PK (2011) Comparative studies on some starchy adsorbents for the uptake of water from ethanol - water mixtures. International Journal of Basic and Applied Sciences 11: 21-31.

14. Okewale AO, Igbokwe PK, Ogbuagu JO (2013) Kinetics and isotherm studies of the adsorptive dehydration of ethanol - water system with biomass based materials. International Journal of Engineering and Innovative Technology 2: 36 - 42.

15. Wu P, Gao H, Sun J, Ma T, Liu Y, et al. (2012) Biosorptive dehydration of tert-butyl alcohol using a starch-based adsorbent: characterization and thermodynamics. Bioresour Technol 107: 437-443.

16. Beery KE, Gulati MK, Eric P, Ladisch MR (1998) Effects of enzyme modification of corn grits and their properties as an adsorbent in a skarstrom pressure swing cycle dryer. Adsorption 4: 321 - 335.

17. American Society for Testing and Materials (1996) Annual Book of ASTM Standard, Volume 15.1, Refractories Carbon and Graphic Products; Activated Carbon ASTM Philadelphia PA.

18. American Society of Testing and Materials (1991) Standard test methods for moisture in activated carbon Philadelphia PA: ASSTM Committee on Standards.

19. Jayanta KB, Mona D, Hassan FM, Gautam N, Kumar A (2012) Fixed - bed column study of textile dye direct blue 86 by using a composite adsorbent. Arch Applied Science Resource 4: 882 - 891.

20. Nwabanne JT (2010) Adsorption and kinetic modeling of heavy metals uptake from waste water effluents. A Ph.D Dissertation in Chemical Engineering Department, Nnamdi Azikiwe University Awka.

21. Sivakumar P, Palanisamy PN (2009) Adsorption studies of basic Red 29 by a non - conventional activated carbon prepared from Euphorbia antiquorum L. International Journal of Chemical Technology Research 1: 502 -510.

22. Kavak D, Ozturk (2004) Adsorption of boron from aqueous solution by sepirolite: II Column studies, II Illsulrararasi. Bor Sempozyumu 23 -25: 495 - 500.

23. Aksu Z, Tezer S (2000) Equilibrium and kinetic modeling of bio-sorption of remazol black B by Rhizopus arrhizus in a batch system: effect of temperature. Process Biochemistry 36: 431 - 439.

24. Quintelas C, Fernandes B, Castro J, Figueiredo H, Tavares T (2008) Bio-sorption of Cr (VI) by three different bacterial species supported on granular activated carbon-A comparative study. Journal of Hazardous Materials 153:799-809.

25. Quintero JA, Cardona CA (2009) Ethanol dehydration by adsorption with starchy and cellulosic materials. Journal of Industrial Engineering Chemical Resources 48: 6783 - 6788.

26. Bertuzzi MA, Vidaurre Castro EF, Armada M, Gottifredi JC (2007) Water vapour permeability of edible starch based films. Journal of food Engineering 80:972 - 978.

27. Ahmad FA, Williams PA, Doublier J, Durand S, Buleon S (1999) Physico - chemical characterization of sago starch. Carbohydrate Polymers 38 : 361-370.

28. Rebar V, Fischbach ER, Apotolopoulos D, Kokini JL (1984) Thermodynamics of water and ethanol adsorption on four starches as model biomass separation systems. Biotechnology Bioengineering 26: 513-517.

29. Lindeboom N, Chang PR, Tyler R (2004) Analytical, biochemical and physic-chemical aspects of granule size with emphasis on small granule starches; A review starch. 56: 89 - 99.

Optimum Concentrate Stream Concentration in $CaCl_2$ Removal from Sugar Solution Using Electrodialysis

Jogi Ganesh Dattatreya Tadimeti, Anusha Chandra and Sujay Chattopadhyay*

Polymer and Process Engineering Department, I.I.T.Roorkee, SRE Campus, UP-247 001, India

Abstract

Batch electrodialysis (ED) with continuous recirculation was carried out to remove CaCl2 from sugar solution at room temperature. Role of limiting process parameters e.g. (i) concentrate concentration, (ii) sugar mass% cum solution viscosity and (iii) ionic conductivity on electrolyte removal rate were investigated. Unsteady state mass balance defining ion transport through the ED cell with its initial and boundary conditions were solved with linearized Nernst-Planck equation. A new Sherwood number expression appropriate to the cell geometry and process parameters (physical properties and conductivity ratios) is developed as $Sh = a.Re^b.Sc^c.\left(\frac{\ddot{E}_{Dil}}{\ddot{E}_{Conc}}\right)^d$ to estimate mass transfer coefficients and ionic flux. Electrolyte removal percentage can't be correctly estimated from ED cell current instead model based estimates of the flux components due to concentration gradient (chemical potential) and applied electric potential could closely predict the experimental data on electrolyte removal. The ratios of electrolytes removal rates (with and without sugar) can be expressed as power function of viscosity ratios as $\frac{percentremoval_c}{percentremoval_{DSC}} = \left(\frac{\mu_{DSC}}{\mu_C}\right)^{0.28}$. This factor of 0.28~0.3 resembles close similarity with Sc number in the Sh number correlation, indicating that the removal rate is a strong function of flow and physical properties of the system. An optimum electrolyte concentration of concentrate, sugar mass% in diluate, and applied electric potential was estimated.

Keywords*:* Electro dialysis; Solution conductivity; Mass transfer coefficient; Ionic flux; Specific energy consumption

Introduction

Presence of excess calcium in the post floculation and precipitation stage of clarified sugar juice creates series of nuisence [1] to sugar industries affecting product quality (listed below). Therefore, removal of it at appropriate stage would drastically reduce operation and maintainence cost and improve product quality. [1]

1. Scale formation in the evaporators
2. Improper crystallization
3. Molasses percentage may increase due to inversion of sugar in alkaline medium
4. Storage is hampered because of hygroscopic nature of these metal ions
5. Excess calcium is not hygienic as well

Application of electrodialysis (ED) is propsoed to remove the dissolved calcium ions (commonly as $CaCl_2$). Electrodialysis (ED) is a polar (cation and anion exchange) membrane based ion removal technique driven by externally applied electric potential as the main driving force. Briefly, the diluate (feed solution) containing electrolyte(s) is placed between two oppositely charged membranes through which the dissociated ions get selectively permeated. $CaCl_2$ (strong electrolyte) removal via electrolydialysis could be quite interesting to explore from industry point of view. Earlier electrodialysis (ED) was applied in sugar industry to recover speciality chemicals e.g. tartrate and malate from grape sugar [2], and in demineralisation of beet sugar syrup, juice and molasses [3,4]. Major technical obstacles arose due to poor membrane stability at higher temperatures commonly maintained to reduce solution viscosity-pumping cost and fouling due to deposition of organic molecules (sugars, proteins etc.). Difference in ionic mobility in solution and membrane causes drop in counter-ion concentration at the membrane surface, which causes concentration polarization and adds up complexity to the existing difficulties [5-11]

In a batch recirculation ED process diluate and concentrate streams get depleted and enriched respectively with time. With increase in electrolyte concentration in the concentrate channel, back diffusion (from concentrate to diluate stream) appears affecting overall ion removal percentage. With a given diluate stream concentration, initially the current density increases with applied electric potential. Finally, it reaches a limiting value at an electric potential where no excess ions are available to transport current in solution, thus, concentration drops to zero (theoretically) at membrane surface. This is called limiting current density. Applying an electric potential above this value would result in electrolysis of water and generating H^+ and OH^- ions. In batch recirculation mode, the diluate stream is recycled back to the ED cell to remove left over ions, therefore, gradual lowering of concentration of diluate stream occurs. Unlike continuous electrodialysis the current density keeps changing with time as is with the electrolyte concentration. Energy wastage due to water spliting is avoided by restricting current density below limiting value. Modeling of batch ED with continuous recirculation becomes more challenging as streams' electrolyte concentration changes with time. Commonly Ohm's law is applied [12-15] to evaluate current based on overall applied electric potential and resistances. But, it fails to predict several experimental current density majorly governed by ionic diffusion process, which limits the

***Corresponding author:** Sujay Chattopadhyay, Polymer and Process Engineering Department, I.I.T.Roorkee, SRE Campus, UP-247 001, India

maximum magnitude of current for an applied voltage. Wen et al. [16] experimentally verified this fact and concluded that rise in ohmic resistance with lowering of diluate concentration is not significant and claimed that back emf (due to concentration polarization across the membrane) generation reduced the effective applied electric potential across the cell pair. Theoretical estimate of back emf becomes difficult due to non-avialability of correct procedure. This ensured that a model probably from first principles would give us appropriate estimation of ion flux. Linearized Nernst-Plank flux equation (formulated from first principles) provided closure prediction of current density and ion flux and can quantitatively estimate drop in ion concentration. The entire batch operation was subdivided into small multiple equilibrium steps involved in irreversible thermodynamics [17].

Kabay et al. [18] carried out batch removal of bivalent Ca^{2+} and Mg^{2+} ions under constant applied DC potentials (3 V,5 V and 10 V) and flowrates (0.6, 1.2 and 1.8 $L.min^{-1}$) and compared energy consumptions. Although time of operation is reduced with applied electric potential but energy consumption has increased. While, flow rate variation did not show much effect on ion removal rate. Several studies [2,18] are reported on Ca^{2+} ion removal using batch recirculation electrodialysis but, role of concentrate concentration on ion removal rate was not emphasized adequately. Herein, we tried to explore few querries: (i) Is there any optimum concentrate concentration for a given diluate composition? (ii) Is there any role of non electrolyte and its concentration (density and viscosity) on ion removal rate and energy consumption? (iii) Can we theoretically model the ion transport and estimate the transport properties? (iv) Is there a generalization possible? To the best of our knowledge, answers to these questions lacks clarity and quantitativeness. Answers to these issues might give a better insight to ion transport mechanism and optimize concentrate compartment.

In cane based sugar industry concentration of sugar in the juice after clarification step (after lime treatment and color removal) commonly reaches ~5% (mass basis). $CaCl_2$ concentration is chosen based on free lime concentration left in clarified juice and the stream entering into evaporator. Role of non electrolyte (sugar), concentrate concentration and $CaCl_2$ viscosity on removal were investigated. A model accounting all experimental observations was used to estimate transport properties involved.

Materials and Methods

Equipment

Electro dialysis setup: The experimental setup layout used for Electrodialysis (ED) application is shown in Figure 1. The ED cell assembly was tailor made in the laboratory and graphite sheets were used as electrodes (0.08×0.04 m^2). An electric DC potential was applied across the cell stack (consisted of two cell pairs containing a concentrate and a diluate channel each.) through a built-in DC source. Voltage and current between two electrodes were continually monitored through a built-in digital voltmeter and ammeter respectively. Cross linked Styrene-divinyl benzene copolymer membranes were obtained from Permionics Ltd. India and used as ion exchange membranes. The properties of the membranes were listed in Table 1. Experiments were performed at a constant applied electric potential of 4 V and a fixed flow-rate for all streams.

Power Supply: The power supply was provided through a voltage stabilizer of 110/220 V AC with 50-80 Hz frequency. The same gave an output voltage 0-20 V DC and had an ammeter that can measure currents from 0-200 mA. Three centrifugal pumps were inbuilt with the system for pumping the solution.

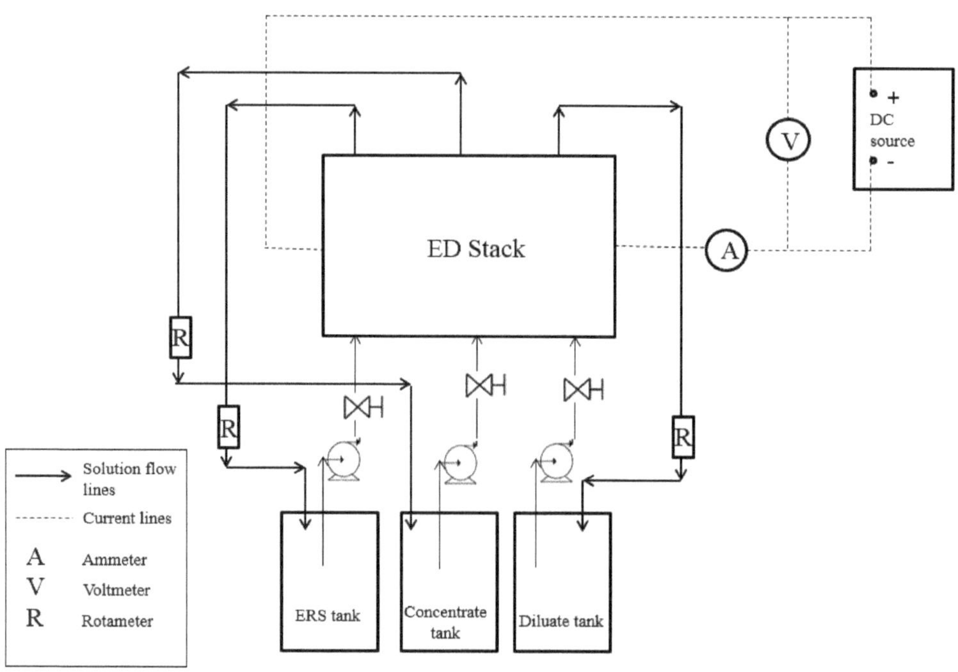

Figure 1: Batch recirculation electro dialysis process showing the storage tanks, pumps, manual control valves, various chambers of the ED cell, electrodes, DC supply source, Voltmeter, ammeter and flow meters used in the experiment.

Conductivity Measurement: Solution conductivity was noted at regular interval through an off line Conductivity and TDS meter CM183 from Elico India.

ED Cell Compartments and Solutions Used: 1000 ml solutions of each stream (concentrate, diluate and Electrode rinsing solution (ERS)) were taken in three tanks (Figure 1) respectively. Each solution tank was connected with its respective flow points of the ED unit through flexi tubing. Solutions were circulated at a constant rate by three centrifugal pumps and the solution flow rates were measured using rotameters connected to each stream. The detailed concentrations of different electrolyte streams were reported in Table 2. Concentrate and diluate solutions were always prepared using deionized water (conductivity <10.0 μS). Electrodialysis experiments were carried out in three sets. (i) DSC1-DSC6 (diluate stream having 25 mol.m^{-3} $CaCl_2$ and 5 mass% sugar and six different concentrate streams having $CaCl_2$ concentrations varying as 5, 10, 20, 25, 30, and 40 mol.m^{-3}), (ii) C1-C6 (diluate stream having 25 mol.m^{-3} $CaCl_2$ and six different concentrate streams with $CaCl_2$ concentrations as 5, 10, 20, 25, 30, and 40 mol.m^{-3}) and (iii) S1-S4 (4 different diluate streams with 25 mol.m^{-3} $CaCl_2$ with amount of sugar varying from 5-20 mass% and concentrate stream having concentration of 25 mol.m^{-3} $CaCl_2$). For convenience DSC4 is renamed as S1 (5 mass% sugar) as both the experiments are identical and other values of sugar concentration are defined as S2 (10 mass% sugar), S3 (15 mass% sugar) and S4 (20 mass% sugar). Detailed information was reported in Table 2. Aqueous solution of NaCl (50 mol.m^{-3}) was used as ERS in all experiments (Table 2). The detailed specifications of ED unit are indicated in Table 3.

Limiting current density (LCD): LCD of the ED cell was estimated using the method reported in the literature [5,15] by plotting cell resistance (V/I) vs. inverse of current (1/I).

Viscosity measurement: Ubbelhood viscometer was fitted in a constant temperature bath to measure solution viscosity at a desired temperature (±1°C). Sugar concentration was varied between 0-20 mass%, while $CaCl_2$ concentration was varied between 0-50 mol.m^{-3} to estimate the solution viscosity.

Modelling of ion transport: Current density and LCD (limiting current density) of an ED cell is a function of series of parameters e.g. flow (cell geometry, flow dynamics, spacer orientation, solution density and viscosity) and physical (ion concentration, transport number and diffusivity) parameters for a given set of membrane pairs. Precise estimation of these parameters and application of Nernst-Plank equation (assuming zero ion concentration on the membrane surface) would give a theoretical estimate of LCD which can be determined experimentally from plot of V/I vs. 1/I [5,6].

Determination of bulk concentration of diluate compartment

The ED process was carried out well below LCD all through.

The concentration of ion can be obtained from unsteady mass balance across diluate, concentrate, and electrode compartments. The unsteady state mass balance equation was formulated based on the assumptions already mentioned in the literature [12].

The mass balance equations for diluate and concentrate compartments in the ED cell are written as [12,13]

$$NV_C^{dil}\frac{dC^{dil}}{dt}=Q^{dil}\left(C_T^{dil}-C^{dil}\right)-\frac{N\eta iA_m}{ZF}+\frac{NA_mD_{AEM}\left(C_{AEM}^{conc}-C_{AEM}^{dil}\right)}{l_{AEM}}+\frac{NA_mD_{CEM}\left(C_{CEM}^{conc}-C_{CEM}^{dil}\right)}{l_{CEM}} \quad (1)$$

$$NV_C^{conc}\frac{dC^{conc}}{dt}=Q^{conc}\left(C_T^{conc}-C^{conc}\right)-\frac{N\eta iA_m}{ZF}+\frac{NA_mD_{AEM}\left(C_{AEM}^{conc}-C_{AEM}^{dil}\right)}{l_{AEM}}+\frac{NA_mD_{CEM}\left(C_{CEM}^{conc}-C_{CEM}^{dil}\right)}{l_{CEM}} \quad (2)$$

Where, V_C^{dil} , V_C^conc are the volumes of diluate and concentrate

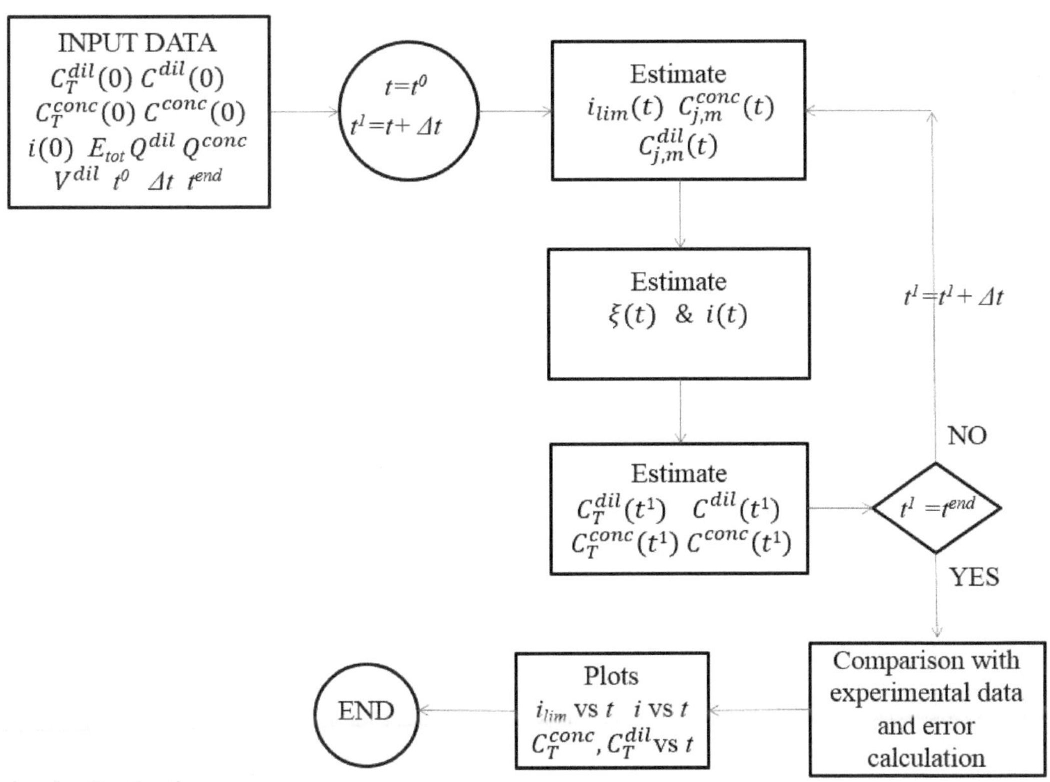

Figure 2: Algorithm explaining steps followed in estimating process parameters using MATLAB program.

Physical Parameters	Cation exchange Membrane	Anion exchange Membrane
Transport number	0.91	0.9
Experimental Resistance (ohm cm^{-2})	2.0-3.5	-----
Max. pressure allowed ($kgcm^{-2}$)	3.0	3.0
Thickness (mm)	0.11 to 0.15	0.09 to 0.11
Max Temperature (°C)	60	60

Table 1: Membrane parameters as obtained from Permionics India Ltd.

Expt. no.	Concentrate	Concentrate flowrate × $10^6(m^3.s^{-1})$	Diluate	Diluate flowrate × $10^6(m^3.s^{-1})$	ERS	ERS flowrate × $10^6(m^3.s^{-1})$	Voltage applied (V)	Time (min)
	SC			Concentrate concentration varied			5% sugar in diluate	
DSC1	Aqueous 5 $mol.m^{-3}$ $CaCl_2$	8.43	25 $mol.m^{-3}$ $CaCl_2$ in 5% sugar solution	8.78	Aqueous 50 $mol.m^{-3}$ NaCl solution	9.15	4	150
DSC2	Aqueous 10 $mol.m^{-3}$ $CaCl_2$ solution	8.43	25 $mol.m^{-3}$ $CaCl_2$ in 5% sugar solution	8.78	Aqueous 50 $mol.m^{-3}$ NaCl solution	9.15	4	150
DSC3	Aqueous 20 $mol.m^{-3}$ $CaCl_2$ solution	8.43	25 $mol.m^{-3}$ $CaCl_2$ in 5% sugar solution	8.78	Aqueous 50 $mol.m^{-3}$ NaCl solution	9.15	4	150
DSC4	Aqueous 25 $mol.m^{-3}$ $CaCl_2$ solution	8.43	25 $mol.m^{-3}$ $CaCl_2$ in 5% sugar solution	8.78	Aqueous 50 $mol.m^{-3}$ NaCl solution	9.15	4	150
DSC5	Aqueous 30 $mol.m^{-3}$ $CaCl_2$ solution	8.43	25 $mol.m^{-3}$ $CaCl_2$ in 5% sugar solution	8.78	Aqueous 50 $mol.m^{-3}$ NaCl solution	9.15	4	150
DSC6	Aqueous 40 $mol.m^{-3}$ $CaCl_2$ solution	8.43	25 $mol.m^{-3}$ $CaCl_2$ in 5% sugar solution	8.78	Aqueous 50 $mol.m^{-3}$ NaCl solution	9.15	4	150
	C			Concentrate concentration varied			No sugar in diluate	
C1	Aqueous 5 $mol.m^{-3}$ $CaCl_2$	8.43	Aqueous 25 $mol.m^{-3}$ $CaCl_2$ solution	8.78	Aqueous 50 $mol.m^{-3}$ NaCl solution	9.15	4	150
C2	Aqueous 10 $mol.m^{-3}$ $CaCl_2$ solution	8.43	Aqueous 25 $mol.m^{-3}$ $CaCl_2$ solution	8.78	Aqueous 50 $mol.m^{-3}$ NaCl solution	9.15	4	150
C3	Aqueous 20 $mol.m^{-3}$ $CaCl_2$ solution	8.43	Aqueous 25 $mol.m^{-3}$ $CaCl_2$ solution	8.78	Aqueous 50 $mol.m^{-3}$ NaCl solution	9.15	4	150
C4	Aqueous 25 $mol.m^{-3}$ $CaCl_2$ solution	8.43	Aqueous 25 $mol.m^{-3}$ $CaCl_2$ solution	8.78	Aqueous 50 $mol.m^{-3}$ NaCl solution	9.15	4	150
C5	Aqueous 30 $mol.m^{-3}$ $CaCl_2$ solution	8.43	Aqueous 25 $mol.m^{-3}$ $CaCl_2$ solution	8.78	Aqueous 50 $mol.m^{-3}$ NaCl solution	9.15	4	150
C6	Aqueous 40 $mol.m^{-3}$ $CaCl_2$ solution	8.43	Aqueous 25 $mol.m^{-3}$ $CaCl_2$ solution	8.78	Aqueous 50 $mol.m^{-3}$ NaCl solution	9.15	4	150
		S				Mass% of sugar varied in diluate		
S1	Aqueous 25 $mol.m^{-3}$ $CaCl_2$ solution	8.43	25 $mol.m^{-3}$ $CaCl_2$ in 5% sugar solution	8.78	Aqueous 50 $mol.m^{-3}$ NaCl solution	9.15	2	150
S2	Aqueous 25 $mol.m^{-3}$ $CaCl_2$ solution	8.43	25 $mol.m^{-3}$ $CaCl_2$ in10% sugar solution	8.78	Aqueous 50 $mol.m^{-3}$ NaCl solution	9.15	4	150
S3	Aqueous 25 $mol.m^{-3}$ $CaCl_2$ solution	8.43	25 $mol.m^{-3}$ $CaCl_2$ in 15% sugar solution	8.78	Aqueous 50 $mol.m^{-3}$ NaCl solution	9.15	6	150
S4	Aqueous 25 $mol.m^{-3}$ $CaCl_2$ solution	8.43	25 $mol.m^{-3}$ $CaCl_2$ in 20% sugar solution	8.78	Aqueous 50 $mol.m^{-3}$ NaCl solution	9.15	8	150

Table 2: Different process variables chosen during ED experimentation.

Parameter	Value	Reference
Temperature, T	308 K	This work
Transport number of cation in the solution,2+,	0.4387	This work [5]
Transport number of anion in the solution,	0.5613	This work [5]
Diffusivity of Ca^{2+} ions in solution at 20°C	7.92×10^{-10} $m^2.s^{-1}$	This work [25,26]
Diffusivity of Cl^- ions in solution at 20°C,	17.7×10^{-10} $m^2.s^{-1}$	[5]
Diffusivity of CaCl2 through AEM	1.8×10^{-11} $m^2.s^{-1}$	[30]
Diffusivity of CaCl2 through CEM	1.8×10^{-11} $m^2.s^{-1}$	[30]
Distance between adjacent membranes	4.7×10^{-3} m	This work
Area of the membrane, Am	32×10^{-4} m^2	This work
Charge on the Calcium ion,2+	+2	This work
Current efficiency,	0.81	This work
Sh number empirical equation constant, a	0.0583 ± 0.0045	This work
Sh number empirical equation constant, b	0.63	This work [24]
Sh number empirical equation constant, c	0.33	This work [24]
Sh number empirical equation constant, d	-0.131	This work

Table 3: Values of different physical parameters used in the model.

compartments (m^3) respectively, N is number of cell pairs, t is time (s), $C_T^d il$ and C_T^{conc} represents diluate concentrations, leaving diluate feed tank and leaving diluate cell compartment ($mol.m^{-3}$) respectively, C_T^{conc} and $Q^d il$ represents diluate concentrations, leaving concentrate feed tank and leaving concentrate cell compartment ($mol.m^{-3}$) respectively, $Q^d il$ and , $Q^c onc$ are the diluate and concentrate streams volumetric flow rate ($m^3.s^{-1}$), η is the current efficiency, t is the current density ($A.m^{-2}$), η is the effective membrane area (m^2).

η can be estimated from the following equation [12,13]:

$$\eta = t_{+,CEM} + t_{-,AEM} - 1 \tag{3}$$

Where, $t_{-,AEM}$ is the transport-number of cations in cation exchange membranes and $t_{-,AEM}$ is the transport-number of anions in anion exchange membrane.

Similarly, unsteady state mass balance around feed tank can be written as

$$\frac{(d(V_T^c onc C_T^c onc))}{dt} = Q^c onc(C^c onc - C_T^c onc) \tag{4}$$

$$\frac{(d(V_T^c onc C_T^c onc))}{dt} = Q^c onc(C^c onc - C_T^c onc) \tag{5}$$

Where V_T^{conc} and V_T^{conc} are volumes of concentrate and diluate feed tanks (m^3) respectively. During electrodialysis water transport occurs across the membranes due to electro-osmosis and osmosis [13]. In the present work volume change (due to water transport) is ignored as there was no major net volume change observed (concentrate and diluate compartments) experimentally.

Determination of the current density

Overall flux equation : The current density was estimated using following assumptions (usually valid for all practical situations) [13]. The solution between the two membranes was assumed, thoroughly mixed to ensure uniform concentration throughout. Stationary membrane surfaces leads to the formation of static boundary layer whose thickness is mainly governed by solution viscosity and existing turbulence. Fluid flow alone can determine the boundary layer thickness [5]. Ion transport due to convection was ignored [19] because the direction of ion transport was perpendicular to the direction of the flow. Trans-membrane pressure was assumed to be zero. Stable boundary layers are formed along the concentrate and diluate compartments of the membrane due to difference in viscosity of feed and concentrate solution. The current density can be estimated from mass balance equation either for cations through CEM or anions through AEM, assuming overall (sum) transport numbers to be unity. The flux of ions passing through the membrane can be expressed by generalized Nernst Plank equation as [5]

$$N_j = -D_j \frac{\partial C_j}{\partial x} - \frac{z_j C_j F D_j}{RT} \frac{\partial \psi}{\partial x} \tag{6}$$

Where, x is the distance measured from boundary layer in contact with the diluate channel towards the membrane, C_j is the diffusivity of ion j (either cation or anion) ($m^2.s^{-1}$), 'C_j is the concentration of ion j ($mol.m^{-3}$), 'R' is the universal gas constant (8.314 $J.mol^{-1}.K^{-1}$), 'T' is the temperature (K), 'z_j' is the charge of diffusing species j, and N_j is the electric potential gradient ($V.m^{-1}$) and F is the faradays constant ($Columb.geqv^{-1}$).

The total molar flux of ion 'j' through the ion exchange membrane, $N_{j,m}$ can be related to the current density, i as

$$N_{j,m} = \frac{t_{j,m} i}{z_j F} \tag{7}$$

The subscript m denotes either AEM or CEM, Z_j is the transport number of ion 'j' in the membrane, i is current per unit area of membrane or current density ($A.m^{-2}$) and Z_j is the charge of the ion.

At steady state $N_{j,m}$ and $N_{j,m}$ are equal, i.e.:

$$\frac{t_{j,m} i}{Z_j F} = -D_j \frac{\partial C_j}{\partial x} - \frac{Z_j C_j F D_j}{RT} \frac{\partial \psi}{\partial x} \tag{8}$$

Assuming that a linear profile of the concentration distribution exists along the boundary layer, the linearized Nernst-Planck equation could be used instead of Eq. (8). Expression for the linearized Nernst-Planck equation when applied in the diluate chamber is [5]:

$$\frac{t_{j,m} i}{z_j F} = \frac{D_j \left(C_{j,b}^{dil} - C_{j,m}^{dil}\right)}{\delta} - \frac{z_j D_j F \xi C_{j,m}^{dil}}{RT} \tag{9}$$

Where, δ is the boundary layer thickness (m), $C_{(j,b)}^d il$ and $C_{(j,m)}^d il$ are concentrations of ions in bulk and at the membrane surface respectively of the diluate compartment, while ξ is the electric potential gradient ($V.m^{-1}$) and is expressed as

$$\xi = \frac{\psi}{\delta} \tag{10}$$

where, δ is the boundary layer thickness.

In Eq. (9) the first part, i.e., $\frac{D_j\left(C_{j,b}^{dil} - C_{j,m}^{dil}\right)}{\delta}$ denotes contribution due to molecular diffusion while the second part i.e., $\frac{z_j D_j F \xi C_{j,m}^{dil}}{RT}$ reflects the effect of voltage on ion-transfer through the membrane.

Boundary layer thickness, δ estimation: δ is estimated using film theory (Eq. (11)) [5,20] and salt mass-transfer coefficient. Salt mass-transfer coefficient is usually determined based on salt diffusivity and suitable mass-transfer correlation, which in-turn is dependent on flow profile and physical properties of the fluids, cell geometry, surface morphology of membranes used in ED cell [5,6,19,21,22].

$$\delta = \frac{D_j}{k} \tag{11}$$

Where, k and k are diffusivity and mass transfer coefficient of diffusing species in solution. Each of these parameters were separately estimated using standard correlations. The mass transfer coefficient was obtained from Sherwood number. [5,19,20] as:

$$Sh = \frac{k.l}{D_j} \tag{12}$$

Where, l is the characteristic length (m). Sherwood number Sh, is expressed as a function of Reynolds number, Re and Schmidt number, Sc [5,20]. The empirical expression of Sherwood number is based on cell geometry and spacer configuration chosen for the present cell is indicated below [20-24]

$$Sh = a.Re^b.Sc^c.\left(\frac{\ddot{E}_{Dil}}{\ddot{E}_{Conc}}\right)^d \tag{13}$$

Where, Sc (Schmidt's number, $\mu/\rho D_j$ is estimated from physical properties (viscosity and density) of the medium while Reynolds number, $\left(\frac{\rho v l}{\mu}\right)$ indicates flow characteristics of the medium [20] and $\left(\frac{\Lambda_{Dil}}{\Lambda_{conc}}\right)$ represent the conductivity ratios of diluate and concentrate, while 'a' 'b' 'c' 'd' are the empirical constants.

Ionic diffusivity is a strong function of hydrodynamic diameter of ions. Assuming infinite dilution ionic diffusivity is estimated using the Nernst-Haskel equation (Eq. 14) [25] as:

$$D^{\circ} = \frac{RT\left[(1/Z)+(1/Z-)\right]}{F^2\left[(1/\lambda_+)+(1/\lambda_{-1})\right]} \quad (14)$$

Where z_+ and z_- denote charges of cation and anion respectively while λ_+ and λ_- denotes limiting ionic conductance in the solvent. Other parameters bearing meaning and units are as reported in the nomenclature.

In general diffusivity of ions from electrolytes e.g. NaCl, $CaCl_2$, LiCl in water are closely related to changes in solution viscosity at constant temperature. Therefore, necessary correction of diffusivity on viscosity is essential. Li et al. [26] experimentally verified that diffusivity of ions in an electrolyte solution is inversely proportional to the solution viscosity.

$$\frac{D}{D^{\circ}} = \frac{\mu^{\circ}}{\mu} \quad (15)$$

Therefore, the estimated diffusivities were corrected due to solution viscosity.

Estimation of Membrane Surface Concentration: The membrane surface concentration of ions is dependent on current density under an applied voltage. As long as the ED operation is executed below limiting current (surface concentration would become zero at limiting current), the surface concentration on either side can be estimated from bulk concentration measurement (diluate/concentrate), current density and limiting current density using Eqs. (16) and (17) [13,27].

$$C^{c}_{(j,m)}onc = C^{c}_{j}onc(1+i/i_{j,lim}) \quad (16)$$

$$C^{dil}_{j,m} = C^{dil}_{j}(1-i/i_{j,lim}) \quad (17)$$

Where, $C^{conc}_{j,m}$ and C^{conc}_{j} are the concentrations of ion *j*, at the membrane surface and in bulk of the concentrate compartment respectively in the ED cell. $C^{dil}_{j,m}$ and C^{dil}_{j} are the concentrations of ion *j*, at the membrane surface and in the bulk of the diluate side respectively in the ED cell.

Estimation of current density and limiting current density ($\mathbf{i}$ and $\mathbf{i_{j,lim}}$): LCD (of a single electrolyte) is estimated from linearized Nernst Planck equation [6,8]:

$$i_{j,\mathrm{lim}} = \frac{C^{dil}_{j,b} D_j Z_j F}{\delta\left(t_{j,m}-t_j\right)} \quad (18)$$

Where, $t_{j,m}$ and t_j are transport numbers of ion *j* in membrane and electrolyte solution respectively.

Considering ion flux in the diluate side of the IEM, Eq. (17) is used to calculate concentration of ion *j*, at the membrane surface of diluate side $C^{dil}_{j,m}$

The current density can be expressed by Eq. (17) after substitution of Eqs. (17) and (18) in Eq. (8).

$$i = \frac{C^{dil}_{j,b} D_j Z_j \xi \dfrac{F}{RT}}{\dfrac{t_{j,m}-t_j}{Z_j F} + \dfrac{\xi\delta\left(t_{j,m}-t_j\right)}{RT} - \dfrac{t_{j,m} Z_j}{F}} \quad (19)$$

Where ξ, the electric potential gradient can be estimated from Nernst equation given below [12,13]

$$\xi = -(2t_{j,m}-1)\frac{RT}{\delta F} ln\left(\frac{\gamma^{dil}_b C^{dil}_{j,b}}{\gamma^{dil}_m C^{dil}_{j,m}}\right) \quad (20)$$

Where γ^{dil}_m and γ^{dil}_b are the mean ionic activity coefficients corresponding to the ions at the wall of IEM and in the bulk of solution respectively within the diluate channel and they are estimated using Debye–Huckel limiting law.

Numerical Estimation of Parameters

The sequence of steps followed in the model to estimate current density and concentration change is described in the flow chart (Figure 2). The differential equations (Eqs. (1), (2), (4) and (5)) were integrated using Euler method using 1 second step size interval. Initial process conditions were obtained from literature. Few crucial parameters and their estimation techniques which were not taken up earlier are presented below.

Experimental current density and concentration of Ca^{2+} ions

The current flowing through the electrolyte solution of the ED cell under an applied voltage was recorded in ammeter. Ca^{2+} ions concentration was estimated from conductivity measurement and using standard calibration chart (mass concentration vs. conductance).

Estimation of mass transfer coefficients

Mass transfer correlations for the electro dialysis process based on the cell geometry were reported earlier by several authors [15,24]. They expressed Sherwood number (Sh) as a power function of Reynolds (Re) and Schmidt number (Sc) determined from their experimentation. In this report we have developed a suitable mass transfer correlation using the LCD and current density expressions that suits all of our experimental results.

LCD was estimated experimentally for the case where both concentrate and diluate concentrations were 25 mol.m^{-3} of $CaCl_2$ in water (Figure 3). Using this LCD value and Eqs. (11), (12) and (18) along with physical properties of the electrolytes reported in Table 3, the mass transfer coefficient was estimated. Which was subsequently

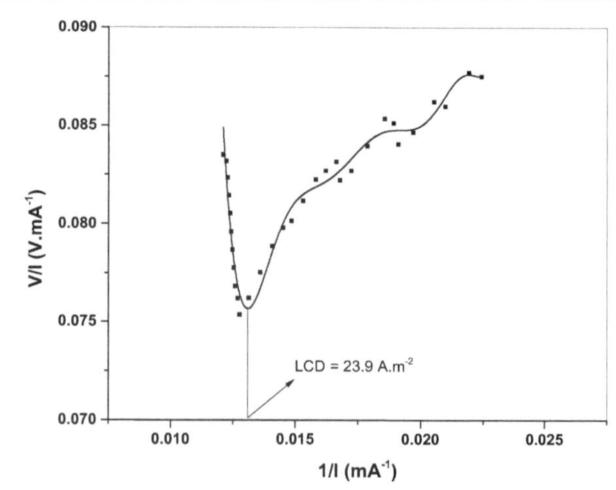

Figure 3: Limiting current estimation from the plot of electrical resistance (V/I) versus 1/I, while the diluate and concentrate streams have 25 mol.m^{-3} $CaCl_2$ and flow rates are 4.93 × 10^{-6} m^3.s^{-1} and 5.27 × 10^{-6} m^3.s^{-1} respectively.

used to evaluate Sherwood number as a function of system flow parameters e.g. Reynolds number, Schmidt number and dimensionless ratio of conductivities Eq. (13). It was found that $Sh.Sc^{-0.33}$ was a linear function of $Re^{0.63}$, where $Sh.Sc^{-0.33} = 0.0583 . Re^{0.63}$.

The constant 0.0583 is a product of parameters a and $\left(\frac{\Lambda_{Dil}}{\Lambda_{Conc}}\right)^d$ For the set with same $CaCl_2$ concentration in concentrate and diluate streams, the ratio of conductivity values is close to unity $\left(\frac{\Lambda_{Dil}}{\Lambda_{Conc}}\right)$ ~1.0. This makes the factor a =0.0583. The factor $a.\left(\frac{\Lambda_{Dil}}{\Lambda_{Conc}}\right)^d$ was optimized to fit each experimental data point (concentration and current density with time). From each of this optimum value, d was obtained knowing $\left(\frac{\Lambda_{Dil}}{\Lambda_{Conc}}\right)$ for any set and is found out to be -0.131. The Sherwood number correlation suitable for the ED cell flow channel geometry was obtained as

$$Sh = (0.0583 \pm 0.0045).Re^{0.63}.Sc^{0.33}.\left(\frac{Ë_{Dil}}{Ë_{Conc}}\right)^{-0.131} \quad (21)$$

The proposed equation (Eq. 21) has been validated and reported in Appendix.

Determination of transport number of ion in solution

Transport number t_j of ion j is the fraction of total current carried by the ion type which is a function of diffusion coefficient and ionic mobility of hydrated species. Difference in hydration ability influences size, diffusivity and mobility of the ion. Thus, ions do not transport current equally in solution. The transport number was estimated from the following equation [5]:

$$t_j = \frac{|z_j| D_j C_j}{\sum_{j=1}^{n} |z_j| D_j C_j} \quad (22)$$

For a binary-ion salt solution, n = 2, j= 1 for cation and j = 2 for anion and transport number is measured.

Determination of current

Initial current density estimation is essential to obtain salt concentration at membrane surface, and to start numerical integration, which may be evaluated either experimentally or from the applied electric potential and solution resistances using Ohms law. The applied electric potential may be expressed as

$$E_{tot} - E_{el} = R_{tot}.I \quad (23)$$

Where, E_{el} is the electric potential drop near the electrodes, R_{tot} is the overall resistance (ohm) of the ED cell and I is the current (A). The overall resistance is the sum of individual values:

$$R_{tot} = R_{conc} + R_{dil} + R_{ERS} + R_{mem} \quad (24)$$

Where, resistance of ERS, concentrate, and diluate channels were estimated either directly from conductivity measurement from extended Kohlrausch-equation [13,28]. The conductivity and the resistance are related as

$$Resistance = \frac{1}{Ë}\frac{L}{A_m} \quad (25)$$

Where, Λ is the conductivity of solution ($S.m^{-1}$), L is the gap between membranes or the compartment thickness (m), A_m is the effective membrane area (m^2).

Determination of the γ_m^{dil} and γ_b^{dil}

γ_b^{dil} and γ_b^{dil} are the mean ionic activity coefficients of salt at membrane and in bulk respectively in the diluate. These were estimated using Debye-Huckel limiting law [28].

$$\log \gamma_{\pm} = \frac{-0.511|z_+ \cdot z_-| I_s^{1/2}}{1 + I_s^{1/2}} \quad (26)$$

Where $\gamma_{\pm}$ is the mean ionic activity of a salt in an electrolyte solution and I_s is the ionic strength defined as,

$$I = 0.5 \sum m_j z_j^2 \quad (27)$$

Where m_j is molality (mol/kg solution) of the ion 'j' having charge E_{sp}.

$CaCl_2$ solution (with 5% sugar) densities were estimated experimentally and these data were linearly regressed (R^2=0.984) and used whenever required (Figure 4).

Determination of specific energy consumption

The specific energy consumption, E_{sp} (kWh.kg^{-1}) was estimated from the following equation

$$E_{sp} = \frac{\int_{t_1}^{t_2} \varepsilon A_m i(t) dt}{M_{CaCl_2} \Delta n_{CaCl_2}(t)} \quad (28)$$

Where, ε is the applied electric potential in volt, M_{CaCl_2} is the area of the membrane in m^2, $i(t)$ is the current density ($A.m^{-2}$) as a function of time, M_{CaCl_2} is the molecular mass of $CaCl_2$(=111.02 g.mol^{-1}) and $\Delta n_{cacl_2}(t)$ is the number of moles of $CaCl_2$ removed from the feed solution at various time interval.

Results and Discussions

Role of Sugar (mass%) and $CaCl_2$ concentration on solution viscosity and influence of temperature

Sugar solution viscosity measurement at constant temperature shows nonlinear (Figure 5) rise with sugar (5 to 20 mass%). Viscosity values were found to be between 0.72 - 1.5 mPa.s with increase in sugar mass%. These values were very much comparable with the literature reported data [29]. Solution viscosity does not show appreciable change with $CaCl_2$ concentration (0-50 mol.m^{-3}) (Figure 6). Viscosity measurement of $CaCl_2$ solution at 20, 25, 32, 37, 42°C show lowering

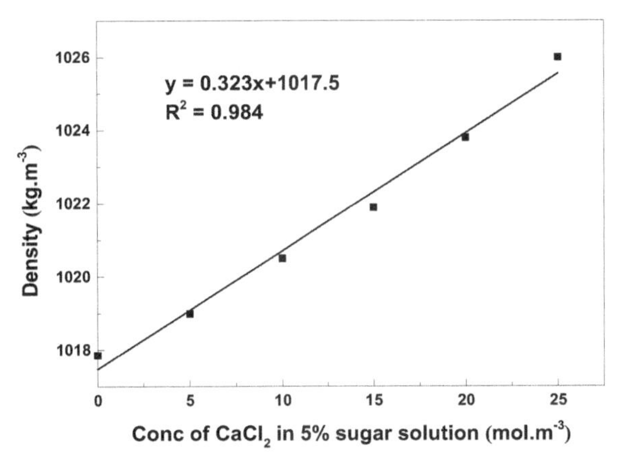

Figure 4: Calibration chart used to estimate solution density (kg.m^{-3}) from molar concentration of $CaCl_2$ (mol.m^{-3}) in 5 mass% sugar solution.

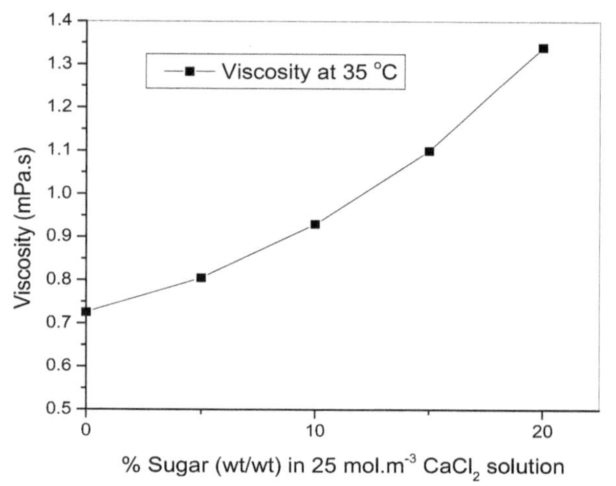

Figure 5: Solution viscosity variation with mass % of sugar containing 25 mol m^{-3} $CaCl_2$ in water at 35°C.

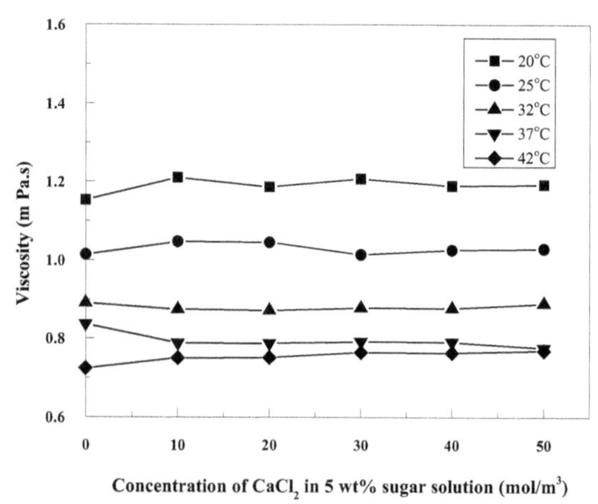

Figure 6: Solution viscosity variation with concentration of $CaCl_2$ in 5 mass % sugar solution at different temperatures.

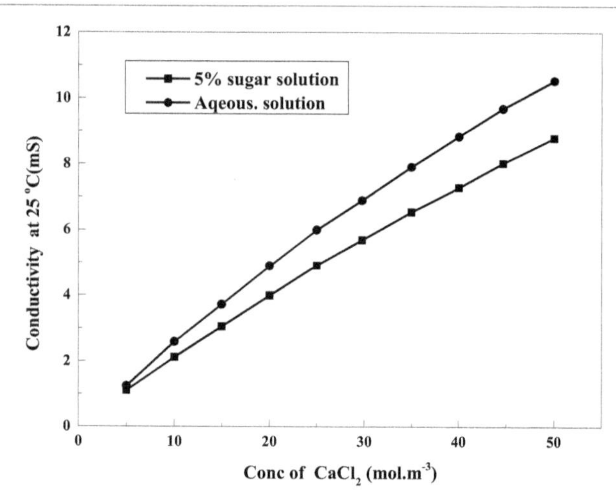

Figure 7: Conductivity variation with concentration of $CaCl_2$ in water with sugar and without sugar.

between 1.18 to 0.6 mPa.s. Almost ~37% lowering in viscosity was recorded (Figure 6).

Effect of Sugar and $CaCl_2$ concentration on electrical conductivity

Figure 7 shows plot of electrical conductivity with $CaCl_2$ concentration, estimated in presence and absence of sugar. Solution conductivity increases almost linearly with rise in $CaCl_2$ concentration (5 to 50 mol.m^{-3}). $CaCl_2$, a strong electrolyte, dissociates completely in solution and increase number of ions per unit volume available for ionic conductance. Addition of sugar dampens the ion conductivity values. This may be due to sugar, a water soluble non-electrolyte, doesn't dissociate and change the number of ions responsible for current carriage, therefore, presence of inert sugar molecules causes increase in crowding in solution.

Experimental results and model prediction

The LCD measurement by plotting V/I vs. 1/I (Figure 3) (diluate and concentrate concentrations are 25 mol.m^{-3} $CaCl_2$) shows a minima at 0.013 mA^{-1} which corresponds to the applied electric potential ~6 V. All experiments were categorically performed at 4 V, hoping current density to remain always below limiting value because, initial concentration of $CaCl_2$ in the diluate stream was kept unchanged for all experiments.

Current density and concentrate concentration variation were noted at regular interval for all the experiments and reported in Figures 8-10. With progress of electrodialysis current density gets lowered (Figure 8) while, concentrate $CaCl_2$ concentration increased (Figure 9). Lowering of the initial $CaCl_2$ concentration in concentrate stream (C6 to C1, Figure 8) for the same diluate stream current density dropped. Similar trend was noted with increase in sugar mass% in the diluate channel (DSC6 to DSC1, Figure 8), but overall current density values were lower than the corresponding non-sugar cases (Figures 8 and 10) shows current density is a strong function of sugar concentration i.e. with increase in sugar mass% (S1, 5% to S4, 20%) the current density drops.

The ion removal percentage estimated from the initial and final $CaCl_2$ concentration of the diluate channel (Figure 11) doesn't show any trend with concentrate concentration. Experiments C5 and DSC2 showed higher ion removal among their corresponding sets C1-C6 and DSC1-DSC6 respectively (Figure 11). This possibly hints at presence of a series of complex reaction taking place in ion transport.

Presence of sugar in the diluate (DSC1-DSC6) stream reduced ion removal rate compared to corresponding non-sugar cases (C1-C6). This may be attributed to tremendous defiance experienced by hydrated calcium ions in presence of inert sugar molecules.

Eqs. (1), (4), (2) and (5) along with initial conditions (Table 3) were solved as per the reported sequence (Figure 2) using MATLAB. The solution gave us theoretical estimate of concentration. Predicted molar concentration of both concentrate and diluate stream shows nice agreement with all experimental data. This is represented by continuous line (Figures 8-10) showing an excellent agreement with the experimental data.

Effect of concentrate stream concentration

Influence of concentrate stream concentration on (i) overall current density (Figure 8), and (ii) ion removal rate (percentage) from diluate solution (Figure 11) were experimentally estimated. Figure 8

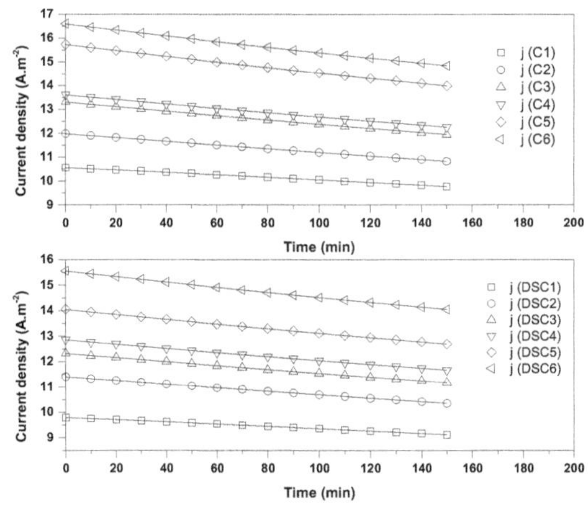

Figure 8: Current density change with time for the experiments C1-C6 and DSC1-DSC6. The symbols represent the experimental data, while the straight lines represent the model predicted values.

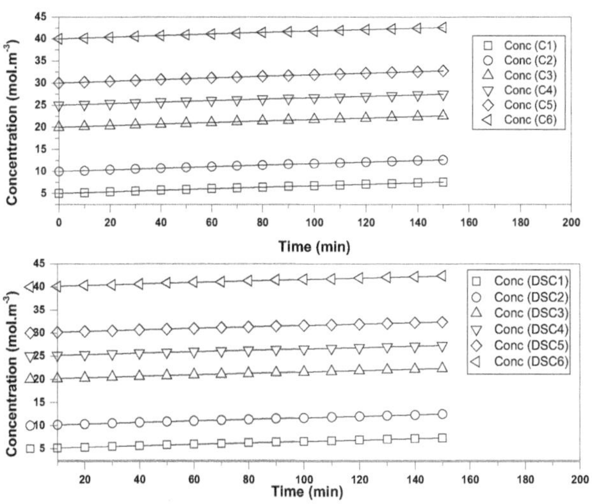

Figure 9: Variation of $CaCl_2$ concentration in concentrate with time for the experiments C1-C6 and DSC1-DSC6. The symbols represent the experimental values, while the straight lines represent the model predicted data.

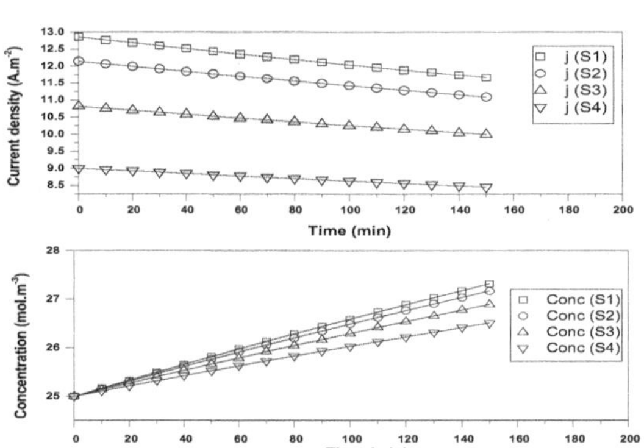

Figure 10: Variation of current density and concentration of $CaCl_2$ in the concentrate with time for the experiments S1-S4. The symbols represent the experimental data, while the straight lines are obtained from model.

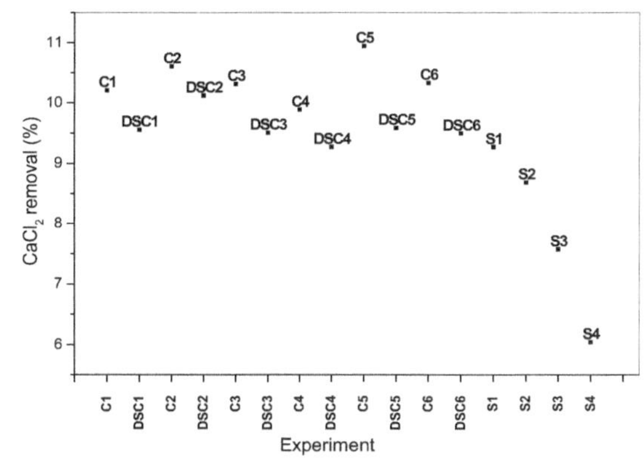

Figure 11: Final % removal of $CaCl_2$ removal for all experiments C1-C6, DSC1-DSC6 and S1-S4.

shows increase in current density with rise in initial concentration of $CaCl_2$, which decreases slowly with time. This is true for both types of diluate with sugar and without sugar. Initially high current density value is probably due to availability of dissociated Ca^{2+} and $2Cl^-$ ions of $CaCl_2$ (a strong electrolyte). With increase in $CaCl_2$ concentration, more such dissociated species appear per unit volume which reduces overall solution resistance and increases current density under a given electric potential. A slow drop in current value appeared due to gradual depletion of ions from diluate compartment. Figure 11 showing percentage removal from diluate stream didn't follow any trend probably due to involvement of several competing factors in overall ion removal. The current density values (Figure 8) don't predict the net ion removal percentage, instead overall ion removal depends on effective or net flux existing in the cell. Effective ion removal rate is due to cumulative addition of two competing processes namely: (i) An external applied electric potential and (ii) chemical potential (or diffusive transport) arising out of concentration gradient across the membrane. The applied electric potential always tries to push Ca^{2+} ions towards the appropriate electrode while, the direction of chemical potential or diffusive transport is decided by the concentration gradient across the membrane. Flux components arising out of an applied electric potential (hence forth defined as current flux), concentration gradient (diffusive flux) and overall (net) flux were estimated from model and presented in Figure 12.

The current flux increased with $CaCl_2$ concentration in the concentrate stream in the sequence C1-C6 and DSC1-DSC6. The trend is similar to the current density behavior indicated before. The diffusive component is favorable for C1 and DSC1 (values are positive), while for C2 and DSC2 it is favorable till initial 40 minutes but thereafter it becomes unfavorable. This leads to crossover of net flux and current flux lines (Figures 12a-12c). Similar crossover doesn't arise with any set, because, the diffusive component remained consistently either positive or negative for the rest of the experiment. Ion removal being a complex process gets influenced by e.g. concentrate concentration, the solution viscosity, conductivity, concentration of non-electrolyte (taken up in the following section) even if the applied electric potential, flow rate and cell geometry were kept unchanged. Concentrate Ca^{2+} concentration being lower than diluate stream the diffusive part becomes favorable for initial 40 minutes and thereafter it gets reversed causing cross over.

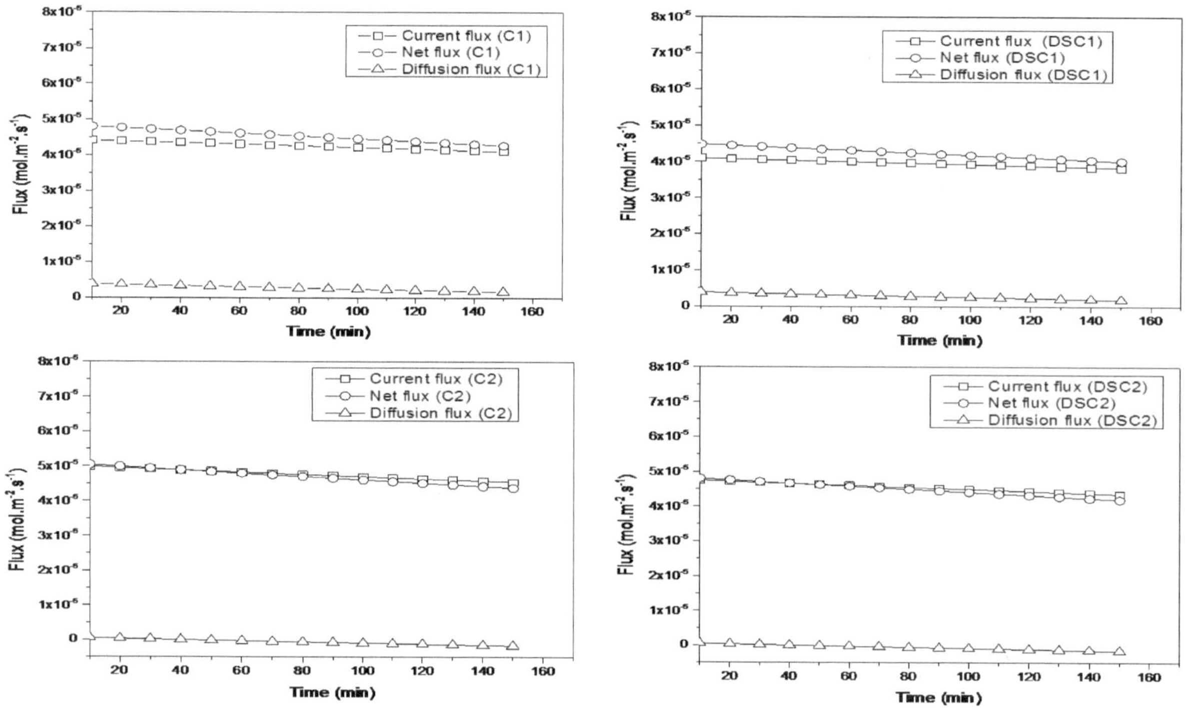

Figure 12: (a): Comparison of current flux, diffusive flux and net flux and their variation with time for experiments C1 - DSC1 and C2 - DSC2.

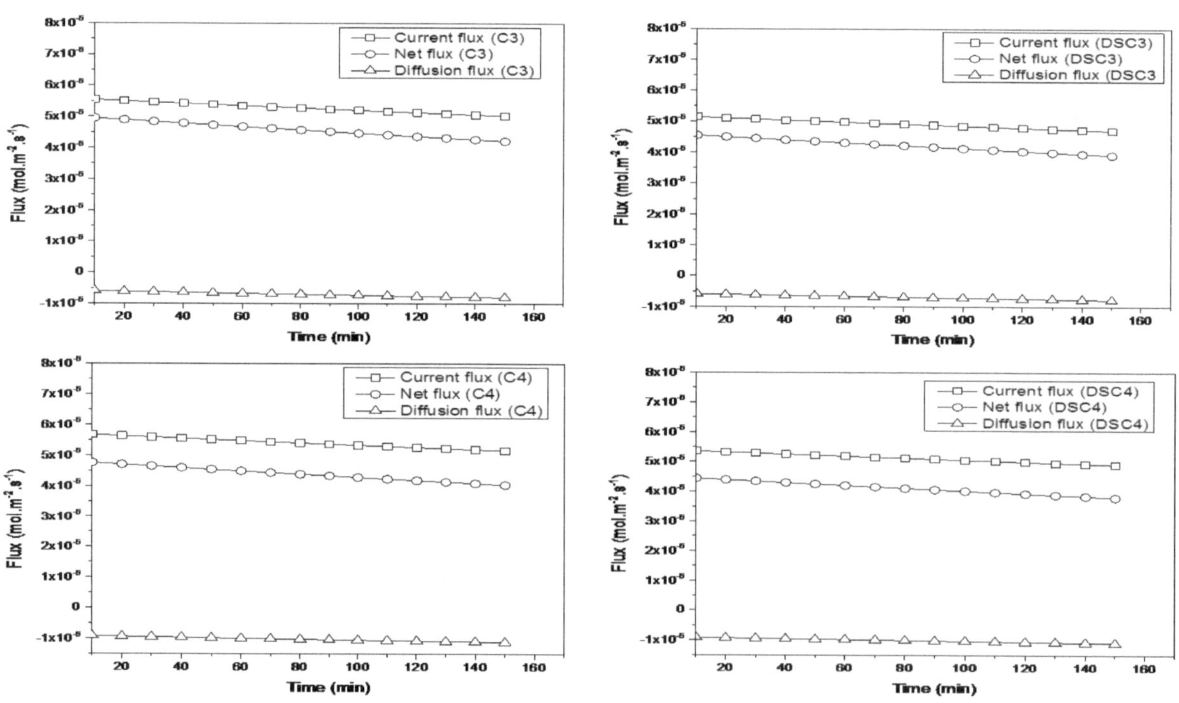

Figure 12: (b): Comparison of current flux, diffusive flux and net flux and their variation with time for experiments C3 – DSC3 and C4 – DSC4.

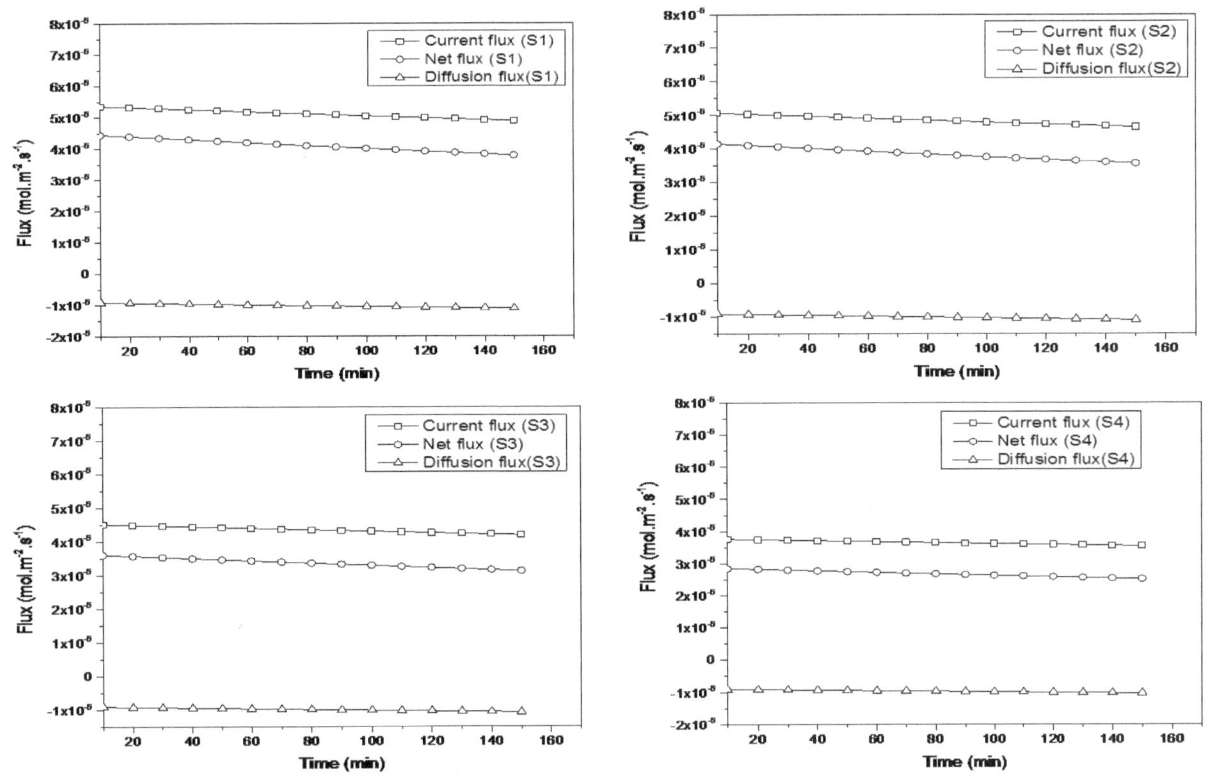

Figure 12: (c): Comparison of current flux, diffusive flux and net flux and their variation with time for experiments C5 – DSC5 and C6 – DSC6.

Investigation on role of sugar concentration in ion removal

Effect of sugar concentration of diluate channel on (i) current density (Figure 10), (ii) concentrate concentration variation (Figure 10), (iii) Ion removal rate (Figure 11) and (iv) net flux Figures 12(d) were experimentally found out.

With increase in sugar concentration the current density drops (Figure 10) thus, current density is maximum with 5% sugar and minimum with 20% sugar and slowly it decreases with time. Gradual lowering in current density between 5% - 20% sugar (S1-S4) may be due to increase in solution viscosity (Figure 5) and reduction in ion conductivity (Figure 7). Thus, hydrated ions experience more resistance under an applied electric potential. Slow drop in current density with time is due to depletion of ions from diluate channel. Gain in concentrate channel concentration supports ion transport from diluate to concentrate channel. The slope of the line indicating ion accumulation rate in concentrate is highest with 5% sugar (S1) while it is the minimum with 20% (S4) sugar concentration. This is due to increase in solution resistance with non-electrolyte sugar molecules. The percent salt removal from diluate stream (Figure 11) supports above fact indicating drop in ion removal rate with increasing sugar concentration.

Estimation of different flux components e.g. current flux, diffusive flux and net flux are presented in Figure 12d. The current flux decreased with increase in the mass% of sugar i.e. in the order of 5% - 20% (S-S4) and thus, effective flux decreases with sugar mass% (Figure 12d).

It is also interesting to note (Figure 13) the ratio of removal rates with sugar and without sugar remains almost constant with concentrate concentration change. Similar constancy is also noted with viscosity ratios and average current density ratios as well. This is possibly due to impermeable (to sugar molecules) nature of ionic membrane used in electro dialysis. This has been confirmed from UV analysis of the concentrate stream before and after operation (Figure 14). No peak is noticed at 400nm (λ_{max} for sugar). Sugar is not an electrolyte and its role is limited to the diluate channel only. The ratios of removal rates (with and without sugar) can be expressed by a simple power function of viscosity ratios as

$$\frac{percent\ removal_C}{percent\ removal_{DSC}} = \left(\frac{\mu_{DSC}}{\mu_C}\right)^{0.28} \tag{29}$$

This factor of 0.28~0.3 resembles close similarity with Sc number in the Sh number correlation, indicating that the removal rate is a strong function of flow and physical properties of the system.

Mass transfer coefficient and specific energy

Local mass transfer coefficients of ions (cation/anion) and specific energy consumption were estimated using Eqs. (12) and (28) respectively and reported in Figure 15. Mass transfer coefficient directly relates the speed of ion transport phenomena and is a strong function of local parameters of the system. Figure 15 shows local mass transfer coefficients 'k' are always higher with non-sugar cases than those with sugar ones. Presence of non-electrolyte (sugar) reduces mass transfer coefficient and increases overall specific energy consumption to execute a desired separation. Experiments C1 and DSC1 are more energy efficient having lowermost specific energy consumption estimates. Favorable diffusional flux and low current values for these sets (C1, DSC1) probably encourage reduction in energy consumption.

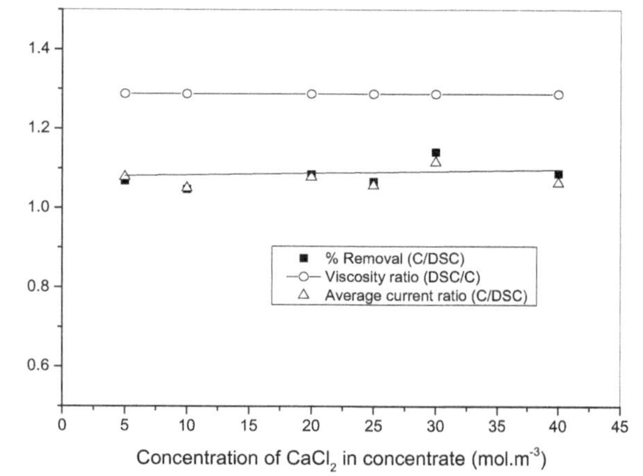

Figure 13: The ratios of % removal (non-sugar/sugar), average current density (non-sugar/sugar) and viscosity (sugar/non-sugar) with change in concentrate concentration.

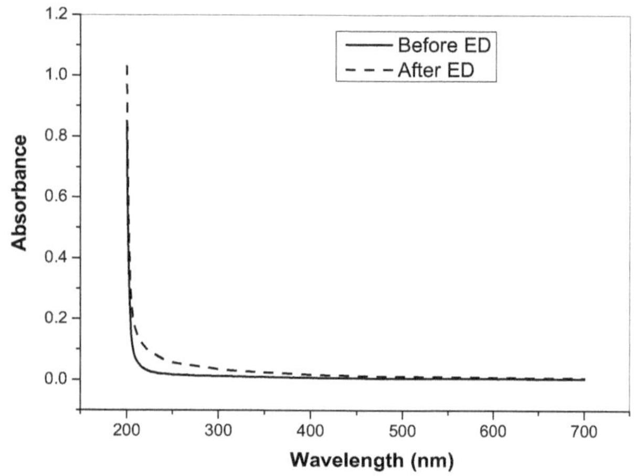

Figure 14: UV absorption of concentrate stream of S1 before and after electro dialysis. Absence of sugar peak (Wavelength, λ=400 nm) ensures the impermeability of sugar molecules through ion exchange membranes.

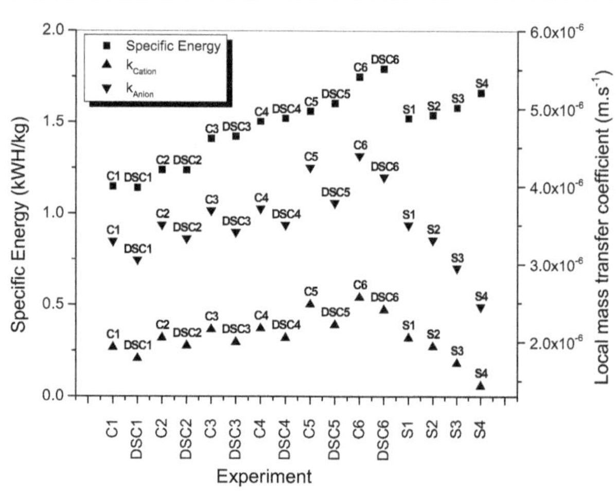

Figure 15: Local mass transfer coefficients of cations, anions and specific energy estimates for experiments C1-C6, DSC1-DSC6 and S1-S4.

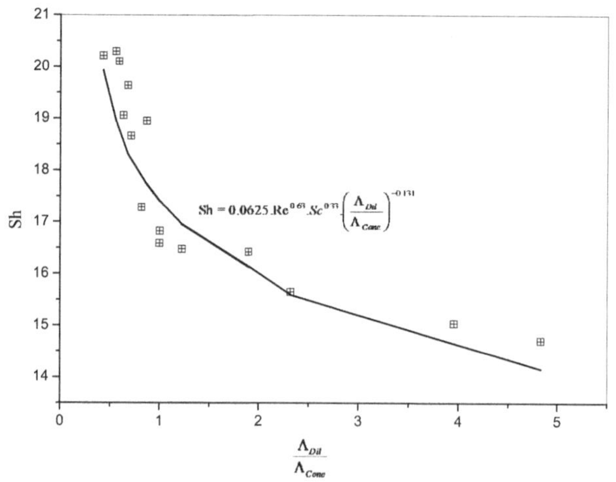

Figure 16: Plot of Sherwood number vs. conductivity ratio of diluate and concentrate to validate the proposed empirical equation (Eq. 21).

Conclusion

Removal of $CaCl_2$ from 5% sugar solution was carried out in batch recirculation electrodialysis mode. All experiments, were performed below LCD to avoid water electrolysis. Role of electrolyte concentration in concentrate channel, sugar mass% in diluate stream in ion removal were evaluated. Concentrate concentration variation doesn't follow any trend to predict ion removal rate. Presence of sugar (in the diluate stream) increases the solution viscosity, reduces ion diffusivity and electrical conductivity thus, lowers ion removal rate compared to those without sugar cases. The linearized Nernst-Plank equation efficiently predicted the experimental data and was extensively used to calculate various fluxes (current, diffusional and net), mass transfer coefficients for ions, specific energy consumption. Increase in concentrate concentration reduced overall cell resistance causing higher current density. Diffusional flux plays a major role in deciding overall or net flux. The experiments C1, C2, DSC1 and DSC2 showed favorable diffusional flux and higher ion removal rate with simultaneous reduction of energy consumption. Therefore, an optimum electrolyte concentration of concentrate, sugar mass% in diluate, and applied external electric potential can be obtained from this apparatus Figure 16.

Acknowledgement

Financial support to execute the experimental work is gratefully acknowledged to IIT Roorkee (No. IITR/SRIC/244/FIG-Sch-A), India.

References

1. Mathur RBL (1978) Handbook of Cane Sugar Technology. Oxford and IBH Publishing Co New Delhi.
2. Smagghe F, Mourgues J, Escudier JL, Conte T, Molinier J, et al. (1992) Recovery of Calcium Tartrate and Calcium Malate in Effluents from Grape Sugar Production by Electrodialysis. Bioresource Technology 39: 185-189
3. Elmidaoui A, Chay L, Tahaikt M, Sahli MAM, Taky M, et al. (2004) Demineralisation of beet sugar syrup, juice and molasses using an electrodialysis pilot plant to reduce melassigenic ions. Desalination 165: 435.
4. Tragardh G, Gekas V (1988) Membrane Technology in the Sugar Industry. Desalination 69: 9-17
5. Geraldes V, Afonso M D (2010) Limiting current density in the electrodialysis of multi-ionic solutions. J Membr Sci 360: 499-508
6. Lee HJ, Strathmann H, Moon SH (2006) Determination Of The Limiting Current Density In Electro-Dialysis Desalination As An Empirical Function Of Linear Velocity. Desalination 190: 43-50

7. Tanaka Y (2005) Limiting Current Density of an Ion-Exchange Membrane and of an ElectroDialyzer. J Membr Sci 266: 6-17

8. Strathmann H (2004) Ion-Exchange Membrane Separation Processes. Elsevier

9. Elmidaoui A, Lutin F, Chay L, Taky M, Tahaikt M, et al. (2002) Removal of melassigenic ions for beet sugar syrups by electrodialysis using a new anion-exchange membrane. Desalination 148: 143-148.

10. Krol JJ, Wessking M, Strathmann H (1999) Concentration polarization with monopolar ion exchange membranes: current-voltage curves and water dissociation. J Membr Sci 162: 145-154.

11. Strathmann H, Krol JJ, Rapp HJ, Eigenberger G (1997) Limiting current density and water dissociation in bipolar membranes. J Membr Sci 125: 123-142

12. Ortiz JM, Sotoca JA, Exposito E, Gallud F, Garcia-Garcia V, et al. (2005) Brackish water desalination by electrodialysis: batch recirculation operation modeling. J Membr Sci 252: 65-75

13. Rohman FS, Othman MR, Aziz N (2010) Modeling of batch electrodialysis for hydrochloric acid recovery. Chem Eng J 162: 466-479

14. Fidaleo M, Moresi M (2010) Application of the Nernst-Planck approach to model the electrodialytic recovery of disodium itaconate. J Membr Sci 349: 393-404

15. Fidaleo M, Moresi M (2005) Modeling of sodium acetate recovery from aqueous solutions by electrodialysis. Biotechnol Bioeng 91: 556-568

16. Wen T, Solt GS, Gao DW (1996) Electrical resistance and coulomb efficiency of electrodialysis (ED) apparatus in polarization. J Membr Sci 114: 255-262

17. Tanaka Y (2005) Irreversible thermodynamics and overall mass transport in ion-exchange membrane electrodialysis. J Membr Sci 281: 517-531

18. Kabay N, Demircioglu M, Ersoz E, Kurucaovali I (2002) Removal of calcium and magnesium hardness by electrodialysis. Desalination 149: 343-349

19. Bird RB, Stewart WE, Lightfoot EN (2002) Transport Phenomenon. John Wiley & Sons.

20. Treybal RE Mass-Transfer Operations. McGraw-Hill publications.

21. Balster J, Yildirim MH, Stamatialis DF, Ibanez R, Lammertink RG, et al. (2007) Morphology and microtopology of cation-exchange polymers and the origin of the overlimiting current. J Phys Chem B 111: 2152-2165.

22. Mischuk NA (1998) Perspective of electrodialysis intensification. Desalination 117: 283-296

23. Grigorchuk OV, Vasileva VI, Shaposhnik VA (2005) Local characteristics of mass transfer under electrodialysis demineralization. Desalination 184: 431-438

24. Isaacson MS, Sonin AA (1976) Sherwood Number and Friction Factor correlations for clectrodialysis systems, with application to process optimization. Ind Eng Chem Process Des Dev 15: 313-321

25. Poling BE, Prausnitz JM, O'Connell JP, The Properties of Gases and Liquids. McGraw-Hill publications

26. Li YH, Gregory S (1974) Diffusion of ions in sea water and in deep sea sediments. Geochimica et Comochimics acta 28: 703-714

27. Barragan VM, Ruiz-Bauza C (1998) Current-Voltage Curves for Ion-Exchange Membranes: A Method for Determining the Limiting Current Density. J Colloid Interface Sci 205: 365-373.

28. Koryta J, Dvorak J, Kavan L (1993) Principles of electrochemistry. John Wiley & Sons, Inc.

29. Nwal Amang D, Alexandrova S, Schaetzel P (2003) The determination of diffusion coefficients of counter ion in an ion exchange membrane using electrical conductivity measurement. Electrochimica Acta 48: 2563-2569

The Effect of Temperature and Pressure on the separation of Calcium Carbonate and Barium Sulphate from a Mixed Sludge

Rukuni TT[1]*, Maree JP[2], Zvinowanda CM[1] and Carlsson FHH[1]

[1]*Department of Environment, Water & Earth Sciences, Tshwane University of Technology, Private Bag X680, 0001, South Africa*
[2]*Incumbent, Rand Water Chair in Water Utilisation, Tshwane University of Technology, Private Bag X680, 0001, Pretoria, South Africa*

Abstract

South Africa is one of the first countries to implement full-scale mine water reclamation to drinking water quality. Reverse osmosis is already being used on full scale for desalination of mine water. However, with increased recycling of mine water, the result has been the increased generation of sludge. The Council for Scientific and Industrial Research (CSIR) has developed the Alkali-Barium-Carbonate (CSIR-ABC) process that can be used for neutralization and desalination of sulphate-rich effluents while recovering valuable by-products from the mixed sludge's produced. A mixture of $BaSO_4$ and $CaCO_3$ sludge is produced as one of the by-products, which preferably needs to be separated into its components prior to thermal treatment. The aim of this study was to separate $CaCO_3$ and $BaSO_4$ from a $CaCO_3$-$BaSO_4$ mixed sludge through dissolution of $CaCO_3$ as $Ca(HCO_3)_2$ in contact with CO_2. Measured quantities of a simulated $CaCO_3$-$BaSO_4$ mixed sludge from the CSIR-ABC process were fed into a reactor vessel containing deionized water and pressurized CO_2 was introduced. The effects of temperature and pressure with time were investigated while monitoring alkalinity, pH and calcium concentration. The findings of this study were: (1) The dissolution rate of $CaCO_3$ was rapid i.e. from 0 to 2000mg/L in the first 20minutes; (2) $Ca(HCO_3)_2$ had a high solubility of about 2600 mg/L when in contact with CO_2 at 1atm., while $BaSO_4$ was almost completely insoluble; (3) The solubility of $Ca(HCO_3)_2$ increased with decreasing temperature and increasing pressure; (4) $CaCO_3$, after conversion to $Ca(HCO_3)_2$, was separated from $BaSO_4$ in a $CaCO_3$-$BaSO_4$ mixed sludge; (5) Visual MINTEQ model is a powerful tool that can be used to predict the solubilities of $CaCO_3$ and $BaSO_4$ when contacted with CO_2.

Keywords: Acid mine drainage; carbonation; Dissolution, Reclamation, Carbon dioxide, Sludge disposal

Introduction

South Africa currently faces both a shortage of water as well as a deterioration of the quality of the available water due to industrial activities such as mining and power generation. The gold mines in Gauteng are expected to decant 345 ML of acid mine drainage per day when open voids have filled up. This water will pollute the surface water resources and cause serious environmental impacts [1]. South Africa was one of the first countries to implement commercial scale drinking water reclamation from mine water using reverse osmosis. The cost of reverse osmosis amounts to R10/m^3 and produces gypsum and ferric hydroxide rich sludge that has to be disposed of at a cost on sludge disposal dumps.

The CSIR-ABC process is an alternative technology and offers the benefit of lower cost as calcium carbonate, lime; barium carbonate and coal are the main process raw materials. It produces minimum amount of sludge from raw materials ($BaCO_3$ and lime) which are recovered from the sludge. Sulphur, a valuable by-product, can also be recovered from the sulphate removed from the mine water. The CSIR-ABC process involves three stages. In the first stage $CaCO_3$ is dosed to raise the pH of the AMD to 7 for removal of free acid, iron (II) as iron (III) and aluminium as aluminium (III). In the second stage lime is dosed to precipitate other metals, including magnesium, as metal hydroxides. Partial sulphate removal is also achieved in these two stages through gypsum crystallization. In the third stage, barium carbonate is dosed for final sulphate removal through $BaSO_4$ precipitation and carbon dioxide is dosed to precipitate calcium as $CaCO_3$. The resultant, mixed $BaSO_4$/$CaCO_3$ sludge is treated in a thermal stage to produce BaS and CaO. As a portion of the sulphide combines with calcium, the BaS yield is never higher than 85%, compared to 96% when only $BaSO_4$ is processed in the thermal stage [2]. It would therefore be beneficial if the mixture of $BaSO_4$ and $CaCO_3$ could be separated prior to thermal treatment. Figure 1a shows the schematic diagram of the CSIR-ABC Process and Figure 1b shows the process flow diagram of the $BaSO_4$-$CaCO_3$ sludge separation process.

The aim of this study was to separate calcium carbonate from barium sulphate, produced in the third stage of the CSIR-ABC Process, through dissolution of $CaCO_3$ with CO_2. By adding CO_2 to a mixed $BaSO_4$-$CaCO_3$ sludge, selective dissolution of $CaCO_3$ occurs according to the following reaction:

$$CO_{2(g)} + CaCO_{3(s)} + H_2O_{(l)} \rightarrow 2HCO_{3\ (aq)}^{-} + Ca^{2+}_{(aq)} \quad \textbf{(1)}$$

As the system becomes enriched in CO_2, the dissolution rate of $CaCO_3$ could be influenced by its saturation level, Ω [3].

$$\Omega = [Ca^{2+}][CO_3^{2-}]/K_{sp} \quad \textbf{(2)}$$

where K_{sp} (T, S, p) is the solubility product (T is temperature, S is solubility and p is pressure) [4] with the pressure dependency adjustments by Millero et al. [5].

$$K_{sp} = [Ca^{2+}][CO_3^{2-}] \quad (3)$$

***Corresponding author:** Rukuni TT, Department of Environment, Water & Earth Sciences, Tshwane University of Technology, Private Bag X680, 0001, South Africa

Note: All the square-bracketed species are stoichiometric concentrations (molality or mol /kg) and disregard any complex formation or ion pairs. It is thus necessary to specify not only the pH scale used in the evaluation of the dissociation constants [6], but also their dependency on ionic strength, temperature, and pressure.

The $Ca(HCO_3)_2$ formed,according to Equation (1), can be precipitated as $CaCO_3$ by lowering the pressure to atmospheric conditions (Equation 2);

$$Ca(HCO_3)_{2(aq)} \rightarrow CaCO_{3(s)} + CO_{2(g)} + H_2O_{(l)} \qquad (4)$$

Materials and Methods

A 2% $BaSO_4/CaCO_3$ slurry was prepared with a $BaSO_4/CaCO_3$ mole ratio of 1 as feedstock. A simulated mixed sludge was prepared using commercial grade $BaSO_4$ (28g) and analytical grade $CaCO_3$ (12 g). The salts were mixed in deionised water (2L). Bottled CO_2 was used for dissolving $CaCO_3$. The composition and the behaviour of the simulated sludge was the same as that of the real sludge due to the fact that all possible impurities (metals) in the real sludge are removed in the pre-treatment stage of the CSIR-ABC process using limestone and lime.;

Batch studies were carried out using a completely-mixed pressurized reactor (Figure 2). It consisted of a 3L reaction vessel equipped with a BirCraft stirrer, temperature sensor, pressure gauge, pH and conductivity control sensors. This unit was designed to withstand pressures of up to 10bar and a maximum temperature of 150°C. The main body of the reactor was constructed from a Class 12unplasticised polyvinyl chloride (uPVC) pipe with a wall thickness of 3.2mm and an internal diameter of 560mm. The length of the pipe was 520mm with uPVC base and top plates that were each 65mm thick. The dosage points were 300mm above the effluent off-take point that was fitted at the bottom of the reactor. Calcium carbonate and barium sulphate were first fed into the reactor from the top and deionized water was pumped into the reaction vessel with a Watson-Marlow pump with continuous stirring. Pressurized carbon dioxide was dissolved in water

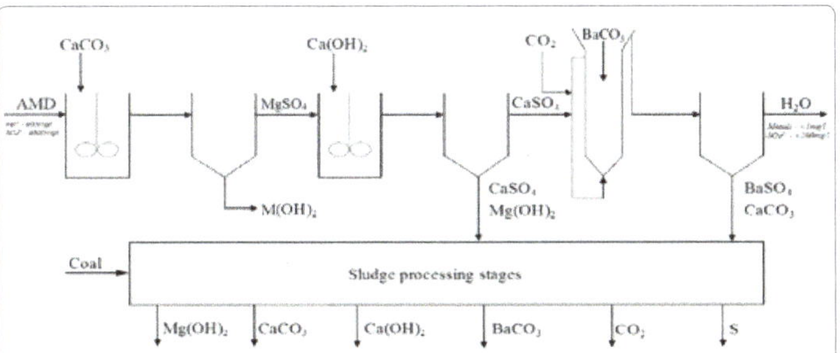

Figure 1a: Schematic diagram of the CSIR-ABC Process for neutralization, metal removal and desalination of AMD.

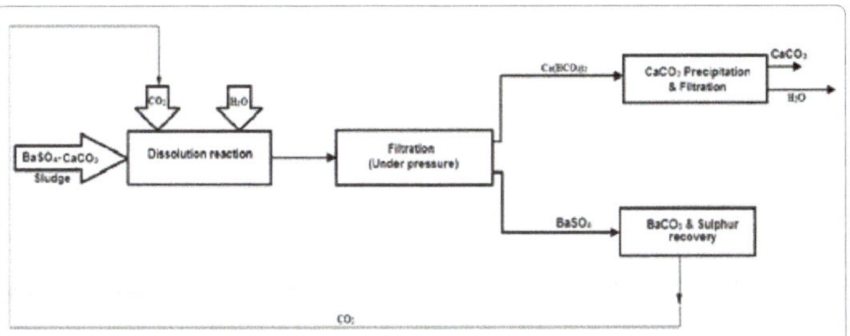

Figure 1b: Schematic flow diagram of the $BaSO_4$-$CaCO_3$ sludge separation process.

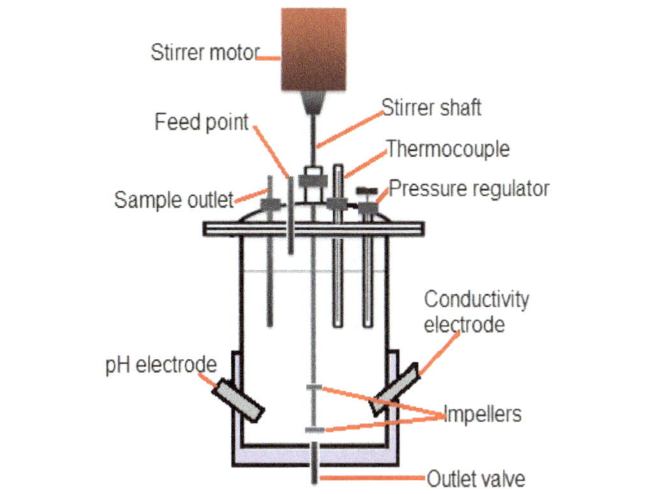

Figure 2: The completely-mixed pressurized reactor.

in the pressurized reactor, to lower the pH to around 6 and increase the pressure to desired levels. A pH control unit was used to control CO_2 dosage by monitoring the pH. At the end of the reaction the solution was filtered under pressure and the filtrate released from the pressure vessel through a valve.

The effect of the following parameters was investigated:

- Time: 0, 10, 20, 40, 60 min (measured).
- Temperature: 0, 10, 25, 45°C (measured and predicted).
- Pressure (predicted).
- $CaCO_3$ concentration: 6, 20 g/L (measured and predicted).
- $BaSO_4$ concentration: 0, 14, 23.3 g/L (predicted).

The default conditions were 0°C, 6 g/L $CaCO_3$, 14 g/L $BaSO_4$. Filtered samples were collected over the time period 0-60 min and analysed for pH, alkalinity, calcium, barium and sulphate. The samples were filtered (0.45μm membrane filter) to remove all solid material.

Sulphate concentration was measured in the effluent water. The sample was conditioned using a glycerine conditioning reagent, and solid barium chloride was added and mixed. The turbidity of the $BaSO_4$ suspension was measured using a turbidity meter and the reading was compared with a calibration curve to determine the sulphate concentration. The minimum detectable concentration is approximately 1mg SO_4^{2-}/L and possible interferences are silica in excess of 500 mg/L and large quantities organic materials that may slow down the precipitation of $BaSO_4$ [7]. Correction for sample colour and turbidity was corrected by running blanks to which $BaCl_2$ was not added.

The total dissolved carbonate species were measured by titrating the sample with standard hydrochloric acid to pH 4.3. The volume of acid was used to calculate the total alkalinity of the sample. From this calculation, the carbonate and hydroxide species concentrations were calculated, and the total carbonate species concentration (as mg/L $CaCO_3$) [7]. The known interferences for this method are soaps, oily matter and precipitates and our samples where filtered to remove all undissolved substances.

Calcium hardness was determined by titration with EDTA (0.02M) using P&R indicator and NaOH buffer. Total hardness was determined with EDTA (0.02M) using Eriochrome Black T indicator and ammonia buffer [7].

Barium concentrations were determined by a Spectro Arcos Inductively Coupled Plasma Optical Emission Spectroscopy (ICP-OES). Filtered samples were diluted using deionised water and analysed while alternating with blanks to eliminate the occurrence of carry-over effect [7]. The instrument's estimated detection limit for barium is 2 µg/L and the upper limit of the effective calibration concentration range is 50 mg/L.

The pH, temperature, electrical conductivity and pressure were measured in situ.

Results and Discussion

Table 1 and Figure 3-6 show the dissolution of $CaCO_3$ (calcite) and $BaSO_4$ (barite) when contacted with CO_2 in water under various conditions. The Figure 3-6 contain measured values while Table 1 contains both measured and predicted values. The predictions were done using a Visual MINTEQ model [8].

Kinetics of $CaCO_3$ dissolution on CO_2-rich water

The dissolution rate was rapid as indicated by the fact that more than 50% of the $CaCO_3$ that dissolved was already in solution when the first sample was collected after 10 min (Figure 3). The dissolution rate is influenced by particle size, pressure and temperature.

Separation of $CaCO_3$ and $BaSO_4$

The mixture of $CaCO_3$ and $BaSO_4$ sludge that is produced by the CSIR-ABC process cannot be separated by solubility differences due to the low solubility of both $CaCO_3$ (1.3 mg/L Ca,) and $BaSO_4$ (5.6 mg/L Ba). As CO_2 is produced as a waste product in the CSIR-ABC process, it was decided to investigate whether $CaCO_3$ could be separated from

Exp No	Feedstock		pH		Temperature(°C)	Pressure (atm)	$CaCO_3$ (aq) (mg/l)		Mg/l Ba
	$CaCO_3$ (s) (g/l)	$BaSO_4$ (s) (g/l)	Predicted	Determined			Predicted	Determined	
	Behavior of various parameters (see Figure 3)								
1.1	20	0		6.06	0	1	1458	2600	
	Comparison of $CaCO_3$ and $BaSO_4$ when in contact with CO_2; pH, Ca and Ba measures and predicted (Figure 4)								
1.2	20	0		6.03	10	1	1228	2050	-
3.2		20		3.87	10	1		50	1.17
	Effect of BaSO4; pH, Ca and Ba measured and predicted (Figure 5)								
1.1	20	0	6.06	6.26	0	1	1458	2600	-
2.1	6	14	6.01	6.01	0	1	1460	2801	2.39
1.2	20	0	6.26	6.29	10	1	1228	2050	-
2.2	6	14	5.94	5.94	10	1	1230	2560	2.77
Exp 1	Effect of temperature on CaCO3; pH and Ca predicted and measured (see Figure 6 and 7)								
1.1	20	0	6.06	6.26	0	1	1458	2600	-
1.2	20	0	6.03	6.29	10	1	1228	2050	-
1.3	20	0	6.01	6.58	25	1	946	800	-
1.4	20	0	6.00	6.07	45	1	665	600	-
Exp 2	Effect of temperature on $CaCO_3$ and $BaSO_4$; pH, Ca and Ba predicted and measured (see Figure 6 and 8)								
2.1	6	14	6.06	6.01	0	1	1460	2810	2.39
2.2	6	14	6.03	5.94	10	1	1230	2560	2.77
2.3	6	14	6.01	5.98	25	1	947	1000	3.40
2.4	6	14	6.00	6.04	45	1	666	600	4.30
Exp 4	Effect of pressure on CaCO3; pH and CaCO3 predicted (see Figure 9)								
4.1	20	0	-	6.25	0	0.5	1108	-	-
4.2	20	0	-	6.06	0	1	1458	-	-
4.3	20	0	-	5.87	0	2	1926	-	-
4.4	20	0	-	5.76	0	3	2271	-	-
4.5	20	0	-	5.68	0	4	2553	-	-
4.6	20	0	-	5.62	0	5	2798	-	-
Exp 5	Effect of pressure on $CaCO_3$ and $BaSO_4$; pH, Ca and Ba predicted								
5.1	20	23.3	6.25	-	0	0.5	1107	-	2.18
5.2	20	23.3	6.06	-	0	1	1457	-	2.39
5.3	20	23.3	5.87	-	0	2	1924	-	2.63
5.4	20	23.3	5.76	-	0	3	2268	-	2.79
5.5	20	23.3	5.62	-	0	5	2795	-	2.74

Table 1: Effect of various parameters on the solubility of $CaCO_3$ and $BaSO_4$ in a CO_2-rich solution.

the almost insoluble $BaSO_4$ by dissolving it as $Ca(HCO_3)_2$ (Equation1), by contacting the sludge mixture with CO_2.

Figure 3 shows that $CaCO_3$ dissolved when contacted with CO_2 at 1atm. Ca and Alkalinity (Alk) values increased rapidly in stoichiometric concentrations from the low solubility of $CaCO_3$ to the high solubility of $Ca(HCO_3)_2$ (2600 mg/L Alk, as $CaCO_3$) and 2650 mg/L Ca (as $CaCO_3$). Over the same period the pH dropped from 8.7 to 6.2. Due to the ease of alkalinity measurements, they were used to monitor the formation of $Ca(HCO_3)_2$ and includes the parameters listed in Equation 5. Figure 3 also shows the ionic balance between the measured calcium concentration and the total alkalinity of the system. As the system became enriched in CO_2, the extent of dissolution decreased as a function of changes in the $CaCO_3$ saturation state to yield both calcium ions and alkalinity (Equation 1).The solubility of $CaCO_3$ decreases with an increase in pH. Therefore, increasing CO_2 input into the reaction lowers the pH and calcium carbonate becomes more susceptible to dissolution. In this study, the pH was lowered to 6 and this enabled CO_3^{2-} to associate with protons, while driving the solubility equilibrium toward the formation of HCO_3^-.

$$Alk = 2[CO_3^{2-}] + [HCO_3^-] + [OH^-] + [H^+] \quad (5)$$

Figure 4 shows the contrast between the solubilities of $CaCO_3$ (2200 mg/L; Ca as $CaCO_3$) and $BaSO_4$ (50 mg/L as Ba). The results showed that $BaSO_4$ does not dissolve when contacted with CO_2 and consequently there were negligible losses of $BaSO_4$ due to dissolution. The effect of $BaSO_4$ on the rate of formation and solubility of $Ca(HCO_3)_2$ was studied. Figure 5 showed that $BaSO_4$ had no effect on the solubility of $Ca(HCO_3)_2$. This is because $BaSO_4$ has a very low solubility The known low solubility of $BaSO_4$ was also confirmed by the low predicted values for Ba^{2+} in solution, when contacted with CO_2 (Table 1).

Effect of temperature and pressure on solubility

In the previous section it was shown that $CaCO_3$ can be dissolved through CO_2 addition without affecting the low solubility of $BaSO_4$. The sludge separation process will be more effective at higher solubility values for $CaCO_3$. Therefore, it was decided to determine the effect of pressure and temperature on the solubility of $CaCO_3$. According to Henry's law, it was expected that the solubility should increase with decreasing temperatures. Figure 6 confirmed the validity of Henry's law [9,10]. In Figure 6, the effect of temperature on the rate of formation and solubility of $Ca(HCO_3)_2$ in the presence of $BaSO_4$ as well as in the absence of BaSO4 was investigated. It was found that $BaSO_4$ had no effect on the solubility of $CaCO_3$ when contacted with CO_2.

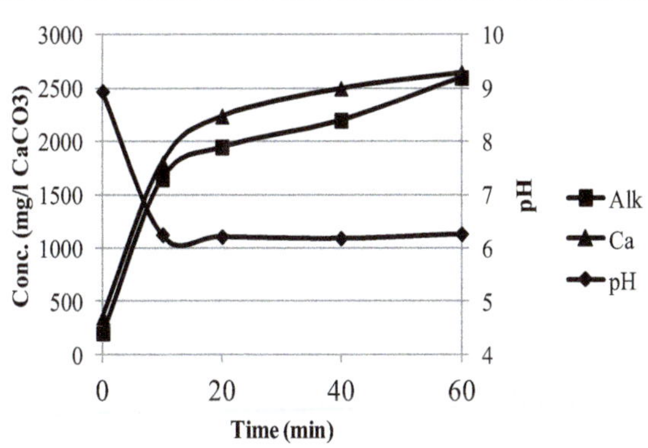

Figure 3: Behaviour of various parameters during contact between $CaCO_3$ and CO_2 (Temp = 0°C; 20 g/L $CaCO_3$).

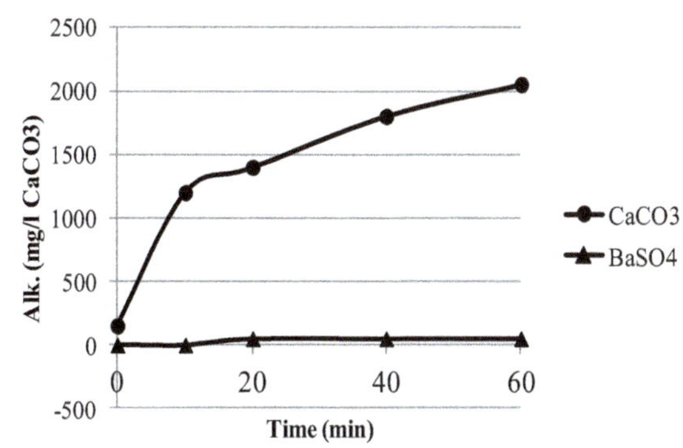

Figure 4: Comparison between dissolution of $CaCO_3$ and $BaSO_4$ in a CO_2-rich solution (10°C; 20 g/L $CaCO_3$; 20 g/L $BaSO_4$).

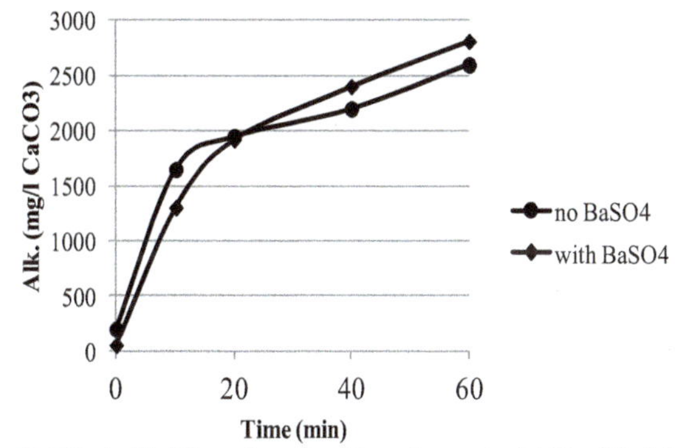

Figure 5: Effect of $BaSO_4$ on the rate of formation and solubility of $Ca(HCO_3)_2$ (20 g/L$CaCO_3$ and absent $BaSO_4$; 6 g/L $CaCO_3$ and 14 g/L $BaSO_4$).

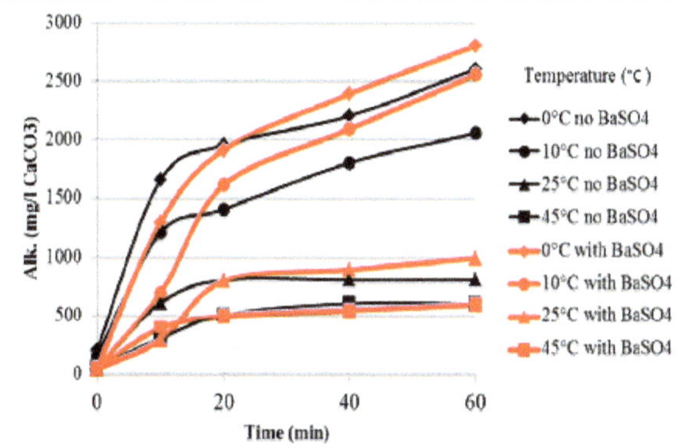

Figure 6: Effect of temperature on the rate of formation and solubilityof $Ca(HCO_3)_2$at low pressure (1atm) in the presence of $BaSO_4$ (6 g/L $CaCO_3$; 14 g/L $BaSO_4$)and under $BaSO_4$-free conditions (20 g/L $CaCO_3$).

Measured versus predicted solubility values

Table 1 shows a comparison of the predicted and determined values for the effect of various parameters on the solubility of $CaCO_3$ and $BaSO_4$ in a CO_2-rich solution. The predictions were done using the Visual MINTEQ (2000) model. The model was designed to simulate equilibrium and speciation of inorganic solutes in natural waters.

Temperature

Figure 7 and 8 (Expt 1) (Expt 2) compare the solubility of $Ca(HCO_3)_2$ at 1atm CO_2 when no $BaSO_4$ and 14 g/L $BaSO_4$, respectively, were present over the temperature range 0-45°C. Both sets of results showed that the solubility of $CaCO_3$ increases with decreasing temperature. The measured values at the lower temperatures (0° and 10°C) were higher than the predicted values. These differences indicate that the Visual MINTEQ (2000) model can be improved further. However, it is a very useful tool as it can predict trends for a large number of compounds.

Pressure

As the results in the above section showed, the Visual MINTEQ(2000) model can predict trends, it was used to predict the effect of pressure on the solubility of $Ca(HCO_3)_2$ in the absence and presence of 23.3 g/L $BaSO_4$. It was noticed that the solubility of $CaCO_3$, when contacted with CO_2, increased with increasing pressure (Figure 9). The model also showed, as in the case of the temperature studies above, that $BaSO_4$ had no influence on the solubility of $CaCO_3$ when contacted with CO_2 (compare Expt 4 with Expt 5, Table 1). The practical implications of these findings are that the solubility of $CaCO_3$, when contacted with CO_2, increases with decreasing temperature and increasing pressure. The CSIR-ABC Process offers a sustainable method for neutralization, metal removal and desalination of acid mine drainage and recovery of by-products from the mixed sludge produced.

Separation rate

In this study, it was found that the dissolution rate of $CaCO_3$ is directly related to the separation rate from the mixed sludge. Figure 10 shows the dissolution rate of the $CaCO_3$ with temperature from a $CaCO_3$-$BaSO_4$ mixed sludge in a CO_2-rich solution. The highest rate of separation was recorded at 0°C. Figure 10 also shows that lower temperatures favour high rates of separation of the $CaCO_3$ from the mixed sludge while increased temperature leads to a significant decrease in the dissolution rate of $CaCO_3$.

Conclusions

The results of this study will contribute to the improvement of the CSIR-ABC Process design to meet the criteria for maximum value of treated water and by-products coupled with reduced running and sludge disposal costs. The conclusions drawn from this work were:

- The dissolution rate of $CaCO_3$ in the presence of CO_2 is fast at low temperature and high pressure;
- $CaCO_3$ has a high solubility of about 2600 mg/l when in contact with CO_2 at 1atm., while $BaSO_4$ is almost completely insoluble;

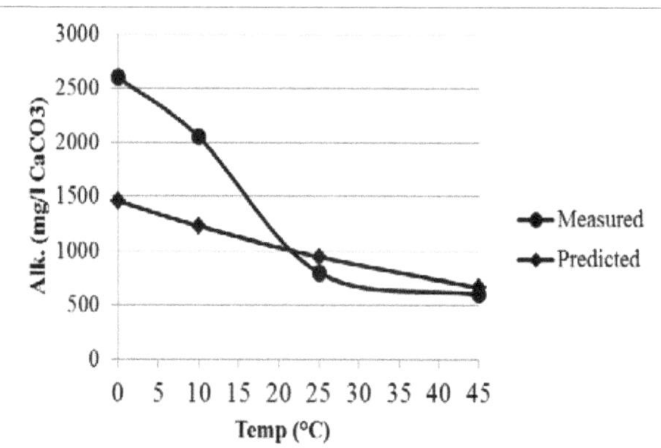

Figure 7: Comparison between measured and predicted values for $CaCO_3$ solubility with temperature when in contact with CO_2 at 1 atm.

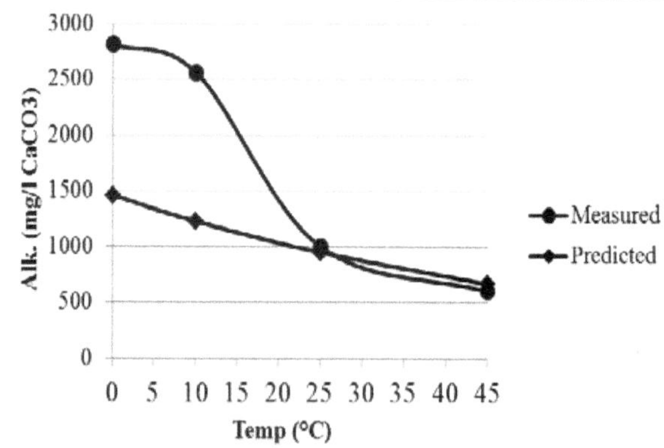

Figure 8: Comparison between measured and predicted values for $CaCO_3$ solubility with temperature in the presence of $BaSO_4$ when in contact with CO_2 at 1 atm.

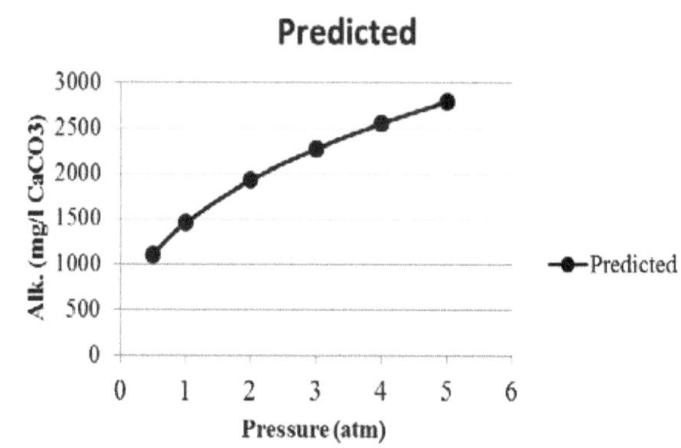

Figure 9: Effect of pressure on the solubility of $CaCO_3$ when in contact with CO_2 at 0°C.

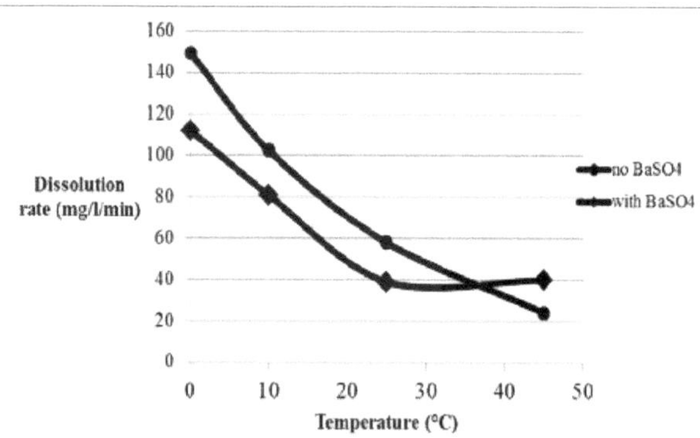

Figure 10: The effect of temperature on the dissolution rates of $CaCO_3$ when contacted with CO_2.

- The solubility of $CaCO_3$ increases with decreasing temperature and decreasing pressure;
- The Visual MINTEQ model was a powerful tool used to predict the solubility of $CaCO_3$ and $BaSO_4$ when contacted with CO_2.
- $CaCO_3$ can be separated from $BaSO_4$ in and $CaCO_3$-$BaSO_4$ mixed sludge.

The cost of this process is low because all the process raw materials (CO_2 and the $CaCO_3$-$BaSO_4$sludge) are waste products of the CSIR-

ABC process (Figure 1a and 1b). It is also foreseen that the cost can be kept low in other applications by producing CO_2 on-site by burning coal and scrubbing the off-gas in water rather than purchasing pure CO_2. From the reaction equilibrium and the results of this study, it is apparent that decreasing the temperature and increasing CO_2 pressure will result in a better conversion of $CaCO_3$ to $Ca(HCO_3)_2$. However, high pressure systems are more complex to run and require sophisticated reactor designs and special construction material which will ultimately increase costs. Because $Ca(HCO_3)_2$ has high solubility when contacted with CO_2 at low temperatures and high pressure, the practical, optimal operation conditions for the dissolution reaction are temperatures close to 0°C and pressures as close as possible to atmospheric depending on the application.

Acknowledgments

Financial support from THRIP and the use of the laboratory facilities at Tshwane University of Technology are gratefully acknowledged.

References

1. (2010) Mine water management in the witwatersrand gold fields with special emphasis on acid mine drainage. Report to the inter-ministerial committee on acid mine drainage.
2. Patrick H, Jannie M, Dolf B (2007) Barium Carbonate Process for Sulphate and Metal Removal from Mine Water. Mine Water and the Environment 26: 14-22.
3. Feely RA, Sabine CL, Lee K, Millero FJ, Lamb MF, et al. (2002) In situ calcium carbonate dissolution in the Pacific Ocean. Global Biogeochem Cycles 16: 1144.
4. Mucci A (1983) The solubility of calcite and aragonite in seawater at various salinities, temperatures, and one atmosphere total pressure. Am J Sci 283: 780-799.
5. Millero FJ (1995) Thermodynamics of the carbon dioxide system in the oceans. Geochim Cosmochim Acta 59: 661-677.
6. Dickson AG (1984) pH scales and proton-transfer reactions in saline media such as sea water. Geochim Cosmochim Acta 48: 2299-2308.
7. Eaton AD, Clesceri LS, Greenberg AE (1995) Standard Methods for the Examination of Water and Wastewater. (19thedn), APHA, AWWA & WEF, New York.
8. http://www2.lwr.kth.se/English/OurSoftware/vminteq/Visual%20MINTEQ_tutorial.pdf
9. Ten Hulscher Th E M, Van Der Velde LE, Bruggeman WA (1992) Temperature dependence of henry's law constants for selected chlorobenzenes, polychlorinated biphenyls and polycyclic aromatic hydrocarbons. Environmental Toxicology and Chemistry 11: 1595-1603.
10. Alaee M, Whittal RM, Strachan WMJ (1996) The effect of water temperature and composition on Henry's law constant for various PAH's. Chemosphere 32: 1153-1164.

Future Scope of Silicone Polymer based Functionalized Nanocomposites for Device Packaging: A Mini Review

Satyajit Gupta[1,2], Praveen C Ramamurthy[1,2*] and Giridhar Madras[2]

[1]Department of Materials Engineering, Indian Institute of Science, Bangalore, India
[2]Center for Nanoscience and Engineering, Indian Institute of Science, Bangalore, India

Abstract

Organic electronic devices, such as conducting polymer and/or perovskite based solar cells are highly promising and emerging, but they are highly unstable in the presence of moisture and oxygen. It is important to protect these devices from these deterrent gases by using a suitable encapsulant and to improve the lifetime of these promising devices. "Silicone polymer" impregnated with suitable functionalized nano fillers/moisture or oxygen scavengers can be potentially used for sensitive organic device encapsulation applications, where a combined property of flexibility, thermal stability, transparency, and low moisture permeation is critical. Suitable fillers can also enhance the thermo-mechanical properties of silicone polymers, which is also important for hermitic sealing applications. Silicone polymer have a wide range of applications in electronic, mechanical, and biological industries, due to their flexibility over a broad temperature window, resiliency, transparency, superior thermal stability, resistant to aging and degradation from sunlight. This polymer also coats a surface conformally, having moisture resistance property and inertness. In this mini-review, functionalized composites based on silicone polymers, which have been developed for the potential organic device encapsulation application, has been discussed. In addition to that, future direction towards the fabrication of new functionalized nanocomposites, based on silicone polymers is also discussed briefly.

Keywords: Curing; Organic devices; Nanocomposite; Functionalization; Silicone polymer

Introduction

Organic devices such as conducting polymer [1,2] based or perovskite [3,4] based solar cells are flexible, lightweight, and more economical than traditional silicon devices, but cannot be commercialized/marketed due to device failure caused by atmospheric moisture/oxygen [5] and hence protection of those devices are very critical. Silicone polymer based composite encapsulants are attractive to this purpose because of its easy processability, wide range of curing temperatures (room temperature to moderately higher temperature), low shrinkage, high thermal stability, low volatility etc. All these properties of silicones partly coincide with the requirements of an ideal device encapsulant. Even though, silicone polymers are moisture resistant, but, the oxygen barrier properties of silicones are not desirable to fulfil organic device requirement. If the barrier properties of silicone polymers are further improved, then it may find an important application towards encapsulation of those devices.

Silicone polymers are an important class of polymer that are used in numerous applications such as drug-delivery vehicle, biomedical implants, lubricants, cosmetics industries and lot more [6-8]. These polymers are flexible, attributed to the Si-O-Si linkages at the back-bone of the polymer chain. This Si-O-Si can flip easily from one configuration to another due to long bong lengths (163 pm) and high bond strengths (~452 KJ mol^{-1}) and low activation energy of Si-O-Si bond rotation, are responsible for its flexibility. The excellent thermal stability also arises out of the siloxane linkages in the structure. Due to the overlap of O '2p' electrons in the vacant Si '3d' orbitals, a 'π' bond character arises, in addition to the strong Si-O 'σ' bond. This increased bond multiplicity of the Si-O-Si linkages makes the polymer thermally stable.

These polymers can be easily modified by chemical reactions and the physical properties can be tailored/altered. Hydrosilylation reaction is one of the important reactions that are used to modify/conjugate other chemical functionality with silicone polymers at the Si-H sites of the "polydimethylsiloxane (PDMS)" backbone. Due to chemical inertness, these are used in several biological applications [9] such as, drug delivery [10] etc. The mechanical properties of silicones are poor, but reinforcement with nanoparticles can improve the mechanical properties to a desired extent. Various nano fillers have been introduced in the composites such as clays, carbon nanotubes, oxide nanoparticles etc. in order to modify/enhance the properties of pristine silicones. Nanofiber reinforced PDMS by using femtosecond laser irradiation have also been carried out [11]. Flexibility and transparency of silicones are also exploited in electronic device packaging. The processability of these polymers are also simple, coupled with the above mentioned physico-chemical properties, leads the polymer for the applications in various fields of material science and technology.

Various oligosiloxane based hybrid materials has also been synthesized for the encapsulation of organic devices [12-14]. Reactive and functionalized nanoparticles are supposed to provide a torturous pathway towards an incoming gas molecule and hence will reduce the permeability to a greater extent (Scheme 1). Recently, it has been shown covalently attached silicones with functionalized nanoparticles like alumina/silica to fabricate composite films fabricated by an in-situ technique, can be used for organic device encapsulation [15-18]. In this context, It is also important to mention that conductive carbon based material such as graphene, carbon nanotubes [19], are also promising for the improvement in the barrier properties [20,21] of polymers.

***Corresponding author:** Praveen C Ramamurthy, Department of Materials Engineering, Indian Institute of Science, Bangalore, India

This mini-review encompasses **(1)** various in-situ strategies to fabricate polymer/nanocomposites with silicones towards organic electronics encapsulation, **(2)** a brief comparison between 'in-situ' method with solution casting and melt-processing technique, and, **(3)** some future directions toward the synthesis of new barrier composites with silicone polymers.

In-situ Curing Technique for the Fabrication of Polymer Nanocomposites

In this section, various methods/techniques have been discussed for the fabrication silicon polymer based nanocomposites. Functionalized silica/alumina has been used as a filler material for silicone polymers. The functionalizations were carried out to make the surface more compatible with the polymer matrix and the reactive surface functional groups were supposed to participate in the curing reaction with the specific functional groups in polymer matrix to enhance the interfacial interaction. The intended inorganic phase was also supposed to reduce the gas permeability through providing a torturous pathway.

Functionalized mesoporous silica as filler materials for silicone polymers

Mesoporous silica nanoparticle (MSN) is an important class of mesoporous material that is used in various applications such as in targated drug delivery as a 'drug loaded vehicle' because of its biocompatibility [22,23], filler material in polymer nanocomposites, active catalyst support, adsorbent, separation and bio-sensing [24] applications etc. Large surface area and pores of these particles has been exploited for various applications. In addition, the surface functionality of these particles can be tuned through functionalization by silane coupling chemistry (Scheme 2). Silanes can be represented as $(R'O)_3SiR$ in general, where R'O- is either EtO- or, MeO- which undergo condensation with the surface Si-OH groups of mesoporous silica surface. This provides different functional groups at the surface. Basically it generates '$Si_{surface\text{-}core}OSiR$', where R is the organic functionality and '$Si_{surface\text{-}core}OSi$' bond generates after the silane coupling at the surface. This provides a platform to conjugate various bio-molecules or polymers. The functionalization makes the surface either polar or non-polar depending on the organic functional groups present.

MSN's with higher pore diameters also serves as a critical filler material for the reinforcement of polymer matrix. Large pore sized mesostructured silica (MSU-J, pore size-5.3) was observed to improve tensile properties like tensile modulus, toughness etc. of rubbery-epoxy polymer [25]. A significant improvement has been observed in the composites as compared to the neat epoxy polymer system. The reinforcement effect in this case was comparable to that of exfoliated organo nanoclay. The improvement can be explained by the segmental confinement of the polymer chains in the mesopores at the inter-phase region of the silica particles [26,27]. This segmental confinement of the polymer chains imparts a profound impact on the properties of the polymer composite materials. Hence the organic inorganic interface is very critical for these composites. The chemical bonding between the polymer and the silica particles also plays a vital role in the enhancement of the thermal [28] and mechanical properties as through these covalent chemical bonding the load transfer to the inorganic filler is possible. Recently it has been reported that mesoporous silica can enhance the thermal stability of polymethylmethacrylate (PMMA), may be due to the radical scavenging activity of silica.

Allyl functionalized silica with internal hydride containing siloxane: Hydrosilylation strategy: In this study, the composites were fabricated by platinum catalyzed hydrosilylation reaction. Internal hydride containing silicone polymer has been covalently grafted over allyl decorated mesoporous silica (MSU-H, 2D hexagonal with a pore diameter of ~7 nm) by hydrosilylation reaction [18]. Mesoporous silica was functionalized by allyltrimethoxysilane, which provides allyl fragments at the particle surface. This organic modification also makes the surface hydrophobic. The grafting was carried out in dry toluene reflux condition to avoid the self-condensation of the silane and this functionalization was determined from FTIR and solid state NMR spectroscopy (^{29}Si and ^{13}C). Two different wt% (0.6 and 1.5) of these functionalized mesoporopus silica have used to fabricate the composites. The curing temperature of this system was estimated from differential scanning calorimetry (DSC), which showed a broad exothermic peak around 135°C. For a curing mixture it is very important to understand the curing behavior as the mixture needs thermal treatment for the application purpose. In addition to DSC studies, rheological experiments were also carried out to understand the curing behavior of the mixture, which showed a self-coupling reaction of 'Si-H' between the same polymer chains at the beginning, followed by the true curing (cross-coupling) i.e. hydrosilylation reaction between the allyl groups of the functionalized nanoparticles surface and the Si-H of the polymer. Energy of activation (E_a) was calculated from DSC by heating at various heating rates by Kissinger method and was found to be 15.9 kJ mol^{-1}. Contact angle measurement of the composite an increase in contact angle with the increase in allyl functionalized silica content (98.7° for lower loadings and 102.5° for the higher loadings). This indicated an increased hydrophobicity, which is important for an encapsulant material. Even though the oxygen barrier properties of the silicone polymers are not good, but the impregnated silica particles were supposed to reduce the oxygen barrier property through torturous pathway. Swelling study of the composites was carried out in dry toluene medium, which is a good solvent for silicones. This study showed that the composite with higher silica loadings is having weight

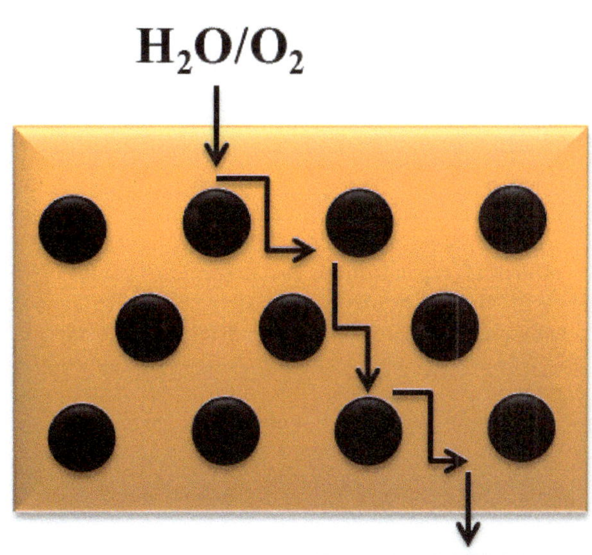

Scheme 1: Reduction in gas permeability due to torturous pathway, created by the inorganic phase

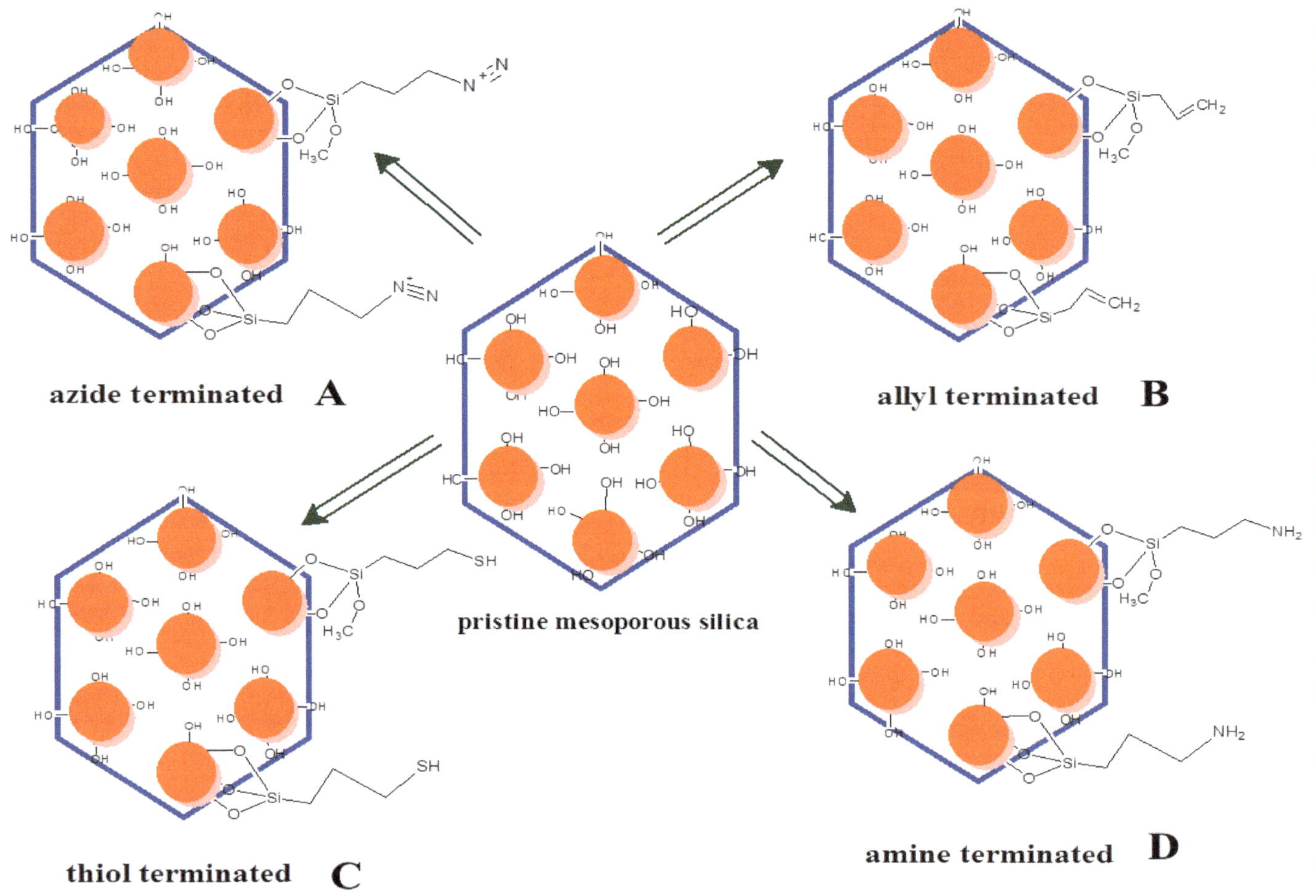

Scheme 2: Various functionalized mesoporous silica to generate different surface functionality and tuning of surface chemistry

swelling ratio (Q_t) lower than the composite with lower filler loadings. This is due to the increased cross-linking density with the increase in the allyl terminated silica loadings of the composite that prevented the permeation of toluene molecules through the composite matrix. Density of the composites also increases with the increase in the silica loadings as observed from density gradient column. This supports that there is a better integrity and cross linking density with increased filler loadings. Nanoindentation and dynamic mechanical analysis (DMA) of the composites indicated an increase in the modulus values with the increase in mesoporous silica loadings. An increase in the co-efficient of thermal expansion was also observed as the filler loading was increased. The composites were observed to be thermally stable up to ~300°C, which is an important property for the composites as it has to protect active and sophisticated device components from harsh environmental conditions. All the properties of the composites indicated that it can be used as a potential encapsulant for organic devices. The curing mixture (liquid polymer dispersed with allyl functionalized silica particles and platinum catalyst) can be easily spin coated or drop casted over the devices and finally cured by thermal treatment to get the in-situ encapsulated device. As compared to the complex, high-cost and laborious multilayered type construction/architecture for encapsulation [29], the 'in-situ' curing technique of encapsulation will be cost effective, easy and a faster method.

Functionalized alumina as filler materials for silicone polymers

Various functionalized (amine and allyl functionalized) alumina particles has been used as a filler material for silicone polymers. The surface functionalization of alumina was carried out by silane coupling chemistry where, '$Al_{surface-core}OSi$' bond formation takes place at the surface, instead of '$Si_{surface-core}OSi$' (as discussed in the section Functionalized mesoporous silica as filler materials for silicone polymers). The composite fabrication with these functionalized nanoparticles with the silicone polymers are discussed below.

(A) Amine functionalized alumina with epoxy-terminated siloxane: *'Amine-Epoxy'* curing strategy: Amine functionalized alumina nanoparticles has been used as a filler in the polymer [poly (dimethylsiloxane) diglycidyl ether terminated] matrix [15] to fabricate composites. Hydride terminated polydimethylsiloxane (PDMS) is used as a precursor for the synthesis of [poly (dimethylsiloxane) diglycidyl ether terminated]. Surface functionalization of the nanoparticles were thoroughly studied through various techniques like, FTIR (Fourier transformed infrared), XPS (X-ray photo electron spectroscopy), TGA (thermogravimetric analysis), Brunauer–Emmett–Teller (BET) isotherm and CHN analysis. Various other functionalities can be also attached onto the hydride terminals of PDMS to get desired properties or for further conjugation with specific surfaces/polymers. In this study, it has been shown that surface amine groups can react with the epoxy terminals of the polymer and forms the flexible composite. Differential scanning calorimetry analysis indicated the curing reaction between amine at the surface and epoxy at the polymer end group, apart from curing of the polymer with the additional curing agent present in the system. This covalent modification is to provide enhanced reinforcement of the nanoparticles with the polymer matrix. The dispersed alumina is expected to provide a torturous pathway for the incoming gas molecules and hence reduces the permeability of incoming gas molecules like moisture vapor/oxygen and hence to form a gas barrier encapsulant, which are critical for organic devices. Atomic force microscopy showed an increased in roughness with the increase in the particle loadings in the matrix. The thermal stability of those composites are found to be high (~305°C) till the optimum loadings of alumina. The optimum loading plays a key role for fabricating a composite as after that important property deteriorates.

(B) Allyl functionalized alumina with hydride terminated siloxane: *Hydrosilylation strategy*: In this study, silicone polymer/functionalized alumina nanocomposites have been also prepared by chemical attachment of the silicone polymer (hydride terminated polydimethylsiloxane) with allyl functionalized alumina through hydrosilylation reaction catalyzed by Karstedt platinum catalyst, where the impregnated nanoparticles in the matrix can reduce the permeability of gases [16]. The alumina surface terminated with allyl groups that make the nanoparticles non-polar and hence compatible in the polymer matrix and cross-links with Si-H of the silicone polymer chains by platinum catalyzed hydrosilylation reaction. It is important to understand the nature of curing, as these materials will be treated thermally for application purpose. The curing behavior of the system was analyzed in details, from non-isothermal rheology and DSC analysis [17]. The mechanistic model for the system was also proposed, based on the two studies. It was observed that an initial self-coupling of the polymer chains occurs first, followed by hydrosilylation reaction occurs in the curing system. Since the oxygen barrier, properties of silicones are not good so the dispersed nano-phase can reduce the permeability of such gases through a torturous pathway. Nano-indentation of the composites indicated the flexible nature of the films. The composites were good electrical insulators and thermally stable up to ~400°C. The FTIR of the composites showed that all the Si-H groups have been reacted. In order to achieve the best performance of the composite material it is important to introduce optimum concentration of fillers in the matrix. The fractured surfaces of the composites reflected the presence of interfacial bindings within the matrix. With the increase in the filler loadings an increase in the contact angle was observed, indicating the increase in hydrophobicity of the matrix. The composites were also stable at two extreme pH values, reflects that the matrix is effectively protecting the alumina nanoparticles from dissolution. This indicates a robust and cross-linked composite matrix.

It is important to note, in this study since hydride terminated silicone polymer have been used for composite fabrication, so the cross-linking density will be much less than the composites (discussed in the section Functionalized mesoporous silica as filler materials for silicone polymers), where internal hydride terminated silicone polymer was used. The improved cross-linking density helps in the improved interfacial interactions within the composites. The increase in cross-linking density is important to reduce the gas permeability further. For all these cases, the liquid polymer dispersed with nanoparticles can be easily spin coated or drop casted over the devices and cured to get an in-situ encapsulated device (Scheme 3).

Advantages of *in-situ* curing over solution casting and melt processing

In this section, a solution casted film fabrication method have been shown, where amine functionalized mesoporous silica particles were used as a filler material and the composite properties are also discussed briefly. Finally the advantages of in-situ curing technique over solution casting and melt-processing for device encapsulation application are compared.

Hybrid organic-inorganic composite films can be fabricated by solution casting [30-32] methodology (Scheme 4B). A solution casted composite film of polyvinyl butyral (PVB) as a base polymer matrix and amine functionalized mesoporous silica as a filler material has been synthesized, where ethanol was used as a solvent [31]. A polyol was also used for the composite fabrication to increase the polar content in the polymer matrix. In this study, mesoporous silica was used as a filler material, which was supposed to reduce the permeability by providing an extended pathway. The surface of the mesoporous silica particles were functionalized with aminopropyltriehoxysilane (APTMS), which generates $-NH_2$ groups at the surface (Scheme 2D). This makes the surface hydrophilic and helps in better mixing in the PVB polymer matrix. The functionalization reduces (~84% reduction was observed) the surface area as compared to the pristine silica, as observed from the BET analysis. Previously it has been shown that amino-propyl functionalized mesostructured silica acts as better reinforcing agent for epoxy polymer [33]. Nanoscale mesoporous silica (NMS)/polyimide nanocomposites have been synthesized [34], in which silica was functionalized with aminopropyltrimethoxysilane. A homogeneous dispersion of nanostructured mesoporous silica was observed in the composites. Due to which substantial improvement in the thermo-mechanical and moisture absorption properties were observed for the composites, as compared to neat polyimide. Homogeneous dispersion is an important criteria for the improvement of the properties of composites. Incorporation of $-Si(CH_3)_3$ modified MCM-48 particles (hydrophobic and compatible to the polymer matrix) in the Polyvinylacetate (PVAc) was also observed to improve mechanical properties resulting composites [35].

In this case the polar $-NH_2$ groups can also help in forming hydrogen bonding with incoming moisture molecules, to trap the moisture. Hence it was anticipated that the synergistic effect of torturosity and the trapping effect can reduce the permeation rate to a greater extent. Three different wt% of functionalized mesoporous silica (0.25 wt %, 0.5 wt %

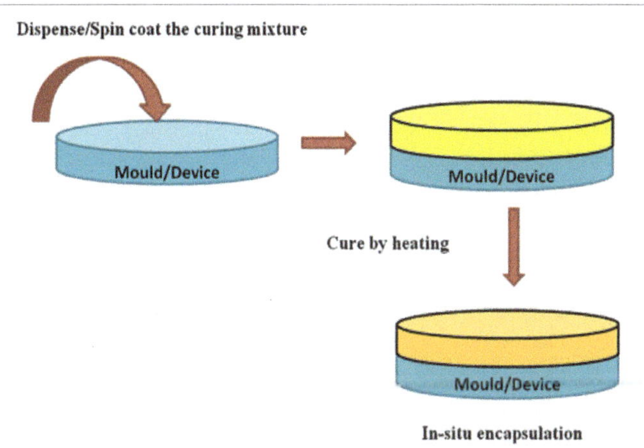

Scheme 3: A model for in-situ cast to encapsulate a device

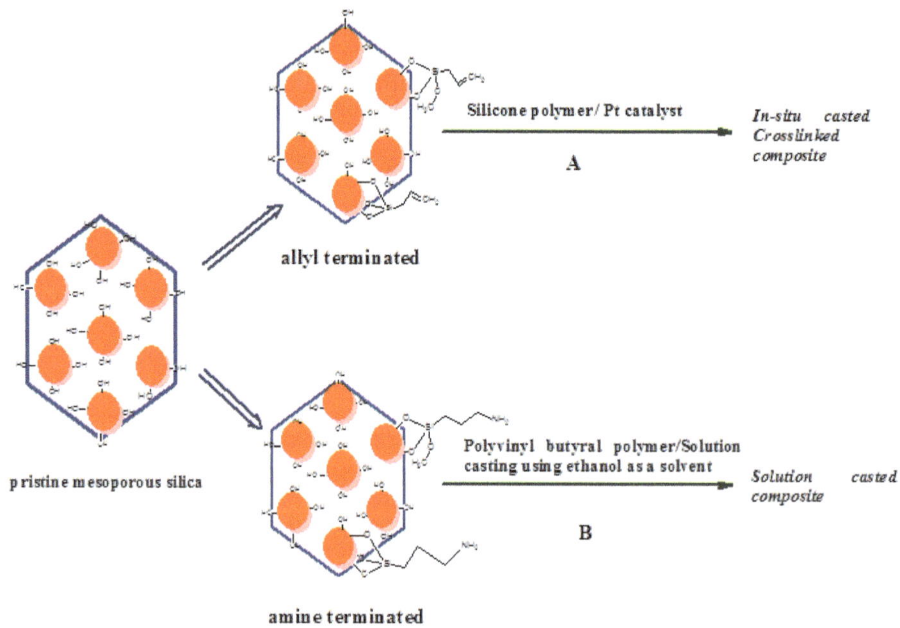

Scheme 4: Composites fabricated by two different strategies, using functionalized mesoporous silica as a filler material

and 1.5 wt %) were intended in the composite matrix. The moisture permeability was determined from the calcium degradation test. This indicated a huge reduction in the moisture permeability with the increase in the increase in silica loadings in the matrix. The permeability value was reduced by an order in case of the composites as compared to the composites. But this reduction was significant till an optimum loading. This amine functionalized silica can also form hydrogen bonding with the polar groups (alcohol and acetate) of PVB, which helps to enhance the interfacial interactions (this was observed for the composites from the fractured surface SEM images which showed higher crack density as compared to the neat polymer matrix). The mechanical property (as obtained from tensile and DMA analysis) and thermal properties (as observed from TGA analysis) showed deterioration. Hence the critical loading or the percolation threshold plays a vital role in fabricating a composite. After the critical loading, aggregation of silica nanoparticles predominates as observed from scanning electron microscopy (SEM) and transmission electron microscopy (TEM) which is responsible for the degradation in the important properties. The composites were used to encapsulate Schottky diodes by sealing with additional glue and tested under accelerated aging conditions. It was observed that the diodes encapsulated with the pristine PVB polymer matrix exhibited almost two times reduction in the current density, as compared to the devices encapsulated with the composite films. This reflects the mesoporous silica loaded composites can be used a gas barrier layer for organic devices.

However, solution casting method suffers from several drawbacks. In-situ curing is a single step process, while solution casting or melt processing needs multiple steps. In solution casting, *step 1* is the evaporation of solvent to get films and *step 2* is the use of a glue to encapsulate the device with the as obtained film (Scheme 5). Unfortunately, solution casting can generate pinholes during the evaporation of the solvents. While in-situ curing method does not use any solvent and hence, less chance of pinhole generation. Melt processed films [36] also requires an additional glue to seal the device.

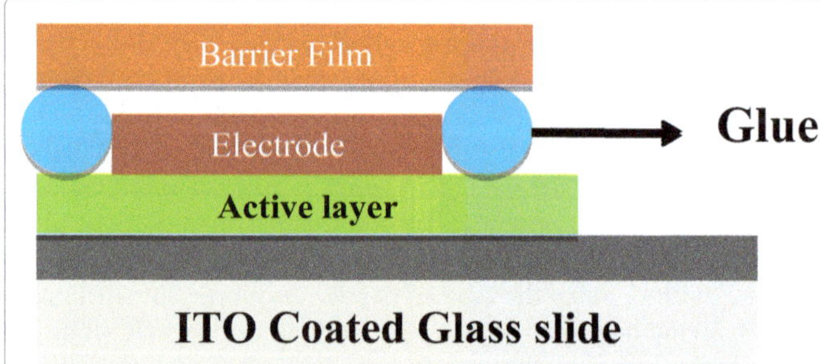

Scheme 5: A model for solution casted or melt processed film to encapsulate a device

Hence, there are several advantages that can take silicone polymers to the forefront towards the advancements of various technologies.

Future Directions towards New Gas Barrier Composites Preparation

Furthermore, the incorporation of oxygen/moisture scavengers in the composites can further reduce the permeability of such composites. It is important to note that silicones are very good moisture resistant, but oxygen barrier properties are not desirable. Hence, in addition to the mixing of nanoparticles, a moisture/oxygen reactive component will reduce to a much greater extent and will be much fruitful for the encapsulation of organic devices.

Several other strategies can be also developed to form the composites such as, '*azide-ene click- chemistry*' and '*thiol-ene click chemistry*'. In which the inorganic surface should have reactive azide, alkene or thiol groups [illustrated in Scheme 2(A) and (C)], while the silicone should have the counter-functional group to couple through click-reaction. Functionalized mesoporous silica is a critical choice as a filler material towards the design and fabrication of composites for various important future applications. Apart from the other potential applications of this important porous material, a suitable and optimized mesoporous silica loaded composite with high gas barrier properties can play a major role in device packaging, in the near future. Until now, the films have been made by using silicones as a base polymer matrix, which are easy to cure. One of the important features regarding 'in-situ curing' method is, it involves no solvent. These composites will further find applications in other potential areas where moisture and oxygen permeation is disadvantageous such as, food/beverage packaging.

Conclusions

Siloxanes are smart material for the purpose of encapsulation because of its easy processability, wide range of curing temperatures, low shrinkage, high thermal stability, low volatility etc. Furthermore, the properties for an encapsulant material such as, (1) hermetic and conformal coating, (2) good chemical, electrical, environmental and thermal stability, (3) compatibility with mass production techniques, and (4) minimum shrinkage of the encapsulant upon curing, coincide with the critical properties of a suitable organic device encapsulant. Unfortunately, the oxygen permeability of silicones is relatively high, so dispersion of nanofillers and oxygen scavengers will be critical for low gas barrier applications. This will open up a new application for silicones in the field of organic device packaging. In comparison to the existent atomic layer deposition or molecular layer deposition route, the in-situ casted silicone polymers having an improved barrier properties will be cost effective, easy to fabricate and will capture the market of organic device packaging industries in near future.

Acknowledgements

S.G acknowledges Council of Scientific and Industrial Research (C.S.I.R), New Delhi for financial support and fellowship. Authors also gratefully acknowledge the financial support from DST No SR/S3/ME/022/2010-(G). This work was also partially supported by the Ministry of Communication and Information Technology under a grant for the Centre of Excellence in Nanoelectronics, Phase II.

References

1. Silva WJ, Kim HP, Yusoff ARM, Jang J (2013) Transparent flexible organic solar cells with 6.87% efficiency manufactured by an all-solution process. Nanoscale 5: 9324-9329.
2. Seo JH, Gutacker A, Sun Y, Wu H, Huang F, et al. (2011) Improved High-Efficiency Organic Solar Cells via Incorporation of a Conjugated Polyelectrolyte Interlayer. J Am Chem Soc 133: 8416-8419.
3. You J, Hong Z, Yang Y, Chen Q, Cai M, et al. (2014) Low-Temperature Solution-Processed Perovskite Solar Cells with High Efficiency and Flexibility. ACS Nano 8: 1674–1680.
4. Luo S, Daoud WA (2015) Recent progress in organic–inorganic halide perovskite solar cells: mechanisms and material design. J Mater Chem A.
5. Pietro RD, Fazzi D, Kehoe TB, Sirringhaus H (2012) Spectroscopic Investigation of Oxygen- and Water-Induced Electron Trapping and Charge Transport Instabilities in n-type Polymer Semiconductors. J Am Chem Soc 134: 14877-14889.
6. Duffy DC, McDonald JC, Schueller OJA, Whitesides GM (1998) Rapid Prototyping of Microfluidic Systems in Poly(dimethylsiloxane). Anal Chem 70: 4974-4984.
7. Leclerc E, Sakai Y, Fujii T (2003) Cell Culture in 3-Dimensional Microfluidic Structure of PDMS (polydimethylsiloxane). Biomed Microdevices 5: 109-114.
8. Cong H, Pan T (2008) Photopatternable Conductive PDMS Materials for Microfabrication. Adv Funct Mater 18: 1912–1921.
9. Abbasi F, Mirzadeh H, Katbab AA (2001) Modification of polysiloxane polymers for biomedical applications: a review. Polym Int 50: 1279-1287.
10. Soroory H, Mashak A, Rahimi A (2013) Application of PDMS-based coating in drug delivery systems using PVP as channeling agent. Iran Polym J 22: 791-797.
11. Selvaraj H, Tan B, Venkatakrishnan K (2011) Synthesis of nanofiber-filled polydimethylsiloxane using ultrafast laser irradiation. J Polym Res 18: 1659-1665.
12. Kim JS, Yang S, Bae BS (2010) Thermally Stable Transparent Sol–Gel Based Siloxane Hybrid Material with High Refractive Index for Light Emitting Diode (LED) Encapsulation. Chem Mater 22: 3549-3555.
13. Jung K, Bae JY, Park SJ, Yoo S, Bae BS (2011) High performance organic-inorganic hybrid barrier coating for encapsulation of OLEDs. J Mater Chem 21: 1977-1983.
14. Yang SC, Kwak SY, Jin JH, Kim JS, Choi Y, et al. (2012) Thermally resistant UV-curable epoxy–siloxane hybrid materials for light emitting diode (LED) encapsulation. J Mater Chem 22: 8874-8880.
15. Gupta S, Ramamurthy PC, Madras G (2011) Synthesis and characterization of flexible epoxy nanocomposites reinforced with amine functionalized alumina nanoparticles: a potential encapsulant for organic devices. Polym Chem 2: 221-228.
16. Gupta S, Ramamurthy PC, Madras G (2011) Covalent Grafting of Polydimethylsiloxane over Surface-Modified Alumina Nanoparticles. Ind Eng Chem Res 50: 6585-6593.
17. Gupta S, Ramamurthy PC, Madras G (2011) Mechanistic overview of the curing behavior of hydride terminated polydimethylsiloxane with allyl functionalized alumina by calorimetry and rheometry. Thermochimica Acta 524: 74-79.
18. Gupta S, Ramamurthy PC, Madras G (2011) Synthesis and characterization of silicone polymer/functionalized mesostructured silica composites. Polym Chem 2: 2643-2650.
19. Moradi O, Yari M, Zare K, Mirza B, Najafi F (2012) Carbon Nanotubes: A Review of Chemistry Principles and Reactions. Fullerenes, Nanotubes and Carbon Nanostructures 20: 138-151.
20. Yoo BM, Shin HJ, Yoon HW, Park HB (2014) Graphene and graphene oxide and their uses in barrier polymers. J Appl Polym Sci 131: 39628.
21. Peng R, Wang Y, Tang W, Yang Y, Xie X (2013) Progress in Imidazolium Ionic Liquids Assisted Fabrication of Carbon Nanotube and Graphene Polymer Composites. Polymers 5: 847-872.
22. Moorthy MS, Park SS, Fuping D, Hong SH, Selvaraj M, et al. (2012) Step-up synthesis of amidoxime-functionalised periodic mesoporous organosilicas with an amphoteric ligand in the framework for drug delivery. J Mater Chem 22: 9100-9108.
23. Park C, Kim H, Kim S, Kim C (2009) Enzyme Responsive Nanocontainers with Cyclodextrin Gatekeepers and Synergistic Effects in Release of Guests. J Am Chem Soc 131: 16614-16615.
24. Slowing II, Trewy BG, Giri S, Lin VSY (2007) Mesoporous Silica Nanoparticles for Drug Delivery and Biosensing Applications. Adv Funct Mater 17: 1225-1236.
25. Park I, Peng HG, Gidley DW, Xue SQ, Pinnavaia TJ (2006) Epoxy–Silica Mesocomposites with Enhanced Tensile Properties and Oxygen Permeability. Chem Mater 18: 650-656.
26. Zhang FA, Lee DK, Pinnavaia TJ (2010) PMMA/mesoporous silica nanocomposites: effect of framework structure and pore size on thermomechanical properties. Polym Chem 1: 107-113.
27. Ji X, Hampsey JE, Hu Q, He J, Yang Z, et al. (2003) Mesoporous Silica-Reinforced Polymer Nanocomposites. Chem Mater 15: 3656-3662.
28. Bershtein VA, Egorova LM, Yakushev PN, Pissis P, Sysel P, et al. (2002) Molecular dynamics in nanostructured polyimide–silica hybrid materials and their thermal stability. J Polym Sci Part B: Polym Phys 40: 1056-1069.
29. Burrows PE, Graff GL, Gross ME, Martin PM, Hall M, et al. (2001) Gas Permeation and Lifetime Tests on Polymer-Based Barrier Coatings. Proc SPIE 4105: 75-83.
30. Gupta S, Sindhu S, Varman KA, Ramamurthy PC, Madras G (2012) Hybrid nanocomposite films of polyvinyl alcohol and ZnO as interactive gas barrier layers for electronics device passivation. RSC Advances 2: 11536-11543.
31. Gupta S, Seethamraju S, Ramamurthy PC, Madras G (2013) Polyvinylbutyral based hybrid organic/inorganic films as a moisture barrier material. Ind Eng Chem Res 52: 4383-4394.
32. Seethamraju S, Ramamurthy PC, Madras G (2013) Flexible poly(vinyl alcohol-co-ethylene)/modified MMT moisture barrier composite for encapsulating organic devices. RSC Adv 3: 12831-12838.

33. Jiao J, Sun X, Pinnavaia TJ (2008) Reinforcement of a Rubbery Epoxy Polymer by Mesostructured Silica and Organosilica with Wormhole Framework Structures. Adv Funct Mater 18: 1067-1074.

34. Cheng CF, Cheng HH, Cheng PW, Lee YJ (2006) Effect of Reactive Channel Functional Groups and Nanoporosity of Nanoscale Mesoporous Silica on Properties of Polyimide Composite. Macromolecules 39: 7583-7590.

35. He J, Shen Y, Yang J, Evans DG, Duan X (2003) Nanocomposite Structure Based on Silylated MCM-48 and Poly(vinyl acetate). Chem Mater 15: 3894-3902.

36. Saravanan S, Gupta S, Ramamurthy PC, Madras G (2014) Effect of silane functionalized alumina on poly(vinyl butyral) nanocomposite films: Thermal, mechanical, and moisture barrier studies. Polymer Composites 35: 1426-1435.

Comparative Studies on Accumulation of Selected Microelements by *Spirulina Platensis* and *Chlorella Vulgaris* with the Prospects of Functional Food Development

Szabolcs Molnár[1]*, Attila Kiss, Diána Virág and Péter Forgó

Regional Knowledge Centre, Eszterházy Károly College, Leányka Street, Hungary

Abstract

Chlorella vulgaris and Spirulina platensis as plausible basic material of perspective functional foodstuffs were in the focus of the research due to the high content of bioactive compounds, favorable growing feature and pronounced microelement accumulation capabilities.

The emphasis was laid on the investigation of bioaccumulation of four microelements (Fe(III), Cu(II), Zn(II), Mo(VI)) in cases of the two abovementioned alga species. Metal binding aspects of algae in our case was not examined for perspective utilisation in environmental protection, but purposively for revealing ways for possible functional food developments.

Both *Spirulina platensis* and *Chlorella vulgaris* proved to be efficient in terms of metal accumulation in two-week-long experiments. With the exception of molybdenum the bioaccumulation capacity of Chlorella exceeded that of Spirulina. Iron (789.7 ± 102.7mg kg^{-1}) and zinc (378.1 ± 5.5mg kg^{-1}) were accumulated to a large extent by Chlorella vulgaris, while bioaccumulation ability of *Spirulina platensis* proved to be the most significant for iron (676.9 ± 27.6mg kg^{-1}) and molybdenum (5.79 ± 1.1mg kg^{-1}).

The biomass of *Spirulina platensis* increased to greater extent than that of Chlorella after the same incubation time in case of the control media, however metal treatment of the media favoured more pronouncedly the growth of biomass of *Chlorella vulgaris*. The biomass of *Spirulina platensis* was diminished when incubated in metal fortified media, with the exception of iron.

As a consequence it might be established that both studied algae species can be suggested for involvement in further functional food developments, since they may inevitably contribute to microelement supply of humans, or to the prevention of iron deficiency anaemia thanks to their large microelement accumulation capabilities.

Keywords: *Chlorella Vulgaris*; *Spirulina Platensis*; Microelements; Bioaccumulation; Growth of Biomass; Functional food; Iron deficiency anaemia

Introduction

Chlorella and *Spirulina* are of outstanding physiological activity including antioxidative, antiflammatory, antimutagenic, antiviral, cardioprotective effects, immune enhancing and anticancer properties [1-3]. As a consequence, elaborations of food-related applications are in the forefront of interest of recent researches, hence such implications demand thorough scientific arguments. Novelty of this paper might be regarded the fact that majority of previous studies on algae species aimed at revealing their fitoremedial properties and metal sorption activities with regard to utilisation in environmental protection. The prospects of possible functional food applications need to be grounded more extensively by purposive researches on physiological and chemical feature of algae.

The outstanding nutritive value and high content of bioactive components of the algae is already well known as several studies (nutritive biological, human clinical) were focused on the analysis of their important components which have beneficial effects on human health like minerals, proteins, lipids and vitamins [4,5]. The exact and detailed composition and the ratio of organic constituents of selected algae have also been revealed [6]. It was found that *Tetraselmis sp.* and *Nannochloropsis oculata*, cultivated in industrial-scale bioreactors, produced 2.33 and 2.44% w/w lipid (calculated as the sum of fatty acid methyl esters) in dry biomass, respectively [7].

Chlorella sp. and *Nannochloropsis salina* cultivated in a lab-scale open pond stimulating reactor grew well and produced 350-500 mgL^{-1} of biomass containing approximately 40% and 16% of lipids, respectively [8]. Several studies were focused on the biosorption of heavy metals (Pb, Cd, Cu, Zn, Ni) by algae species in order to eliminate toxic elements from the environment [9]. The studied algae were suitable for the elimination of toxic heavy metal levels. Regular analysis and monitoring of the metal content of algae showed the extension of the metal pollution of a given area [10]. Biological materials can bound to cellular surfaces via the process called biosorption or bioaccumulation can take place inside the cell [11,12]. Biosorption of metal ions may be accomplished via complex mechanisms, such as ion exchange, complex formation, electrostatic interactions and precipitation. Correlation was observed between the content of trace elements or other biologically active substances of the growth medium and the composition of the algae biomass [13,14] The capability of metal uptake depends on several factors like the growing circumstances (temperature, pH), the level of available nutrients and microelements, metal concentration of the growth medium, the amount of the alga in the solution (biomass) and the biosorption capacity of the studied species [15].

***Corresponding author:** Szabolcs Molnár, Regional Knowledge Centre, Eszterházy Károly College, 3300 Eger, Leányka Street 6, Hungary

By taking advantage of the high bioaccumulation characteristics of algae, their nutritive value can considerably be improved making sensible the concept of plausible functional food applications.

Biosorption capacity of seaweeds from Baltic sea was also studied in order to get information about the accumulated amounts of metals with the prospect of utilisation as feed supplements for livestock [16]. According to our knowledge it might regarded as one of the first attempts to find food or feed related applications for plants suitable for metal accumulation. Our research was carried out with similar objectives.

The majority of nutrition supplements comprise trace elements as inorganic salts but the bioavailability of inorganic forms is not so pronounced. The bioavailability of organic forms is larger and more effective than that of the inorganic forms [17,18]. Application of organic iron complexes signifies a new trend in the clinical practice during the treatment of iron deficiency anaemia [19-22] Studies on dietary application of *Spirulina platensis* [13] related the bioaccumulation of iodine, zinc and selenium. It was observed that the uptaken metals are in organic form in the *Spirulina*. Functional yoghurt has also been produced by the application of this alga with enhanced metal-content.

Several kinds of alga based products appeared on the market as food supplements in the previous decades, but the elaboration of functional, algae containing products started just in the 90's in the European Union [23-25]. The annual growth rate for the sale of functional foods is about 14% at world level, thus these algae containing functional products with enriched metal content would enter an expanding market.

Our objective in this study was to broaden the scope of possible alimentary application of specific algae species (*Chlorella vulgaris* and *Spirulina platensis*) by revealing their bioaccumulation characteristics in terms of 4 important microelements. These examinations should provide a good basis for elaboration of new functional food prototypes based on metal enriched alga biomass. This endeavour might be regarded as the most pronounced novelty and relevance of our study, as formerly environmental protection related utilisations of algae species were in the forefront of researches.

Molybdenum binding capabilities of alga species have not been examined previously, therefore our studies might convey new knowledge to the current state-of-the art of that segment of science.

Most of the previous food analytical studies aim at revealing amino acid, fatty acid and protein profiles of various alga species, while our work focuses on microelement accumulation aspects. In this paper binding efficiency of the studied microelements was studied under various conditions and in different concentrations. Biosorption characteristics of iron has been more thoroughly scrutinised than that of the other microelements, since iron deficiency occurs the most frequently in humans among microelement nutrition disorders.

Materials and Methods

Organisms and applied chemicals

The algae culture of *Spirulina platensis* M2 was obtained from Soley Institute (El Sobrante, California), while wild-type of *Chlorella vulgaris* was obtained from the Plant and Soil Protecting Station (Gödöllő, Hungary). The algae cultures were stored in sterile test-tubes on slant agar at 4oC in dark. All the applied reagents and chemicals (analytical purity) were purchased from Sigma-Aldrich (Darmstadt, Germany), and Merck (Darmstadt, Germany).

Bioaccumulation of metals by *Spirulina platensis* and *Chlorella vulgaris* from growth medium with diverse microelement content

Growing media: The composition of the growth medium in the case of *Spirulina platensis* was in accordance with Zarrouk's medium [26], while for *Chlorella vulgaris* the OECD 201, AAP (US, EPA, 2011) conditions were applied. [27]

Four metals (Fe(III), Cu(II), Zn(II), Mo(VI)) were chosen for experiments. Enhancement of metal content of the growing media was performed separately for each microelement. Five different solutions were prepared in case of each microelement: one control solution and 4 solutions with diverse metal concentrations [28]. The metal content of the control solution (C) was in accordance with the recipe of OECD and Zarrouk guideline, as follows: [$FeCl_3$*$6H_2O$]: 0.16 mgl^{-1}, [$CuCl_2$*$2H_2O$]: 0.012 μgl^{-1}, [ZnCl2]: 0.00327 mgl^{-1}, [Na_2MoO_4*$2H_2O$]: 0.00726 mg l^{-1}. The metal concentration was separately and stepwise enhanced to tenfold of the control's Fe(III), Cu(II), Zn(II) and Mo(VI) content. These solutions were signed as: Fe10X, Cu10X, Zn10X, Mo10X. The metal solutions were prepared from $FeCl_3$ x $6H_2O$, $CuCl_2$ x $2H_2O$, $ZnCl_2$, Na_2MoO_4 x $2H_2O$ (Table 1.).

In order to establish the extent of maximum adsorption (saturation) of iron, and its affect on the biomass production, the iron content was separately and stepwise increased: 5x (C_{Fe5x}=0.165mg l^{-1}), 10x (C_{Fe10x}=0.33mg l^{-1}), 25x (C_{Fe25x}=0.825mg l^{-1}), 50x (C_{Fe50x}=1.65mg l^{-1}) and 100x (C_{Fe100x}=3.3mg l^{-1}), compared to that of the control ($C_{Control}$=0,03.3mg l^{-1}) (OECD 201, Zarrouk's) (Table 1).

Alga suspensions with 10 mgl^{-1} concentration were prepared in Erlenmeyer-flasks (500 ml) by the application of the above described growing media. The experiments were performed in triplicates (n=3) and after the incubation time the whole amount of the gained alga was analysed.

Alga growing apparatus: The alga growing apparatus consists of a horizontal glass surface whereon the Erlenmeyer-flasks were placed. Special lamps (Aqua-Medic Gmbh., Bissendorf, Germany) were applied in order to induce alga growing, constant illumination of 160 μE m^{-2}s^{-1} was provided by 24 W T5 light bulbs. The illuminating periods were set in accordance with the natural conditions by automated switch. Fresh air was pumped into the solutions through plastic tubes in order to avoid the generation of alga film layer on the wall of the flasks. The growing period was 2 weeks; the temperature was 25°C in case of *Chlorella* and 32°C in case of *Spirulina*.

Sample preparation

After the two-week long incubation time the alga solutions were centrifuged for 15 minutes at 6000 rpm than the supernatant was removed and the alga was washed with distilled water (conductivity was lower than 0.055 mS) and centrifuged again. The water was removed from the centrifuge tubes and the alga was dried at 50°C for 48 hours. The amount of the produced dry alga was established gravimetrically.

Growth media	Fe^{3+}	Cu^{2+}	Zn^{2+}	Mo(VI)
Control*	33μgl^{-1}	0.004 μgl^{-1}	1.6 μgl^{-1}	2.9 μgl^{-1}
Enhanced media (Fe 10X)	330μgl^{-1}	0.004 μgl^{-1}	1.6 μgl^{-1}	2.9 μgl^{-1}
Enhanced media (Cu 10X)	33μgl^{-1}	0.04 μgl^{-1}	1.6 μgl^{-1}	2.9 μgl^{-1}
Enhanced media (Zn 10X)	33μgl^{-1}	0.004 μgl^{-1}	16 μgl^{-1}	2.9 μgl^{-1}
Enhanced media (Mo 10X)	33μgl^{-1}	0.004 μgl^{-1}	1.6 μgl^{-1}	29μgl^{-1}

*The metal content of the control media in accordance with OECD 201 (Chlorella), Zarrouk media (Spirulina).

Table 1: Applied metal concentration in the growth media.

Prior to establishment of the metal content of the samples the organic matter was decomposed with the application of microwave accelerated reaction system (MARS, CEM Corporation, Metthews, USA) in the presence of 8 mlcc. HNO_3 and 2 ml of cc. H_2O_2. The working conditions of the decomposing system were as follows: 800 W power, 220°C, 30 bars for 45 min.

Sample preparation was performed in triplicates (n=3) from the three parallel growth media with the same metal-composition.

Measurement conditions

The metal content of the samples was measured by atomic absorption spectrometer SpectrAA 50B Varian (Mulgrave Virginia, Australia). During the measurements flame (FAAS) and electrothermal atomization (ET-AAS) were applied (GTA-100, Varian, Mulgrave Virginia, Australia) with deuterium background correction.

For both AAS techniques (flame and electrothermal) Varian SpectrAA hollow cathode lamps were used. For quantitative analysis CertiPUR standard solutions (Merck, Darmstadt, Germany) were applied. The measurement parameters are summarized in Table 2. The obtained results were calculated from the values of three parallel measurements and were expressed in mg kg^{-1} dried alga.

Calculations

Data for the calculations (both EF, WR) gained from the results of measured metal content of 30 samples prepared as follows. Five samples were prepared in triplicates with four different the compositions of the metals (with a control sample) as it is demonstrated in Table 1, so 15 samples were prepared per algae species.

Enrichment factor (EF) was calculated according to the following ratios.

$$EF = \frac{C_E}{C_C}$$

C_E = Microelement concentration of dry alga grown in the media with enhanced metal content

C_C = Microelement concentration of dry alga grown in the control media

Weight ratio (WR) was calculated according to the following

$$WR = \frac{W_E}{W_C}$$

W_E = Weight of dry alga grown in the media with enhanced metal content

W_C = Weight of dry alga grown in the control media

The theoretical maximum of the adsorbing (Fe^{3+}) capacity was calculated from the average iron content (three performed measurements) of the set of six samples by the application of parametrical fit (n=18). The calculations were performed by the application of the following formula:

	Method	Wavelength (nm)	Conc. Range	Dynamic interval
Fe^{3+}	FAAS	248.3	0.1876 3 ppm	0.0039-0.2101
Cu^{2+}	FAAS	324.7	0.125-1 ppm	0.0138-0.1569
Zn^{2+}	FAAS	213.9	0.125-2 ppm	0.0239-0.2735
Mo(VI)	ET-AAS	313.3	0.015-0.045 ppb	0.0099-0.0528

Table 2: Measurement conditions.

Results and Discussion

Comparison of nutritive values of algae's different growing parameters

Food industrial utilization the studied algae should be based on the accurate knowledge on their nutritive characteristics [29]. Therefore their basic nutritive parameters and components were revealed as it is summarized in Table 3.

In comparison with previous results [7,8] it can be stated that the lipid content of the studied samples are in compliance with the global trends. The protein level of *Spirulina* samples are 50% higher than that of *Chlorella*, as it was formerly revealed. All the other examined nutritive parameters have not displayed great alteration when comparing them with previous studies.

Comparison of bioaccumulation capacity of *Spirulina platensis* and *Chlorella vulgaris*

It is of great importance to establish that the actual extent of metal fortification of the alga species in respect of being able to tailor practical application.

Metal-accumulating capacities of *Chlorella vulgaris* and *Spirulina platensis* were compared for 4 distinctive microelements in cases of growing media of 10 times fortified metal concentration. It can be stated that the extent of bioaccumulation of the microelement studied proved to be more pronounced for *Chlorella vulgaris*, with the exception of molybdenum. *Spirulina platensis* accumulated Mo(VI) significantly more efficiently than *Chlorella vulgaris* under the experimental conditions (Figures 1 and 2).

Copper and zinc contents of *Chlorella vulgaris* exhibited great variability when grown in metal fortified media. Ten times more copper and zinc were analysed in *Chlorella* grown in the metal-treated media in comparison with the control medium (Zn 10X: 378.1 mg kg^{-1}, Cu 10X : 103.1 $mgkg^{-1}$ dry biomass). Equal amount of copper was uptaken by *Spirulina* from the control medium and the copper fortified medium (Cu: 10.03 $mgkg^{-1}$, Cu 10X: 10.50 mg kg^{-1} dry biomass). Almost four times more zinc was uptaken by *Spirulina* from the Zn-enriched growth medium than from the control medium (C: 13.1 $mgkg^{-1}$, Zn 10X: 47.2 $mgkg^{-1}$ dry biomass). If the metal accumulation feature is compared of the two species, Chlorella uptakes ten times more zinc and copper than *Spirulina*.

Marked difference was observed in the extent of iron accumulation of the two studied algae grown in the control medium. After two-week-long incubation period the iron concentration of *Spirulina* was 300.5 mg kg-1, while *Chlorella* accumulated five times less metal (66.3 $mgkg^{-1}$ dry biomass). On the contrary, when examining algae grown in iron fortified media the extent of iron uptake of *Chlorella* proved to be more considerable than that of *Spirulina* (*Spirulina* Fe10X: 676.9 $mgkg^{-1}$, Chlorella Fe 10X: 789.7 $mgkg^{-1}$ dry biomass).

		Nutritive Value	
		Alga grown in normal medium	Alga grown in normal medium with 10 times enhanced microelement content
Total Lipid	*Spirulina p.*	9.38 ± 0.38%	8.87 ± 0.28
	Chlorella v.	15.11 ± 1.22	12.34 ± 0.97
Sodium	*Spirulina p.*	1.72 ± 0.05	1.92 ± 0.10
	Chlorella v.	1.11 ± 0.09	1.44 ± 0.71
Total Carbohydrate	*Spirulina p.*	22.88 ± 3.39	19.98 ± 2.88
	Chlorella v.	19.25 ± 2.33	18.23 ± 1.92
Protein	*Spirulina p.*	59.87 ± 3.66	61.22 ± 4.28
	Chlorella v.	41.37 ± 4.55	43.71 ± 3.87

Table 3: Nutritive value of the samples.

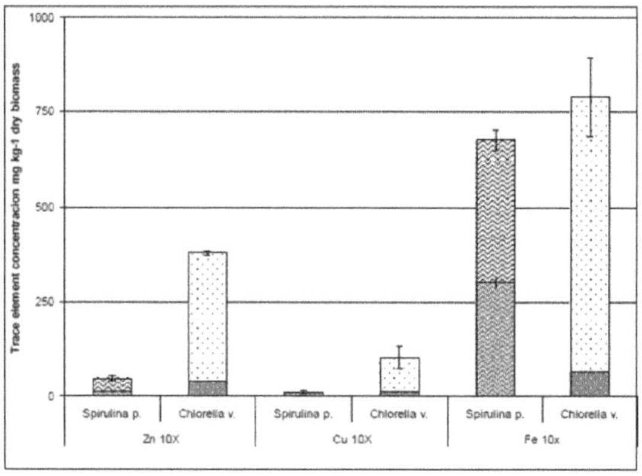

Figure 1: Bioaccumulation of zinc, Cu, iron by Spirulina and Chlorella from growing media with 10 times fortified metal content ■: Content media; □: Growth media with enhanced trace element level.

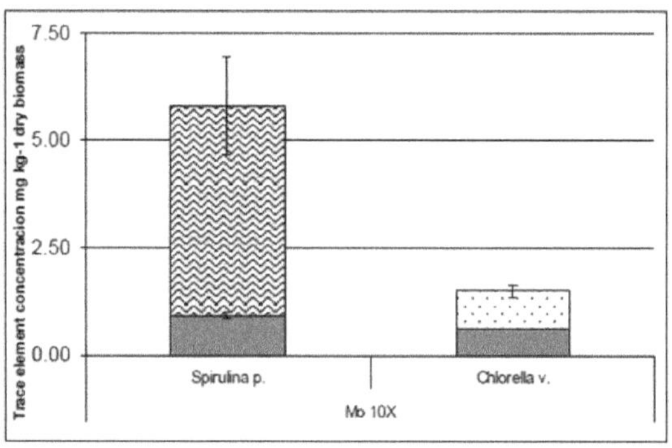

Figure 2: Bioaccumulation of Mo by Spirulina and Chlorella from growing media with 10 times fortified Mo(VI) ■: Content media; □: Growth media with enhanced trace element level.

Extent of bioaccumulation of molybdenum in *Spirulina* was found to be four times larger than in *Chlorella* (*Spirulina* Mo10X: 5.79 mgkg^{-1}, *Chlorella* Mo10X: 1.51 mg kg^{-1} dry biomass).

Comparison of iron accumulation of *Spirulina platensis* and *Chlorella vulgaris*

Figure 3 shows that by enhancement of the iron content of the growing media (5X, 10X, 25X, 50X, 100X) the amount of the accumulated iron in *Spirulina platensis* slowly and continuously increased. The extent of iron accumulation of *Chlorella* was significantly increased when applying sample with 50 times fortified iron concentration. In case of higher concentration of iron less amounts of algae biomass were generated.

When comparing the iron accumulation capacity of the two studied alga species it can be stated that *Spirulina platensis* accumulated greater amount of iron from the control growth medium than *Chlorella vulgaris*. In cases of samples involving growing media with enhanced iron concentration, *Chlorella* adsorbed significantly larger quantity of iron than *Spirulina*.

The calculated theoretical maximum of adsorbing capacity of the biomass was 419.17 mg/100 g dried alga (*Chlorella vulgaris*) and 133.57 mg/100 g dried alga (*Spirulina platensis*), respectively.

Alga mass (biomass) increment

After the two-week-long incubation period the dried biomass of the two studied algae, grown in media with different microelement content, was compared to one another and to their control as well (Figure 4).

The biomass of *Chlorella* grown in the medium with ten times enhanced Zn(II) concentration was approximately three times larger than the biomass of *Chlorella* grown in the control medium (Figure 4). Zinc, molybdenum and copper enrichment of the growing media had adverse effect on the growth of *Spirulina*, as its biomass was significantly lower than the biomass of the control sample after two weeks incubation. Iron fortification of the growing media resulted in a definite increase of the biomass of *Chlorella*, while that of *Spirulina* was not affected. Molybdenum fortification hardly influences the biomass of *Chlorella*, while significantly lower biomass of *Spirulina* was measured than in the control one. The enhanced copper-content of the growing media exerted negative effects on the growth of both studied algae species.

Since our major interest was directed towards the accurate estimation of the extent of biosorption of iron, it was of high importance to reveal its effect on the increase of alga biomass. The biomass of *Spirulina* and *Chlorella* grown in control media were compared to the ones which were grown in media with enhanced iron concentrations (5x, 10x, 25x, 50x) (Figure 5).

It was observed that the amount of biomass of *Chlorella* was found

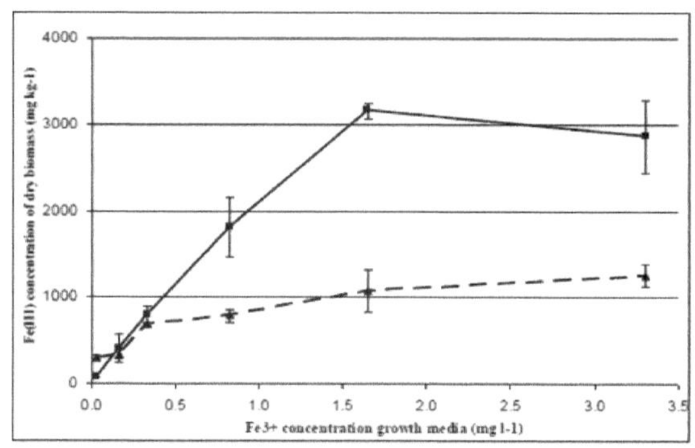

Figure 3: Biosorption of iron (Fe3+) by Spirulina (—▲—) and Chlorella (—■—).

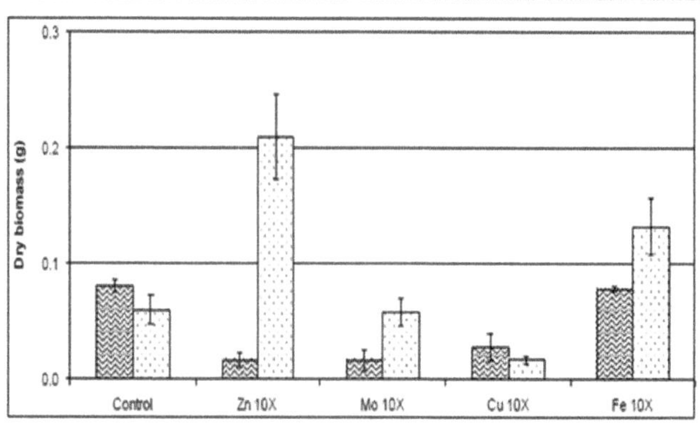

Figure 4: Biomass of spirulina (▧) and Chlorella (□) grown in media with diverse microelement contant after two-week-long incubation.

to be roughly four times larger in comparison with the control sample in case of applying growth medium with 5 times increased Fe (III) content, and three times bigger in case of Fe10X medium. The iron enrichment of the growing media had hardly affected the biomass of *Spirulina* in cases of applying Fe5X and Fe10X growing media; however more significant increase in the iron concentration of the media (Fe25X, Fe50X and Fe100X) led to moderate enhancement in the amount of biomass. On the contrary the biomass of *Chlorella* was dramatically lower at these iron levels (Fe25X, Fe50X and Fe100X).

Therefore the iron can be considered as one of the limiting factors of the growth of *Chlorella* as increase of the biomass reacts sensitively to the alterations in the iron content of the media. Beyond a certain iron concentration further biomass production is hindered. Biomass growth of *Spirulina* is not so adversely affected by the increasing iron concentration of the medium, even though slight biomass enhancement is observed in the previous cases.

Comparison of enrichment factors and weight ratios

The enrichment factor was outstanding in case of *Chlorella* for Fe(III); Cu(II) and Zn(II) as well as in case of *Spirulina* for Mo(VI). In case of *Chlorella* Fe(III) and Zn(II) the weight ratio was also remarkable (Table 4).

Practical application

Both micoalgae species might be suitable as row material of plausible functional food application; however *Chlorella* exhibits more favourable feature in respect of metal binding capability and efficiency, than *Spirulina*. As a consequence, *Chlorella* can be advised for manufacturing use of microalga-based functional food products fortified with distinctive microelements, if microelement enrichment is the major scope of the development. In contrary to the previous arguments, *Spirulina* might be more favoured to *Chlorella* in case of production just Fe-fortified products, as the biomass growth of *Spirulina* proved to be much more pronounced when applying excessively iron-enriched media (Fe25X, Fe50X, Fe100X). In any other cases *Chlorella* displayed more beneficial biomass growing characteristics, however the high reproduction rate and effective biomass production of *Spirulina* make this species also appropriate for involvement in functional food production in a quick, cost-effective and environmentally friendly way.

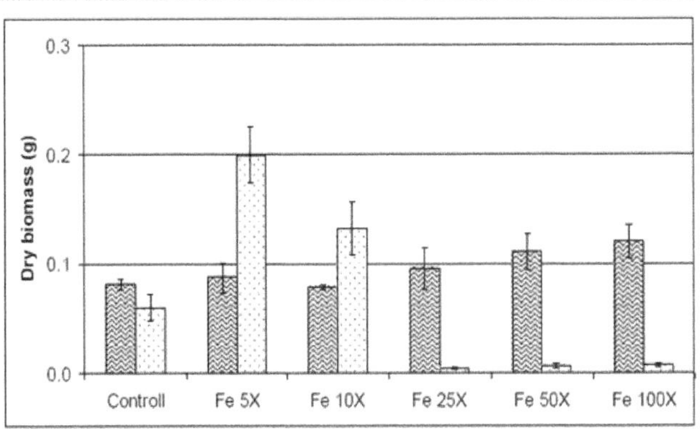

Figure 5: Biomass of spirulina (▨) and Chlorella (□) grown in media with differently enhanced iron content.

	Fe^{3+}		Cu^{2+}		Zn^{2+}		Mo(VI)	
	S.p	*C.v*	*S.p*	*C.v*	*S.p*	*C.v*	*S.p*	*C.v*
Enrichment factor	2.25	11.92	1.05	8.89	3.61	9.14	6.30	2.40
Weight ratio	0.97	2.19	0.35	0.29	0.20	3.49	0.20	0.97

Table 4: Enrichment factor and weight ratio.

In Table 5 the amounts of algae are presented necessary to be consumed in order to ensure 15% of the relevant RDA-s in cases of the studied 4 microelements [26].

Conclusions

It might be concluded that metal enriched microalgae can be considered as highly efficient tools of microelement supply, as it involves considerable amounts of metals in bioavailable forms (organic bonds). Due to the pronounced metal accumulation feature of the two studied species, they are suitable for being involved in possible functional food developments.

If the metal accumulation feature is compared of the two species, *Chlorella* uptakes ten times more zinc and copper than *Spirulina*.

It was found that the extent of bioaccumulation of the studied microelement proved to be significantly more pronounced for *Chlorella* vulgaris than for *Spirulina platensis*, with the exception of molybdenum.

Chlorella displayed more beneficial biomass growth characteristics than *Spirulina platensis* when incubated in media with enhanced microelement content in cases of Zn(II), Mo(VI) and Cu(II).

Iron can be considered as one of the limiting factors of the growth of *Chlorella* as increase of the biomass reacts sensitively to the alterations in the iron content of the media. Biomass growth of *Spirulina* is not so adversely affected by the increasing iron concentration of the medium.

It was established that significantly less amount of algae has to be consumed to comply with the physiological needs of some relevant microelements, if they are incubated in metal fortified media.

The abovementioned results confirm the relevance of development of microelement enriched products based on specifically selected and grown microalga species. The most efficient ways and conditions of metal bioaccumulation have been established laying the foundations for functional food application in the future.

Acknowlegdement

The authors thank the National Office for Research and Technology for financial support of this work (TÁMOP-4.2.2.A-11/1/KONV-2012-0008 and TÁMOP-4.2.3-12/1/KONV-2012-0025).

References

1. Watanuki H, Ota K, Malina ARAS, Tassaka KT, Sakai M (2006) Immunostimulant effects of dietary Spirulina platensis on carp, Cyprinus carpio. Aquaculture 258: 157-163.
2. Mendiola JA, Jaime L, Santoyo S, Reglero G, Cifuentes A, et al. (2007) Screening of functional compounds in supercritical fluid extracts from Spirulina platensis. Food Chem 102: 1357-1367.
3. Rodrigez-Garcia I, Guil-Guerro JL (2008) Evaluation of the antioxidant activity of

		Amount of Algae (g)	
		Alga grown in normal medium	**Alga grown in normal medium with 10 times enhanced microelement content**
Mo(VI)	*Spirulina p.*	40.74	6.47
	Chlorella v.	59.51	24.81
Zu^{2+}	*Spirulina p.*	172.13	47.71
	Chlorella v.	54.37	5.95
Cu^{2+}	*Spirulina p.*	20.94	20.01
	Chlorella v.	18.10	2.04
Fe^{3+}	*Spirulina p.*	6.99	3.10
	Chlorella v.	31.69	2.66

Table 5: Amount of algae necessary for intzake of 15% of RDA of 4 microelements.

three microalgal species for use as dietary supplements and in the preservation of foods. Food Chem 108: 1023-1026.

4. Vonshak A (1997) Spirulina platensis (Arthrospira): Physiology, cell-biology and biotechnology. Taylor and Francis Ltd, London 213-226.

5. Ajeesh M, Bohra CPN, Gupta C, Rajasekaran C (2009) Spirulina as "functional food". New Biotech 255: 5285.

6. Tokusoglou Ö, Üunal MK (2006) Biomass nutrient profiles of three algae: Spirulina platensis, Chlorella vulgaris, and Isochrisis galbana. J Food Sci 68: 1144-1148.

7. Makri A, Bellou S, Birkou M, Papatrehas K, Dolapsakis NP, et al. (2011): Lipid synthesized by micro-algae grown in laboratory- and industrial- scale bioreactors Engineering in Life Sciences 11: 52-58.

8. Bellou S, Aggelis G (2013) Biochemical activities in Chlorella sp. and Nannochloropsis salina during lipid and sugar synthesis in a lab-scale open pond simulating reactor. Journal of Biotechnology 164: 318-329.

9. Romera E, Gonzales F, Ballester A, Blazquez ML, Munoz JA (2007) Comparative study of biosorption of heavy metals using different types of algae. Biores Techn 98: 3344-3353.

10. Topcuoglu S, Guven KC, Balkis N, Kirbasogulu C (2003) Heavy metal monitoring of marine algae from the Turkish Coast of the Black Sea, 1998-2000. Chemosphere 52: 1683-1688.

11. Chojnacka K, Chojnacki A, Górecka H (2005) Biosorption of Cr3+, Cd2+, and Cu2+ ions by blue-green algae Spirulina sp. : kinetics, equilibrium and the mechanism of the process. Chem 59: 75-84.

12. Chojnacka K (2010) Biosorption and bioaccumulation- the prospects for practical applications. Envir Internat 36: 299-307.

13. Varga L, Szigeti J, Ördög V (1999) Effects of a Spirulina platensis biomass enriched with trace elements on combinations of starter culture strains employed in the dairy industry. Milchwissenschaft 54: 247-248.

14. Yuan LZ, Wang CG, Zhou BC (2008) Effects of iron on growth and lipid accumulation in Chlorella vulgaris. Bioresour Technol 99: 4717-4722.

15. Lovley DR (2000) Environmental Microbe-Metal Interactions. ASM Press, Washington, DC.

16. Chojnacka K (2007) Using biosorption to enrich the biomass of seaweeds from the Baltic Sea with microelements to produce mineral feed supplement for livestock. Biochem Engineering J 39: 246-257.

17. House WA (1999) Trace element bioavailability as exemplified by iron and zinc. Field Crops Res 60: 115-141.

18. Worms I, Simon DF, Hassler CS, Wilkinson KJ (2006) Bioavailability of trace metals to aquatic microorganisms: importance of chemical, biological and physical processes on biouptake. Biochimie 88: 1720-1731.

19. Sas G, Nemesánszky E, Brauer H, Scheffer K (1984) On the therapeutic effects of trivalent and divalent iron in iron deficiency anaemia. Arzneimittelforschung 34: 1575-1579.

20. Umbreit JN, Conrad ME, Moore EG, Latour LF (1998) Iron absorption and cellular transport: The mobilferrin/paraferritin paradigm. Semin Hematol 35: 13-26.

21. Naude S, Clijsen S, Naulaers G, Daniels H, Vanhole C, et al. (2000) Iron supplementation in preterm infants: A study comparing the effect and tolerance of a Fe2+ and a nonionic Fe III compound. J Clin Pharmacol 40: 1447-1451.

22. Schümann K, Ettle T, Szegner B, Elsenhans B, Solomons NW (2007) On risks and benefits of iron supplementation recommendations for iron intake revisited. J Trace Elem Med Biol 21: 147-168.

23. Plaza M, Cifuentes SA, Ibanez E (2008) In the search of new functional food ingredients from algae. Trends Food Sci Technol 19: 31-39.

24. Sloan AE (2002) The top ten functional food trends the next generation. Food Technol 56: 32-56.

25. Gouveia L, Batista AP, Sousa I, Raymundo A, Bandarra NM (2008) Food Chemistry Research Developments. Hauppauge USA 2: 2-37.

26. OECD (2011) Guidelines for testing of chemicals proposal for updating guideline 201. Freshwater Alga and Cyanobacteria, Growth Inhibition Test.

27. Dietary Reference Intakes (DRIs) Codex Alimentarius Hungaricus.

28. Raoof B, Kaushik BD, Prasanna R (2006) Formulation of a low-cost medium for mass production of Spirulina. Biomass Bioenerg 30: 537-542.

29. Lau TC, Chan MW, Tan HP, Kwek CL (2013) Functional Food: A Growing Trend among the Health Conscious. Asian Social Science 9: 198.

Permissions

All chapters in this book were first published in JCEPT, by Omics International; hereby published with permission under the Creative Commons Attribution License or equivalent. Every chapter published in this book has been scrutinized by our experts. Their significance has been extensively debated. The topics covered herein carry significant findings which will fuel the growth of the discipline. They may even be implemented as practical applications or may be referred to as a beginning point for another development.

The contributors of this book come from diverse backgrounds, making this book a truly international effort. This book will bring forth new frontiers with its revolutionizing research information and detailed analysis of the nascent developments around the world.

We would like to thank all the contributing authors for lending their expertise to make the book truly unique. They have played a crucial role in the development of this book. Without their invaluable contributions this book wouldn't have been possible. They have made vital efforts to compile up to date information on the varied aspects of this subject to make this book a valuable addition to the collection of many professionals and students.

This book was conceptualized with the vision of imparting up-to-date information and advanced data in this field. To ensure the same, a matchless editorial board was set up. Every individual on the board went through rigorous rounds of assessment to prove their worth. After which they invested a large part of their time researching and compiling the most relevant data for our readers.

The editorial board has been involved in producing this book since its inception. They have spent rigorous hours researching and exploring the diverse topics which have resulted in the successful publishing of this book. They have passed on their knowledge of decades through this book. To expedite this challenging task, the publisher supported the team at every step. A small team of assistant editors was also appointed to further simplify the editing procedure and attain best results for the readers.

Apart from the editorial board, the designing team has also invested a significant amount of their time in understanding the subject and creating the most relevant covers. They scrutinized every image to scout for the most suitable representation of the subject and create an appropriate cover for the book.

The publishing team has been an ardent support to the editorial, designing and production team. Their endless efforts to recruit the best for this project, has resulted in the accomplishment of this book. They are a veteran in the field of academics and their pool of knowledge is as vast as their experience in printing. Their expertise and guidance has proved useful at every step. Their uncompromising quality standards have made this book an exceptional effort. Their encouragement from time to time has been an inspiration for everyone.

The publisher and the editorial board hope that this book will prove to be a valuable piece of knowledge for researchers, students, practitioners and scholars across the globe.

List of Contributors

Xiao-Juan Jin
MOE Key Laboratory of Wooden Material Science and Application, Beijing Key Laboratory of Lignocellulosic Chemistry, MOE Engineering Research Center of Forestry Biomass Materials and Bioener, Beijing Forestry University, Beijing, China

Yue Mei Zhu
MOE Key Laboratory of Wooden Material Science and Application, Beijing Key Laboratory of Lignocellulosic Chemistry, MOE Engineering Research Center of Forestry Biomass Materials and Bioener, Beijing Forestry University, Beijing, China

Yahya Chetouani
Département Génie Chimique, Rue Lavoisier, 76821, Mont Saint Aignan Cedex, France

Abdouelilah Hachimi
Research Institute of Catalysis and Environment of Lyon (IRCELYON), UMR 5256 CNRS, Claude University, Bernard Lyon I, Villeurbanne, France

Laouedj Nadjia
DR. Moulay Tahar University, Saida, Algeria

Elaziouti Abdelkader
LCPCE Laboratory, Faculty of sciences, Department of industrial Chemistry, University of the Science and the technology of Oran (USTO M.B), BP 1505 El M'naouar 31000 Oran, Algeria

Bekka Ahmed
LCPCE Laboratory, Faculty of sciences, Department of industrial Chemistry, University of the Science and the technology of Oran (USTO M.B), BP 1505 El M'naouar 31000 Oran, Algeria

AK Minocha
Environmental Science and Technology Division, Central Building Research Institute, Roorkee- 247667(UA) & NIPER Mohali, Punjab-160062, India

Manish Kumar Goyal
Environmental Science and Technology Division, Central Building Research Institute, Roorkee- 247667(UA) & NIPER Mohali, Punjab-160062, India

Nguyen Tan Dzung
HCMC University of Technical Education, 01-Vo Van Ngan Street, Thu Duc District, Viet Nam

Nguyen Quoc Dzung
HCMC University of Technical Education, 01-Vo Van Ngan Street, Thu Duc District, Viet Nam

Trinh Van Dzung
HoChiMinh City University of Technology, Viet Nam National University, HoChiMinh City

Le Xuan Hai
HoChiMinh City University of Technology, Viet Nam National University, HoChiMinh City

Salimah Shafii
Faculty of Industrial Science and Technology, University Malaysia Pahang, 26300, Kuantan, Pahang, Malaysia

Wu Lihua
Faculty of Industrial Science and Technology, University Malaysia Pahang, 26300, Kuantan, Pahang, Malaysia

M. Ridzuan Nordin
Faculty of Industrial Science and Technology, University Malaysia Pahang, 26300, Kuantan, Pahang, Malaysia

Liew Kong Yong
Faculty of Industrial Science and Technology, University Malaysia Pahang, 26300, Kuantan, Pahang, Malaysia

Mahlatse M. Sekhula
Department of Environmental, Water & Earth Sciences, Faculty of Science, Tshwane University of Technology, 175 Nelson Mandela Drive, Arcadia Campus, Pretoria, South Africa

Jonathan O. Okonkwo
Department of Environmental, Water & Earth Sciences, Faculty of Science, Tshwane University of Technology, 175 Nelson Mandela Drive, Arcadia Campus, Pretoria, South Africa

Caliphs M. Zvinowanda
Department of Environmental, Water & Earth Sciences, Faculty of Science, Tshwane University of Technology, 175 Nelson Mandela Drive, Arcadia Campus, Pretoria, South Africa

Nana N. Agyei
Department of Chemistry, University of Limpopo, MEDUNSA Campus, Pretoria, South Africa

Abdul J.Chaudhary
Institute for the Environment, Brunel University, Kingston Lane, Uxbridge, Middlesex, UB8 3PH, UK

Abarasi Hart
School of Chemical Engineering, University of Birmingham, Edgbaston, Birmingham, B15 2TT, UK

Yuan-Cheng Cao
Key Laboratory of Optoelectronic Chemical Materials and Devices of Ministry of Education, Jianghan University, Wuhan 430056, China

Wenjun Wei
Key Laboratory of Optoelectronic Chemical Materials and Devices of Ministry of Education, Jianghan University, Wuhan 430056, China

Jiyan Liu
Key Laboratory of Optoelectronic Chemical Materials and Devices of Ministry of Education, Jianghan University, Wuhan 430056, China

Mariia Artus
Precarpathian National University named after V. Stephanyk 57, Shevchenko str., 76025 Ivano-Frankivsk, Ukraine

Ivan Kostiv
State Enterprise Scientific-Research Gallurgi Institute 5a, Fabrychna str., 77300 Kalush, Ukraine

N. Jhider
Applied Thermodynamics and Environment National engineering Research Unit, School of Gabes, Gabes, Tunisia

M. Bagane
Applied Thermodynamics and Environment National engineering Research Unit, School of Gabes, Gabes, Tunisia

Hossein Faghihian
Department of Chemistry, Shahreza Branch, Islamic Azad University, Iran

Leila Riazi
R&D Department, Iranian Chemical Industries Investment Company, Isfahan, Iran

Phung Kim Le
Department of Chemical Engineering, University of Technology, Ho Chi Minh City, Vietnam

Kien Anh Le
Institue of Tropical and Environmental, District10, Ho Chi Minh City, Vietnam

N. Abdennebi
National School of Engineers of Gabes (ENIG), University of Gabes, University City, Gabes, Tunisia

N. Hajji
Laboratory of Energy and Environment, ENIG, University City, Gabes, Tunisia

Nisha Yadav
Forest Ecology and Environment Division, Forest Research Institute, Dehradun, India

Rajdeo Kumar
Forest Ecology and Environment Division, Forest Research Institute, Dehradun, India

Dr. (Mrs.) Laxmi Rawat
Forest Ecology and Environment Division, Forest Research Institute, Dehradun, India

Sarita Gupta
CSWCRTI, Dehradun, India

M. Selatnia
Laboratoire de commande des processus Ecole Nationale Polytechnique d'Alger, BP 182 Avenue Hassen Badi, El Harrach, Algeria

R. Illoul
Laboratoire de commande des processus Ecole Nationale Polytechnique d'Alger, BP 182 Avenue Hassen Badi, El Harrach, Algeria

MS Boucherit
Laboratoire de commande des processus Ecole Nationale Polytechnique d'Alger, BP 182 Avenue Hassen Badi, El Harrach, Algeria

Rodrigues LF
Pontifical Catholic University of Rio Grande do Sul-PUCRS, Center of Excellence in Research and Innovation in Petroleum, Mineral Resources and Carbon Storage CEPAC, Brazil

FS Goudinho
Pontifical Catholic University of Rio Grande do Sul-PUCRS, Center of Excellence in Research and Innovation in Petroleum, Mineral Resources and Carbon Storage CEPAC, Brazil

DO Laroque
Pontifical Catholic University of Rio Grande do Sul-PUCRS, Center of Excellence in Research and Innovation in Petroleum, Mineral Resources and Carbon Storage CEPAC, Brazil

LV Lourega
Pontifical Catholic University of Rio Grande do Sul-PUCRS, Center of Excellence in Research and Innovation in Petroleum, Mineral Resources and Carbon Storage CEPAC, Brazil

R. Heemann
Pontifical Catholic University of Rio Grande do Sul-PUCRS, Center of Excellence in Research and Innovation in Petroleum, Mineral Resources and Carbon Storage CEPAC, Brazil

JMM Ketzer
Pontifical Catholic University of Rio Grande do Sul-PUCRS, Center of Excellence in Research and Innovation in Petroleum, Mineral Resources and Carbon Storage CEPAC, Brazil

J. Boken
Department of Physics, Banasthali Vidyapith, Rajasthan, India

S. Dalela
Department of Pure and Applied Physics, University of Kota, Kota, Rajasthan, India

CK. Sharma
Assistant Professor, Department of Bioscience & Biotechnology, Banasthali University, Rajasthan, India

D. Kumar
Department of Chemistry, Banasthali Vidyapith, Rajasthan, India

R. U. Owolabi
Fountain University, Industrial & Environmental Chemistry Department, Osogbo Osun State, Nigeria

N. A. Osiyemi
Ahmadu Bello University, Chemical Engineering Department, Zaria, Kaduna, Nigeria

M. K. Amosa
Covenant University, Chemical Engineering Department, Ota, Ogun State, Nigeria

M. E. Ojewumi
Fountain University, Industrial & Environmental Chemistry Department, Osogbo Osun State, Nigeria

Magda A. Akl
Chemistry department, Faculty of Science, Mansoura University, Mansoura, Egypt

AbdElFatah M. Yousef
Chemistry department, Faculty of Science, Mansoura University, Mansoura, Egypt

Sameh AbdElnasser
Chemistry department, Faculty of Science, Mansoura University, Mansoura, Egypt

Ashish Chauhan
Department of Chemistry, Dr. B. R. Ambedkar National Institute of Technology, Jalandhar 144 011 (Pb) India

Priyanka Chauhan
Department of Chemistry, Dr. B. R. Ambedkar National Institute of Technology, Jalandhar 144 011 (Pb) India

Balbir Kaith
Department of Chemistry, Dr. B. R. Ambedkar National Institute of Technology, Jalandhar 144 011 (Pb) India

AK Minocha
Central Building Research Institute, Roorkee and NIPER, Mohali-Punjab, India

Manish Kumar Goyal
Central Building Research Institute, Roorkee and NIPER, Mohali-Punjab, India

Gabriela Rigo Roso
Department of Food Science and Technology, Federal University of Santa Maria, Roraima Avenue, 1000, 97105-900, Santa Maria, RS, Brazil

Maria Isabel Queiroz
School of Chemistry and Food, Federal University of Rio Grande, Itália Avenue, 96201-900, Rio Grande, RS, Brazil

Nívia Maria Streit
School of Chemistry and Food, Federal University of Rio Grande, Itália Avenue, 96201-900, Rio Grande, RS, Brazil

Cristiano Ragagnin de Menezes
Department of Food Science and Technology, Federal University of Santa Maria, Roraima Avenue, 1000, 97105-900, Santa Maria, RS, Brazil

Leila Queiroz Zepka
Department of Food Science and Technology, Federal University of Santa Maria, Roraima Avenue, 1000, 97105-900, Santa Maria, RS, Brazil

Eduardo Jacob- Lopes
Department of Food Science and Technology, Federal University of Santa Maria, Roraima Avenue, 1000, 97105-900, Santa Maria, RS, Brazil

Hoang N. Nguyen
Department of Chemical Engineering, College of Engineering, De La Salle University, 2401 Taft Avenue, 1004 Manila, Philippines

Pag-asa D. Gaspillo
Department of Chemical Engineering, College of Engineering, De La Salle University, 2401 Taft Avenue, 1004 Manila, Philippines

Julius B. Maridable
Department of Chemical Engineering, College of Engineering, De La Salle University, 2401 Taft Avenue, 1004 Manila, Philippines

Roberto M. Malaluan
Department of Chemical Engineering Technology, School of Engineering Technology, Iligan Institute of Technology, Mindanao State University, Andres Bonifacio Avenue, Tibanga, 9200 Iligan City, Philippines

Hirofumi Hinode
Department of International Development Engineering, Graduate School of Science and Engineering, Tokyo Institute of Technology, 2-12-1-I4-1, Ookayama, Meguro-ku, Tokyo 152-8550, Japan

Chris Salim
Department of International Development Engineering, Graduate School of Science and Engineering, Tokyo Institute of Technology, 2-12-1-I4-1, Ookayama, Meguro-ku, Tokyo 152-8550, Japan

Ha K.P. Huynh
Department of Inorganic Chemistry, Faculty of Chemical Engineering, Ho Chi Minh City University of Technology, 268 Ly Thuong Kiet Street, District 10, Ho Chi Minh City, Vietnam

Jean Paul Sinumvayo
School of Biotechnology, Jiangnan University, Wuxi, P. R. China

Nestor Ishimwe
School of Biotechnology, Jiangnan University, Wuxi, P. R. China

Ali Fakhri
Department of Chemistry, Shahre-Qods Branch, Islamic Azad University, Tehran, Iran

Saeideh Adami
Department of Chemistry, Shahre-Qods Branch, Islamic Azad University, Tehran, Iran

Naruporn Monmaturapoj
National Metal and Materials Technology Center, 114 Thailand Science Park, Pathumthani 12120, Thailand

Witoon Thepsuwan
National Metal and Materials Technology Center, 114 Thailand Science Park, Pathumthani 12120, Thailand

Suda Wanakitti
National Metal and Materials Technology Center, 114 Thailand Science Park, Pathumthani 12120, Thailand

Charusporn Mongkolkachit
National Metal and Materials Technology Center, 114 Thailand Science Park, Pathumthani 12120, Thailand

Katanchalee Mai-ngam
National Metal and Materials Technology Center, 114 Thailand Science Park, Pathumthani 12120, Thailand

Sawinee Ngernpimai
National Metal and Materials Technology Center, 114 Thailand Science Park, Pathumthani 12120, Thailand

Wattana Klinsukhon
National Metal and Materials Technology Center, 114 Thailand Science Park, Pathumthani 12120, Thailand

Chureerat Prahsarn
National Metal and Materials Technology Center, 114 Thailand Science Park, Pathumthani 12120, Thailand

Abraham Tamir
Department of Chemical Engineering, Ben-Gurion University of the Negev, Beer-Sheba, Israel

L. Karimi
Department of Chemistry, Faculty of Science, University of Zanjan, Zanjan, Iran

S. Nouroozi
Department of Chemistry, Faculty of Science, University of Zanjan, Zanjan, Iran

MH Rasoulifard
Department of Chemistry, Faculty of Science, University of Zanjan, Zanjan, Iran

Delia Teresa Sponza
Department of Environmental Engineering, Engineering Faculty, Dokuz Eylül University, Tınaztepe Campus, 35160, Buca, İzmir, Turkey

Rukiye Oztekin
Department of Environmental Engineering, Engineering Faculty, Dokuz Eylül University, Tınaztepe Campus, 35160, Buca, İzmir, Turkey

Farah Zemani
Laboratoire d'Energie et Propulsion Navale, Faculté de Génie Mécanique, Université des Sciences et de la Technologie Mohamed, Boudiaf d'Oran, B.P. 1505 Oran El- M'naouar, 3-1000 Oran, Algeria

Amina Sabeur-Bendehina
Laboratoire d'Energie et Propulsion Navale, Faculté de Génie Mécanique, Université des Sciences et de la Technologie Mohamed, Boudiaf d'Oran, B.P. 1505 Oran El- M'naouar, 3-1000 Oran, Algeria

Mustapha Boussoufi
Laboratoire d'Energie et Propulsion Navale, Faculté de Génie Mécanique, Université des Sciences et de la Technologie Mohamed, Boudiaf d'Oran, B.P. 1505 Oran El- M'naouar, 3-1000 Oran, Algeria

AO Okewale
Department of Chemical Engineering, Federal University of Petroleum Resources, P.M.B., 1221, Effurun, Nigeria

PK Igbokwe
Department of Chemical Engineering, Nnamdi Azikiwe University, P.M.B., 5025, Awka, Anambra State, Nigeria

KA Babayemi
Department of Chemical Engineering, Anambra State University, P.M.B., 02, Uli, Anambra State, Nigeria

Jogi Ganesh Dattatreya Tadimeti
Polymer and Process Engineering Department, I.I.T.Roorkee, SRE Campus, UP-247 001, India

Anusha Chandra
Polymer and Process Engineering Department, I.I.T.Roorkee, SRE Campus, UP-247 001, India

Sujay Chattopadhyay
Polymer and Process Engineering Department, I.I.T.Roorkee, SRE Campus, UP-247 001, India

TT Rukuni
Department of Environment, Water & Earth Sciences, Tshwane University of Technology, Private Bag X680, 0001, South Africa

JP Maree
Incumbent, Rand Water Chair in Water Utilisation, Tshwane University of Technology, Private Bag X680, 0001, Pretoria, South Africa

CM Zvinowanda
Department of Environment, Water & Earth Sciences, Tshwane University of Technology, Private Bag X680, 0001, South Africa

FHH Carlsson
Department of Environment, Water & Earth Sciences, Tshwane University of Technology, Private Bag X680, 0001, South Africa

Satyajit Gupta
Department of Materials Engineering, Indian Institute of Science, Bangalore, India
Center for Nanoscience and Engineering, Indian Institute of Science, Bangalore, India

Praveen C Ramamurthy
Department of Materials Engineering, Indian Institute of Science, Bangalore, India
Center for Nanoscience and Engineering, Indian Institute of Science, Bangalore, India

Giridhar Madras
Center for Nanoscience and Engineering, Indian Institute of Science, Bangalore, India

Szabolcs Molnár
Regional Knowledge Centre, Eszterházy Károly College, Leányka Street, Hungary

Attila Kiss, Diána Virág
Regional Knowledge Centre, Eszterházy Károly College, Leányka Street, Hungary

Péter Forgó
Regional Knowledge Centre, Eszterházy Károly College, Leányka Street, Hungary